W9-CTH-244

2ND EDITION

PRINCIPLES OF WATER RESOURCES

HISTORY, DEVELOPMENT, MANAGEMENT, AND POLICY

2ND EDITION

PRINCIPLES OF WATER RESOURCES

HISTORY, DEVELOPMENT, MANAGEMENT, AND POLICY

THOMAS V. CECH

WILEY JOHN WILEY & SONS, INC.

Cover Photo: Niagara Falls, Ontario. The roar of falling water is deafening when you stand at the overlook of Horseshoe Falls, seen at the bottom of the photo. A cool, thick mist rises from the cascading water and towers above the falls, easily visible from passing aircraft. It's a humbling experience—the power of Niagara beneath your feet—as it churns past the deck of the *Maid of the Mist* (visible at the lower left) and eventually into Lake Ontario.

ACQUISITIONS EDITORS	Ryan Flahive, Jerry Correa
ASSOCIATE EDITOR	Denise Powell
EDITORIAL ASSISTANT	Christine Cordek
MARKETING MANAGER	Clay Stone
PRODUCTION EDITOR	Barbara Russiello
TEXT/COVER DESIGNER	Madelyn Lesure
SENIOR PHOTO EDITOR	Sara Wight
ILLUSTRATION EDITOR	Sandra Rigby
ILLUSTRATIONS	Hadel Studio
MAPS	Cartographics
COVER PHOTO	© Alan Schein Photography/Corbis Images

This book was set in 10/12 Sabon by Matrix Publishing Services and printed and bound by Malloy Lithography. The cover was printed by Lehigh Press.

This book is printed on acid free paper. ∞

ISBN 0-471-48475-X (Main Book)
ISBN 0471-65810-3 (Wiley International Edition)

Printed in the United States of America

10 9 8 7 6 5 4 3 2

For my parents, Theofil & Agnes Cech

Epiphany

My fingertips becoming numb
I realize now where I am
submerged between two islands
waiting for another breath to come.
A pencil-thin line
dividing the Basque sky and the sea
disappears as I stare longer
into heavenly whiteness,
forgetting the tingle in my legs.
This is not an option.
The magnitude of my companion
has distanced the destination
drawing the journey closer.

With every playful brush
I am tossed forward like a seal
rising in and out of the water
to reveal itself.
The game continues
until my body releases
floating safely on the surface
of cupped hands,
the faint murmur of fisherman's boats,
displaceable in distance and depth,
makes all other sound obsolete.
I breathe out
as shallow grains of sand
blue like the color of my skin
bring me back.

—Denise C. Powell

ABOUT THE AUTHOR

I grew up on a small family farm near Clarkson, Nebraska, in the eastern part of the state. From an early age, I was intrigued by water. My great-grandparents, Vaclav and Katherine Cech (pronounced *check*), homesteaded the farm in 1879 and built a two-story frame house next to a dry creek bed. We never bothered to name the creek, which only flowed intermittently since it was so small and dry most of the time. My sister Judy, brother Jerry, and I crossed Maple Creek everyday on our walk to our one-room country school—District 14. It was a typical, white clapboard Nebraska rural school of the 1960s, and it had an oil furnace, wooden floors, and a ceramic water jug since we had no indoor plumbing. The water jug was filled every morning at the Novotny farm just down the road.

The Cech farm relied on the Nebraska wind and a Chicago-made windmill for drinking water. The windmill pumped water through a system of buried pipes to a cement-lined vault buried underground. The water storage vault, called a cistern, was a marvel of engineering to a young Nebraska farm kid. Once a year we let all the faucets in the house run until the cistern was emptied. Then, my older brother took a wooden ladder and climbed down into the cistern to clean out the silt and sand that had settled to the bottom. We used a corn scoop shovel and a metal bucket for the job. The bucket had a rope tied to the handle, and my job was to pull each bucket-full of dirty water and sand out of the cistern, and get rid of it above ground. The entire process fascinated me. How did my grandparents know to drill a well at that spot to find water? Who laid out the network of buried pipes on our farm? How did the groundwater get beneath our farm in the first place?

When I grew older, I crossed the Platte River many times near Grand Island, Nebraska, on my trips home from college at Kearney State. The Platte River was not very deep, but its broad sweep across Nebraska impressed me. If I was lucky, my drives along the Platte River coincided with the annual sandhill crane migration that announced spring on the prairie.

As an adult, I've worked in the field of water resources in Utah, Colorado, and Nebraska. My academic background includes a B.S. in Education from the University of Nebraska at Kearney and a Master's in Community and Regional Planning from the University of Nebraska—Lincoln. I've been involved in water development and conservation, environmental issues, groundwater management, lobbying at the state and federal levels, and in education. I have contemplated writing a book about water for years.

In some ways, the idea for this textbook began on my daily walk to the little one-room country school on the Nebraska prairie. Perhaps something special in the drinking water had a part in it. My classmate, Sandra Novotny, lived on the farm where we fetched the daily bucket of drinking water for school. Today, Sandra is president of Nova Environmental Services in Kensington, Maryland, and is a consultant for the U.S. Environmental Protection Agency, among other clients. Dr. Glenn Cada, who lived a few miles west of the school, is now one of the country's leading authorities on salmon migration along the Columbia River. Glenn works for the Oakridge National Laboratory in Oakridge, Tennessee, and his review and comments in Chapter 12 are most appreciated. Lorraine Smith, formerly Lorraine Tuma, lived just over the hill to the east of our windmill and is now the Town Clerk for the Town of Clarkson, Nebraska. Lorraine helped the community develop a wellhead protection program (discussed in Chapter 11) to protect the quality of local drinking water supplies. And finally, my brother, Leroy (the one who cleaned the bottom of the cistern and then went on to receive his Master's degree from the University of California—Davis) is gaining a new appreciation for water resources management and policy after reviewing the manuscript for this textbook endless times, in between trips to the family cabin on Birch Lake just outside Ely, Minnesota.

Working in the field of water resources is fantastic. I hope you learn a great deal from this textbook—about water, about our world, and about yourself.

Tom Cech

PREFACE

The study of water resources is a fascinating, but too often frustrating, process. It's fascinating because it involves a wide variety of disciplines such as mathematics, science, geography, geology, biology, political science, meteorology, and even psychology. Water resources management includes the construction of physical features, such as dams and other storage projects, to conserve water during wet periods for later use. It can take the form of cooperative legal agreements, negotiated over many years, between neighbors, states, or countries to share scarce water resources. Water management even involves volunteer community groups that inventory a watershed to protect a local drinking water supply.

The study of water resources can also be very frustrating. Strange terminology, incomprehensible data, diverse viewpoints, and wide ranging, complex topics can quickly become overwhelming. Too often, water resources "experts" neglect to provide straightforward, understandable explanations of water issues for the layperson. News reports often provide only "sound bite" reporting to gain interest and do not have the time or space needed to provide solid background information on a particular water resources issue. Unfortunately, this occurs quite often since it is difficult to break down complex water resources topics into short, understandable, and interesting explanations. This textbook is an attempt to change that paradigm.

How should one proceed with the study of water resources? First, it's important to understand the historical context of both simple and complex water resources issues. Questions we might ask are: How did early civilizations obtain water for personal needs, irrigation, navigation, and hydropower? What conflicts existed between these early water users? What techniques were used to construct these water resources projects?

Second, what are the natural physical processes of water? How do climate and weather patterns affect the distribution of water resources around the world? How do surface and groundwater processes work, and what is the interaction between surface water and groundwater?

Third, what are the primary components of water quality? What constituents affect human health and the health of ecosystems? How is water treated for human consumption and then treated once again before being released back to rivers, lakes, and ponds?

Fourth, since dams and irrigation projects are major water uses around the world, what construction and management methods are used to build and operate these facilities? What impacts, both positive and negative, do such water projects have on the surrounding area? Why are dams and irrigation projects important to the survival of a civilization? What role do these facilities play in environmental degradation?

Fifth, what water agencies exist at local, state, regional, and federal levels to manage water resources? What are their responsibilities, how do they interact, and how do they overlap? Do our neighbors in Mexico and Canada follow the U.S. system of water management, or have they adopted different methods?

Sixth, what aspects of the environment are at risk due to water shortages, poor water quality, or existing water development projects? How are endangered species being protected? What economic and societal issues surround the protection of these species?

And finally, how are water conflicts around the world resolved? And how will we solve the need for additional clean water supplies in the twenty-first century?

As you read through this textbook, keep in mind the role that geography, climate, technology, and population growth play in all aspects of water use. These complex interrelationships provide the basic underpinnings for the history, development, management, and policy of water resources around the world.

ACKNOWLEDGMENTS

Writing a college textbook is enlightening, challenging, and exhilarating at various stages of the process. Many individuals provided support, guidance, and critical review of the text as it evolved from an idea in a file cabinet to the bound textbook in your hands. The outstanding staff at John Wiley & Sons, Inc. has been excellent to work with on this project. In particular, my special thanks go to Anne E. Smith, Publisher, Ryan Flahive and Jerry Correa, Acquisitions Editors, Denise Powell, Associate Editor, Christine Cordek, Editorial Assistant, Tom Kulesa, New Media Editor, Sandra L. Rigby, Senior Illustration Editor, Sara Wight, Photo Editor, Elyse Rieder, Photo Researcher, Barbara Russiello, Production Editor, Clay Stone, Marketing Manager, and Madelyn Lesure, Cover and Text Designer. I also wish to thank Betty Pessagno for her work as copy editor, and Pam Whiteley, customer service representative, and all at Matrix Publishing who typeset the pages of this book.

I would like to thank the following people who have reviewed this second edition to make it an even better book.

Grace M. Cech, *James A. Michener Library, University of Northern Colorado, Greeley, Colorado*
Dr. James E. Hairston, *Agronomy and Soils Department, Auburn University, Auburn, Alabama*
Dr. Robert M. Hordon, *Department of Geography, Rutgers University, New Brunswick, New Jersey*
Dr. William H. Hoyt, *Department of Earth Sciences, University of Northern Colorado, Greeley, Colorado*
Dr. James W. Jawitz, *University of Florida, Soil and Water Science Department, University of Florida, Gainesville, Florida*
Dr. Marie Livingston, *Department of Economics, University of Northern Colorado, Greeley, Colorado*
Ruth Kline-Robach, *Institute of Water Research, Michigan State University, Lansing, Michigan*
Dr. John D. Skalbeck, *Department of Biology and Geology, University of Wisconsin—Parkside, Kenosha, Wisconsin*
Dr. Michael Vorwerk, *Biology Department, Westfield State College, Westfield, Massachusetts*

I gratefully acknowledge the detailed work of the following who reviewed the entire manuscript of the first edition for accuracy and content:

Dr. Philip Chaney, *Department of Geology and Geography, Auburn University, Auburn, Alabama*
Dr. Burrell Montz, *Department of Geography and Regional Development, Binghamton University, Binghamton, New York*
Gary G. Peterson, *Pima County Flood Control District and The University of Arizona, Tucson, Arizona*
Dr. Robert M. Hordon, *Department of Geography, Rutgers University, Piscataway, New Jersey*
Dr. William Hoyt, *Department of Earth Sciences, University of Northern Colorado, Greeley, Colorado*
Grace Cech, *James A. Michener Library, University of Northern Colorado, Greeley, Colorado*
Dr. Beverly Wemple, *Department of Geography, University of Vermont, Burlington, Vermont*
Dr. Ellen Marsden, *School of Natural Resources, University of Vermont, Burlington, Vermont*
Leroy Cech, *Burnsville, Minneosta*

The following individuals reviewed chapters or smaller segments of the manuscript and contributed greatly to the accuracy and breadth of topics found within the textbook:

Dr. Kenneth Hopkins, *Department of Earth Sciences, University of Northern Colorado, Greeley, Colorado*
Dr. J. David Aiken, *Department of Agricultural Economics, University of Nebraska–Lincoln, Lincoln, Nebraska*
Dr. Mel Goldstein, Meteorologist, *WTNH—Channel 8, New Haven, Connecticut and Western Connecticut State University, Danbury, Connecticut*
John Miriovsky, Director, *Lincoln Water System, Lincoln, Nebraska*
Geoffrey C. Ryan, Acting Director, *Bureau of Public Affairs, New York City Department of Environmental Protection, New York, New York*
Bill Irvine, Agricultural Specialist, *St. Paul, Nebraska*
Eric Tharp, *Legislative & Public Affairs, Department of Water and Power, Los Angeles, California*
Robert Walsh, External Affairs Officer, *Hoover Dam, Bureau of Reclamation, Department of Interior, Boulder City, Nevada*
Diana Cross, Regional Public Affairs Officer, *Bureau of Reclamation, Department of Interior, Boise, Idaho*
Jim Mumford, Regional Dam Safety Coordinator, *Bureau of Reclamation, Department of Interior, Boise, Idaho*
Jenny Cech, *University of Colorado, Boulder, Colorado*
Jeff Buettner, Communications Officer, *Central Nebraska Public Power & Irrigation District, Holdrege, Nebraska*
Dave Lyngholm, Power Manager, *Grand Coulee Dam, Bureau of Reclamation, Department of Interior, Grand Coulee, Washington*
Craig Sprankle, Public Affairs Officer, *Grand Coulee Power Office, Grand Coulee Dam, Bureau of Reclamation, Department of Interior, Grand Coulee, Washington*
Peter Soeth, Public Affairs Specialist, *Bureau of Reclamation, Department of Interior, Denver, Colorado*
Robert A. Bank, Engineering Manager, *U.S. Army Corps of Engineers, Department of Defense, Washington, D.C.*
Dr. Marty Reuss, Chief Historian, *U.S. Army Corps of Engineers, Department of Defense, Washington, D.C.*

Dr. Ted Nelson, Director of Navigation, *Tennessee Valley Authority, Knoxville, Tennessee*
April Cech, student, *McGill University, Montreal, Quebec*
Cynthia Dyballa, Acting Chief, *Washington Policy Staff, Bureau of Reclamation, Department of Interior, Washington, D.C.*
Glenn G. Patterson, Hydrologist, *U.S. Geological Survey, Department of Interior, Reston, Virginia*
Mitch Snow, Chief of Media Services, *U.S. Fish & Wildlife Service, Department of Interior, Washington, D.C.*
Dr. Dwight T. Pitcaithley, Chief Historian, *National Park Service, Department of Interior, Washington, D.C.*
Dr. Janet McDonnell, Bureau Historian, *National Park Service, Department of Interior, Washington, D.C.*
Eric Janes, *Rangelands, Soil, Water, Air Group, Bureau of Land Management, Department of Interior, Washington, D.C.*
Sharon Kliwinski, *Water Resources Division, National Park Service, Department of Interior, Washington, D.C.*
Dr. Douglas Helms, Senior Historian, *Natural Resources Conservation Service, U.S. Department of Agriculture, Washington, D.C.*
Jack Frost, Special Studies Coordinator, *Natural Resources Conservation Service, U.S. Department of Agriculture, Washington, D.C.*
Warren Harper, *U.S. Forest Service, Department of Interior, Washington, D.C.*
John Paquin, *Office of Energy Projects, Federal Energy Regulatory Commission, Department of Energy, Washington, D.C.*
Amy Zimmerling, *U.S. Environmental Protection Agency, Washington, D.C.*
Holly Harrington, *Office of Public Affairs, Federal Emergency Management Agency, Washington, D.C.*
Dr. Brit Storey, Senior Historian, *Bureau of Reclamation, Department of Interior, Denver, Colorado*
Melissa Cech, student, *University of Colorado, Boulder, Colorado*
Ian Derk, student, *Colorado State University, Fort Collins, Colorado*
Laurie Van Leuven, *Public Information/Water Conservation Office, Highline Water District, Kent, Washington*
Willis Emmons, District Manager, *Kennebunk Sewer District, Kennebunk, Maine*
Ralph Pentland, Canadian Co-Chairman, *Upper Great Lakes Plan of Study Team, International Joint Commission, Ottawa, Quebec*
Tom Knutson, General Manager, *Farwell Irrigation District, Farwell, Nebraska*
Kelly Fackel, Community & Public Relations Manager, *Miami Conservancy District, Dayton, Ohio*
Dr. Mark J. W. Bamberger, Water Resources Information Coordinator, *Miami Conservancy District, Dayton, Ohio*
Emmett Egr, Information/Education Coordinator, *Papio-Missouri River Natural Resources District, Omaha, Nebraska*
Wayne A. Bossert, Manager, *Northwest Kansas Groundwater Management District No. 4, Colby, Kansas*
C.E. Williams, General Manager, *Panhandle Groundwater Conservation District, White Deer, Texas*
Don Avila, *Public Information Office, The County Sanitation Districts of Los Angeles County, Whittier, California*
Alisa Richardson, *Department of Environmental Management, State of Rhode Island, Providence, Rhode Island*
John Manning, *Department of Environmental Management, State of Rhode Island, Providence, Rhode Island*
Ann Pesiri Swanson, Executive Director, *Chesapeake Bay Commission, Annapolis, Maryland*
Richard Opper, Executive Director, *Missouri River Basin Association, Lewistown, Montana*
Dr. Irina Cech, *School of Public Health, The University of Texas, Houston, Texas*
Dr. Ernst M. Davis, *School of Public Health, The University of Texas, Houston, Texas*
Sandra K. Novotny, President, *Nova Environmental Services, Kensington, Maryland*
Dr. Jim Loftis, *Department of Chemical and Bioresource Engineering, Colorado State University, Fort Collins, Colorado*
Gary Litherland, *Commissioner's Office, Chicago Water Department, Chicago, Illinois*
Harold Gorman, Executive Director, *Sewerage & Water Board of New Orleans, New Orleans, Louisiana*
Lorraine Smith, *Town of Clarkson, Clarkson, Nebraska*
Lynn O'Neil, *Sierra Club Books, Sierra Club, San Francisco, California*

Ben Beach, *Wilderness Society, Washington, D.C.*

Tim McLean, Senior Editor, *National Wildlife Federation, Reston, Virginia*

Dr. Keith McKnight, Manager of Conservation Programs, *Ducks Unlimited, Memphis, Tennessee*

Tom Graff, California Regional Director, *Environmental Defense, Oakland, California*

Kate Brauman, *Membership and Public Education, Natural Resources Defense Council, New York, New York*

Dr. Glenn Cada, *Environmental Sciences Division, Oak Ridge National Laboratory, Oak Ridge, Tennessee*

Kelly Mills, Staff Geologist, *Texas Natural Resource Conservation Commission, Austin, Texas*

Bob Kerr, Director, *Pollution Prevention Assistance Division, Georgia Department of Natural Resources, Atlanta, Georgia*

Jeremy Berkoff, Consultant Economist, *Cambridge, United Kingdom*

Dr. Thomas Naff, Director, *Middle East Research Institute, University of Pennsylvania, Philadelphia, Pennsylvania*

Dr. Aaron T. Wolf, *Department of Geosciences, Oregon State University, Corvallis, Oregon*

Dennis Nelson, Executive Director, *Western Watercourse and National Project WET, Montana State University, Bozeman, Montana*

Sue McClurg, Chief Writer, *The Water Education Foundation, Sacramento, California*

Andrew Stone, Executive Director, *American Ground Water Trust, Concord, New Hampshire*

Chad Pregracke, *Mississippi River Beautification and Restoration Project, East Moline, Illinois*

Susan Seacrest, President, *The Groundwater Foundation, Lincoln, Nebraska*

Cindy Kreifels, Executive Director, *The Groundwater Foundation, Lincoln, Nebraska*

Others who provided assistance include:

Melissa Alverson, *Tennessee Valley Authority, Knoxville, Tennessee*

Nancy E. Mayberry, *U.S. Army Corps of Engineers, New Orleans District, New Orleans, Louisiana*

Adele Merchant, *U.S. Army Corps of Engineers, Northwestern Division, Portland, Oregon*

Andrew Senti, *Bureau of Land Management, Lakewood, Colorado*

Tiffany Butler, *Orange County Water District, Fountain Valley, California*

Doris Wemhoff, *Lindsay Manufacturing Company, Lindsay, Nebraska*

Jessica Morales, *South Florida Water Management District, West Palm Beach, Florida*

Robert Schaefer, *Los Angeles Department of Water and Power, Los Angeles, California*

Michael Laffin, *Nova Scotia House of Assembly, Halifax, Nova Scotia*

J. Michael Jess, Associate Director, *Conservation and Survey Division, University of Nebraska, Lincoln, Nebraska*

Dr. Gerry Saunders, *Department of Biological Sciences, University of Northern Colorado, Greeley, Colorado*

Bud Summers and Phong Hoang, *City of Columbia Water Works, Columbia, South Carolina*

Sven-Erik Skogsfors, Chief Director, *Stockholm International Water Institute, Stockholm, Sweden*

Dr. Dean Pennington, Executive Director, *YMD Joint Water Management District, Stoneville, Mississippi*

Joe Puglia, The PRGroup, *New Orleans, Louisiana*

I thank everyone for their attention to detail and their insightful suggestions for improving the manuscript. Every effort has been made to provide a sound and accurate text for students of water resources.

Thomas V. Cech

CONTENTS

BOOK COMPANION WEBSITE

Our Instructor Companion Site offers a wealth of teaching material for instructors and additional research guidance for students. Our assets include:

Instructor Resources:

- **PowerPoint LECTURE SLIDES** *Chapter-oriented slides including text art*
- **SAMPLE EXAMS** *Includes midterm exam and final exam*
- **STUDENT RESEARCH PAPERS** *Two sample student research papers for reference*

Student On-line Resources:

- **ANNOTATED WEBLINKS** *Useful weblinks selected to enhance chapter topics and content*

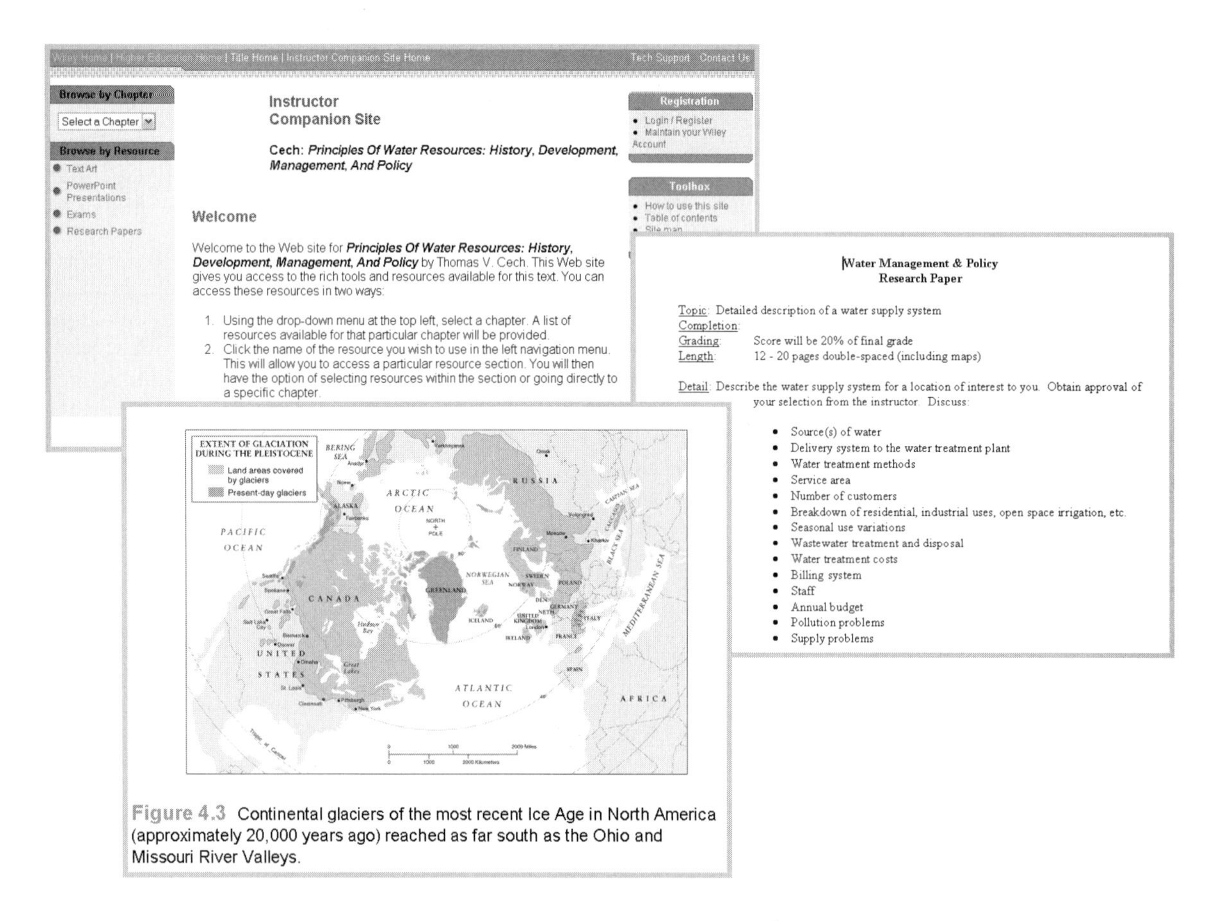

Go to **www.wiley.com/college/cech** to access these instructor and student resources.

A NOTE TO THE STUDENT

From space, the view of planet Earth is quite misleading. Our blue planet has approximately 75 percent of its surface covered by oceans, rivers, and lakes. However, almost 98 percent of that water is too salty to drink. Of the remaining 2 percent, less than half is found in rivers, lakes, or groundwater. And of that small fraction of freshwater, much of it is polluted from human activity.

The human body is like the surface of the Earth since almost 75 percent of our body weight is composed of water. We can live about 40 days without food before starving to death but only about three days without water before we die of dehydration. Humans require an abundant and clean supply of water to survive.

Worldwide, the use of freshwater is increasing. The world's population continues to increase—and so does water usage per capita. A 1996 report published in the journal *Science* estimated that humans use 54 percent of all freshwater found in rivers and streams on Earth.[1] A 1997 report from the Johns Hopkins University School of Public Health predicts severe water shortages for more than one-third of the world's human population by 2025.[2]

It took from the beginning of human existence until 1800 for the world population to reach approximately 1 billion people. By 1900, that figure increased to about 1.7 billion. In 1999, an unimaginable threshold of 6 billion people was reached. Where will our abundant and clean freshwater come from in the future? Students like you and others will be needed to work in the field of water resources to plan, develop, and implement water resources management projects to meet future needs. This course could be the beginning of a career in water resources.

Residents of more-industrialized countries such as the United States, Canada, Australia, the United Kingdom, Western Europe, and Japan to name a few are extremely fortunate to have safe, dependable, and relatively abundant supplies of freshwater. Through the years, extensive water delivery systems and water treatment facilities have been constructed by local, state, and federal agencies to serve the needs of citizens in these countries. This is not generally the case in less-industrialized countries. Many residents on islands in the Caribbean live in extreme poverty and daily must carry drinking water from community wells or other water distribution points to their homes. Numerous African countries have polluted water supplies—rivers and lakes that receive raw sewage from undersized or outmoded wastewater treatment plants.

Infant mortality rates in some of these countries are incomprehensible—an estimated 6000 children die every day from diseases associated with unsafe water and poor sanitation and hygiene. Approximately 3 million Africans die each year from waterborne illnesses. Unsafe water and sanitation cause up to 80 percent of all diseases in less-industrialized countries. According to author and global water expert Peter Gleick, the sanitation systems of ancient Rome would be a significant improvement for almost 3 billion (50 percent) of the world's population today. This is why your study of water resources is so important.

What is being done to improve the supply of water around the world? In China, the Three Gorges Dam is being constructed to provide hydropower for electricity, flood protection, and improved commercial shipping. The cost of this project, the world's largest dam, is great. Besides the financial expenses, more than 1 million residents in the inundation area will have to be relocated. Cultural sites, ancestral burial grounds, and productive farmland will be lost forever. And yet, the Chinese government and many residents are confident this project will propel the region into the twenty-first century (see Chapter 7).

In the Middle East, wars have been fought for centuries, and recently the lack of water has led to skirmishes, loss of life, and political unrest. In the United States, particularly in Texas, Colorado, Nebraska, Georgia, and California, water conflicts are present and growing annually. Drought, competing land uses, urbanization, and water entrepreneurs purchasing water for economic gain are at the epicenter of this conflict (see Chapter 14).

[1]Postel, S.L., G.C. Daily, and P.R. Ehrlich, "Human Appropriation of Renewable Fresh Water." *Science* 271: 785, February 9, 1996.

[2]Henricsen, D., Robey, B., and Upadhyay, U.D., "Solutions for a Water-Short World." Population Reports, Series M., No. 14, Baltimore, Johns Hopkins School of Public Health, Population Information Program, December 1997.

What can you do to make a difference? The future of water resources planning and management will require intelligent, dedicated, and visionary individuals. Students interested in biology, geography, engineering, law, chemistry, geology, environmental studies, watershed science, forestry, and other related fields will be in great demand. Federal, state, and local water agencies will be hiring such people for streamflow analysis, water quality sampling, ecosystem restoration, long-range planning projects, project management, GIS mapping, statistical and economic analysis, biological surveys, political analysis and lobbying, project management and construction, report writing, data acquisition and analysis, and public outreach. This course will benefit you in each of those job opportunities and could lead to a very fulfilling career in water resources.

Let's begin.

CHAPTER 1

HISTORICAL PERSPECTIVE OF WATER USE AND DEVELOPMENT

Life must be lived forward, but understood backward.

Søren Kierkegaard (1813–1855)
Danish Philosopher

Throughout history, the development, management, and policy of water resources have evolved in a variety of ways. In the arid Middle East, for example, elaborate irrigation projects were constructed thousands of years ago to raise food and fiber. The region between the Tigris and Euphrates rivers was known as the Fertile Crescent in large part because of the abundance provided by ancient irrigation projects. In China, canals in use today were built during royal dynasties around 600 B.C. and earlier to transport people, cargo, and armies. Later, similar construction techniques were used in Western Europe to develop elaborate water transportation networks.

For centuries waterwheels were used to divert water for crops, to provide water for fountains in royal gardens, to grind grain, and to supply drinking water. The technology of waterwheels, developed in Greece and Rome approximately 2000 years ago, transformed the economy of Western Europe by A.D. 1100.

Much of our knowledge of ancient hydraulic, or water-based, civilizations has been obtained from ruins, artifacts, and artwork—the remains of a Roman aqueduct in Italy, a bas-relief sculpture from Mesopotamia, or a temple in China. Ancient societies prospered when water supplies were properly managed; conversely, poor water management brought a decline in the health and well-being of citizens and, in extreme cases, even death to an entire civilization.

In this chapter, pay close attention to the historical context of the early hydraulic civilizations. What similarities existed between irrigation in Egypt, Babylonia, Spain, and Mexico? What role did the *qanat* play in the settlement of desert regions around the world? How important were canals in the settlement and prosperity of China and Western Europe? We can learn important lessons from these ancient civilizations, and can gain insight into present-day issues and opportunities regarding water development, management, and policy.

Knowledge of the past helps to anticipate the future.

Thucydides (460 B.C.–400 B.C.), Greek Philosopher

DRINKING WATER FOR EARLY CIVILIZATIONS

Water is the basis of life on Earth and the foundation of all civilizations. To show its importance in their culture, the ancient Persians listed "water" as the first word in their dictionary, calling it *ab*. The Egyptian civilization used a wavy line to represent the word "water." This symbol later became the Hebrew letter *mem* (representing *mayim* or water) and eventually the Latin letter M. (1)

Long before these early civilizations flourished, our Stone Age ancestors lived in caves and other camps that were close to sources of drinking water, such as springs and lakes. Wild game often congregated near these watering holes and provided a source of food. The needs of early people were quite basic—food, water, and shelter. As time passed and human populations increased, prehistoric communities tended to form near lakes in central Africa and along rivers in the Middle East, northern China, and India (Figure 1.1).

In some regions of the world, variable precipitation forced humans to relocate to wetter areas that had more reliable food supplies. The early Somalians of eastern Africa were nomadic because of their constant search for water and grass for their cattle herds. Between periods of drought and continual rains, this nomadic culture traveled great distances across deserts to reach greener pastures. Whenever possible, the Somalians dug groundwater wells by hand, at regularly spaced intervals along their desert routes, to provide drinking water for caravans of nomads and cattle. Groundwater wells in the desert provided a reliable source of drinking water for their own use and later served as the foundation for the development of small desert communities. Eventually, larger cities developed around these underground water sources in the African desert.

Why were ancient well diggers able to find groundwater in the middle of an African desert? The most probable answer is that during the Pleistocene epoch, when glaciers covered portions of North America and Europe, the climate in northern Africa and the Middle East was relatively wet. Modern satellite imagery of the region shows evidence of ancient riverbeds that have long since been covered by blowing desert sands. Ancient water from that geologic time remains underground even today.

SIDEBAR

Camels were the primary method of transportation across deserts because they could live without water for over a month. However, if a nomad miscalculated traveling distances, or if a desert groundwater well suddenly went dry, a caravan could run out of drinking water. This created a deadly situation, leaving the traveler with few options for survival. At this point, nomads were forced to resort to the *uusmiirad,* a procedure that involved extracting water from the stomach and intestines of a slaughtered camel, and then drinking it to survive. It was the ultimate sacrifice for the animal, but it meant life for the nomad. (2)

Groundwater systems became more elaborate in many parts of the world with the development of **qanats** (underground water delivery systems, from a Semitic word meaning "to dig"). Since around 1000 B.C., *qanats* have been constructed in southwest Asia, North Africa, and the Middle East to tap into reliable sources of groundwater. A *qanat* consists of a mother well connected to long, underground delivery tunnels stretching to nearby communities. A *qanat* mother well is dug by hand, generally near the foothills of a mountain range, to tap into plentiful sources of groundwater. A gently sloping tunnel is constructed from the mother well to villages and fields at lower elevations. Gravity provides the means to move groundwater from the mother well to lower elevations along the tunnel system. Vertical shafts are constructed along the sloping delivery tunnel to allow numerous access points. *Qanats* vary in length from 25 to 28 miles (40 to 45 km) and have depths up to 400 feet (122 m). (3)

In Afghanistan, Pakistan, and western China, a groundwater system with a mother well and delivery tunnel is called a *karez*. The word *falaj* or "unfailing springs" is used in Oman, while in Morocco and Cyprus it is called a *foggaras*. *Qanats* were probably one of the greatest hydrologic achievements of the ancient world, allowing communities to develop in locations without reliable water supplies. Today, most of the world's *qanats* are found in Iran (Figure 1.2). (4)

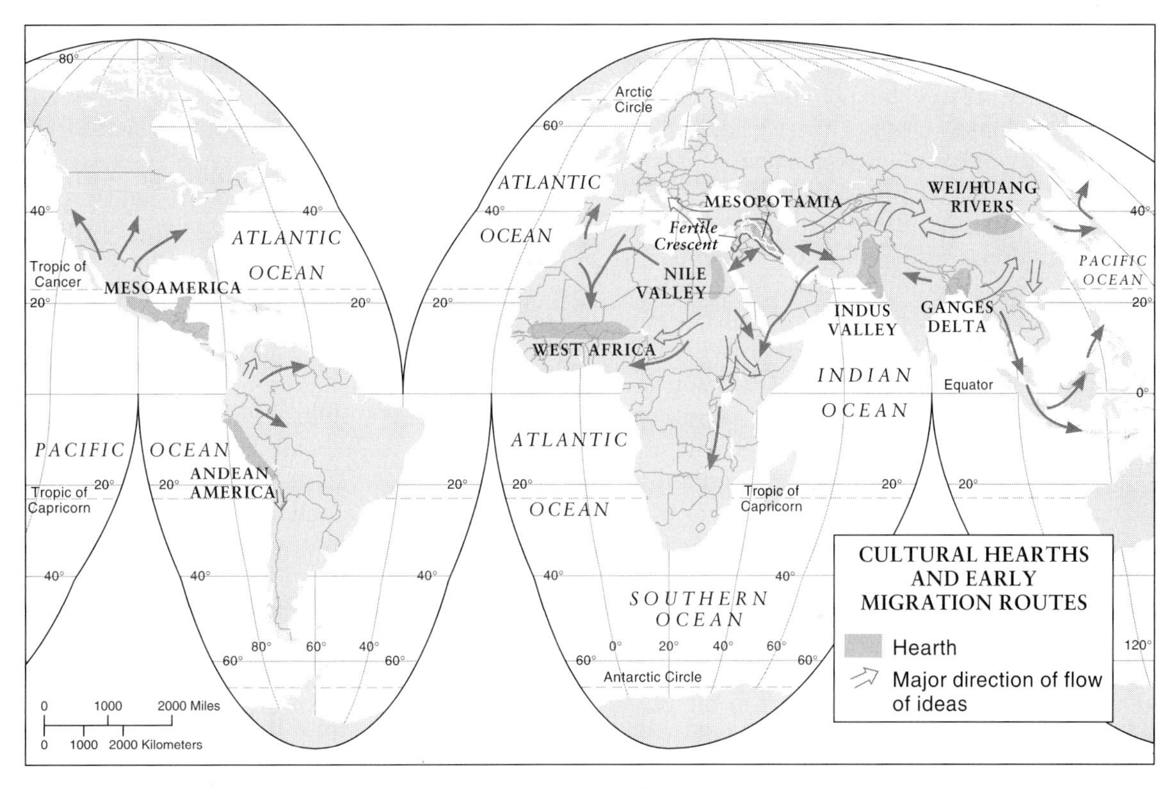

FIG. 1.1. Centers of early civilizations—Water was a necessity in these dry regions to provide food and fiber for survival. *Geography: Realms, Regions, and Concepts,* 10th Edition, by Harm de Blij and Peter O. Muller. Copyright © 2002 H. J. de Blij and John Wiley & Sons, Inc. This map was originally produced in color. Adapted and reprinted by permission of John Wiley & Sons, Inc.

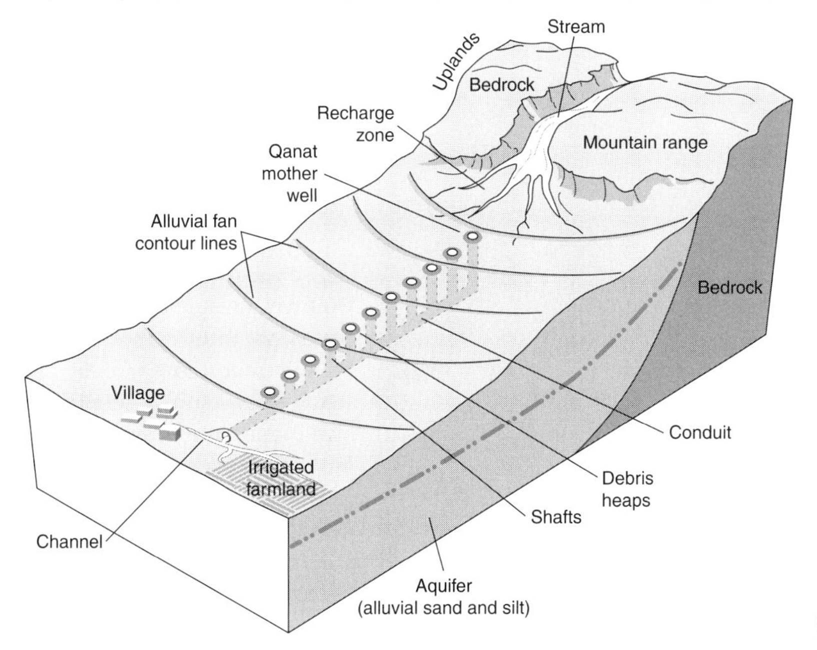

FIG. 1.2. Typical *qanat* system.

Qanats are still the traditional source of water in the Middle East and parts of China. In northern Iraq, they are used today to provide water to some of the oldest cities of the world, like Sulaimaniya (population approximately 400,000). Iran has over 22,000 *qanats* that supply 75 percent of all water currently used in the country. Spanish explorers transferred the technology to northern Chile where five hand-dug groundwater tunnels are still used today in the nitrate mining region of the Atacama Desert. (5)

Digging underground tunnels for a *qanat* was very dangerous work, and cave-ins were common. Generally, an underclass of the local population was forced to construct and maintain the water delivery system. Small boys were often used to dig in the cramped and more confined areas during construction, and the loss of life was shocking. On some construction projects, workers wore their funeral clothes as they dug in case the earth above the tunnel collapsed. This eliminated the need for co-workers to dig out a buried worker to provide a proper burial. (6)

The ancient Romans also developed extensive water delivery systems to their cities. Surface water and groundwater were stored in cisterns (underground reservoirs lined with clay or dug in limestone formations) at higher elevations near a city. Water from these underground storage reservoirs was distributed by gravity through a network of pipes to public fountains and baths, as well as to a few private citizens. The distribution pipes were made of lead or baked clay. Wastewater from the water system was generally returned to a river through sewer pipes buried beneath city streets. The first Roman **aqueduct** (an elevated water delivery system made of stone) was built in 312 B.C., and by 300 B.C. there were 14 aqueducts in Rome (see Figure 1.3) delivering 400,000 gallons (over 1.5 million l) of water daily to Roman citizens. (7) Excess water from these delivery systems was used to power the city's fountains and flush sewage into the Tiber River. (8)

SIDEBAR

Sextus Julius Frontius (A.D. 30–104) was the first water commissioner of Rome. In A.D. 97, he wrote a two-volume work, *On Aqueducts,* in which he described the advances in aqueduct construction since ancient times:

Will anybody compare the idle Pyramids, or those other useless, though much renowned, works of the Greeks, with these aqueducts, with these many indispensable structures? (9)

FIG. 1.3. Crossing of two Roman aqueducts in Via Latina southeast of Rome. These aqueducts were described as *Romanae providentiae magnitudinis que primitiae* (the first fruits of Rome's foresight and greatness) by author Raffaele Fabretti in 1680. This painting is by Zeno Diemer and is on display at the Deutsches Museum, Munich, Germany.

The construction of aqueducts and other water delivery structures allowed Roman cities to grow in size and population, and reduced the amount of time required for individuals (usually women) to obtain daily water supplies for the home. Water pipelines, aqueducts, and other delivery features became a symbol of a maturing civilization. Little or no concern was given to wastewater disposal as long as wastes were removed from homes, streets, and gutters. Natural cleaning processes in local rivers were able to purify some human wastes if the population density in an area was not too great.

The Romans also altered natural water resources features by digging canals to drain lakes and marshes. In A.D. 52, the Roman emperor Claudius constructed a canal to Lake Fucinus, a naturally formed lake near Rome. His plan was to grow crops on the fertile soil left behind in the dried lake bed. The canal included a 3-mile-long (5 km) and 7-foot-diameter (2 m) tunnel dug by hand through a mountain. In Damascus, the capital of present-day Syria, Roman workers constructed six canals to divert water from the Barada River for use in the city. A dividing point of all six canals was called "the parting of the streams" and was a popular recreation spot for Roman citizens. (10)

The Roman Empire also developed extensive water delivery systems in France, Italy, the Netherlands, and Great Britain. In A.D. 1236, a system of pipelines was constructed in London to carry water from the Thames River and nearby springs to residents in the city. In 1619, the New River Company (a privately owned company) delivered water throughout London, making it the first time that every home in a city received water through a network of pipes. By the end of the 1800s, most towns and cities in Great Britain had municipal water systems in place. (11) Some English communities privatized their public water systems (which were developed through the payment of taxes) by selling stock to fund the development of private water companies. In other locations, carts went door to door to deliver water to English homes.

A CLOSER LOOK

The search for water dominated the activities of early civilizations. Early inhabitants of central Africa, northern China, and India lived in regions where drinking water supplies were adequate for small numbers of people. However, as populations increased, human settlements relocated into areas that received less (or erratic amounts) of precipitation. This placed greater pressure on communities to develop more reliable sources of drinking water.

The Romans created the most extensive water delivery systems of the ancient era with the construction of aqueducts, although the *qanat* system in the Middle East was also quite elaborate. Why did ancient civilizations alter the natural environment in search of water supplies? What role did climate and geography play in decisions to dig wells, to construct elaborate aqueducts, or to drain lakes and marshes to create additional cropland? Did other alternatives exist during ancient times to meet the water needs created by population growth and limited water supplies?

Today, drinking water sources vary greatly around the world. For example, cities such as New Orleans, Louisiana, and Izmit, Turkey, use rivers as their drinking water sources. Paris, France, uses a combination of water from rivers and groundwater. Chicago, Illinois, obtains water from Lake Michigan, while Reykjavik, Iceland, uses groundwater that does not require any treatment because of its high quality.

Major drinking water issues today often revolve around water quality concerns. Much of the Earth's freshwater resources found in lakes, rivers, and groundwater have been contaminated to some degree, either naturally or by humans, and is not safe to drink without water treatment. We'll discuss these issues in detail in Chapter 5 and Chapter 11.

EARLY IRRIGATION AND FLOOD-CONTROL PROJECTS

Irrigation projects are found throughout the world today but were also prevalent thousands of years ago along major rivers of the world, such as the Nile, Indus, and Yangtze. Without irrigation, food was scarce during dry periods, and civilizations could not survive.

EGYPT

When the Nile inundates the land, all of Egypt becomes a sea, and only the towns remain above water, looking rather like the island of the Aegean.

King Herodotus, Egypt, 500 B.C.

Irrigation was vital for the permanent settlement of extreme northeastern Africa since the region receives only about 1 inch (2.5 cm) of precipitation annually. The ancient Egyptians relied on monsoonal rains in the mountains of Ethiopia to the south to bring floodwater and fertile sediments to the Nile River Valley of Egypt. During seasonal floods, water spilled over the banks of the swollen Nile and naturally irrigated adjacent lands. These floods were sometimes so extensive that ships didn't bother to follow the river but instead floated across the countryside as if on an ocean.

As the Egyptian population increased and more food was required, dikes and irrigation canals were constructed to direct floodwaters across more land in the Nile Valley. King Scorpion (circa 3200 B.C.) called the initial cutting of ground for a new canal the "Day of Breaking the River" (Figure 1.4). The development of each new irrigation canal signified the growing power and wealth of the Egyptian civilization through increased food production, the payment of additional taxes to the king or queen of Egypt, and a general improvement in economic prosperity.

To supplement the floodwaters of the Nile River, Egyptians used lifting devices (probably imported from Mesopotamia) to withdraw groundwater to irrigate crops. These ancient tools included the **shadouf,** a lever system with a bucket attached to a long pole on a pivot; the **tambour,** an auger-type device attributed to Archimedes and sometimes called the "Archimedes Screw"; and the **saqia** waterwheel, an elaborate animal-powered waterwheel (probably introduced from Persia) with multiple buckets to lift water (see Figure 1.5). These were the earliest irrigation "pumps" invented and are still in use today in the Middle East. (12)

FIG. 1.4. This earthen jug depicts King Scorpion cutting the first sod of an irrigation canal, a ceremony that continued into the nineteenth century in Egypt (now in the Ashmolean Museum, Oxford, England).

CHINA

In China, the Huang He River (also called the Yellow River) was the cradle of Chinese civilization. The Huang He has a long history of devastating floods and repeated attempts by Chinese dynasties to control, or conquer, the "raging beast." Precipitation in the Huang He River Basin ranges between 8 and 24 inches (20 and 60 cm) per year. This wide variation in precipitation leads to variable water needs throughout the country. In some locations the diversion of water for irrigation is a major concern, while flood control is practiced in other areas. Flooding has been a major problem along the Huang He for centuries, with thousands of lives often lost during single flood events.

Levees (earthen and rock dikes) were constructed over 2500 years ago along smaller branches of the Huang He River to control floods. However, larger flood-control projects were rarely attempted on the main river channel because the water flow was too great to control with the dam construction technology of the time.

Yu the Great was the first manager of Chinese waters and later became emperor of China (Figure 1.6). Around 2280 B.C., Emperor Yao asked Yu to construct dams, dikes, and other waterworks along the Huang He River to protect and enhance life for his citizens. Yu the Great was so successful in reclaiming land and controlling floods that after the death of Emperor Shun (Yao's successor), Yu became the new emperor of China. Later, Emperor Yu also became the head

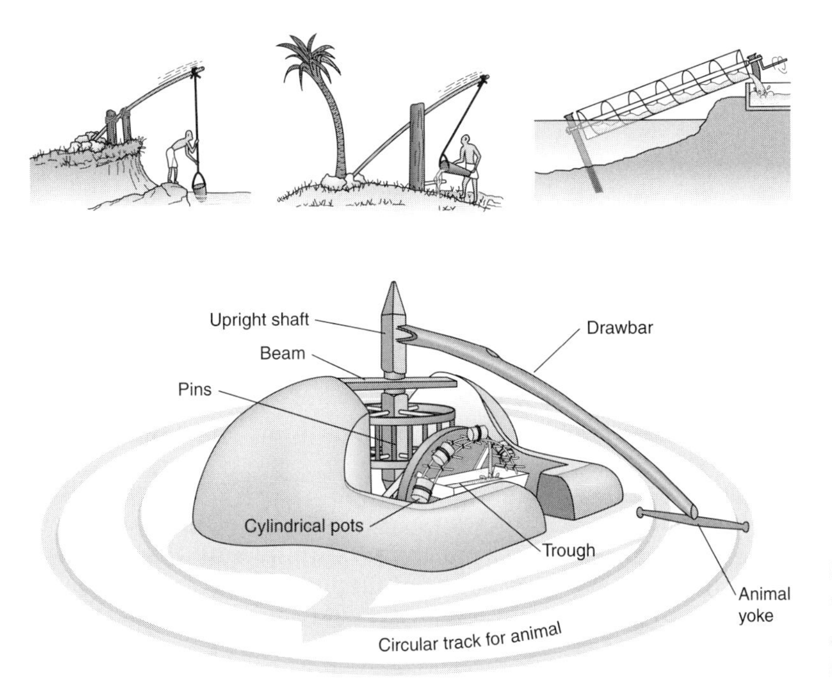

FIG. 1.5. Ancient water-lifting devices. Top left and top center: *Shadouf;* Top right: *Tambour* (Archimedes screw); Bottom: *Saqia.*

FIG. 1.6. This temple to Yu the Great is located at the foot of Kuaiji Mountain in Shaoxing City, Zhejiang Province, China. The three characters, from top to bottom on the temple, read *Da Yu Ling—Da* meaning "big," "great," or "majesty," *Yu* for Emperor Yu's name, and *Ling* meaning "mausoleum." The three Chinese characters together mean "The Majesty Yu's Mausoleum." Emperor Yu led the fight against floods throughout China and gained eternal respect from the people of his country.

of all water engineers in his country. Emperor Yu the Great is still honored in China today.

In addition to floods, food shortages were common in ancient China, so government-sponsored irrigation projects were developed to feed the growing population. Around 560 B.C., the Cheng State Irrigation Canal was completed for use in the north-central region of modern Honan. Around 300 B.C. the Changshui and Cheng Kuo canals were completed to irrigate extensive areas around Honan and Shensi. (13)

During the Chai Dynasty in about 300 B.C., the Ministry of Public Works for the Chinese government was responsible for water management, repair of levees, cleaning of irrigation ditches, drainage of floodwaters, and storage of water in reservoirs. Chinese dynasties were considered "good" or "bad" based on the maintenance and improvements they made to the country's irrigation systems. (14)

During the Ming Dynasty (A.D. 1368–1644), the chief water administrator in China advocated building huge stone levees along the Huang He River to channelize floodwaters. He theorized that confining the flow of water in the river channel would increase the velocity of floodwater and scour the riverbed to a greater depth. It was hoped that this greater capacity in the river channel would ultimately reduce flooding. Today, the same theory is being debated along the Platte River of central Nebraska.

Recent proponents of the concept argue that increased flows in the Platte will accelerate scouring of the river channel. This, in turn, will enhance habitat for the endangered whooping cranes that stop to rest along the Platte River Valley during their annual migration between the Gulf Coast of Texas and Canada. Opponents argue that the scouring concept is simply an attempt to claim additional water supplies from the neighboring upstream states of Wyoming and Colorado. (This controversy will be discussed further in Chapter 12.)

For centuries, many water managers along the Huang He River in China tried to alleviate flooding by "using nature to control nature." Some believed that natural river processes such as erosion, flooding, and silt deposition should be allowed to occur without interference from humans. It was argued that only small-scale levee projects should be used to control natural flooding events. Chinese water managers who followed this philosophy constructed small flood-control projects such as long, low levees that used natural sediment deposits (including silt, gravel, sand, and other earthen materials) found within the river channel. This use of natural materials to hold a river in its channel was thought to be a wise practice since it didn't greatly interfere with nature.

On the other hand, followers of other Chinese philosophies tried to control or "conquer" nature and used larger, more restrictive construction techniques. In the twentieth century, for example, the government of the former Soviet Union waged "a war against nature" to control the floods of the Huang He River. The communist philosophy was to "force" the river to serve useful purposes for Chinese residents. Russian planners, civil engineers, and hydrologists (those who apply scientific and mathematical principles to solve a water-related problem) devised elaborate flood-control and irrigation projects throughout the country. Contrasting philosophies caused conflict, and debates over proper water management methods continue today in China and around the world. This issue is discussed further in Chapter 7 in a guest essay regarding construction of China's Three Gorges Dam.

SIDEBAR

Levees were used not only for flood protection in China, but also as instruments of war between feudal states. Between 404 B.C. and 221 B.C., dikes were used to divert floodwaters onto an opponent's fields and villages. Loss of life was extensive, and eventually treaties were signed to ban this practice of hydraulic warfare.

THE MIDDLE EAST

The ancient Sumerian culture had two very unpredictable rivers—the Tigris and Euphrates (located in present-day Turkey, Syria, and Iraq). Numerous irrigation projects were constructed, and the science of flood protection was well developed by Sumerian rulers. Taxes were collected from irrigators, and extensive laws were adopted to properly operate and maintain their irrigation systems.

The Assyrians, located in portions of modern-day Turkey, Iran, Iraq, and Syria in approximately 2400 B.C., also created extensive irrigation laws. Various rules were developed for water use that was obtained from precipitation, groundwater, and water stored in underground cisterns. All irrigators had to share in the work of removing sand and gravel that was deposited in irrigation canals by slow-moving water. In addition, irrigators had to minimize water contamination, assist with canal repairs after floods, and ensure that other water users at the end of the ditch received their fair share of water. Irrigators unwilling to cooperate with these rules were either beheaded or stoned to death. (15)

Around 500 B.C., rainfall harvesting was developed in the Middle East to channel surface water runoff for irrigation of crops. Stone walls were constructed to divert precious precipitation directly to crops or into underground cisterns for drinking water. These methods are still used today in the Judean Desert of Israel.

INDIA, SPAIN, PORTUGAL, AND SOUTH AMERICA

A water-harvesting system similar to the methods used today in Israel was developed in the Thar Desert of western India. Stone walls, cisterns, dams, water holes, and tanks captured enough stormwater to allow thousands of people to live in the desert. Roman and Moorish invaders of Spain and Portugal brought irrigation techniques to the Iberian Peninsula around A.D. 800. **Acequias,** or irrigation ditches, were developed as well as **acenas** (water mills) and **charcas** (reservoirs). In South America, stone-walled terraces have been found that were drained with elaborate ditch systems on steep hillsides of the Andes Mountains. In Peru, the lost city of Machu Picchu (Figure 1.7) stands as a monument to the engineering skills of the ancient Incas. Similar technology was used in Bolivia, Ecuador, Colombia, and Suriname.

NORTH AMERICA

Large-scale irrigation began in the present United States with the efforts of the **Hohokam Indians** in approximately A.D. 800. Canals 30 to 60 feet (9 to 18 m) wide diverted water from the Salt River and irrigated land near the present site of Phoenix, Arizona. Almost 300 miles (483 km) of canals irrigated 250,000 acres (101,175 ha) of desert land and provided food for 200,000 people. (16)

Around A.D. 950 the **Anasazi Indians** developed community irrigation projects in the desert lands of southwest Colorado. Small reservoirs were used to collect surface water runoff during rainstorms. Upstream water channels were lined with rocks, soil, and brush to divert water into reservoirs for drinking water or onto fields for

FIG. 1.7. The ancient Inca ruins at Machu Picchu in Peru still contain remnants of an expansive drainage network. It is located on a high mountain ridge at an elevation of 7,999 feet (2,438 m) in the Andes Mountains. Machu Picchu had a permanent population of about 300, but was abandoned around A.D. 1540. See http://www.waterhistory.org/histories/machu/ for an excellent overview of water supply and drainage at Machu Picchu.

crop irrigation. Other methods used small rows of strategically placed rock dams (today called check dams) to flood small fields of vegetables, beans, and corn. (17)

In the fifteenth century, Spanish settlers migrated north from Mexico to modern-day California, Arizona, and New Mexico. Spanish missionaries constructed many small irrigation canals to raise crops near their churches. Some of this irrigation technology was obtained from the ancient Aztecs of Mexico, while other techniques were replicated from farms in Spain and Portugal. The Aztec civilization had an impressive system of aqueducts that delivered irrigation and drinking water to their communities, including the Aztec capital of Tenochtitlán, at the present site of Mexico City. The Aztecs used rock and mortar for these aqueducts, and even designed dual water delivery systems so that one could be cleaned while the other provided an uninterrupted water supply to residents.

Another major irrigation effort began in North America in 1847 when **Brigham Young** (1801–1877) and 1500 Mormon followers settled in the Salt Lake Valley of Utah Territory, a dry region that receives only 15 inches (38 cm) of average annual precipitation. Fortunately, over 40 inches (102 cm) of water falls as snow during the winter months in the nearby Wasatch Mountains. Mormon settlers constructed diversion dams across river channels and diverted runoff from melting snow into irrigation canals (see Figure 1.8). These early dams were not elaborate and were typically made of logs, rocks, and brush (sticks and weeds) that were similar to methods used by the Anasazi Indians centuries earlier. Irrigation canals were constructed with the use of horses pulling plows followed by workers with picks, shovels, and their bare hands. (18)

Construction of an irrigation canal was not a simple task. A canal with too steep a slope would allow water to move too fast. This often led to erosion problems that could wash out the side of an earthen canal. A canal that was too flat caused irrigation water to flow too slowly or to pool and not move at all in low spots. Canals were usually constructed with a fall of about 2 feet per mile (0.4 m/km). Some early settlers in the Salt Lake Valley used the water surface in a tea cup as a "level" to guide canal diggers. (19)

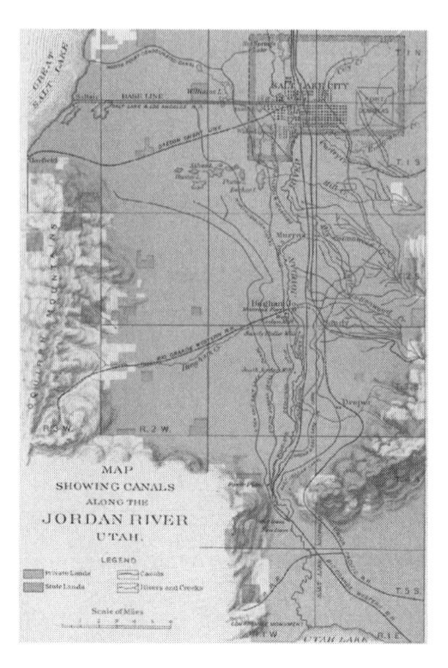

FIG. 1.8. This map was made for a 1903 U.S. Department of Agriculture report and shows the primary irrigation ditches in the Salt Lake City, Utah, area. The Mormon leader, Brigham Young, was largely responsible for the development of these water delivery systems throughout the Salt Lake Valley.

During the initial years of settlement, competition for irrigation water in the Salt Lake Valley was limited because the Mormon settlers were part of a patriarchal society. That is, the welfare of the community generally came before the needs of the individual, and that philosophy was enforced by their leader, Brigham Young. (It was also quite clear to these early settlers that survival depended on cooperation to grow food in the arid valley.) Water resources were treated in the same manner and were shared during times of scarcity. In later years, this concept of equal rights for water use was replaced throughout much of the West by a rigid water allocation system based on a strict priority system of water use (see Chapter 7).

In 1862, the U.S. Congress passed the **Homestead Act,** a landmark piece of federal legislation that opened the floodgates of development in the West. This law allowed anyone over the age of 21 to acquire ownership of 160 acres (65 ha) of land if they lived on it for a period of five years and made improvements such as constructing a house or barn, cultivating crops, or planting trees. The cost of the land was only $1.25 per acre ($3.09 per ha). As more people moved to the arid West, water for irrigation became a critical issue. In areas with flowing rivers, settlers constructed irrigation canals to grow food for growing communities. In other locations, groundwater wells were dug by hand, and windmills were used to pump it up to the land surface.

A CLOSER LOOK

"Obtaining water loomed as the settlers' first and most pressing problem. Before the 1880s, no one attempted to farm any distance from a stream. But the great majority of newcomers staked claims wherever they could, relying on nature and neighboring wells until the time came when they could dig their own. Lacking special drills and often the horsepower to run them, settlers dug into the ground with tools as basic as picks and shovels. A descent of forty feet or more to reach water was not unusual. In the meantime, settlers hauled water from some distant source, usually a well located in town or on a neighbor's claim. In the absence of windmills, they pumped or raised water by hand, a time-consuming and strenuous task.

"From hard experience, high plains farmers soon learned to appreciate a dependable water supply. With a well and a windmill, they could keep stock; and stockraising proved a crucial hedge against failure. By combining crop culture with keeping small numbers of horses and cattle, a cow for milk and butter, and a few pigs and chickens, a family had a better chance of hanging onto their land through the droughts that lay ahead." (20)

Katherine Harris, *Long Vistas—Women and Families on Colorado Homesteads*. Reprinted with permission of the University Press of Colorado, Niwot, Colorado.

SIDEBAR

Sears, Roebuck & Company had an extensive line of windmills in the late 1800s to pump groundwater. The rotor blades were turned by the wind and rotated a gear-driven shaft that pumped a sucker rod. The up-and-down movement of the rod forced groundwater to the land surface. Windmills offered in the Sears, Roebuck & Company catalog ranged from the "direct-stroke" wooden model to the "high-geared" steel line. Most models had 8-foot (2.4-m) rotors on top of towers that were 30 to 40 feet (9 to 12 m) high.

Around 1870 **Horace Greeley,** an editor for the *New York Tribune,* promoted the phrase "Go West, Young Man," to encourage settlement in the western United States. Through newspaper editorials and speeches around the country, Greeley expressed his strong belief that the United States needed to settle the vast regions west of the Mississippi River. The timing for his promotion was ripe—the Civil War had recently ended in 1865, and construction of the transcontinental railroad between Sacramento, California, and Omaha, Nebraska, had been celebrated in 1869. Greeley organized a group of settlers and investors from the East, and encouraged them to settle at the confluence (junction) of the South Platte and Cache la Poudre rivers in northern Colorado. In many ways, Greeley's group hoped to replicate the success of Brigham Young and his followers who settled in the Salt Lake Valley 23 years earlier. The geography of Horace Greeley's settlement (today called Greeley, Colorado) was in many ways similar to the site selected by Brigham Young—irrigation water was required to grow crops in the arid climate; adjacent streams had ample water supplies when snow melted from nearby mountains; and fertile land was available for crop production if adequate water was applied.

Construction of an irrigation canal (also called an irrigation ditch) began immediately as Horace Greeley's settlers arrived at their new community. (Greeley stayed in New York City to attend to newspaper duties.) Horses and mules were used to pull metal scrapers called *slips* to construct the 15-mile-long (24 km) Greeley Irrigation Company Ditch. At the time, this effort was the largest community irrigation project in the United States. Over the next 20 years, scores of irrigation ditches were constructed in the South Platte River Basin of Colorado, and the area prospered, becoming one of the top agricultural producing regions in the country.

In later years, and into the twentieth century, irrigation projects expanded across the United States. Settlers in arid regions relied on irrigation water to produce crops such as corn, wheat, oats, sugar beets, alfalfa, and vegetables. Some crops, such as corn, oats, and alfalfa, were fed to livestock on farms to be used for food or to sell to buyers for cash. Sugar beets were produced for sale to local sugar factories for refining, while vegetables were grown for sale to local urban markets. Settlers in the West without adequate irrigation water were forced to move back to more humid regions (areas of greater precipitation) east of the Mississippi River, or moved into town to work in other professions. Entire farming communities were abandoned in later years when crops died as a result of inadequate water supplies. Drought during the 1930s drove thousands of settlers from farms and ranches to seek new opportunities in California, Oregon, and Washington. Local economies of the abandoned areas were devastated, and many lives were financially ruined.

Irrigation continues to be extensively used around the world today. Between 1970 and 2000, the Earth's population increased by 2.3 billion people. This growing demand for food and fiber places great stress on food production and delivery systems around the world. Conflicts are increasing over water supply needs between irrigators and urban centers, over the need to use fertilizer and other chemicals to enhance crop production while not polluting rivers, lakes, or groundwater, and over the high production costs of operating irrigated farms. These issues will be discussed further in Chapter 5, Chapter 6, and Chapter 14.

POLICY ISSUE

Ancient civilizations developed irrigation water projects to provide food for growing communities. Common crops produced were grains such as wheat and corn, vegetables, and hay for livestock such as horses and cattle. Without supplemental water supplies, crops withered and died owing to inadequate moisture during the growing season. Irrigation meant food and life.

As technology improved, mechanical devices were invented to irrigate lands that were formerly at too high an elevation or were too far from a stream to receive water. The *shadouf, tambour,* and *saqia* were early methods used to lift water to fields that were higher than nearby water supplies. Dams and other methods of water diversion were later constructed to change the course of streams to deliver irrigation water to additional lands.

Irrigation has its drawbacks. Land irrigated year after year can become unproductive through a buildup of salts naturally found in the soil. Sediments (silt, gravel, and other earthen materials) can accumulate behind dams and can eventually make these structures unusable. Centuries of irrigation can destroy the ability of soils to produce adequate crops. When this loss of productive land occurs, an entire civilization may collapse, as has happened over the ages. Could early inhabitants in dry climates have developed alternatives to irrigation, or is it inevitable that humans will alter the natural environment by developing irrigation projects when food is needed?

EARLY WATER TRANSPORTATION DEVELOPMENT

EGYPT AND GREECE

Inland waterways such as rivers and canals have been used around the world for centuries to transport people and goods. In ancient Egypt, the movement of commerce relied on navigation of the Nile River, its tributaries, and artificial canals since few roads existed in this desert region. Approximately 80 canals were constructed in the Nile River Valley by various dynasties for navigation and irrigation. In Greece, a canal was constructed for navigation across the Isthmus of Corinth to connect the Aegean and Ionian seas. However, opposition by Greek merchants along the longer, more circuitous water route nearly stopped the project before it was completed. The merchants were afraid that business would suffer due to a loss of customers sailing past their establishments. They were right.

CHINA

Chinese commerce relied primarily on river transportation along the Huang He and Yangtze rivers. Combined, these two massive inland waterways allowed navigation on over 5000 miles (8000 km) of mainland China. Since these rivers flowed west to east, a north-south "Grand Canal," the Da Yunhe, was constructed to allow the transport of freight, grain, and military troops between river valleys of the Wei, Yangtze, Hai, and Huang He.

The Da Yunhe Canal extends nearly 1000 miles (1600 km) between Beijing to the north and Hangzhou in the southern part of the country (Figure 1.9). The canal maintains a depth of approximately 10 feet (3 m) and has an elevation change of only 138 feet (42 m) over its entire course. However, boats must contend with steep gradients of 20 to 30 feet (6 to 9 m) in a few locations. Inclined planes (ramps) were constructed in ancient times out of stone, and boats were dragged up or down these steep sections by horses and gangs of men. The ramp had a relatively smooth surface that extended the length of the unnavigable reach (section) of the canal. In other locations, water from rivers and lakes was released into the canal, and sluice gates (wooden barriers) were used to maintain adequate water levels for navigation.

In some areas it was not possible to dig the Da Yunhe Canal to the depth necessary to maintain a gradual elevation change. As technology

FIG. 1.9. The Grand Canal (in Chinese, Da Yunhe or "great transport river") is the world's longest artificial canal and was constructed during the fourth and fifth centuries B.C. Peasants did most of the work, and nearly half of the six million men who worked on the canal died during construction. Today, the Da Yunhe Canal connects the major Chinese cities of Beijing in the north and Hangzhou to the south.

improved, crude locks were constructed to replace inclined planes at some locations. **Locks** are basically a series of steps, or step chambers, that can be filled or emptied with water. Once a vessel enters a lock, water-tight doors are closed, and sluices (wooden doors) are opened to release water into or out of the lock.

Filling a lock with water elevates a boat to a higher elevation; releasing water from the lock lowers the elevation of the vessel. When the desired height is reached, lock doors are opened and a boat continues on its journey. Occasionally, flash locks were used to provide a quick release, or flash, of water. This temporary discharge of water increased the amount of water in a section of a canal or small stream, raised the elevation of the water surface, and allowed a vessel to pass to a higher elevation along the Da Yunhe.

A section of the Da Yunhe Canal was extended to Beijing around A.D. 600 to supply northern armies with food and other supplies. This helped improve China's defenses and also increased economic trade between the northern and southern regions of the country. The Chinese economy relied heavily on canal navigation because it allowed more grain to reach urban markets. Increased grain sales also meant more taxes paid to the Chinese government. (21) The Da Yunhe Canal is still the longest artificial waterway in the world and represents an engineering feat comparable to the Great Wall of China.

EUROPE

Numerous canals were built throughout Western Europe. The construction of the first navigation canal in France, the Canal de Briare, was completed around 1610 to link the Loire and the Seine rivers. It was approximately 27 miles (43 km) in length. The Canal du Midi was completed in 1681 during the reign of Louis XIV and was the largest construction project in Europe up to that time. France developed approximately 3000 miles (4800 km) of artificial canals that interconnected with 4600 miles (7400 km) of navigable rivers. On the Thames River in England, sluice gates were often used to maintain adequate water levels for navigation. These gates channeled water to one side of a river to increase depths for navigation. (22)

Freight transportation along waterways was extensive in the region known today as the Czech Republic. As early as the sixth and seventh centuries, corn was transported on rivers from Bohemia to be traded with the Saxons for weapons and cloth. In the eleventh century, landowners along the Czech waterways revolted when excessive tolls were charged for the transport of beer and field machinery (cutting sickles). All barge traffic was stopped along the route until the ruler of Bohemia, Charles of Luxemburg, ordered the opening of Bohemian waterways to free passage. (23)

A CLOSER LOOK

Leonardo da Vinci (1452–1519) was a great water engineer and military planner. Around 1500, he developed plans to improve navigation and irrigation in the Arno River Valley around Milan and Florence. A byproduct of this scheme was the drying up of the river at Pisa to destroy the city's water supply, thereby winning a war that had been going on for 10 years. Da Vinci designed canals, sluice gates, locks, and diversion dams to implement his plan for war and peace (Figure 1.10).

Niccolo Machiavelli, a government official in Florence (who would become famous for the Machiavellian method of politics of corruption and ruthlessness), was da Vinci's cohort in the water scheme. If successful, the Florentine economy would have flourished, da Vinci would have been rich, and Machiavelli would have been a political heavyweight in Italy. However, the plan was not successful for a variety of reasons. Da Vinci went on to other interests, and Machiavelli was ousted from public office. (24)

UNITED STATES

George Washington (1732–1799) was a strong proponent of canal construction in the original colonies of the United States. Through his travels, he had seen extensive inland waterways in England and France, and knew the great benefits they would provide for the economy of the young nation. At the end of the Revolutionary

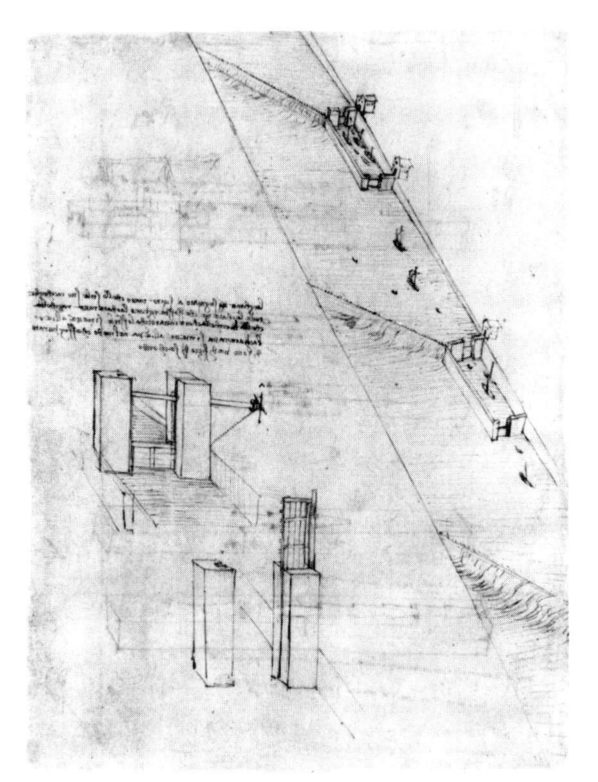

FIG. 1.10. These sketches are based on Leonardo da Vinci's studies of navigation on rivers with irregular flow rates. The left side of the drawing shows two sketches of dropdown sluice gates designed to allow boats to go either upstream or downstream. The annotation explains how the gates are operated by means of a winch. The drawing is from *Il Codice Atlantico di Leonardo da Vinci nella biblioteca Ambrosiana di Milano,* Editore Milano Hoepli, 1894–1904. The original is at the Biblioteca Ambrosiana in Milan, Italy.

War, he took a six-week trip across the Allegheny Mountains to develop recommendations for a canal from the Ohio River to the Potomac. Washington later organized a company to build the Chesapeake and Ohio Canal, but it was only partially completed. Portions of the canal can still be seen in Washington, D.C., as part of the National Capital Parks System.

In the early nineteenth century, the United States had only 100 miles (160 km) of canals. However, the **Gallatin Report**, presented to Congress in 1808, fueled dreams of westward expansion. Albert Gallatin, U.S. secretary of the Treasury, completed a comprehensive review and survey of all existing transportation routes in the United States. His report to Congress recommended that the federal government make a strong commitment to construction of new roads and canals to the West, with an emphasis on a canal/river link between the Hudson River and the Great Lakes.

SIDEBAR

De Witt Clinton, mayor of New York City and later governor of the State of New York, was a strong proponent of the Erie Canal, a project that began in 1817 and was completed by 1825. It was 363 miles (584 km) in length and connected Lake Erie near Buffalo to the Hudson River at Albany. However, not everyone was in favor of the $7 million canal. Although "Clinton's Ditch" was ridiculed as a potential money loser, freight and passenger traffic repaid the project in only 10 years. Today, only pleasure craft use the Erie Canal, and tolls paid by travelers on the New York Thruway (I-90) underwrite the operation and maintenance of the State Canal System.

Shortly after the Gallatin Report was presented to Congress, the country went "canal crazy." The state governments of Virginia, Pennsylvania, Maryland, New York, Ohio, New Jersey, Indiana, and Illinois entered the canal-building business and developed waterways such as the Erie, Pennsylvania Main Line, and the Chesapeake and Ohio canals. These new transportation routes helped open markets to farmers and ended the isolation of "western" areas of the United States. Small towns grew as emigrants streamed west of the Allegheny Mountains, employment increased, and the nation's economy strengthened. The Erie Canal cut travel time between New York City and Buffalo, New York, from 20 days to 6 and reduced the cost of moving freight from $100 to $5 per ton. (25) By the end of the 1800s, the United States had over 4000 miles (6400 km) of navigable canals.

The Mississippi River also played a major role in the history of navigation in the United States.

In the 1700s, keelboats and flatboats were the typical mode of transportation on the Mississippi until the introduction of steamboats in the 1800s. The first steamboat to travel on the Mississippi River was probably the *New Orleans,* built by Robert Fulton and Robert Livingston in 1810 in Pittsburgh. (26)

Obstacles and other natural barriers were common along the Mississippi River during the 1800s. Floods, erosion, and variations in streamflow all combined to make navigation on the river quite dangerous (see Figure 1.11). Hidden sandbars and deadly "snags" (a Scandinavian word meaning "sharp") were a constant worry and led to a refinement of terms to describe different types of snags in the Mississippi:

Planter A tree trunk buried in the sand on one end but floating free at the other end.

Raft A snag that almost completely blocked a channel.

Sawyer A snag balanced so delicately that a breeze caused it to bob up and down.

Sleeper A water-logged tree floating just beneath the surface of the water. (27)

SIDEBAR

A new obstacle to navigation came along in 1856 when a railroad bridge was constructed across the Mississippi River between Davenport, Iowa, and Rock Island, Illinois. Shortly after completion, the steamer *S.S. Effie Afton* rammed a pier (bridge foundation) and severely damaged the bridge. Immediately, the steamer company sued the Chicago & Rock Island Railroad Company for obstructing navigation on the river. The local court agreed and called the bridge a public nuisance. The railroad company appealed the ruling and took the case all the way to the U.S. Supreme Court where the decision was reversed. The Court issued a landmark ruling in favor of the railroad company after listening to the compelling arguments of their attorney from Illinois, Abraham Lincoln.

Steamboats of the 1800s could travel along the Mississippi River from New Orleans north to St. Louis and beyond on the main stem of the Mississippi, to the northeast on the Ohio River, or northwest along the Missouri River into Montana. Navigation provided a critical supply of materials to western cities and outposts until the construction of railroads took business away from the riverboats. Powerful railroad executives and politicians helped secure favorable government contracts for the construction of railroads around the country. This provided direct competition to navigation interests for the transportation of materials and led to a steady decline in the number of riverboats operating along the Mississippi River and its tributaries.

Even now in the twenty-first century, water transportation on rivers and lakes continues to be a very important mode of transportation around the world. Grains, textiles, cotton, lumber, fuel, and innumerable other commodities are shipped daily on intricate networks of rivers and lakes in many parts of the world. Unfortunately, conflicts over water flows and the needs of other water users and the natural environment are becoming more frequent. See Chapter 7 for more information on present-day water transportation issues.

EARLY HYDROPOWER DEVELOPMENT

The energy contained in moving water has been used to power some of the earliest machines made by humans. The first waterwheels were developed to grind grains such as corn and wheat for human consumption. Later, improved waterwheel technology was utilized to power factories that produced textiles, wood products, and metal machines. More recently, modern versions of the waterwheel have generated electricity when constructed within the hydropower facilities of a dam.

In Mesopotamia and ancient Egypt, the **noria,** or Egyptian waterwheel, was used to divert water from a stream or groundwater well for irrigation. This waterwheel had a chain pump connected to a series of earthen pots rotated by a wheel. Water was lifted by the pots and then spilled into an irrigation channel. If the flow of the stream increased, more jars were added to

FIG. 1.11. River obstructions have plagued navigation on the Ohio, Missouri, and Mississippi rivers and their tributaries since the 1800s. Due to the economic depression of 1838, no federal snag removal efforts were made between 1839 and 1842. During that time, hundreds of steamboats sank and caused enormous problems for the transportation of people and materials. This photo is of the U.S. Army Corps of Engineers snagboat *R.E. DeRussy*, which was built in Albany, New York, in 1867. The vessel had two hulls and a windlass for hoisting snags and wrecks from the Ohio River and its tributaries. Note the workers with axes standing in the pile of trees in the foreground. Snag removal was very labor intensive.

the waterwheel. If water flow decreased, jars were removed to allow the *noria* to rotate more efficiently.

Waterwheels were used in Greece as early as 100 B.C. to grind grain. Water from a stream flowed into a horizontal paddle wheel that contained numerous buckets mounted on a vertical shaft. The weight of water in the buckets caused the waterwheel to rotate. This turned a shaft attached to a millstone that ground corn or wheat for bread or other food items. This simple technology, called milling, spread quickly and was used in China by A.D. 100. (28)

The City of Hama in present-day Syria received its water from the River Orontes. River water was lifted by huge waterwheels and then emptied into enormous stone aqueducts. They were waterproofed with a 1-foot-thick (0.3 m) layer of primitive concrete made from lime, sand, and broken limestone. Millions of limestone blocks formed the archways across valleys.

Around A.D. 300 the Romans improved waterwheel design by using a horizontal drive shaft attached to a vertical wheel. This new shape allowed the buckets to be placed just below the surface of a flowing stream. Again, the weight of the water in the buckets turned the waterwheel.

By 1086, there were 5000 mills operated by water power in England. By 1800, the number had grown to over 500,000 mills throughout Europe. Many included a **mill raceway,** or canal, that diverted water from a river to provide an efficient and steady flow of water for a waterwheel. Mills powered by waterwheels ground corn and wheat, powered bellows and hammers to make iron, ground ingredients to make paper, cut wood, crushed olives for oil, drilled gun barrels, and powered textile factories. (29)

Conflicts over waterwheel diversions and river navigation occurred as early as A.D. 1000. In England, waterways were under the jurisdiction of the Lord High Admiral. Since most watermills required the construction of a dam across a river to supply water to a raceway, conflicts with navigation traffic would inevitably occur. The dam at Chester was installed to power mills in the city, but the dam had to be constructed across the River Dee. When royal command ordered it removed to restore navigation, the business community in the city objected. The

Lord High Admiral agreed with the merchants and refused the order. The dam at Chester still stands today. (30)

The mills on the Blackstone River in Rhode Island and Massachusetts are excellent examples of U.S. watermills (Figure 1.12). The Blackstone has a drop in elevation of over 400 feet (122 m) between Worcester, Massachusetts, and Providence, Rhode Island, and a massive 31-foot (9.4-m) drop at Woonsocket, Rhode Island. This reach of the Blackstone River became a hotbed for waterwheel development, and Woonsocket became one of the largest textile manufacturing centers in the United States in the 1800s.

Conflict between navigation and watermills in the United States was common. If a miller needed to divert water for a watermill at the same time a boat needed to pass by, water in a river had to be shared. In addition, the miller was also required to pay the navigator for lost water, or "flash," since navigation was slowed down, or "injured." The law recognized that river navigation existed long before the miller arrived. It was commonly agreed that the earlier (or "prior") water use for navigation had a priority over water needs at a watermill.

FIG. 1.12. Slater Mill and Dam are located on the Blackstone River in Pawtucket, Rhode Island, at the head of Narragansett Bay. This early mill was constructed by Samuel Slater in 1793 and was the first cotton mill in the United States to use mechanical spinning machines. By 1800, the town of Pawtucket had 29 cotton mills, and by 1830, there was a dam located every mile along the Blackstone River and its tributaries. Today, the Slater Mill Historic Site is a popular tourist attraction as the birthplace of the Industrial Age in America.

The first recorded industrial use of the Niagara River, a boundary between Ontario and New York, was in 1759. A small canal was dug to power a waterwheel for a sawmill. In 1875, the Niagara Falls Hydraulic Power & Manufacturing Company built a canal 35 feet (11 m) wide and 8 feet (2.4 m) deep to divert water from the Niagara River, above the Falls, to sawmill sites below. The 150-foot (46-m) drop in elevation provided massive amounts of water energy to turn the waterwheels. A few years later, those same wheels generated electricity. (31)

Thomas Edison's development of the incandescent light bulb created the demand for cheap, plentiful power sources to turn electrical generators. The first hydropower station for Edison's electric system was a waterwheel constructed on the Fox River in Wisconsin in 1882. The Westinghouse Company installed the first electric generators along Niagara Falls near Buffalo, New York, and later at other locations around the United States and Canada. By the early 1900s, hydropower provided over 40 percent of all electricity used in the United States, and by 1945 it supplied 75 percent of the electrical needs for the Pacific Northwest. Today in Canada, 60 percent of all electricity is generated with hydropower, while worldwide the figure is approximately 15 percent. (32) See Chapter 7 to learn more about modern power production at Hoover and Grand Coulee dams in the western United States.

POLICY ISSUE

The invention of the waterwheel was an extremely important event in history. One of its earliest uses was to divert water from a river or a groundwater well for irrigation. By 100 B.C. waterwheels were used to grind grain, and the technique quickly spread to other civilizations. Centuries later, the waterwheel was redesigned to operate mills around the world. Significant events related to the development of waterwheels are included in Table 1.1.

TABLE 1.1 Selected Water Development Events, 3200 B.C.–A.D. 2004

Year[a]	Event	Present Location
B.C.		
3200	King Scorpion proclaims "Day of Breaking the River"	Egypt
2280	Yu the Great constructs various waterworks	China
1000	*Qanats* constructed	Middle East
560	Cheng State Irrigation Canal completed	China
500	Water harvesting developed	Middle East
500	Dikes and levees constructed for flood control	China
312	First Roman aqueduct built	Italy
100	Waterwheels used to grind grain	Greece
A.D.		
100	Waterwheels used to grind grain	China
300	Romans improve design of waterwheels	Italy
800	Hohokam Indians develop irrigation	Arizona
800	Irrigation introduced by Romans and Moors	Spain
950	Anasazi Indians use irrigation	Colorado
1000	Conflict between boats and waterwheel in Chester	England
1086	5000 mills operated by waterwheels	England
1400	Spanish settlers migrate north to California, Arizona, and New Mexico	United States
1500	Leonardo da Vinci proposes irrigation/navigation system at Milan	Italy
1610	Canal de Briare completed	France
1619	Network of pipes deliver water to every home in London	England
1800	Waterwheel development rapidly increases	Rhode Island
1808	Gallatin Report presented to Congress	Washington, DC
1810	Steamboat *New Orleans* constructed in Pittsburgh	Pennsylvania
1825	Erie Canal completed	New York
1847	Mormons develop irrigation in the Salt Lake Valley	Utah
1856	Abraham Lincoln argues Mississippi River navigation case	Washington, DC
1862	U.S. Congress passes Homestead Act	Washington, DC
1870	Horace Greeley promotes irrigation in Greeley	Colorado
1882	Hydropower plant developed for Edison electrical system	Wisconsin
1945	Hydropower provides 75 percent of electrical needs in Pacific Northwest	United States
2004	Hydropower provides 60 percent of all electricity supplies	Canada

[a] Some dates are approximate.

Why was there such a tremendous application of this hydropower technology? What role did population growth play in this transfer of technology? Was it inevitable that navigation would come into conflict with this new "use" of water from rivers? What if the navigation industry had been highly organized prior to the construction of waterwheels in England? What could they have done to prevent infringement on their industry? Would this have been a proper protection of a prior (or "senior") water user, or should the new milling technology have been given unrestricted use of water as long as society benefited from this new water use? What rights should an existing industry have over a new water use from a river? Finally, how should the environment be protected within this debate? These issues will be explored in Chapter 8 and Chapter 12.

CHAPTER SUMMARY

Water use has evolved over thousands of years from basic human needs to complex technological innovations. Early civilizations focused on the need for food, shelter, and drinking water. Later, as populations multiplied and food demands increased, irrigation was developed. Improved food supplies led to larger urban population centers that required greater supplies of water for drinking needs and sanitation requirements. The Romans were one of the earliest civilizations to construct extensive aqueducts and other urban water delivery systems. Navigation flourished along natural and artificial waterways to transport goods, soldiers, and other material. Hydropower was widely used to grind grain and later for other manufacturing purposes. Conflict between navigation and milling industries occurred frequently in some locations.

As the population of the world expanded, irrigation, navigation, and milling industries expanded. In the United States, canal construction for navigation was extensive in eastern regions of the country, while irrigation was generally confined to the western territories. Why? Chapter 2 will attempt to answer that question.

QUESTIONS FOR DISCUSSION

1. Describe the earliest water uses by humans.

2. Discuss the development of *qanats*.

3. Explain why groundwater exists today in desert regions of the Middle East.

4. Why did early civilizations develop irrigation projects?

5. Discuss the development of navigation in the United States.

6. Discuss the allocation of water between mill owners and navigators. Within this context, do you agree that the prior (or "senior") user of water (navigation) should obtain priority over the "junior" water user (mill owners who diverted water from a stream and often constructed dams across rivers)?

7. Abraham Lincoln was a staunch supporter of construction of the transcontinental railroad. Almost a century earlier, George Washington had lobbied heavily for the construction of canals to improve the economy of the country. What recent U.S. presidents have used water to promote (or discourage) economic development? in what way?

8. Are you aware of any ongoing conflicts between navigation and other water uses in your community or region?

9. Discuss the evolution of waterwheels. What types of products were produced with this technology?

10. Are irrigation, navigation, or hydropower generation controversial topics in your area? If so, what are some of the issues under debate? What water resources agencies are involved?

KEY WORDS TO REMEMBER

acenas p. 9
acequias p. 9
Anasazi Indians p. 9
aqueduct p. 4
Brigham Young p. 10
charcas p. 9
Gallatin Report p. 15
Hohokam Indians p. 9
Homestead Act p. 10
Horace Greeley p. 11
locks p. 14
mill raceway p. 17
noria p. 16
qanats p. 2
saqia p. 6
shadouf p. 6
tambour (Archimedes screw) p. 6

SUGGESTED RESOURCES FOR FURTHER STUDY

READINGS

Biswas, Asit K. *History of Hydrology.* Amsterdam: North-Holland Publishing Co., 1970.

De Villiers, Marq. *Water.* Boston: Houghton Mifflin Co., 2000.

Harris, Katherine. *Long Vistas—Women and Families on Colorado Homesteads.* Niwot: University Press of Colorado, 1993.

Heat-Moon, William Least. *River-Horse: Across America by Boat.* New York: Penguin Books, 1999.

Hillel, Daniel. *Rivers of Eden—The Struggle for Water and the Quest for Peace in the Middle East.* New York: Oxford University Press, 1994.

Masters, Roger D. *Fortune Is a River—Leonardo da Vinci and Niccolo Machiavelli's Magnificent Dream to Change the Course of Florentine History.* New York: Free Press, 1998.

Michener, James A. *The Source.* New York: Random House, 1965.

Sandoz, Mari. *Old Jules.* Lincoln: University of Nebraska Press, 1962.

White, Gilbert F., David J. Bradley, and Anne U. White. *Drawers of Water—Domestic Water Use in East Africa.* Chicago: University of Chicago Press, 1972.

VIDEOS

Far & Away. Directed by Ron Howard, Universal City Studios, 1992. 2 hrs., 20 min.

Oklahoma. Directed by Fred Zinnemann, 20th Century-Fox Studios, 1955. 2 hrs., 25 min.

"Aqueducts, Man Made Rivers of Life." Modern Marvels Series, The History Channel. 50 min.

"The Erie Canal." Modern Marvels Series, The History Channel. 50 min.

"Niagara Power Plant." Modern Marvels Series, The History Channel. 50 min.

WEBSITES

"WaterHistory.org," International Water History Association, http://www.waterhistory.org/, January 2004.

"The Grand Canal," Thinkquest Team, http://www.library.thinkquest.org/20443/grandcanal.html, June 2001.

International Water History Association, http://www.iwha.net/, January 2004.

"The History of Hydropower Development," U.S. Department of Interior, Bureau of Reclamation, http://www.usbr.gov/power/edu/history.htm, January 2004.

"The Roots of the City," http://www.mexicocity.com.mx/history1.html, January 2004.

"Molinos Nuevos (Museo Hidraulico) in Murcia, Spain," WaterHistory.org, http://www.waterhistory.org/histories/murcia/, January 2004.

REFERENCES

1. Hillel, Daniel. *Rivers of Eden—The Struggle for Water and the Quest for Peace in the Middle East.* New York: Oxford University Press, 1994, 23.
2. Laitin, David D., and Said S. Samatar. *Somalia—Nation in Search of a State.* Boulder, CO: Westview Press, 1987, 6–7.
3. Biswas, Asit K. *History of Hydrology.* Amsterdam: North-Holland Publishing Co., 1970, 26–29.
4. Pearce, Fred. "Ancient Lessons from Arid Lands." *New Scientist* 132, No. 7 (December 1991), 47.
5. Ibid.
6. Ibid., 48.
7. Kerr, Senator Robert S. *Land, Wood and Water.* New York: Fleet Publishing Corp., 1960, 27–28.
8. De Villiers, Marq. *Water.* Boston: Houghton Mifflin Co., 2000, 57.
9. Biswas, 93.
10. Payne, Robert. *The Canal Builders.* New York: Macmillan Co., 1959.

11. White, Gilbert F., David J. Bradley, and Anne U. White. *Drawers of Water—Domestic Water Use in East Africa.* Chicago: University of Chicago Press, 1972, 3.

12. Hillel, 61.

13. Greer, Charles. *Water Management in the Yellow River Basin of China.* Austin: University of Texas Press, 1979.

14. Biswas, 8–9.

15. Carr, Donald E. *Death of the Sweet Waters.* New York: W.W. Norton, 1966, 23.

16. Kerr, 125–126.

17. Yandell, Michael D., et al. *National Parkways—Mesa Verde, Canyon de Chelly & Hovenweep.* Casper, WY: World-Wide Research and Publishing Co., 1987, 42–43.

18. Kerr, 129–130.

19. Ibid., 130.

20. Harris, Katherine. *Long Vistas—Women and Families on Colorado Homesteads.* Niwot: University Press of Colorado, 1993, 35–36.

21. "The Grand Canal," Thinkquest Team, http://www.library.thinkquest.org/20443/grandcanal.html, June 2001.

22. Calvert, Robert. *Inland Waterways of Europe.* London: George Allen & Unwin, 1963, 62.

23. Ibid., 175–176.

24. Masters, Roger D. *Fortune Is a River—Leonardo da Vinci and Niccolo Machiavelli's Magnificent Dream to Change the Course of Florentine History.* New York: Free Press, 1998.

25. The Western Writers of America. *Water Trails West.* Garden City, NY: Doubleday & Co., 1978, viii.

26. Ibid., ix.

27. Ibid., 44.

28. "History of Energy-Conversion Technology—Water Wheels." *Encyclopaedia Britannica,* http://www.britannica.com, September 17, 2000.

29. "Inventions and Discoveries." Encyclopedia, http://www.lineone.net/encyclopedia/technology/c-inventions1-d.html, June 2001.

30. Reynolds, John. *Windmills & Watermills.* New York: Praeger Publishers, 1970, 17–19.

31. "The History of Hydropower Development," U.S. Department of Interior, Bureau of Reclamation, http://www.usbr.gov/power/edu/history.htm, January 2004.

32. "Niagara Falls—History of Power," http://www.law?.com/,tans?/power.html, September 17, 2000.

CHAPTER 2

THE HYDROLOGIC CYCLE, CLIMATE, AND WEATHER

THE HYDROLOGIC CYCLE

CLIMATE AND WEATHER

All streams flow into the sea, yet the sea is never full. To the place the streams come from, there they return again.

Ecclesiastes 1

Ancient civilizations were curious about the world around them but had limited scientific knowledge regarding the natural processes of the oceans, land, and atmosphere. This led to some very interesting theories regarding the natural environment.

Thales (636 B.C.–546 B.C.), pronounced **Thay**'leez, is generally considered the founder of Greek science (Figure 2.1). A merchant, seafarer, natural philosopher, and weather observer who lived in the Greek city of Miletus, Thales sought to determine the essence or substance of all matter. He theorized that everything in the universe originated and ended as water.

Through his observations of the weather and his time at sea, Thales reasoned that all things were made of water. Why else would the gods drop water from the heavens, allow it to flow into rivers and eventually to the oceans, and require that all living things consume water every day? It was an interesting philosophy that was shared by many in Greek society.

In many ways, Thales' philosophy of water described the hydrologic cycle. This natural cycle of precipitation, runoff, storage, and evaporation greatly affects all living things and alters the Earth's surface every day. The water in the hydrologic cycle changes the landscape, creates weather patterns, and provides life. One could still argue today that water is at the center of all things on Earth.

For thousands of years, humans have tried to manipulate the hydrologic cycle through the construction of dams and canals to control the movement of water in the hydrologic cycle. Chapter 1 presented several examples of civilizations that developed elaborate urban and agricultural water delivery systems for drinking water, irrigation supply, and other uses. In this chapter, the role of the hydrologic cycle, climate, and weather will be considered as it relates to settlement patterns and water availability around the world.

FIG. 2.1 Bust of Thales of Miletus, one of the Seven Wise Men of ancient times.

THE HYDROLOGIC CYCLE

The movement of water between the land, oceans, and the atmosphere is called the **hydrologic cycle** (Figure 2.2). This natural process is driven by solar energy from the Sun. Moisture circulates from the Earth into the atmosphere through evaporation and then back to the Earth as precipitation. Water is not created or destroyed in this process but simply changes form and location.

TABLE 2.1 Total Water Supplies in the Hydrologic Cycle

Location of Storage	Total	Percent of Freshwater
Total water on Earth	100.0	
Sea water	97.5	
Total freshwater	2.5	100.0
Ice caps and glaciers		74.0
Groundwater		25.6
Lakes, rivers, soil moisture, and atmosphere		0.4

Source: Barbara W. Murck and Brian J. Skinner, *Geology Today: Understanding Our Planet* (New York: John Wiley & Sons, 1999). Reprinted with permission of John Wiley & Sons, Inc.

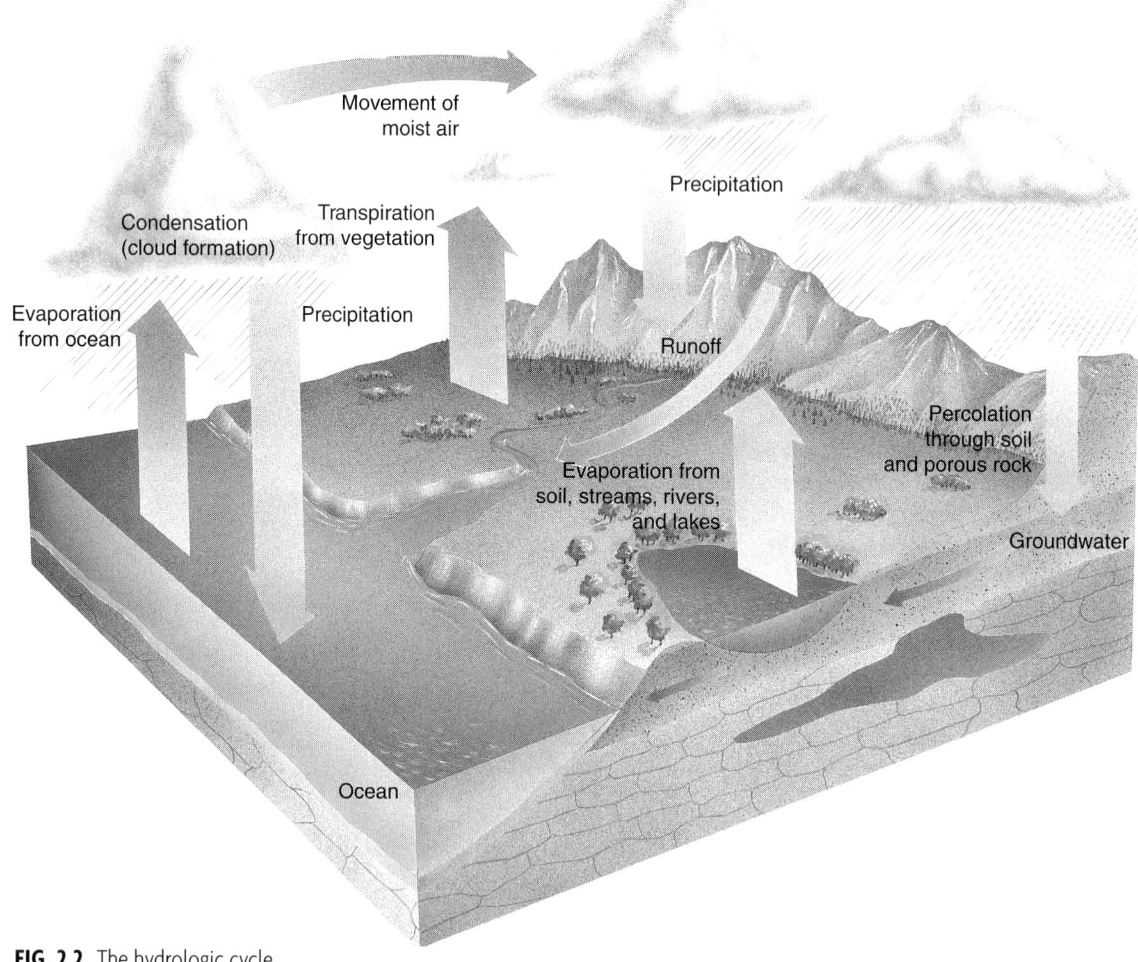

FIG. 2.2 The hydrologic cycle.

The oceans of the world contain over 97 percent of all water found in the Earth's hydrologic cycle (see Table 2.1 and Figure 2.3). All but a fraction of the remaining water is found either in frozen polar ice caps, where it may have been stored as ice for thousands of years, or as groundwater beneath the land surface. Rivers, lakes, ponds, wetlands, and moisture in the atmosphere contain less than 1 percent of the freshwater on Earth as shown in Table 2.1.

The hydrologic cycle contains five key components:

1. precipitation
2. runoff
3. surface and groundwater storage
4. evaporation/transpiration
5. condensation

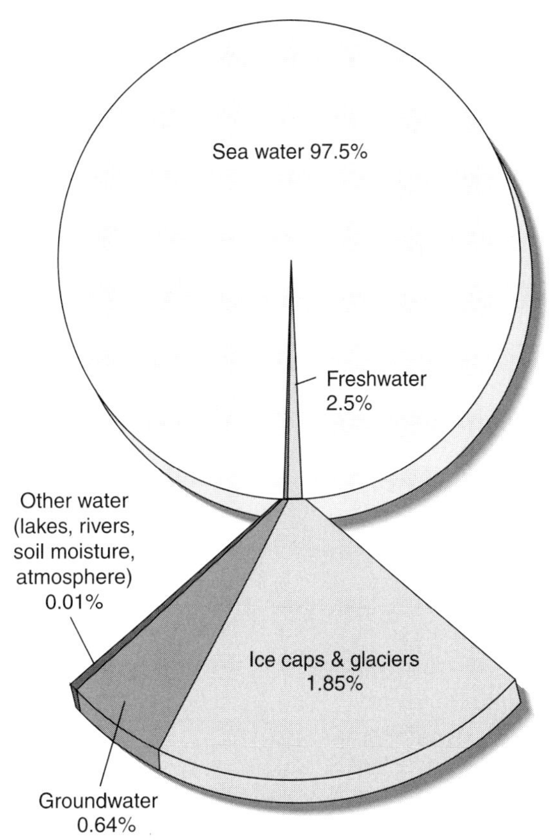

FIG. 2.3 Distribution of the total water in the hydrologic cycle. Less than one percent is available to humans since most water is sea water, frozen, or inaccessible in soil moisture or the atmosphere.

PRECIPITATION

Precipitation occurs when atmospheric moisture becomes too great to remain suspended in clouds. Under proper conditions, small moisture droplets undergo a process of coalescence, or joining together, and fall in the form of rain, snow, sleet, hail, or **virga** (rain that evaporates before reaching the ground). Once it reaches the Earth's surface, precipitation can become surface water runoff, surface water storage, glacial ice, water for plants, groundwater, salt water in the oceans, or may evaporate and return immediately to the atmosphere.

Ocean evaporation provides approximately 90 percent of the Earth's precipitation. However, living near an ocean does not necessarily imply increased rainfall. Southern California and the island of Aruba, near Venezuela, are examples of relatively dry regions adjacent to an ocean or sea. Aruba receives only 17 inches (43 cm) of precipitation per year and relies on reservoir storage and desalination of salt water from the Caribbean Sea for its water supply. San Diego, located along the shores of the Pacific Ocean in southern California, has an average annual precipitation of only 10 inches (25 cm) and receives very limited rainfall during the summer. The reasons for such natural climatic phenomena will be discussed later in this chapter.

Table 2.2 presents a wide range of average annual precipitation quantities in western U.S. cities such as Phoenix, Portland, Denver, and Los Angeles, and in eastern cities such as Atlanta and Bangor. However, Portland, Oregon, is an anomaly, for it receives more annual precipitation, on average, than Chicago, or the combined average annual totals of Phoenix, Denver, and Los Angeles. Why? (Check your hypothesis when climate and weather are discussed later in this chapter.)

TABLE 2.2 Average Annual Precipitation for Selected U.S. Cities

City	Amount	
	Inches	Centimeters
Phoenix, Arizona	7.6	19.3
El Paso, Texas	8.6	21.8
Los Angeles, California	11.9	30.2
Denver, Colorado	15.4	39.1
Salt Lake City, Utah	15.6	39.6
Fargo, North Dakota	19.6	49.8
Dallas, Texas	35.0	88.9
Chicago, Illinois	35.8	90.9
Portland, Oregon	36.3	92.2
Columbus, Ohio	37.8	96.0
Seattle, Washington	38.1	96.8
Kansas City, Missouri	38.2	97.0
Bangor, Maine	40.5	102.9
New York City, New York	41.5	105.4
Atlanta, Georgia	49.8	126.5
Memphis, Tennessee	52.7	133.9
Miami, Florida	59.0	149.9

Source: National Climate Data Center, National Oceanic and Atmospheric Administration, U.S. Department of Commerce, Asheville, North Carolina; see http://www.ncdc.noaa.gov/ol/climate/globalextremes.htm, September 2001.

Figure 2.4 and Table 2.3 present average annual precipitation totals from around the world. Notice the low rainfall amounts (less than 11 in., or 28 cm) for Cairo, Ahmadi, Riyadh, Damascus, Tehran, and Amman. The Tigris, Euphrates, and Nile rivers flow through these **arid** regions (areas that receive less than 10 in., or 25 cm, of average annual precipitation) and provided the irrigation water the early civilizations needed to survive. Without this supplemental water, crops such as wheat would have died since grains require approximately 15 inches (38 cm) of moisture during the growing season. The precipitation component of the hydrologic cycle (or the lack of it) played a major role in settlement patterns, water use, and the construction of irrigation projects in this desert region.

Tables 2.4 and 2.5 show the erratic nature of the hydrologic cycle. Notice that the island of Hawaii has both the high and low precipitation extreme records. Elevation is the key reason to Hawaii's extremely variable precipitation and will be explained later in this chapter.

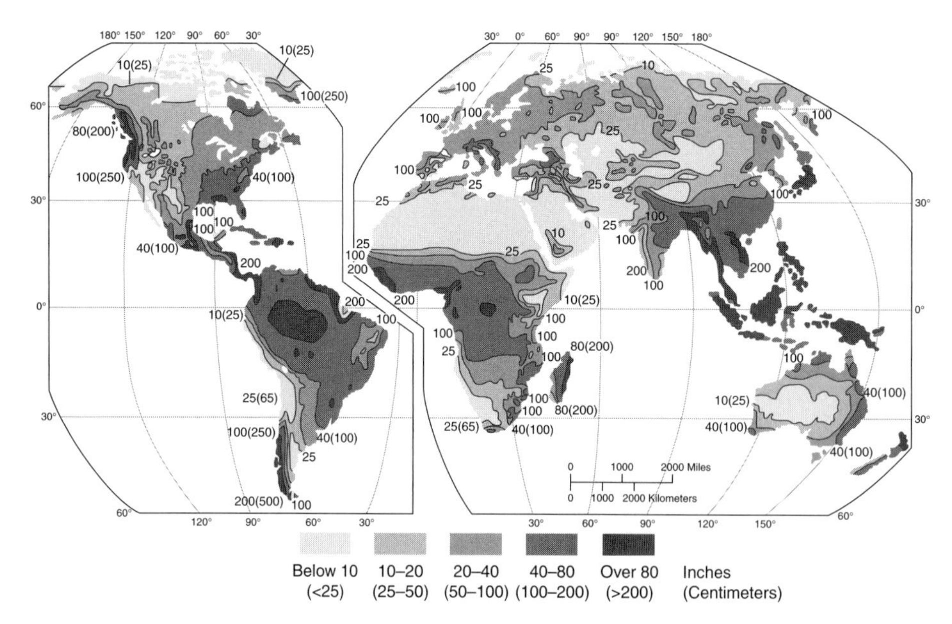

FIG. 2.4 Distribution of mean annual precipitation around the world. (Numbers in parenthesis are in centimeters.)

TABLE 2.3 Average Annual Precipitation for Selected World Cities

City	Amount		City	Amount	
	Inches	Centimeters		Inches	Centimeters
Cairo, Egypt	1.0	2.5	Nairobi, Kenya	29.9	75.9
Ahmadi, Kuwait	3.9	9.9	Rome, Italy	31.2	79.2
Riyadh, Saudi Arabia	4.4	11.2	Reykjavik, Iceland	32.2	81.8
Damascus, Syria	7.4	18.8	Johannesburg, South Africa	33.3	84.6
McMurdo, Antarctica	8.0	20.3	Perth, Australia	34.1	86.6
Tehran, Iran	9.5	24.1	Montreal, Quebec	40.8	103.6
Amman, Jordan	10.7	27.2	Rio de Janeiro, Brazil	43.4	110.2
Madrid, Spain	17.3	43.9	Brazzaville, Congo	54.0	137.2
Prague, Czech Republic	18.9	48.0	Tokyo, Japan	60.0	152.4
St. Petersburg, Russia	21.2	53.8	Yaounde, Cameroon	62.9	159.8
London, Great Britain	23.2	58.9	Jakarta, Indonesia	71.7	182.1
Paris, France	23.9	60.7			
Beijing, China	25.0	63.5			

Source: National Climate Data Center, National Oceanic and Atmospheric Administration, U.S. Department of Commerce, Asheville, North Carolina; see http://www.ncdc.noaa.gov/ol/climate/globalextremes.html, September 2001.

TABLE 2.4 Low Annual Precipitation Extremes around the World

Location	Amount		Elevation		Years of Record
	Inches	Centimeters	Feet	Meters	
Arica, Chile	0.03	<.08	95	29	59
Wadi Halfa, Sudan	<0.1	<.25	410	125	39
Amundsen-Scott South Pole Station, Antarctica	0.8	2.0	9186	2800	10
Batagues, Mexico	1.2	3.0	16	5	14
Aden, Yemen	1.8	4.6	22	7	50
Mulka, South Australia	4.05	10.29	160	49	42
Astrakhan, Russia	6.4	16.3	45	14	25
Puako, Hawaii, United States	8.9	22.6	5	2	13

Source: National Climate Data Center, National Oceanic and Atmospheric Administration, U.S. Department of Commerce, Asheville, North Carolina; see http://www.ncdc.noaa.gov/oa/climate/globalextremes.html, July 2003.

TABLE 2.5 High Annual Precipitation Extremes around the World

	Amount		Elevation		
Location	Inches	Centimeters	Feet	Meters	Years of Record
Lloro, Colombia	523.6*	1329.9	520	158	29
Mawsynram, India	467.4	1187.2	4597	1401	38
Mt. Waialeale, Kauai, Hawaii, United States	460.0	1168.4	5148	1569	30
Debundscha, Cameroon	405.0	1028.7	30	9	32
Quibdo, Colombia	354.0	899.2	120	37	16
Bellenden Ker, Queensland, Australia	340.0	863.6	5102	1555	9
Henderson Lake, British Columbia, Canada	256.0	650.2	12	4	14
Crkvica, Bosnia-Hercegovina	183.0	464.8	3337	1017	22

*Estimate.

Source: National Climate Data Center, National Oceanic and Atmospheric Administration, U.S. Department of Commerce, Asheville, North Carolina; see http://www.ncdc.noaa.gov/oa/climate/globalextremes.html, July 2003.

Measuring Precipitation Precipitation measurement is a very important tool in understanding the amount of water available for human and plant use. A variety of methods are available to obtain data on the amount, location, and intensity of precipitation events. This information allows growers, municipal water providers, crop scientists, forest managers, and others to adjust water use patterns.

FIG. 2.5 The world's oldest surviving rain gages were invented in the fifteenth century in present-day Korea, and were typically placed near government facilities in provinces and cantons of the country to monitor precipitation. Results were made known to the king and the court for planning purposes. The gage in this photo is called Chuk-u-gi, and is the oldest in the world. It is from the early Joseon Dynasty in A.D. 1441.

Simple rain gages have been used in India, China, and Korea for more than a thousand years (Figure 2.5). In 300 B.C. India used rain gages to determine tax collections. Periods of high rainfall meant good crops and higher taxes, whereas low rainfall meant poor crops and a tax break from the government. The rain buckets and precipitation tubes used in ancient times were not much different from some of the instruments used today. Modern measuring devices range from simple plastic and glass tubes available at hardware stores to elaborate "weigh and tip" bucket gages used by the U.S. National Weather Service (NWS). Accurate precipitation measurement requires the placement of rain gages away from trees, buildings, and other features that could interfere with rainfall. Networks of rain gages can be distributed across a wide area to determine the quantity of rainfall over a large geographic region.

Doppler Radar provides a sophisticated and accurate method of precipitation measurement. The name is derived from the Doppler Effect, which was first mathematically described in 1842 by Christian Doppler (1803–1853), an Austrian physicist. Doppler studied the apparent change in sound frequencies as an object, such as a train whistle, approached and then passed a stationary person. His concept of frequency and sound movement is used in Doppler Radar to estimate rainfall amounts based on the intensity of radar echoes. (1) This modern technology sends out continual electromagnetic waves that bounce off suspended water droplets in the air and return as an electronic signature back to a computer. The distribution of droplets provides data on storm intensity and precipitation amounts. Scientists are now using Doppler Radar data from previous storms to estimate historic precipitation amounts. This information provides spatial, gridded, historic precipitation estimates that are not available with rain gage monitoring networks.

In some regions of the world, snow depth measurements are a very important aspect of monitoring precipitation. Water frozen and stored as snowpack in mountainous areas, or in drifts or blankets of snow at lower elevations, is measured to determine the amount of liquid water available. In some areas, this snowmelt will provide water supplies for farms and communities at lower elevations during spring and summer months.

The U.S. National Weather Service and other agencies around the world measure snowpack. Measurements are typically obtained during winter and spring months to determine the depth and water content of snow at various locations. A **snow tube** is used to collect a vertical snow sample from the top of the snow down to the land surface. The metal tube has a sharp edge that cuts through frozen layers of ice and snow when rotated and pushed down. The depth of snow is measured, and then the tube and snow contents are weighed to determine water content.

Snow cores are taken along an established line, called a **snow course**, to take into account variations in snow depth and characteristics. The **water equivalent** is the amount of liquid that would result from melted snow, and it varies based on snow density, water volume, and snow depth. Monthly reports are often prepared to show water users and managers current snow depth and water-equivalent data. This data is usually compared to that of previous years and long-term averages.

Snow pillows are also used to determine the water content of snow. The pillows are made of stainless steel and are rectangular-shaped with sides of approximately 4 feet by 5 feet (1.2 m by 1.5 m) and a thickness of approximately 1 foot (0.3 m). As snow accumulates, internal pressure on antifreeze fluid inside the pillow increases. This increase in pressure is then electronically recorded and transmitted via satellite to monitoring stations at lower elevations. This eliminates the need for scientists to hike into remote mountainous regions during the winter (see Figure 2.6).

FIG. 2.6 Snow pillow station at a mountain station in Utah. A very interesting website on snow pillow data, entitled "Seasonal Automatic Snow Pillow (ASP) Plots for B.C.," can be found at http://www.elp.gov.bc.ca/rib/wat/rfc/river_forecast/snowp.htm. The website is operated by the Aquatic Information Branch of the Ministry of Sustainable Resource Management, Government of British Columbia, Canada.

SIDEBAR

Water managers at the Natural Resources Conservation Service in Denver (an agency within the U.S. Department of Agriculture) monitor snowpack on a monthly basis in all watersheds of Colorado and produce monthly reports. These data are used to predict streamflows during dry summer months and are available to municipal water managers, irrigators, and other water users for planning purposes. Go to ftp://ftp.wcc.nrcs.usda.gov/data/snow/update/co.txt for more information.

RUNOFF

Runoff is the amount of water that flows across the land surface after a storm event. Chapter 1 described the efforts of the ancient Anasazi Indians to capture runoff during rainstorms by diverting and channeling water into small lakes or onto fields for irrigation. Runoff played an extremely important role in the Anasazi's ability to live in the harsh, arid climate of southwest Colorado.

Climate, terrain, precipitation intensity, and volume play a large role in surface water runoff. The Amazon River Basin in South America has tremendous runoff volume due to high precipitation rates, high humidity, and a massive drainage area. The Los Angeles Basin in southern California also has high runoff volume but generally only for short time durations due to infrequent but intense rainfall events, hard and impervious soils, and extensive developed areas of concrete and rooftops. In contrast to these examples, the Sandhills of central Nebraska generally have low runoff volumes owing to lush ground cover of native grasses, gently rolling terrain, and very porous (sandy) soils that allow runoff to seep underground (see Figure 2.7).

Land use has a significant effect on surface water runoff. Barren land surfaces hinder water seepage into the soil and cause runoff to move rapidly downhill. Dense vegetative cover slows surface water flow and allows increased seepage rates. Urban areas with paved streets and parking lots, sidewalks, and roof tops prevent seepage and increase runoff. Areas downstream of urban areas often experience increased streamflows

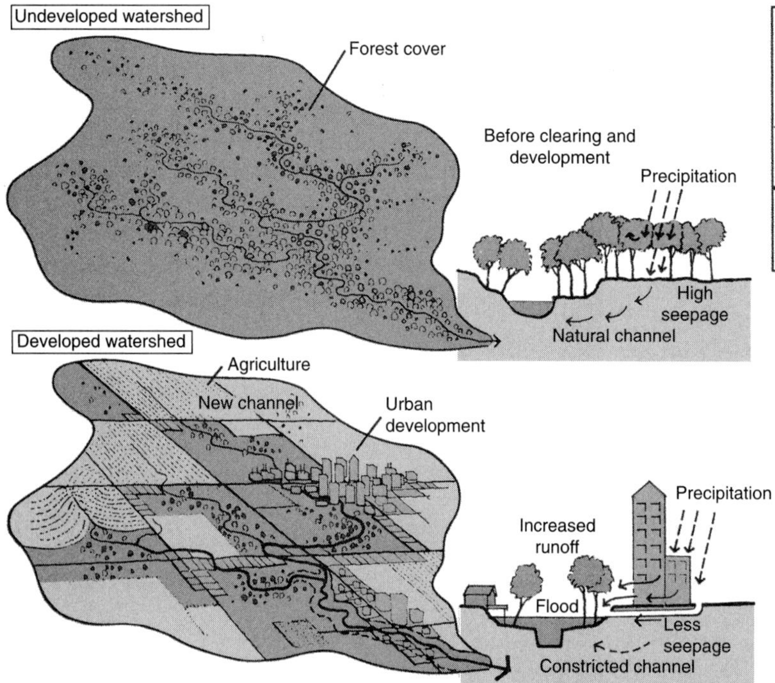

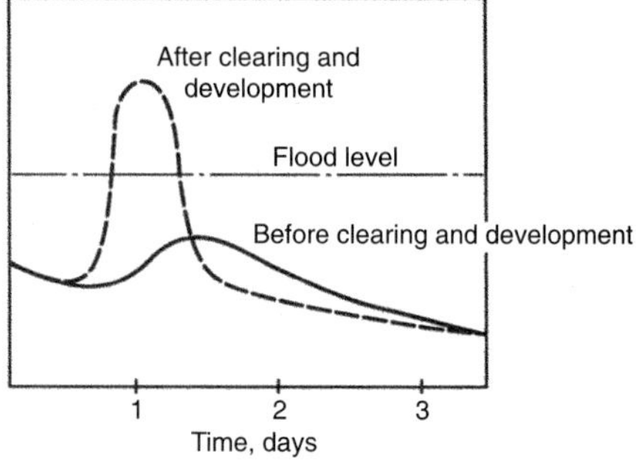

FIG. 2.7 Runoff and streamflow can change dramatically when land use patterns change from undeveloped vegetation to agriculture or urban development. The graph shows a corresponding change in streamflow from a storm event. Note the tremendous change in streamflow levels before and after clearing and development. What problems will this cause?

after major storm events. Measurement of runoff will be discussed in Chapter 3, and relationships between runoff and groundwater will be discussed in Chapter 4.

SURFACE AND GROUNDWATER STORAGE

Lakes and Reservoirs Lakes and reservoirs are important components of the hydrologic cycle. **Lakes** are large bodies of inland water generally formed by glacial activity or surface water runoff, whereas **reservoirs** are either natural or artificial water bodies used to store, regulate, and control water. Both lakes and reservoirs serve as collection points for storage of surface water runoff and groundwater seepage, lose water to evaporation, and can replenish the flow in streams.

Lakes and reservoirs can be created by landslides, tectonism (movement of geologic formations), glaciation, river action, animal activity (beavers), meteorite impact, volcanism, and human activity (the construction of dams). Two of the largest and deepest lakes in the world were created by tectonism. Lake Baikal in Russia is the largest and deepest freshwater lake on Earth, with a depth of 5250 feet (1600 m). It contains as much freshwater as all of the Great Lakes of North America combined—roughly 20 percent of the world's total surface freshwater. Lake Tanganyika, in eastern Africa and bordered by Zambia, Tanzania, Burundi, and the Democratic Republic of Congo (Zaire) is the longest lake in the world (416 mi, or 670 km) and the second deepest at 4823 feet (1470 m). Lake Tahoe, on the border of California and Nevada, was shaped and landscaped by scouring glaciers. It is the second deepest lake in the United States and the tenth deepest in the world, with a maximum depth measured at 1645 feet (501 m). Crater Lake in Oregon, the deepest lake in the United States, has a maximum depth of 1932 feet (589 m) and was created by an ancient volcano named Mount Mazama.

Many lakes in the Northern Hemisphere were created by glacial ice that moved southward out of Canada during the Ice Age and then melted and receded as global warming occurred. The Finger Lakes in upstate New York, Great Slave Lake in the Northwest Territories of Canada, and the Great Lakes of North America were created by this type of glacial action.

SIDEBAR

According to the Canadian Hydrographic Service, Department of Fisheries and Oceans, a century of water-level records of the Great Lakes indicate no regular, predictable cycle. The maximum variation of lake levels ranges from 3.9 feet (1.2 m) in Lake Superior to over 5.9 feet (1.8 m) in the other lakes. (2)

Lakes and reservoirs rely on precipitation, snowmelt, and, in some cases, groundwater infiltration and glacial melt as sources of water. Water stored in lakes is lost through evaporation, groundwater recharge, and outflow. Many lakes and reservoirs are closely monitored to determine water storage volumes. Chapter 3 will discuss the natural functions of lakes, while Chapter 7 will describe the construction of dams to create water storage reservoirs.

CASE STUDIES

Following are two brief examples of natural and human-caused events that have significantly altered water levels in lakes around the world.

Example: Great Salt Lake of Utah

Great Salt Lake is a terminal lake, with no outlet river extending to an ocean. The lake is shallow for its size—about 70 miles (113 km) long and 30 miles (48 km) wide, but only approximately 40 feet (12 m) deep. The lake bottom is so gently sloped that any increase in inflow causes a broad surface area to be inundated with water.

In 1983, water levels in Great Salt Lake began to rise due to very high precipitation and snowmelt in the region. Since it has no outlet, water levels in the lake responded dramatically to increased surface water runoff and reached record levels in June 1986 and March–April

1987. Lake levels increased 12 feet (3.7 m), causing floodwaters to inundate Interstate Highway 80 and parts of Salt Lake City on the south side of the lake. (3)

In response, the State of Utah implemented the West Desert Pumping Project to lower water levels in Great Salt Lake. Three large pumps were installed and pumped water from the lake onto the Bonneville Salt Flats to the west, creating the Newfoundland Evaporation Basin. The first pump was turned on in April 1987 and ran until June 1989 (see Figure 2.8). The combination of pumping, evaporation, and reduced inflow caused lake levels to drop, with about 2 feet (0.6 m) of the decline due to pumping. The pumps were operated for two years at a cost of over $60 million. (4)

During the late 1800s, to coax them to settle in the West, American pioneers often heard the phrase "Rain follows the plow." Land speculators argued that crop production would generate more evaporation and transpiration, and so would cause precipitation to increase. Scientists noted that water levels in Great Salt Lake rose after Mormon settlers began irrigating nearby lands. Others believed that enlarged water surfaces, created by the construction of reservoirs, would also increase evaporation and precipitation. Both claims were overblown and misleading, causing great human misery when drier times prevailed.

Example: The Aral Sea of Uzbekistan

The Aral Sea is dying. Once-prosperous fishing villages along this Central Asia water body are now 60 miles (96 km) from the shore, with water so salty that all fish have died. It is estimated that the entire sea will disappear by 2010.

In about 1965, officials of the former Soviet Union tapped this large body of water (the fourth largest lake in the world at the time) for irrigation. The sea covered 26,255 square miles (68,000 km^2) and had an average depth of 52 feet (16 m). Fish production was tremendous, but Soviet officials wanted to develop irrigation to boost the regional economy. Although planners were aware that the size of the water body would shrink, no one was prepared for the magnitude of change that resulted (Figure 2.9).

Today, the Aral Sea is less than 50 percent its previous size and contains less than 25 percent of its volume in 1960. The resulting concentration

FIG. 2.8 This photo was taken during the first pump test of the Great Salt Lake West Desert Pump Plant in 1987. Great Salt Lake is the largest lake in the Western Hemisphere without an outlet to the sea, and is a remnant of ancient Lake Bonneville. Changing climatic conditions cause long-term fluctuations in lake levels in Great Salt Lake, located just northwest of Salt Lake City, Utah.

FIG. 2.9 The Aral Sea is located between Kazakhstan and Uzbekistan, both former Soviet Socialist Republics. During the communist period, large-scale irrigation projects were developed that dramatically reduced flows into this inland sea. Fishing villages that were once located on the shores of the Aral Sea are now 60 to 90 miles (96 to 144 km) from water. In addition, the increased concentration of pollutants in the Aral Sea has totally destroyed the fishing industry, and many fishing vessels were simply abandoned.

of salts and chemicals has killed all plant and animal life, and has destroyed a productive fishing industry, with tens of thousands of people abandoning their fishing livelihoods. A dry, desert landscape has replaced the retreated water surface, and dust storms, warmer local temperatures, and increased winds are common.

The World Bank has launched a program to develop drinking water supplies and to restore some plant and animal life in the area. However, 95 percent of the surface runoff that historically entered the Aral Sea is now diverted for other uses. The local climate of this region has been moderated with hotter, shorter summers and longer, colder winters. Wind-scattered dust from the dry lake bed carries for great distances and has increased respiratory illnesses and other diseases.

Will the Aral Sea ever return to its previous condition? Unfortunately, experts agree that will never happen. Even so, the independent republics of Kazakhstan, Kyrgyzstan, Tajikistan, Turkmenistan, and Uzbekistan have signed cooperative agreements to attempt to develop a regional water management system to return some water to the Aral Sea. (5)

Wetlands Wetlands play a very important role in the storage of water within the hydrologic cycle. A **wetland** can be described as an area of standing water, usually shallow, that contains cattails or other hydric (water-loving) plants. Wetland scientists and the U.S. Fish & Wildlife Service adopted a formal definition of wetlands in 1979 after several years of review. In a report titled *Classification of Wetlands and Deepwater Habitats of the United States* (Cowardin et al., U.S. Department of the Interior, Fish & Wildlife Service, Washington, D.C., 1979) wetlands are defined as:

> *. . . lands transitional between terrestrial and aquatic systems where the water table is usually at or near the surface or the land is covered by shallow water. . . . Wetlands must have one or more of the following three attributes: (1) at least periodically, the land supports predominantly hydrophytes (plants that have adapted to life in soils that are often flooded or saturated with water); (2) the substrate is predominantly undrained hydric soil; and (3) the substrate is nonsoil and is saturated with water or covered by shallow water at some time during the growing season of each year.*

Wetlands are often found along rivers, lakes, deltas, estuaries, and swamps. These marshy areas provide excellent habitat for wildlife, can induce groundwater recharge, reduce erosion during floods, and provide temporary and permanent storage areas for surface water runoff. In addition, a wetland can improve water quality. Wetlands will be discussed in detail in Chapter 12.

Groundwater Groundwater storage is another important component of the hydrologic cycle. Porous soils allow surface water to seep downward into the soil and underlying geologic material under the force of gravity. Groundwater can be found beneath the land surface in sand and gravel, rocks, fine clay material, and cracks in large rocks. Surface water can also become groundwater through seepage from streams, lakes, wetlands, and salt water. Groundwater moves under the force of gravity through geologic material to lower elevations until it reaches an underground barrier such as clay or rock. Groundwater can eventually reach the land surface at a lower elevation as a spring, or it can infiltrate into a stream, lake, wetland, or ocean. Groundwater properties and movement will be discussed in Chapter 4.

EVAPORATION

Evaporation is the process of liquid water converting into vapor, through wind action and solar radiation, and then returning to the atmosphere. Evaporation occurs from open bodies of water, such as lakes, rivers, and the ocean, and from land surfaces. Evaporation rates are extremely important to determine water availability or water loss rates in a region. These rates can be measured by filling special pans with water and then recording daily loss rates.

The English astronomer-mathematician Sir Edmond Halley (1656–1742) conducted experiments in about 1701 to estimate evaporation from the Mediterranean Sea. He set out several small pans filled with water during hot summer days and calculated the amount of water that evaporated from the Mediterranean in a single day. He then estimated the daily flow of freshwater from contributing rivers and determined a net loss of water to the system. Evaporation, Halley theorized, would lead the Mediterranean Sea to become saltier since the net water lost from the Mediterranean exceeded freshwater inflow. (6)

Future research proved Halley correct in his theory that the deeper Mediterranean waters were saltier than the adjacent Atlantic Ocean. Halley continued his work with the hydrologic cycle and evaporation, and later discovered the comet that carries his name today.

The small pans that Halley used to estimate evaporation in the Mediterranean are very similar to devices used today. The U.S. National Weather Service and other scientific agencies around the world use **Class A Evaporation Pans** to determine evaporation rates. These galvanized steel pans have a diameter of 4 feet (1.2 m) and a depth of 10 inches (25.4 cm). Generally, a pan is filled to a depth of 8 inches (20.3 cm) and is refilled whenever the water level drops to 7 inches (17.8 cm). Water levels in the pan are recorded daily. Errors in measurement can occur if an animal drinks water from the pan, if wind causes water to splash out, or if heating of the metal increases evaporation. The error in evaporation readings, caused by excess energy conducted through the walls of the pans, can be corrected by using a **pan coefficient.** This adjusting factor will lower the pan evaporation rate to more closely reflect actual values; it varies based on the average temperature and elevation of a region.

Evaporation from reservoirs can be significant, particularly in arid regions such as western portions of the United States, northern Mexico, Australia, and the Middle East. Along the Front Range of Colorado, water levels in lakes can drop between 2 to 4 feet (0.6 to 1.2 m) every year. In Nevada, Lake Mead loses over 3 percent of its stored water annually to evaporation. The U.S. Geological Survey has calculated that 70 percent of Georgia's 50 inches (127 cm) of annual precipitation returns directly to the atmosphere through evaporation. Of the remaining 30 percent, approximately 20 percent will become surface runoff and 10 percent will seep into the soil and become Georgia groundwater. (7)

In the 1940s, H. L. Penman worked for the British Army and developed the **Penman Equation** to estimate evaporation rates from plants and soil. (8) This mathematical formula includes the effects of sunshine duration, solar radiation, wind speed, temperature, and vapor pressures on evaporation rates. However, many of the data inputs necessary for this equation are not generally gathered at monitoring sites and must be estimated by researchers. Modified several times since its creation, it is now called the Modified Penman Equation.

Sublimation, an important part of the evaporation process, allows snow or ice to change directly into water vapor, or vice versa, without going through the normal melting process. Snow and ice sublimate (convert between a solid and gaseous state) during the winter and reduce available water supplies for downstream water users. Wind, temperature, and elevation variations can cause the loss of up to 52 percent of available water supply in snowpack in any given season due to sublimation. (9)

Transpiration occurs when water molecules exit living plant tissue (stomata) and enter the atmosphere. In areas of abundant rainfall, transpiration is fairly constant, with variations occurring primarily in the length of each plant's growing season. However, transpiration in dry areas varies greatly by root depth. Shallow-rooted plants often wither and die owing to a lack of moisture, but deep-rooted plants, such as alfalfa *(mendicago sativa)* or cottonwood trees *(Populus deltoids),* will continue to transpire water because of roots that can tap into deeper groundwater supplies. Deep-rooted, water-loving plants are called **phreatophytes** and are often found along river corridors or in areas where the depth to groundwater is not excessive.

Evapotranspiration (ET) includes all evaporation from water and land surfaces, as well as transpiration from plants. ET rates vary greatly throughout the year and are primarily dependent on temperature, wind, and atmospheric moisture conditions. C. Warren Thornthwaite (1899–1963) introduced the concept of **potential evapotranspiration** in the late 1940s. (10) Potential ET is the amount of water in a plant that will be lost if there is never a water deficiency in the soil for use by the plant. This theory recognizes that a plant can consume only so much water during a complete growing season. The application of additional water to a plant, through precipitation or irrigation, cannot be utilized by a plant and will simply percolate downward past its roots. Thornthwaite's concept was very important because it developed the maximum water requirements of various crops. The concept of potential evapotranspiration is still widely used around the world.

A CLOSER LOOK

Many scholars consider C. Warren Thornthwaite the most outstanding American climatologist of the twentieth century. His research into potential evapotranspiration and the water balance model (which compares water need to water supply) was first published in 1948. Previously, Thornthwaite was a geography professor at the University of Oklahoma; later he became chief of the Climatic and Physiographic Division of the Soil Conservation Service in the U.S. Department of Agriculture.

In 1946, he joined Seabrook Farms in New Jersey as an irrigation consultant. While at Seabrook, the U.S. Army and Air Force entered into a contract to have Thornthwaite conduct experiments on evapotranspiration. The military was interested in his research because it believed Professor Thornthwaite could provide valuable weather forecasting. This information could influence vehicular movement on unpaved surfaces, the prediction of fog and wind conditions, and chemical warfare forecasting. (11)

The **Blaney-Criddle Method** was developed by the U.S. Department of Agriculture in the 1950s and is also used to estimate evapotranspiration (ET) rates. This mathematical method uses crop ET rates based on differing water needs during the growing season as plant size increases.

Lysimeters can be used to directly measure the evapotranspiration rates of various types of vegetation. A soil-filled tank is buried at ground level and filled with the type of soil and vegetation to be tested, such as bluegrass or corn. All water inputs and outputs are weighed using the following equation:

$$ET = Si - Sf + P + I - D$$

where ET = evapotranspiration for the test period
Si = volume of soil moisture at the beginning of the experiment
Sf = volume of soil moisture at the end of the experiment
P = amount of precipitation entering the lysimeter
I = amount of irrigation water entering the lysimeter
D = amount of moisture drained from the soil

In some locations, satellite imagery can be used to determine the amount of water consumed by irrigated crops based on color and heat radiated by various crops. ET rates can be computed based on this infrared photography from space.

Consumptive use (CU) is the amount of water transpired and retained within a plant or animal. For example, the consumptive use of corn includes the amount of water transpired during a growing season plus the volume of water stored in the corn stalk and ears.

$$CU = ET + Sc$$

where CU = consumptive use for a growing season
ET = evapotranspiration rate for the entire growing season
Sc = amount of water stored in the plant tissue

Consumptive use is an extremely important measurement of water use. Since consumptive use water is lost to the local surface and groundwater system, water managers and engineers often go to great lengths to determine the precise consumptive use quantities of various crops and animals.

CONDENSATION

Condensation is the cooling of water vapor until it becomes a liquid. This process begins when water vapor in the atmosphere rises in the air and cools. As this occurs, the water vapor undergoes a change of state into liquid or ice. If other atmospheric conditions are present, this process can form clouds at higher elevations, or fog if close to the Earth's surface. As the droplets collide, they merge and form larger droplets. This process is very noticeable on plants as they collect dew in the morning, or when you take a can of soda out of a refrigerator.

In summary, the hydrologic cycle represents the continuous movement of water from the atmosphere to the land surface, from the land, rivers, lakes, and groundwater to the oceans, and from the oceans back to the atmosphere. The movement of water through the hydrologic cycle distributes moisture around the world through precipitation, runoff, storage, evaporation and transpiration, and condensation. Predictable hydrologic patterns and changes to those patterns, caused by natural events or human activities, combine to create our climate and weather around the world.

CLIMATE AND WEATHER

OVERVIEW

In Chapter 1, we discussed human efforts to obtain water for drinking, irrigation, navigation, and hydropower. A common theme in that chapter was the attempt to transport water to other locations by constructing canals, aqueducts, groundwater wells, and diversion dams. The question that naturally arises is: Why didn't these early civilizations simply move to wetter regions with more reliable sources of water?

Early settlers did not consider mass migration to more hospitable climates for several reasons. First, more humid regions may have been located far from existing communities. For many people the process of relocating to unfamiliar settings, far from known sources of food and water, was probably unthinkable.

Second, some areas with adequate water already had settlements with cultures very different from those of drier climates. An influx of "outsiders" looking for food, water, and jobs may not have been welcomed. Third, religion, wealth, family tradition, and fear most likely played important roles in decisions to remain in arid locations. Why else would people remain in dry, inhospitable areas? Do these same factors exist today?

CLIMATE

Climate is defined as the average weather condition of a location and can generally be predicted for centuries into the future. The climate of Egypt is hot and dry, although a few occasional wet and cool days are possible. The climate of southeastern Canada and northeastern United States is generally hot and humid during the summer months but frigid with ice and snow in winter. Several factors affect the climate of a region, but the three primary ones are air currents, ocean currents, and the tilt of the Earth's axis.

Air Currents **Global trade winds** (the word *trade* means "direction" or "course") are persistent winds caused by changes in air temperature and the rotation of the Earth. George Hadley (1685–1758), an English mathematician, pointed out in 1735 that global winds were caused by the uneven distribution of the Sun's heat on the Earth. Hadley theorized that, since more thermal energy reached the equator than the poles, global winds would be generated as the air moved between those two regions to reach equilibrium. (12)

If the Earth were a nonrotating sphere, the air circulation would be in a north-south (pole to equator) pattern so that the poles would be areas of low-pressure and the equator an area of high-pressure zones. (Atmospheric pressure zones are discussed later in this chapter.) The Earth's rotation disrupts this simple atmospheric pattern through a process called the **Coriolis effect.** Moving objects, such as the wind, are deflected to the right in the Northern Hemisphere and to the left in the Southern Hemisphere.

The Coriolis effect is named after Gustav Gaspard de Coriolis, a nineteenth-century French scientist who first explained why weather systems rotate. According to Newton's first law, a moving body that is not affected by a force will continue in a straight line. However, Coriolis deduced that since the Earth and its atmosphere make one rotation every 24 hours, and since the atmosphere at the equator has to rotate faster than the atmosphere at the poles, the force of the Earth's rotation would deflect air to the right in the Northern Hemisphere and to the left in the Southern Hemisphere. (13)

Air also moves between the equator and the poles due to temperature gradients. As a result, a low-pressure zone of convergence, called the **intertropical convergence zone (ITCZ),** is created at the equator. Air also piles up at latitudes 30°N and 30°S, creating two belts of high pressure around those latitudes. Air in the high-pressure zones sinks back to the ground, with some air flowing back toward the poles but most returning back toward the equator.

Global air circulation patterns can be readily seen by the distribution of deserts. Approximately 25 percent of the Earth's land mass (outside the polar regions) is comprised of arid (desert) lands (described earlier as locations that receive less than 10 in., or 25 cm, of average annual precipitation). Substantial regions of **semiarid** climate (annual precipitation of between 10 and 20 in., or between 25 and 51 cm) are found adjacent to many of the deserts in the world (Figure 2.10).

Ocean Currents Ocean currents are broad, slow drifts of sea water in the ocean created by prevailing surface winds, ocean tides produced by the gravitational pull of the moon, the Coriolis effect caused by the Earth's rotation, and differences

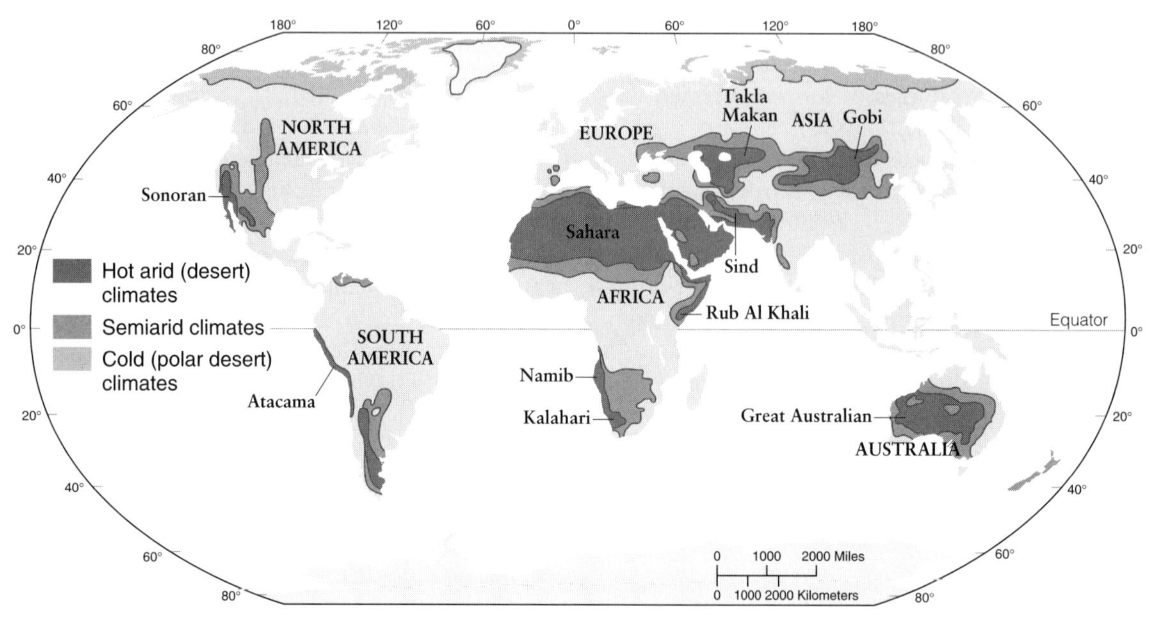

FIG. 2.10 Arid and semiarid regions of the world. Compare the desert locations shown above to the centers of early civilizations shown in Figure 1.1 in Chapter 1.

in water density effected by temperature or salinity variations. Ocean currents affect, and often temper, regional climates through the transfer of thermal energy. The climate in Iceland, for example, is surprisingly mild due to the warm **gulf stream** currents off its shores that originate near the equator (see Figure 2.11).

Surface ocean currents are generally wind-driven and occur in all of the world's oceans. Moving air pushes the surface of sea water and can create a current of water from 165 to 330 feet (50 to 100 m) deep. Prevailing trade winds, discussed earlier, determine the direction of these fairly shallow ocean currents. Examples of large surface currents are the Gulf Stream, the North Atlantic Current, the California Current, the Atlantic South Equatorial Current, and the Westward Drift.

Associated with these wind-driven currents are counter-surface and underlying currents. The Coriolis effect on water of the oceans is similar to its effect on the atmosphere. Water in the oceans near the equator must move faster with the Earth's rotation than does water near the poles. This causes surface ocean currents to be deflected to the right in the Northern Hemisphere and to the left in the Southern Hemisphere. The currents eventually come into contact with landmasses that deflect them again, creating giant oceanic current circles called **gyres** (pronounced *jires*).

Tidal currents, created by rising and falling tides, are especially relevant to coastal environments worldwide. The rise and fall of tides do not occur simultaneously around the world, so that some locations experience higher ocean-surface elevations while other regions have lower ocean-surface levels. This difference in water-surface elevations on the ocean creates currents as sea water attempts to create a common water level.

Vertical and ocean-bottom currents are created mainly by density differences in temperature and salinity. Cold, salty waters from the polar regions sink to the ocean bottom and move toward the opposite poles where they again surface. This is known as *upwelling* and is very important to

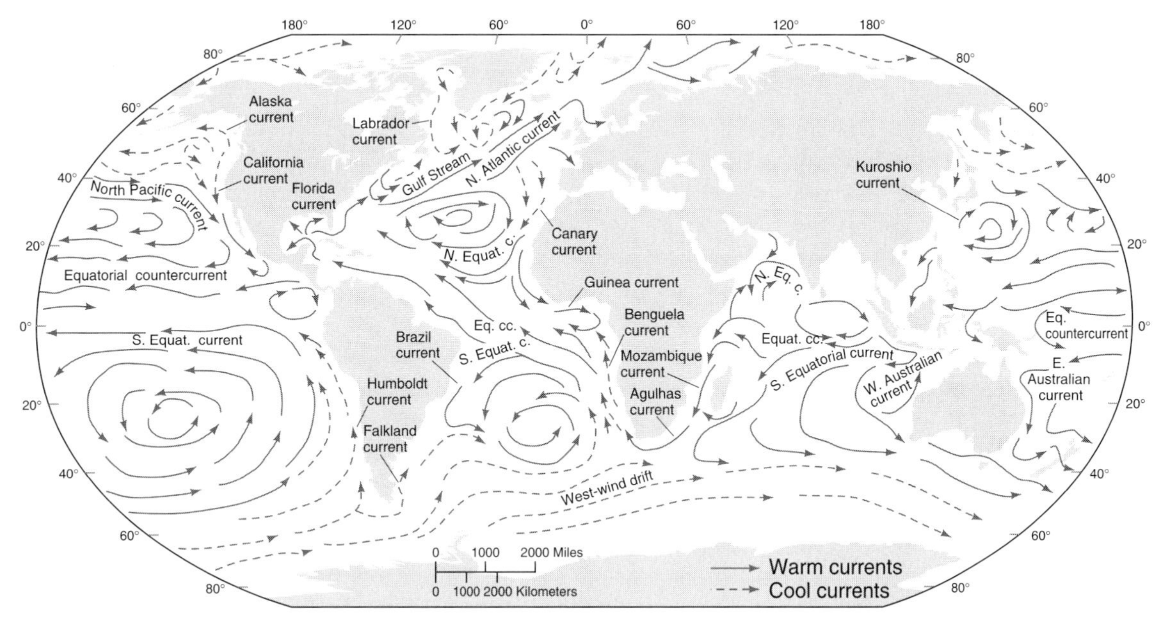

FIG. 2.11 Ocean currents have distinctive patterns that curve to the right (clockwise) in the Northern Hemisphere and to the left (counterclockwise) in the Southern Hemisphere. Continents deflect the westward flow of currents in the Atlantic and Pacific oceans and force warm, equatorial waters toward the poles.

marine life in the movement of food and nutrients. These displaced cooler waters are then replaced by the underlying bottom waters.

A CLOSER LOOK

Clipper ships first arrived in the Americas in the late 1840s and created a new era of sailing vessels for the transport of goods around the world. Speed records were constantly broken on long voyages, and knowledge of trade winds and ocean currents gave a great advantage over rival shipping companies.

During the 1840s, Matthew Fontaine Maury (1806–1873), a Tennessee farm boy who sailed around the world as a midshipman in the United States Navy, raised this science to a high level of sophistication. After one visit home while on leave, he was thrown off the top of a stagecoach and broke his leg. It never healed properly, and the Navy assigned him to a desk job at its Depot of Charts and Instruments in Washington, D.C.

In the Depot's vault, Maury discovered a collection of thousands of ship's logs, including almost every one since the first voyage of a U.S. Navy vessel. The charts included the weather and sea conditions for every month of the year for oceans around the world. Maury and his staff compiled these hundreds of thousands of bits of information. Later, he persuaded the Navy to standardize observations taken of weather, winds, currents, and other hydrological and meteorological data from their entire fleet. This led to the creation of Maury's published work *Wind and Current Charts* in 1847.

Sea captains were initially skeptical of the charts created by a low-level, land-bound Navy officer who hadn't been to sea in years. However, in 1848 a merchant by the name of Jackson challenged Maury's work and decided to follow the suggested courses on the charts. To Jackson's astonishment, by following Maury's charts he cut 35 days off a round trip between Baltimore, Maryland, and Rio de Janeiro, Brazil. Word spread quickly as other naval officers reported similarly reduced travel times on other voyages.

Eventually, Maury and his staff charted all the oceans of the world and offered specialized information on trade winds, monsoons, water-surface temperatures, storms, currents, and even the distribution of whales. In 1854, Maury published *Physical Geography of the Sea* and soon earned the name "Pathfinder of the Seas." For his efforts he received honors from intellectual societies and European royalty, and the gratitude of clipper captains around the world. (14)

A modern-day ocean current charting exercise is occurring as scientists track the epic voyages of 29,000 potential data points. In January 1992, 20 containers of rubber bath toys—rubber ducks, frogs, beavers, and turtles—were washed overboard from a cargo ship in the Pacific Ocean midway between Hong Kong and Tacoma, Washington. All were presumed lost at sea. However, in spite of windstorms, Arctic ice floes, and constant weathering, the flotilla of rubber bath toys have voyaged nearly halfway around the world, through the Bering Strait, across Arctic waters, and into the Atlantic Ocean off North America.

Six rubber toys first washed up on beaches near Sitka, Alaska, in November 1992, ten months after falling overboard in the Pacific storm. Remarkably, approximately 400 of the rubber toys have been found along the coasts of Alaska, Hawaii, Washington, and Vancouver Island and the Queen Charlotte Islands off British Columbia during the first years of their epic voyage. By 1995, they were passing through the Bering Strait but became frozen in Arctic icepack for several years. It's expected that other castaways will later be found along the coasts of Iceland and the United Kingdom. Oceanographers are tracking the location of the beached toys and are using that data to improve our knowledge of prevailing winds and ocean currents. It's a method that Matthew Maury would have found remarkable. See http://www.thefirstyears.com (the company that originally commissioned the shipment of rubber bath toys) for updated information.

Tilt of the Earth's Axis The seasons of the Earth, as well as temperature variations between the equator and the poles, are created when the Earth spins around an axis that is tilted 23.5 degrees from perpendicular to the plane of the ecliptic, i.e., the imaginary surface that contains the Earth's orbit around the Sun. If the Earth were stood straight up on its axis, the equator would always face the Sun and there would be no change in seasons.

A second minor factor that contributes to the change in seasons is the Earth's elliptical orbit around the Sun. This orbit causes the distance between the Earth and the Sun to vary slightly each day, so that we're 3 million miles (5 million km) closer to the Sun in early January than in early July. This distance change causes the Earth to receive slightly more thermal energy in January (during the Southern Hemisphere's summer) and less overall energy during the Northern Hemisphere's summer in July.

MONITORING CLIMATE CHANGE

Each year, the Earth's climate is recorded in tree rings, ice, coral, and sediment cores around the world. The study of tree rings is called **dendrochronology** (from *dendro,* Greek for "tree," and *chronology,* for the science that deals with time and dates). Trees form one growth ring per year, which varies in size, density, and chemical characteristics. Each ring reflects variations in annual climate, including precipitation and temperature.

Andrew E. Douglas (1867–1962) was an astronomer who became the acknowledged father of dendrochronology. Through his work with the Harvard College Observatory in Massachusetts in the late 1800s, he hoped to find evidence of past Sunspot events by studying the growth rings of coniferous trees. In the early 1900s, he began work on his theory in the forests of northern Arizona and later joined the faculty at the University of Arizona. In 1914, he presented a paper on his research to the Carnegie Institution in Washington, D.C., and from there, the science of dendrochronology evolved. (15)

Ice can provide a similar climatic record in regions where trees cannot grow, such as at the polar icecaps. Oxygen isotopes found in extracted ice cores from glaciers or polar icecaps can reveal climate changes over thousands of years. (Ice cores from depths of over 2 mi [3.2 km] were recently collected in Greenland and Antarctica.) Ice core temperature, dust, air bubbles, acidity, and oxygen isotopes can provide fairly detailed and accurate climate history. The Greenland and Antarctica ice cores provide climatic history going back 250,000 and 500,000 years, respectively. (16)

Ocean coral also provides information regarding historical climate data. Scientists have discovered that coral reefs contain a record of ocean-surface temperature changes for many thousands of years. Growth bands on coral provide the data for this research. Researchers are now working to correlate ocean-surface temperature records of El Niño and La Niña cycles to predict future climate changes.

Finally, sediment core analysis has also been used to assess climatic change. In Russia, for example, the study of Lake Baikal sediments began in 1993. Sediment cores of more than 320 feet (98 m) in length were collected and analyzed to reveal the climatic, environmental, and geologic history of that particular region as far back as 5 million years. (17) Drilling in ocean sediments has recorded global climate changes back to 200 million years ago.

WEATHER

Weather is the state of the atmosphere at a given time and place. It is a short-term event and generally is not reliably predicted for more than a few days because the chaotic nature of the atmosphere makes such prediction difficult. Weather is generally determined by five variables: (1) temperature, (2) air pressure, (3) humidity, (4) heating, and (5) wind speed and direction.

Temperature Carbon dioxide, water vapor, methane, ozone, and nitrous oxide create the Earth's "life-maintaining blanket." These five gases are commonly called **greenhouse gases** because they absorb infrared radiation being emitted from the surface of the Earth, much like a glass-covered greenhouse. Changes in the amounts of greenhouse gases, however, can lead to changes in temperature and eventually to changes in climate.

Ocean temperatures play a significant role in global weather patterns. Anyone who has gone to the beach off the coast of Maine, Rhode Island, or Nova Scotia in the summer has experienced surprisingly cold ocean waters (often no more than 53°F [12°C] during the summer months). This is contrasted by ocean temperatures above 80°F (27°C) off the coast of Florida. **Isotherms** (lines connecting points of equal temperatures) show bands of ocean temperatures that generally parallel the equator. The warmest waters in August (approximately 82°F, or 28°C) are found in a zone generally between 30°N and 10°S latitude where solar radiation is the greatest. By January, the zone of maximum solar radiation moves to the Southern Hemisphere, followed by higher ocean temperatures. (18)

The ocean has a high capacity to hold heat and moderates coastal temperatures in many regions of the world. Land temperature extremes have ranged from 136°F (58°C) in the Libyan Desert of northern Africa to −126°F (−88°C) in central Antarctica (a variation of 262°F, or 146°C). By contrast, recorded ocean temperatures have varied by only 69°F, or 38°C. The variance has been from 97°F, or 36°C, in the Persian Gulf and 28°F, or −2°C, in polar seas. (19)

Air and land surface temperatures also play a major role in climate, regional weather patterns, and aquatic ecosystems. The vertical motions caused by the heating of oceans, land surfaces, and air will be discussed later in this chapter in the section titled "Heating."

Air Pressure The Earth's **atmosphere** is composed of invisible and odorless gases and suspended particles that surround our planet. Since air pressure decreases with altitude, the density of the atmosphere (the amount of air per unit volume) decreases as altitude increases. Variations in air pressure are measured by a **barometer.**

The invention of the mercury barometer is generally attributed to the Italian physicist Evangelista Torricelli (1608–1647), a student of Galileo. In 1643, Torricelli constructed a glass bulb with a neck "two cubits" long. (A cubit varies in length but is generally 17 to 21 in., or 43 to 53 cm.) The tube was filled with mercury and inverted into a basin containing more mercury. (Originally, Torricelli used water in his model, but he finally had to use a 60-ft, or 18.3-m, glass tube. He later used the much heavier mercury in his experiment and reduced the height of the tube to only about 3 ft, or 0.9 m.) Torricelli deduced from this experiment that air pressure on the mercury surface in the bowl must be holding up the column of mercury inside the glass tube. Day-to-day measurements of Torricelli's invention showed slight variations in air pressure outside the glass tube filled with mercury. This led to the discovery by Torricelli and others of air pressure changes and its relation to weather changes. (20)

The highest barometric pressure ever recorded on the Earth's surface was 1083.8 millibars (32.00 in., or 81.28 cm) at Agata, Siberia, on December 31, 1968. The temperature was −50°F (−46°C), with the extreme cold and dense air contributing to the high reading. The lowest sea-level pressure recorded was 870 millibars (25.69 in., or 65.25 cm) on October 12, 1979, approximately 1000 miles (1609 km) east of the Philippine Islands inside Typhoon "Tip." The lowest recorded atmospheric pressure in the United States was 892 millibars (26.35 in., or 66.93 cm) inside a 1935 Labor Day hurricane that crossed the Florida Keys. The absolute lowest pressure on Earth is probably inside a tornado. (21) The average barometric pressure, called the *standard atmosphere at sea level,* is 30.17 inches or 76.63 cm of mercury.

Humidity The composition of air varies in different locations at the Earth's surface owing to the presence of **aerosols** and water vapor. Aerosols are very small liquid droplets, or tiny solid particles, that remain suspended in the air. Liquid aerosols can exist in fog, while solid particles such as very small ice crystals, smoke particles from fires, dust, volcanic emissions, and industrial pollutants can be widespread.

Humidity is the amount of water vapor that exists in the air in a given location (referred to as absolute humidity). A hot humid day in the tropics can have as much as 4 percent of the air by volume as water vapor, while a cold winter day in other regions may have less than 0.3 percent water vapor present. (22)

Evaporation causes water vapor to enter the atmosphere and then change back into a liquid form as part of the hydrologic cycle. When the number of water molecules that evaporate (change from a liquid to a gas) equals the amount changing from a gas to a liquid in the atmosphere, the atmosphere is said to be saturated (the **dew point** or **frost point** has been reached). If this water vapor content is exceeded, precipitation will occur in the form of rain, snow, hail, or sleet.

Heating Have you ever pumped air into a bike tire and noticed that the pump became hot as you pushed down on the handle? Some of the heat you felt was generated by friction between the leather sleeve moving inside the pump cylinder. However, much of the temperature increase was caused by the compression of air inside the bike pump. In contrast, have you ever released air from a tire and noticed how cool it felt?

These two conditions—the heat generated from compressed air and the coolness of expanding air—are examples of **adiabatic processes** (from the Greek word *adiabatos,* meaning "no passage"). This term comes from the process of heating or cooling without the addition or subtraction of heat from an outside source. When air is compressed, the result is an increase in temperature. When compressed air expands, the result will be a decrease in temperature.

Warm air is less dense than cool air and will rise in the atmosphere. However, as the air pressure decreases and elevation increases, rising air will expand and cool. If the opposite occurs and cool air begins to sink in the atmosphere, the temperature will increase as the air is compressed at lower elevations. Rising and expanding unsaturated air will drop in temperature at a constant rate of 1°F per 183 feet (1°C per 100 m). Falling and contracting parcels of unsaturated air will increase in temperature at the same rate. These air temperature changes, caused by compression and expansion based on elevation change, are called the **adiabatic lapse rate.** (23)

The adiabatic process often creates clouds, which are composed of aerosols and/or ice crystals, and form when air rises and becomes saturated through the process of adiabatic cooling. Clouds can be classified on the basis of form and height. On the basis of form, clouds are grouped into two major classes: globular masses, or *cumuliform* clouds, and layered, or *stratiform* clouds. Clouds are categorized into four families by height: low, middle, high, and clouds with vertical development. (24)

Cumuliform clouds are puffy, round, individual clouds that form when hot, humid air rises due to convection heating (the transfer of heat to the atmosphere from the land surface). Variations occur in cumulus clouds (stratocumulus, cumulonimbus, altocumulus, and cirrocumulus) as **convection heating** increases the elevation of warm, humid air movement.

Stratiform clouds are best known by the sheet or blanket appearance that extends across the entire sky. These cloud formations reach a certain level of water vapor content and then spread laterally instead of vertically. Nimbostratus, altostratus, and cirrostratus clouds vary from the dense, low cloud blanket to the higher elevation, "wispy" appearance cloud formations. Cirrus clouds are the highest stratus clouds in the sky and look like filaments or feathers. These are composed entirely of ice crystals.

Clouds are generally formed by combinations of five processes:

1. Density lifting caused by warm, low-density air rising due to convection heating (heating from the Sun) and then displacing cooler air located at higher elevations.
2. Frontal lifting caused by warm, moist air moving over cooler air. This is called a **warm front** and often causes the buildup of clouds and the generation of precipitation. A warm front is shown on a weather map by a line with semicircles extending into the cooler air. The term *front* is used because Norwegian meteorologists in World War I compared the clash of air masses above the Earth to the battle lines, or fronts, of warring armies of the time on the land surface. (25)
3. Frontal lifting caused by cool air moving under warmer air. Cold air will move in as a **cold front** and will displace warmer, less dense air by pushing it upward. This may cause cloud formation and precipitation. A cold front is designated on a map by a line with triangle-shaped points extending into the warmer air mass. (Occasionally, weather fronts will become parallel with little or no movement. These are called **stationary fronts** and are shown with triangular points on one side and semicircles on the other side of the weather front.)
4. **Orographic lifting** created when warm, moist air is forced by wind over high natural features such as mountains (Figure 2.12). Some

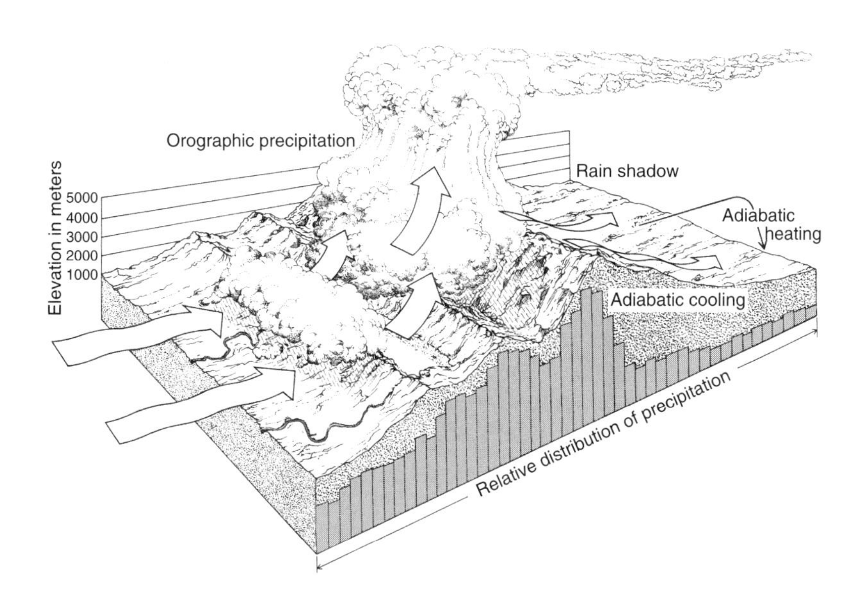

FIG. 2.12 Orographic rainfall—Rain and snowfall are heaviest on the windward slopes as compared to the dry climates of leeward, or downwind, slopes. See Figure 2.13 for the result of this process.

of the highest amounts of precipitation occur in regions that experience orographic lifting—the western coast of Tasmania in Australia, the Owen Stanley mountain range in New Guinea, and the Cascade Range in Washington, Oregon, and British Columbia.

5. Convergence lifting caused when two air masses collide and are both forced upward. Weather patterns along the Florida Peninsula are a good example of this process. Air flows toward land from both the Gulf of Mexico on Florida's west coast and from the Atlantic Ocean on the east coast of the state. The two air masses collide over the state and force some air to rise, resulting in frequent afternoon thunderstorms and rain showers.

All in all, the single basic reason for cloud formation and precipitation, or the lack of it, is related to vertical motions. The upward motion of air can enhance the growth of clouds, the formation of ice crystals, and the precipitation process. If air is sinking or subsiding, relative humidity will decrease. (The air warms while the water content remains the same.) This will cause clouds to dissipate, and precipitation will be sparse. When air rises, clouds form, and a good field of droplets becomes available for coalescence. Ice nuclei then form in the colder portion of a growing cloud as part of the precipitation process. Much of what we see as weather and climate can be explained by these vertical motions. The violence of the weather is directly related to their strength, and meteorologists would love to be able to predict those vertical motions with accuracy. (26)

A CLOSER LOOK

On July 26, 1959, Lt. Col. William H. Rankin, a U.S. Marine Corps pilot, was on a routine flight between Massachusetts and South Carolina. He was flying an F8U Crusader supersonic jet at 47,000 feet (14,326 m) above a well-developed cumulonimbus cloud system, or thunderhead, when he was forced to eject due to a seized engine (a condition created when pistons expand so much due to overheating that they will not move in the cylinders, probably caused by low oil pressure).

Lt. Col. Rankin fell through the extremely cold (−70°F, or −57°C) interior of the thunderhead cloud. Violent updrafts inside the storm changed his normal 10-minute descent into a 40-minute ride of terror. Lightning flashed around him and at times seemed to pass through his body, while continual claps of thunder almost broke his eardrums. Torrential rain and pelting hailstones struck his body, which was protected only by a summer flight suit and helmet. The pilot was terrified that he might drown while still in the thunderhead due to the incredible amount of precipitation inside the cloud.

Ejection into the low atmospheric pressure at 47,000 feet caused his internal organs to swell in size, creating horrific pain. Lt. Rankin's parachute at times oscillated like a pendulum and at other times deflated from violent downdrafts of air. The harrowing aerial experience ended when he landed in a pine tree near Rich Square, North Carolina. He is the only person known to have penetrated the entire length of a severe thunderstorm and survived. (27)

As mentioned earlier, the Cascade Mountains of Oregon are a topographic barrier that produce a large rain shadow over the eastern (downwind) region of the state. Portland, on the west (windward) side of the mountain range, receives approximately 36 inches (91 cm) of precipitation annually. By contrast, the City of Pendleton, Oregon, on the east side of the Cascades, is in a rain shadow and receives only 12 inches (30 cm) per year (Figure 2.13). A **rain shadow** is an area of low precipitation created when warm, moist air runs into a barrier such as a mountain range, which causes orographic lifting and heavy precipitation on the windward side. Low precipitation occurs on the downwind side of such a range. This is the main reason why Portland receives more precipitation than Chicago, or more than Los Angeles, Phoenix, and Denver combined.

Coastal deserts can also occur along the western shores of some continents. Cold, bottom waters of an ocean often meet landmasses, causing maritime air temperatures to cool. Fog can often be the result since the air retains too little moisture to create precipitation. Coastal deserts in

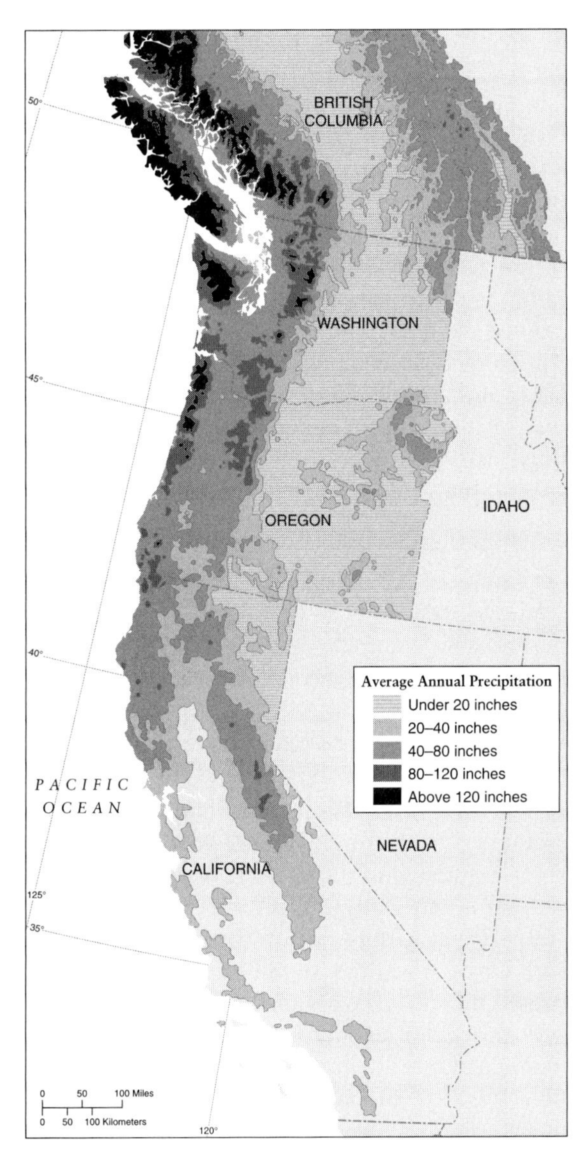

FIG. 2.13 Average annual precipitation in Oregon, Washington, California, and British Columbia. Note the dramatic distribution of orographic rainfall.

Peru, Chile, and southwestern Africa are among the driest regions on Earth.

Heating caused by El Niño and La Niña events can also drastically alter weather around the world. **El Niño** (Spanish for "the Little Boy" or "Christ Child" since it often arrives around Christmas in South America) is a natural phenomenon that occurs when a warm surface ocean current forms in the eastern Pacific Ocean off the coast of Ecuador and Peru. This warm current prevents cold, deeper water from reaching the ocean surface. El Niño is also associated with the southward displacement of the intertropical convergence zone (ITCZ) called the **southern oscillation,** a variable pattern of air pressure between the eastern and western tropical Pacific.

Strong El Niño events often produce floods in Ecuador, Peru, Cuba, and the southern United States, and drought in Australia, Indonesia, the Philippines, and southern Africa. El Niño can also generate more frequent and intense ice storms in southeastern Canada, while generally reducing Atlantic tropical storms and hurricanes.

El Niño is on an approximate cycle of two to seven years and has been monitored in South America for centuries. Peruvian farmers have predicted El Niño rainfall in December by observing the brightness of stars in the Pleiades Constellation in June. Brighter stars meant more rain during the growing season. It's been hypothesized that El Niño conditions in the ocean off the coast of Peru generated high cirrus clouds during the summer that veiled starlight from the constellation. (28)

La Niña ("the Little Girl") is another natural phenomenon caused by cooler water in the equatorial Pacific Ocean. It also leads to flood and drought conditions across the world, but the locations of these weather patterns tend to be opposite from the effects of El Niño.

Wind Speed and Direction **Wind** is the horizontal movement of air caused by differences in air pressure. Air is constantly moving from locations of high pressure to areas of low pressure to eliminate differences in air pressure caused by solar heating. There is always some vertical movement associated with this air movement. Air pressure gradients are determined by **isobars** (lines on a map that connect locations of equal atmospheric pressure) measured by a barometer. The area between isobars determines the air pressure gradient in a location on the ground. If isobars are close together, wind speed will increase as air moves from a region of high pressure to a nearby region of lower pressure. An area where isobars are far apart will experience little or no wind.

Air near the ground will flow inward from all directions into an area of **low barometric pressure** and will develop an inward spiral motion (counterclockwise) in the Northern Hemisphere. Air near the ground will flow in an outward spiral motion (clockwise) from a **high barometric pressure** area. In the Southern Hemisphere the opposite is true; a low-pressure system will spiral clockwise, while a high-pressure system will move counterclockwise.

The spiral pattern of air flow can be seen almost daily on satellite photos or weather maps. Air rotating around a low-pressure system, in a counterclockwise motion, is labeled with an *L* for Low on a weather map. Air rotating in a clockwise motion around a high-pressure system is labeled *H* for High on the same map.

Finally, during the winter, it's not uncommon for the temperature to be 65°F (18°C) in Mississippi but near freezing on the same day in Ohio, only a few hundred miles to the north. These substantial temperature variations arise because of differences in pressure gradients that create **jet streams.** These are relatively narrow ribbons of very strong winds at higher altitudes caused by moving air that tries to even out areas between high- and low-pressure systems. High winds aloft have been compared to streams of water and are generally strongest during winter months.

WEATHER MODIFICATION

Everybody talks about the weather, but nobody does anything about it.

Mark Twain, *Hartford Courant*, August 27, 1897

Weather modification is any change in the weather caused by humans. For centuries, the idea of "controlling" the weather has been an intriguing idea. Efforts have been made to increase rainfall, reduce the size and duration of hailstorms, dissipate fog, reduce the intensity of lightning, and calm the winds of hurricanes. However, rain enhancement and hail suppression have shown the most promise.

One of the first U.S. weather modification programs began in the 1890s. Investigators noted that in the major battles of the Civil War of the 1860s, rainfall seemed to occur after artillery had been fired. The U.S. Congress appropriated $9000 for the U.S. Department of Agriculture to conduct tests near Washington, D.C., and in west Texas. Cannons were fired into the sky, dynamite was detonated, and hydrogen-filled balloons were released into the atmosphere, but the results were inconclusive. (29)

SIDEBAR

In the early 1900s, San Diego was in the middle of a severe drought. In desperation, the city hired the self-proclaimed rainmaker J.S. Stingo. His efforts coincided with torrential rains and flash flooding until irate citizens ran him out of town. The movie *The Rainmaker* (1956), starring Burt Lancaster, is based on this incident.

The first scientific breakthrough in weather modification occurred in 1946 when V. J. Schaeffer, a scientist with the General Electric Company (GE) in Schenectady, New York, accidentally discovered that adding dry ice (frozen carbon dioxide) into a deep-freeze full of supercooled fog caused the moisture to quickly freeze and drop as snow particles. Schaeffer later found that silver iodide produced a similar effect. Before long, airplanes were dropping silver iodide into supercooled clouds above New England to increase the growth of ice crystals. (On one flight, researchers even carved a big GE into the clouds.) (30)

This process of introducing nucleating agents such as silver iodide (AgI), ammonium iodide (NH_4I), or dry ice into clouds to enhance precipitation or to reduce hail is called **cloud seeding.** Cloud seeding causes water vapor in a cloud to freeze, converting it from a gaseous to a solid state. Other moisture is then induced to form around this nucleus until these water droplets become heavy enough to fall as precipitation.

Ground-based generators or specially equipped aircraft deliver the nucleating agents into clouds.

Silver iodide is often introduced into a cloud by burning silver iodide flares. The smoke particles furnish the nuclei for ice crystal formation. A cloud temperature of 10°F (−12°C) is ideal, although a range of 25°F to −22°F (−4°C to −30°C) is acceptable. Cloud temperatures warmer than 25°F (−4°C) are not sufficiently supercooled, while temperatures below −22°F (−30°C) already contain ice crystals since spontaneous nucleation occurs at −40°F (−40°C). (31)

Concerns have been expressed over the environmental effects of using ammonium or silver iodide if large quantities accumulate on the ground after seeding efforts. Measurements have been made of the change of concentrations of silver iodide found in farm ponds, rivers, and other areas where cloud seeding is practiced. Only minor concentrations have been found, but widespread use of silver iodide could result in elevated local concentrations in precipitation and runoff. (32)

It is very difficult to evaluate the effectiveness of precipitation enhancement projects. Since the atmosphere does not generate identical clouds, a complex statistical dilemma is created in determining the value of such cloud seeding programs. One method is to compare the amount and extent of rainfall in a seeded area to a region outside the area, using several different cases. This type of analysis has shown an increase in precipitation of 10 to 20 percent. (33)

POLICY ISSUE

Although cloud seeding efforts have been conducted for over a century, many people adamantly oppose the concept because cloud seeding, they believe, increases the likelihood of damaging hailstorms. Reductions in precipitation downwind of seeded storms have also been attributed, opponents argue, to enhanced rainfall upwind. Some scientists and water resource managers are skeptical of the benefits of cloud seeding since many scientific studies are inconclusive.

Some objectors argue that cloud seeding is unethical because it interferes with natural hydrologic processes. Supporters argue that cloud seeding works and that efforts should continue regardless of political boundaries. In 1972, the United Nations Conference on the Human Environment held in Stockholm, Sweden, urged establishment of an advisory committee to consider the international concerns of weather modification.

If you were the program manager for a cloud seeding experiment, how would you deal with these types of allegations to gain support for your program? How could you design a precipitation monitoring network to support your project?

SIDEBAR

The costliest hailstorm in U.S. history occurred in Denver, Colorado, in 1990, with damages of $625 million. In 1984, a hailstorm in Munich, Germany, caused $1 billion in damages. The largest documented hailstone fell in Aurora, Nebraska, in 2003. It measured 7.0 inches (17.8 cm) in diameter and had a circumference of 18.75 inches (47.6 cm)—almost the size of a soccer ball. The hailstone is now at the National Center for Atmospheric Research in Boulder, Colorado, where it will be preserved indefinitely. (34)

FLOODS

Floods, a most deadly natural disaster, have become more common and severe due to human influence and development. Paved surfaces, road embankments, and wetlands removal can increase runoff volumes in specific locations. A flood can be caused by summer thunderstorms, melting snow, ice jams in swollen rivers, or seasonal monsoons. Storms can be totally unexpected and come with little warning, such as summer floods experienced in Tucson, Arizona, or in New Zealand. By contrast, floods can be predictable and closely monitored, like flooding along the Mississippi River in the United States or the Red River in south-central Canada, after a winter and spring of excess moisture.

Every year thousands of floods occur around the world, with little or no warning. Such an event, called a **flash flood,** happened in the Black Hills of South Dakota on June 9, 1972, when 15 inches (38.1 cm) of rain fell in only five hours, sweeping away more than 200 people in the unexpected floodwaters generated by the storm. In Calama, Chile, virtually no rain fell for 400 years in the city of 150,000 people located in the middle of the Atacama Desert. Then a midafternoon deluge hit on February 10, 1972, creating mudslides and catastrophic floods.

On the evening of July 31, 1976, near Estes Park, Colorado, over 12 inches (30.5 cm) of rain fell in just four hours above the mountainous and narrow Big Thompson River Canyon. Thousands of streamside motorists, residents, and campers were caught by surprise. Surface water runoff was so great that all stream measurement devices were washed out by a 20-foot (6-m) wall of water that crashed down the dark canyon. Loss of life was widespread as 144 people died in the flood. The death toll would have been much higher if Colorado State Patrol Sergeant Willis Hugh Purdy had not driven his patrol car down the canyon, ahead of the racing wall of water, to warn unsuspecting residents and campers. The heroic Sergeant Purdy died in the flood. Today, a plaque in his honor is located in the Big Thompson Canyon.

Flash floods are brutally swift, massive, and destructive. However, larger floods have occurred with several days and even weeks of warning from meteorologists and government authorities. In 1993, massive snowmelt and above-normal precipitation in the upper Mississippi River Basin, associated with the 1992–1993 El Niño, led to enormous amounts of surface water runoff during that spring and summer (Figure 2.14). Iowa, Missouri, and Illinois, as well as other basin states, experienced record water levels that covered substantial portions of those states. Government agencies worked around the clock for weeks in an effort to construct levees and dikes, built of earth or sandbags, to keep floodwaters away from communities. In some cases these efforts were successful, but in other locations entire cities were inundated.

In the Northern Territory of Australia, cyclone season begins in early November and runs to the end of April. (Cyclones are called hurricanes or typhoons in other countries.) Rainfall and flooding can be extreme but are not unexpected during cyclone season. See http://www.bom.gov.au/info/cyclone for excellent information from the Australian Bureau of Meteorology.

DROUGHT

Author John Steinbeck wrote in *East of Eden,* "And it never failed that during the dry years the people forgot about the rich years, and during the wet years they lost all memory of the dry years. It was always that way." (35) Steinbeck wrote of the brutally dry years in Oklahoma in the 1930s and described the hardships created for farmers and business people across the American Midwest.

According to the U.S. National Weather Service, **drought** is "a period of abnormally dry weather which persists long enough to produce a serious hydrologic imbalance." The severity of a drought depends on the degree of moisture deficiency, the duration of the dry weather, and the size of the affected area. Drought is very different from aridity. **Aridity** is a permanent climatic condition in a region, whereas drought is a temporary lack of moisture.

Drought is relative to a particular location. Years without rain in the Atacama Desert, located in the rain shadow of the Andes Mountains of northern Chile, would simply represent the normal aridity of the region. However, two months without moisture in Seattle, Washington, or Montreal, Quebec, could be classified as a serious drought. Droughts generally have no specific beginning or end, cover large areas, and cause little structural damage. Drought can sometimes be predicted through data obtained on developing El Niño and La Niña events.

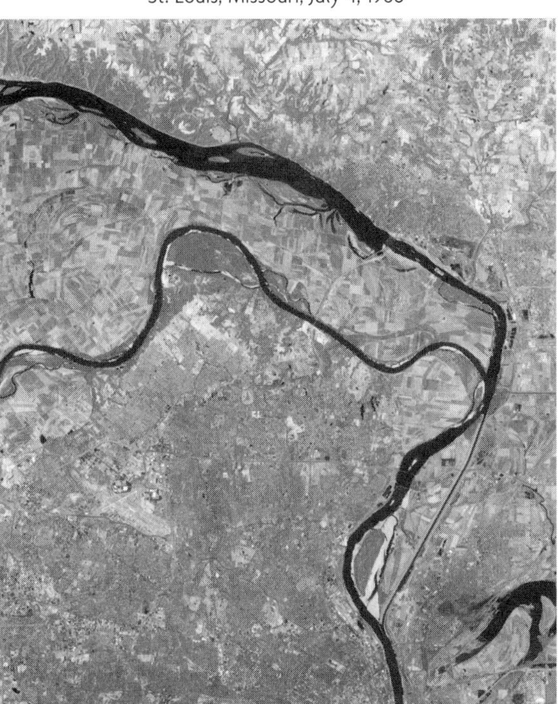

FIG. 2.14 These satellite images show the confluence of the Missouri and Mississippi rivers near St. Louis, Missouri. A normal flow pattern is shown in the photo at the left where the area between the two rivers is dry and cultivated for crops. The photo on the right shows the rivers out of their banks because of weeks of torrential rains in the region. Farmland and communities were inundated with floodwaters, causing billions of dollars in damages.

Drought can be defined in several different ways:

Meteorological Drought Departure from normal precipitation. (Annual precipitation over 5 in. [12.7 cm] per year in some parts of northern Africa would be considered a very wet period.)

Hydrological Drought Below-normal surface and groundwater supplies.

Agricultural Drought Inadequate water supplies to grow a particular crop.

Socioeconomic Drought Inadequate water supplies to serve local residents.

Several **drought indices** exist to measure the severity of below-average moisture in a region. A simple method used quite often is "Percent of Normal," a calculation that meets the needs of TV weather forecasters and general audiences. All audiences easily understand when a meteorologist says "We're 10 percent above normal in precipitation for the year," or "Precipitation is 20 percent below normal in the western part of the province."

The Palmer Drought Severity Index, or **Palmer Index,** was the first comprehensive drought index developed in the United States. (36) The U.S.

FIG. 2.15 Humans have greatly altered the habitat of Australia. In 1859, Thomas Austin imported 24 rabbits from England and released them on his property for sport hunting. With no natural predators, the animals multiplied rapidly; today Australia has more than 200 million rabbits. This extreme overpopulation has created severe stress on the landscape. Here thirsty rabbits converge on a waterhole during a drought.

Department of Agriculture uses this index to determine when to provide emergency drought assistance. It was developed by W. C. Palmer in 1965 and works best in large areas of uniform topography.

This index does not consider human alterations to the natural water balance, such as reservoir storage. An extremely wet region would have a Palmer Classification of 4.0 or more, whereas an extreme drought would be −4.0 or less (see Table 2.6). Weekly Palmer Index maps can be viewed at http://www.drought.noaa.gov/palmer.html.

TABLE 2.6 Palmer Classifications

4.0 or more	Extremely wet
3.0 to 3.99	Very wet
2.0 to 2.99	Moderately wet
1.0 to 1.99	Slightly wet
0.5 to 0.99	Incipient wet spell
0.49 to −0.49	Near normal
−0.5 to −0.99	Incipient dry spell
−1.0 to −1.99	Mild drought
−2.0 to −2.99	Moderate drought
−3.0 to −3.99	Severe drought
−4.0 or less	Extreme drought

A **Surface Water Supply Index (SWSI)** provides a weighted index for a river basin using information on snowpack, reservoir storage levels, precipitation, and streamflow.

Drought is a naturally occurring event throughout the world (see Figure 2.15). The Dust Bowl of the 1930s was probably the worst drought in recorded U.S. and Canadian history. In those years, dust storms occurred weekly and piled drifts of blown soil deposits up to 25 feet (7.6 m) high. Topsoil was blown thousands of miles from dry farmland in the Central Plains out to ships in the Atlantic Ocean. Crops withered away, livestock died, and grasshoppers arrived like the plague. Fortunately, land-use practices have changed since then to protect barren ground susceptible to blowing.

POLICY ISSUE

Extended drought in California from 1986 to 1993 forced residents to drastically reduce water use. Restrictions were implemented to conserve water, as a result of which lawns and crops withered and died from lack of water. A drought hit much of the eastern United States and south-central Canada in 2002, and peaked in the western regions of the two countries in 2003 when below-normal precipitation caused extensive crop loss and ravaging forest fires. Fines of up to $1000 and a six-month jail term were imposed in some locations for watering a lawn or washing a car. Other methods have been used to combat drought, such as in 1988 when Alabama Governor Guy Hunt led a statewide prayer for rain. Wet weather arrived the next day and continued for weeks.

How do you prepare for drought? Government officials can encourage residents to plant xeriscape (low water-use) plants instead of bluegrass for lawns. Emergency water allocation programs, based on priority of use, can be developed. Many cities have contingency plans for water use restrictions during periods of water shortages. Even and odd watering days are often used in drier climates, based on the last digit of a home address, to restrict lawn watering and car washing. In extreme conditions, outdoor water use is prohibited.

Drought planners often consider techniques used in arid climates. Many residents in Phoenix, Arizona, have crushed rock lawns instead of bluegrass to reduce water use. Drip irrigation is used to reduce the amount of water needed to grow crops in southern California. Restrictions on water use are common in these dry regions. What drought contingency plans have been used in your area? Is it appropriate for government agencies to implement water rationing only during drought periods, or should government-mandated water conservation programs be a way of life?

CHAPTER SUMMARY

The hydrologic cycle is a never-ending, naturally occurring feature of our Earth and is powered by energy from the Sun. Through the processes of precipitation, runoff, storage, evaporation and condensation, water is recirculated around the world. Between natural processes of the atmosphere, the land surface, groundwater, and the oceans, the Earth's water is constantly recycled.

The locations of wet and dry regions on our planet are determined by a variety of factors, but primarily by air movement and ocean currents. The Earth's rotation generates patterns of air and ocean water movement that create extremes in precipitation around the world. Orographic barriers, such as mountain ranges in Washington, Oregon, British Columbia, and Chile, can also create extremely dry local climates.

Weather is created by the same factors that generate climate, but is also affected by changes in atmospheric temperature and barometric pressure. Warm and cold fronts can greatly influence daily and weekly weather patterns. Some weather events can be predicted by observing the development of El Niño or La Niña events, while other conditions are almost unpredictable.

QUESTIONS FOR DISCUSSION

1. Name the four primary components of the hydrologic cycle.
2. What role does runoff play in the hydrologic cycle?
3. What is the difference between climate and weather?
4. Why are some regions of the world much drier than other areas?
5. Discuss the role of wetlands within the hydrologic cycle.
6. Discuss orographic lifting and its effects on precipitation variability in the western United States and Canada.
7. Identify several methods used to obtain historical climate data.
8. Explain how an El Niño event develops off the coast of Peru.
9. In the Southern Hemisphere, do ocean current gyres go clockwise or counterclockwise? Why?
10. Why is atmospheric pressure important in predicting weather?
11. What is the difference between aridity and drought?
12. Find the current Palmer Drought Index for a selected region. What classification currently exists, and why?

KEY WORDS TO REMEMBER

adiabatic lapse rate p. 42
adiabatic process p. 42
aerosols p. 42
arid p. 26
aridity p. 48
atmosphere p. 41
barometer p. 41
Blaney–Criddle Method p. 35
Class A Evaporation Pan p. 34
climate p. 37
cloud seeding p. 46
cold front p. 43
condensation p. 36
consumptive use (CU) p. 36
convection heating p. 43
Coriolis effect p. 37
cumuliform clouds p. 43
dendrochronology p. 40
dew point (frost point) p. 42
Doppler Radar p. 28
drought p. 48
drought indices p. 49
El Niño p. 45
evaporation p. 34
evapotranspiration (ET) p. 35
flash flood p. 48
global trade winds p. 37
greenhouse gases p. 41
gulf stream p. 38
gyres p. 38
high barometric pressure p. 46
humidity p. 42
hydrologic cycle p. 24
intertropical convergence zone (ITCZ) p. 37
isobars p. 45
isotherms p. 41
jet stream p. 46
La Niña p. 45
lake p. 31
low barometric pressure p. 46
lysimeters p. 35
orographic lifting p. 43
Palmer Index p. 49
pan coefficient p. 34
Penman Equation p. 35
phreatophytes p. 35
potential evapotranspiration p. 35
precipitation p. 25
rain shadow p. 44
reservoir p. 31
runoff p. 30
semiarid p. 37
snow cores p. 29
snow course p. 29
snow pillows p. 29
snow tube p. 29
southern oscillation p. 45
stationary front p. 43
stratiform clouds p. 43
sublimation p. 35
Surface Water Supply Index (SWSI) p. 50
Thales p. 23
transpiration p. 35
virga p. 25
warm front p. 43
water equivalent p. 29
weather p. 41
weather modification p. 46
wetland p. 33
wind p. 45

SUGGESTED RESOURCES FOR FURTHER STUDY

READINGS

Frisinger, H. Howard. *The History of Meteorology to 1800*. New York: Science History Publications, 1977.

Goldstein, Mel. *The Complete Idiot's Guide to Weather*. New York: Alpha Books, 1999.

Rankin, William H. *The Man Who Rode the Thunder*. Englewood Cliffs, NJ: Prentice-Hall, 1960.

Skinner, Brian J., Stephen C. Porter, and Daniel B. Botkin. *The Blue Planet: An Introduction to Earth System Science*. 2nd ed. New York: John Wiley & Sons, 1999.

Williams, James Thaxter. *The History of Weather*. Commack, NY: Nova Science Publishers, 1999.

WEBSITES

National Center for Atmospheric Research, Homepage, May 2003. http://www.ncar.ucar.edu

National Oceanic and Atmospheric Administration, National Climate Data Center, Homepage, May 2003. http://www.ncdc.noaa.gov/ol/ncdc.html

United Kingdom Meteorological Office, Homepage, May 2003. http://www.meto.govt.uk

National Oceanic and Atmospheric Administration, "NOAA La Niña Page," May 2003. http://www.elnino.noaa.gov/lanina.html

"Western Kansas Weather Modification Program," May 2003. http://pta6000.pld.com/hailman

Texas Natural Resource Conservation Commission, "The Potential of Cloud Seeding for Augmenting Rainwater in Semi-Arid West Texas," George W. Bomar, May 2003. http://www.lib.ttu.edu/playa/text94/playa24.htm

Dartmouth Flood Observatory, Homepage, May 2003. http://www.dartmouth.edu/artsci/geog/floods/index.htm

U.S. Department of Agriculture, "Drought Information," May 2003. http://drought.fsa.usda.gov

National Drought Mitigation Center, University of Nebraska-Lincoln, Homepage, May 2003. http://www.drought.unl.edu/index.htm

U.S. National Weather Service, "All About Droughts," May 2003. http://www.nws.noaa.gov/om/drought.htm

U.S. Army Corps of Engineers, "National Drought Atlas," May 2003. http://www.iwr.usace.army.mil/iwr/atlas/Atlasintro.htm

VIDEOS

The Grapes of Wrath, John Ford, Director, 20th Century-Fox, 1940. 2 hrs., 9 min.

"Weather Predictions," Modern Marvels Series, The History Channel. 50 min.

The Rainmaker, Joseph Anthony, Director, Paramount Pictures, 1956. 2 hrs., 15 min.

"Fatal Flood," Chana Gazit, Director, *American Experience Series*, PBS Home Video, 2001. 60 min.

REFERENCES

1. Eagleman, Joe R. *Meteorology: The Atmosphere in Action*. Belmont, CA: Wadsworth Publishing Co., 1985, 54.

2. Canadian Hydrographic Service, Department of Fisheries and Oceans, http://chswww.bur.dfo.ca/danp/fluctuations.html, July 2001.

3. Utah Geological Survey. "Commonly Asked Questions about Utah's Great Salt Lake and Ancient Lake Bonneville," http://www.ugs.state.ut.us/pi-39/pi39pg1.htm, July 2001.

4. Campbell, Robb. "Earthshots: Great Salt Lake, Utah," http://geochange.er.usgs.gov/sw/changes/anthropogenic/gsl, U.S. Geological Survey, Washington, DC, July 2001.

5. Skinner, Brian J., and Stephen C. Porter. *The Dynamic Earth: An Introduction to Physical Geology*. New York: John Wiley & Sons, 2000, 5.

6. Halley, E. "An Account of the Evaporation of Water." *Philosophical Transactions of the Royal Society*, London, 18, 1694, 183–190.

7. Water Resources of Georgia Homepage, U.S. Geological Survey, Atlanta, GA, http://go.water.usgs.gov, September 2001.

8. Penman, H. L., "Natural Evaporation from Open Water, Bare Soil, and Grass." *Proceedings of the Royal Society,* London, England, Series A, 193, 1948, 120–146.

9. Male, D. H., and D. M. Gray. "Snowcover Ablation and Runoff." *Handbook of Snow.* New York: Pergamon Press, 1981.

10. Thornthwaite, C. W. "Report of the Committee on Transpiration and Evaporation." *Transactions of the American Geophysical Union,* vol. 25, pt. 5, 1943–1944.

11. Mather, John R., and Marie Sanderson. *The Genius of C. Warren Thornthwaite, Climatologist-Geographer.* Norman: University of Oklahoma Press, 1996.

12. Frisinger, H. Howard. *The History of Meteorology to 1800.* New York: Science History Publications, 1977, 125–128.

13. Skinner, Brian J., Stephen C. Porter, and Daniel B. Botkin. *The Blue Planet: An Introduction to Earth System Science.* 2nd ed. New York: John Wiley & Sons, 1999, 248–249.

14. Whipple, A.B.C., and the Editors of Time-Life Books. *The Seafarers: The Clipper Ships.* Alexandria, VA: Time-Life Books, 1980, 41–45.

15. Fritts, H. C. *Tree Rings and Climate.* New York: Academic Press, 1976, 4–7.

16. Williams, James Thaxter. *The History of Weather.* Commack, NY: Nova Science Publishers, 1999, 140.

17. "Lake Baikal—A Touchstone for Global Change and Rift Studies," http://marine.usgs.gov/factsheets/baikal, Marine and Coastal Geology Program, U.S. Geological Survey, Washington, DC, July 2001.

18. Skinner et al., *The Blue Planet,* 246.

19. Ibid.

20. Frisinger, 68–69.

21. Goldstein, Mel, *The Complete Idiot's Guide to Weather.* New York: Alpha Books, 1999, 45.

22. Skinner et al., *The Blue Planet,* 273.

23. Ibid., 283.

24. Strahler, Alan, and Arthur Strahler. *Physical Geography.* 2nd ed. New York: John Wiley & Sons, 2002, 142–143.

25. Ibid., 286.

26. Personal communication with Dr. Mel Goldstein, Meteorologist at WTNH-TV, New Haven, Connecticut, and Director of the Western Connecticut State University Weather Center, Danbury, Connecticut, September 10, 2001.

27. Rankin, William H. *The Man Who Rode the Thunder.* Englewood Cliffs, NJ: Prentice-Hall, 1960.

28. National Oceanic and Atmospheric Administration, "NOAA La Niña Page," http://www.elnino.noaa.gov/lanina.html, July 2001.

29. Humphreys, W. J. *Rain Making and Other Weather Vagaries.* Baltimore, MD: Williams & Williams Co., 1926, 32.

30. Posey, Carl A. *The Living History Book of Wind & Weather.* Pleasantville, NY: Reader's Digest Association, 1994, 170.

31. Eagleman, 277.

32. Ibid.

33. Ibid., 283–284.

34. "Largest Hailstone in U.S. History Found," http://news.nationalgeographic.com/news/2003/08/0804_030804_largesthailstone.html, National Geographic Society, January 2004.

35. Steinbeck, John. *East of Eden.* New York: Viking Press, 1963, 5–6.

36. Palmer, W. C. "Meteorological Drought." Research Paper No. 45. U.S. Department of Commerce, U.S. Weather Bureau, Washington, DC, 1965.

CHAPTER 3

SURFACE WATER HYDROLOGY

The rapids beat below the boat
Deep in the heart of the land
Feel the pulse of the river in the pulse of your throat
Deep in the heart of the land.

"Veins in the Stone" © Lynn E. Noel 1993.
In *Voyages: Canada's Heritage Rivers*
(St. John's, Newfoundland, CANADA: Breakwater Books, 1995).
Crosscurrents Music, http://homepage.mac.com/lynnoel
Reprinted with Permission

Leonardo da Vinci (1452–1510) spent much of his life studying human anatomy. Through his dissections and observations, da Vinci developed comparisons of the body to natural features of the world (Figure 3.1). He compared human arteries to rivers on Earth, the pulsing of blood through the body to the flow of mountain streams, and arterial bleeding to river flooding.

Hydrology, the study of moving water, is somewhat similar to the science of medicine. Rivers deliver the life-blood of water, analogous to the veins and arteries of the human body, throughout the world. Like blood, water cleanses, nourishes, and gives life to plants and animals. Like medicine, water is a fascinating field of study and involves water engineers, hydrologists, and other scientists who investigate water movement, its interaction with landforms, chemical processes, and how water affects other natural systems on Earth. Similarly, medical doctors, nurses, and other health professionals study the movement of blood through the body, its interaction with glands, chemical processes, and how blood affects other physiological processes in the human body. Da Vinci was a keen observer of these similarities in his study of anatomy and hydrology.

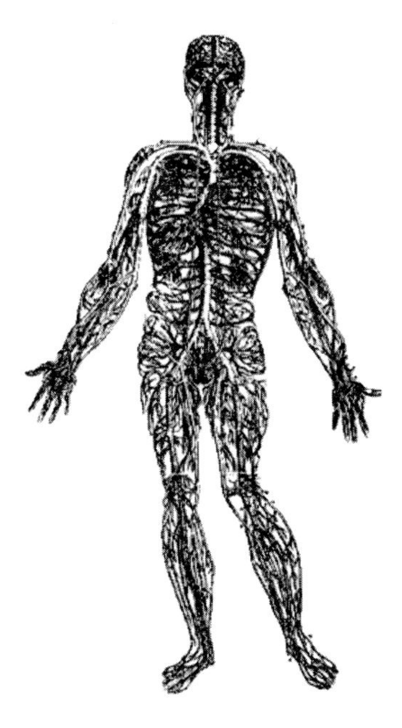

FIG. 3.1 Leonardo da Vinci was fascinated by the similarities between the organization of rivers on the surface of the Earth and the human circulatory system. He developed numerous sketches of both natural systems.

WHAT IS SURFACE WATER HYDROLOGY?

Chapter 2 presented the hydrologic cycle processes of precipitation, runoff, storage, evaporation, and condensation. **Surface water hydrology** is the study of moving water found in rivers, open channels, and runoff flowing across the open land surface. Many ancient cultures utilized the science of hydrology to create sophisticated practices to control or capture moving surface water. This was especially true for cultures in arid settings such as the Anasazi Indians of southwest Colorado, the Sumerians along the Tigris and Euphrates rivers, and the Egyptians along the Nile River (see Chapter 1).

According to the dictionary, a *stream* is a "flow of running water, large or small," whereas a *river* is a "large stream of water." Most people use these terms interchangeably to denote a body of running water of any size. However, a stream is generally considered to be smaller than a river, a creek smaller than a stream, and a brook even smaller. Rills form during precipitation events and gather downhill to form a brook which, if it grows, creates a creek. In this textbook, river and stream will be used to denote a flow of running water, large or small.

WATERSHEDS

The total land area that drains surface water to a common point (or common body of water) is called a **watershed** (also called a *river basin, drainage basin,* and *catchment*). Watersheds can be as small as a parcel of ground that drains into a pond or as large as the 1.26 million square miles (3.26 million km^2) in the United States and Canada that drain into the Mississippi River and its tributaries (see Figure 3.2). The world's largest watershed, the Amazon River Basin, is located in South America and empties into the Atlantic Ocean. Although it contains only 2 percent of the global land area, the Amazon delivers almost 20 percent of the global river discharge to the ocean.

States, provinces, and countries generally contain several watersheds. Canadian watersheds, for example, drain into the Pacific, Atlantic, and Arctic oceans. In Colorado, surface water located west of the Continental Divide of the Rocky Mountains flows toward the Pacific Ocean, whereas surface water east of the divide flows toward the Gulf of Mexico and the Atlantic Ocean. In eastern North Dakota and western Minnesota, water located in the Red River Basin flows north through Manitoba, into Lake Winnipeg, and northward into Hudson Bay. Water in western North Dakota flows south to the Mississippi River and ultimately into the Gulf of Mexico near New Orleans, Louisiana.

Table 3.1 shows the 10 largest watersheds in the world. The quantities under the column "Average Discharge" are in cubic feet per second (cfs) and cubic meters per second (cms). These are rates of flow and are common units of water measurement. (These units will be explained later in this chapter but are presented now to show relationships between drainage areas and average water discharge.) Note that the average discharge of the Amazon River is nearly 10 times greater than that of the Mississippi.

Table 3.1 also shows that the drainage areas of the Zaire River in central Africa and the Mississippi River in the United States are almost identical. Yet, the average discharge is over double in the Zaire River Basin. What climatic factors discussed in Chapter 2 would probably account for this difference?

DELINEATING A WATERSHED

A watershed is delineated by a ridge or drainage divide that marks the boundary of the drainage basin and can be easily identified on topographic maps. All surface water runoff below a ridge line will flow downhill within the watershed. The incline of terrain is generally downhill toward the

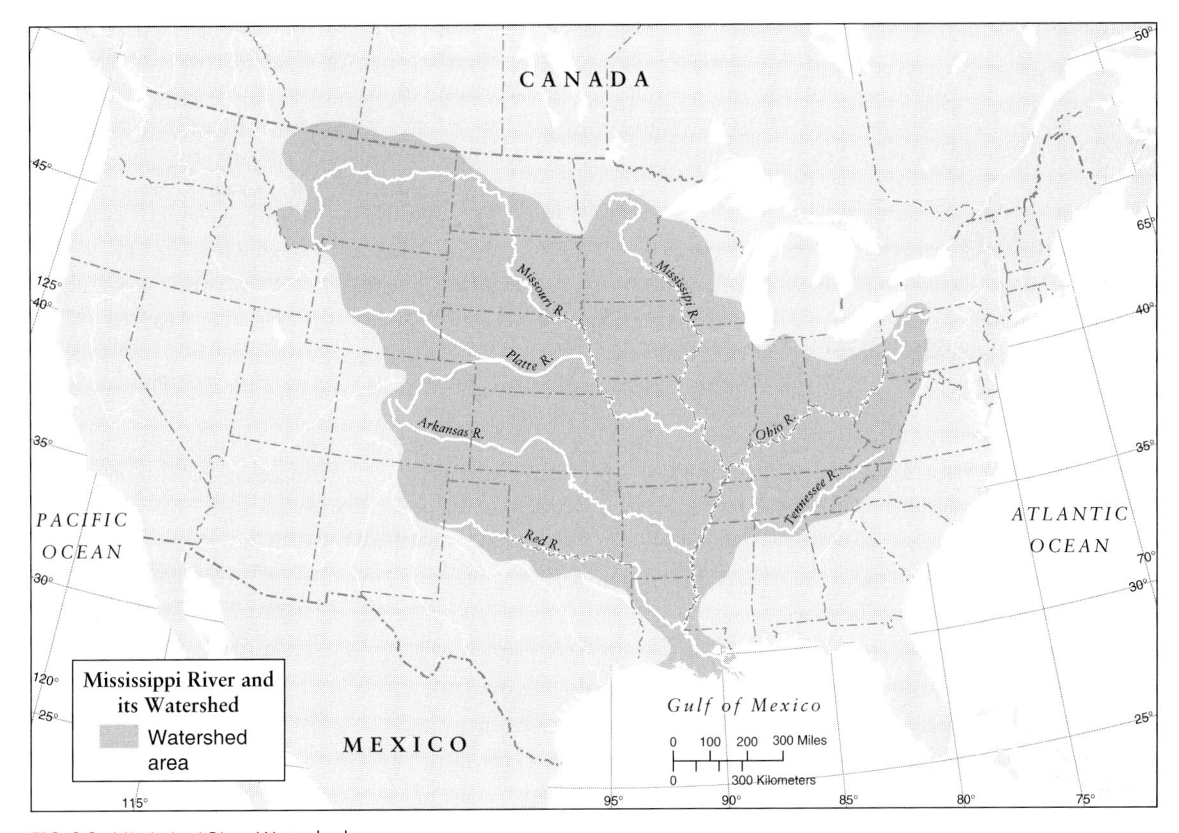

FIG. 3.2 Mississippi River Watershed

TABLE 3.1 World's Largest Drainage Basins

River Basin	Drainage Area (1000 mi²)	Drainage Area (1000 km²)	Average Discharge cfs*	Average Discharge cms*
Amazon, South America	2380	6160	6,183,750	175,100
Zaire (Congo), Africa	1480	3830	1,413,430	40,000
Mississippi, United States	1260	3260	649,820	18,400
Parana-La Plata, South America	1090	2820	526,500	14,910
Yenisei, Russia	1000	2590	627,560	17,770
Lena, Russia	970	2510	568,900	16,110
Yangtze (Chang Jiang), China	750	1940	1,008,480	28,560
Ganges-Brahmaputra, India	570	1480	1,087,990	30,810
Orinoco, South America	380	980	1,232,510	34,900
Mekong, Vietnam	310	800	526,500	14,910

*cubic feet per second and cubic meters per second.

Source: Adapted from *The World in Figures* by Victor Showers (Toronto: John Wiley & Sons, 1973).

main channel of a river. The boundaries of a watershed can be delineated by first locating the lowest point, or watershed outlet, on a topographic map. Then, higher elevations can be followed until a ridge, or high point, is identified.

A CLOSER LOOK

Topography maps are invaluable tools for geographers, planners, engineers, hikers, and others. Since 1879, the U.S. Geological Survey (USGS) has developed **topographic (topo) maps** that provide information on slope, elevation, distance, and physical features for the entire United States. USGS topographic maps (also called quad sheets) present land surface information at various scales of measurement. **Scale** is the relationship between the size of a map feature and its actual dimensions on the ground. USGS maps are drawn in many scales, but 1:24,000 is the most common. The first number of the scale represents the units on the map (i.e., 1 in.), and the second number represents the units on the ground (i.e., 24,000 in.). This means that a distance of 1 inch on a map, with a scale of 1:24,000, equals 24,000 inches (2,000 feet, or 0.8 km) on the land surface being displayed.

Contour lines follow a constant elevation above sea level. A USGS topographic map generally has intervals of 20 feet (6 m) between contour lines to show changes in elevation. In a relatively level area, the contour lines on a topographic map will be fairly straight with large distances (perhaps miles or kilometers) between them. On the other hand, contour lines may almost touch in a mountainous location because the change in elevation (gradient) is very steep and abrupt (see Figure 3.3).

The USGS has maps available through the Internet at http://mcmcweb.er.usgs.gov/topomaps/ordering_maps.html. Topo maps can be obtained by calling 1-800-USA-MAPS, or by writing to the following address: USGS Information Services, Box 25286, Denver Federal Center, Denver, Colorado 80225.

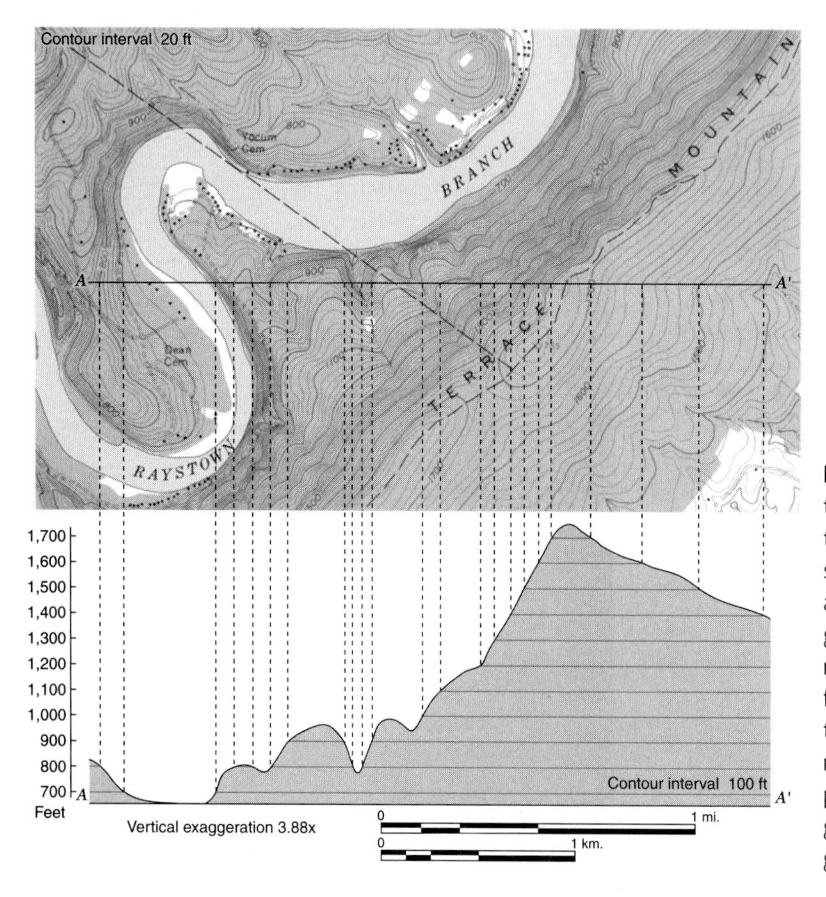

FIG. 3.3 Hills, valleys, and slopes of a topographic map. The steep slopes on the map are represented by closely spaced contour lines. Relatively level areas are shown by contour lines with greater distances between contours (or no contour lines as seen along the surface of the river). The profile below the contour map is exaggerated to make the differences in elevation prominent. Could you determine the general gradient of a river with a topographic map? How?

Another good source of topographical maps is through Topozone.com. Additional information regarding topographic maps can also be found in the Appendix.

Three simple rules can be followed when trying to determine watershed boundaries on a map:

1. Surface water generally flows at right angles across contour lines on a map.
2. Ridges are indicated by the highest elevation contour line in an area.
3. Drainages are indicated by contour lines pointing upstream.

Once the boundaries of a watershed have been determined, several watershed parameters can be computed such as size, maximum and minimum elevations, shape, slope, and drainage patterns. Surface water flows can also be predicted based on various potential precipitation events. Hydrologists—people who study and measure moving water—are also concerned with the aspect and orientation of a watershed. The **aspect** of a watershed is the direction of exposure of sloping lands, whereas **orientation** is the general direction of the main portion of a river as it moves down a watershed. A river with an east-west orientation will probably have slopes that are generally north-south in aspect.

POLICY ISSUE

Watersheds provide a basic geographic unit for water resources planning activities. Water quality plans, streamflow calculations, and flood projections are generally based on watershed size, land use, and other physical features. In 1972, the Nebraska Unicameral Legislature used watersheds to reorganize hundreds of local units of government (such as soil conservation districts, weed control districts, and flood-control districts). In place of these single-purpose agencies, 24 natural resources districts (NRDs) were created to manage surface water and groundwater, soil conservation, weed control, and many other natural resources planning, administration, and development functions. Nebraska's NRDs will be discussed in more detail in Chapter 10.

OVERLAND FLOW

Rain that falls on the land surface within a watershed will immediately move in one of three general directions. First, rain may evaporate back into the atmosphere as described in Chapter 2. Second, precipitation may **percolate**, or seep, down into the soil and eventually become groundwater. (The processes of groundwater movement will be discussed in detail in Chapter 4.) Third, rain may move along the land surface as runoff during and after a storm event. Runoff water that is moving toward a river or stream is called **overland flow.** Some overland flow may become stored in small ponds, wetlands, or lakes before reaching a flowing stream. Overland flow rates and volumes are very dependent on precipitation rates, duration of a storm event, and the spatial distribution of precipitation.

A second feature of surface water runoff, called **interflow,** occurs when precipitation percolates just below the land surface and moves in the same direction as overland flow. Interflow moves in subsurface materials at a slower rate than moving water on the surface and will arrive at a river later than overland flow. A heavy 45-minute downpour typically generates a more rapid overland flow than a calm, soaking shower over a 48-hour period that generates interflow. This variation in surface water runoff is an important reason why accurate precipitation measurement, discussed in Chapter 2, is so important in water resources management.

Both overland flow and interflow are greatly affected by human development. Hard surfaces, such as parking lots, roads, and rooftops act as funnels to drainage pathways that ultimately empty into rivers, streams, ponds, and lakes. Impervious barriers created by development also inhibit interflow and percolation into the soil. By contrast, forests, cultivated ground, open space, parks, and other vegetated areas are relatively porous and slow the runoff of precipitation and promote percolation into the soil.

RIVERS

Rivers contain less than 0.01 percent of the Earth's water but originate in several possible sources. Some are fed by springs or small streams coming together to create larger rivers of water. Some originate in lakes, such as the Mississippi River at Lake Itasca in northern Minnesota, or Egypt's Nile River which begins near Lake Tama in the highlands of Ethiopia. Other rivers, such as the Colorado River in the Rocky Mountains of Colorado, begin as trickles of melted snow water.

COMPONENTS OF A RIVER

A river consists of a main channel and all tributaries that flow into it. The beginning of a river is called its **headwaters,** or source. **Tributaries** are smaller streams that combine to form larger streams and ultimately rivers. When viewed from above or on an aerial or satellite photo, tributaries often look like the branches of a tree. The site at which a tributary joins the main river channel is called the **confluence** of a river. **Upstream** denotes a location toward the headwaters of a river or tributary, whereas **downstream** is toward the direction of a confluence with a larger stream, mouth, or other end point of a river. The imaginary line that connects the deepest points of a river channel is called the **thalweg** of a river.

Overland flow and interflow are important sources of water for many rivers. After a rainstorm, water levels in rivers often rise and cause some water to percolate into the banks (sides) of a river (called **bank storage**). As local conditions become drier, this bank storage will slowly move back into the river as flow decreases. Porous riverbanks allow significant amounts of storm runoff to be temporarily held as bank storage and can reduce the threat of flooding downstream.

The zone beneath a river is called the **hyporheic** (from the Greek meaning "to flow beneath") **zone** and varies in depth depending on the composition and size of a river bottom. The hyporheic zone can extend beneath a riverbed to a depth of a few inches (less than 10 cm) or down to several feet (1 m or more). Oxygen is the limiting factor that determines what type of organisms can survive in this area, with the colder waters of high altitudes supporting the most life. One researcher in Germany counted more than 365,000 organisms per square foot (3.9 million/m^2) beneath a German stream, while Canadian biologists found more invertebrates living in the hyporheic zone of an Ontario river than in the sand, gravel, and other materials on the river bottom itself. (1)

RIVER MORPHOLOGY

Geomorphology is the study of forces that shape the surface of the Earth. The greatest force in the alteration of land is moving water, and its effects can easily be seen along rivers. Rivers develop many features after years of traveling the same course. A "young" river has a V-shaped valley, with some rivers having almost vertical walls and swift-flowing water (see Figure 3.4). As a river continues toward its mouth and seeks a base level, the slope of the river channel generally decreases. Eventually, the river valley widens and becomes more U-shaped where silt and sand have created a wide plain caused by previous floods.

The width and depth of a river increase as it proceeds downstream. This occurs due to increased volumes of water and erosion. An individual can step across the Mississippi River at its source at Lake Itasca in northern Minnesota, where this mighty river is only 4 inches (10 cm) deep. However, one would have to be an excellent swimmer to cross the approximately half-mile-wide (0.8 km) and 200-foot-deep (61 m)

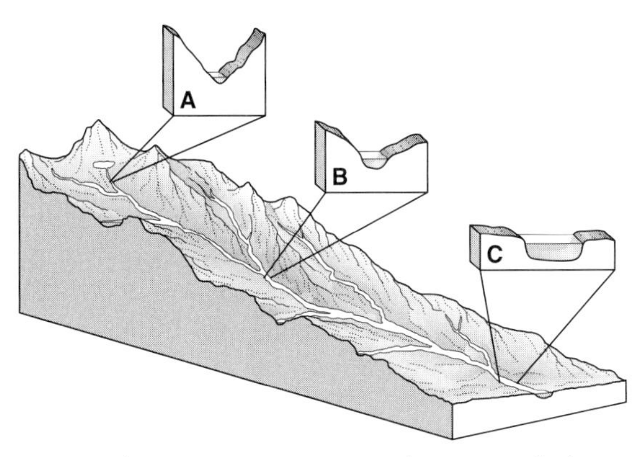

FIG. 3.4 Changes in stream properties along a watershed are denoted by changes in channel width and depth. Cross sections A, B, and C represent changing gradients and discharge as a river system flows through a watershed toward a confluence or to the ocean. Notice how the river system changes from a V-shaped channel to a U-shaped channel as it moves downstream.

channel of the Mississippi River at New Orleans before it enters the Gulf of Mexico (Figure 3.5).

In 1802, Scottish geologist John Playfair described the formation of rivers with the following statement, which today is known as Playfair's Law:

Every river appears to consist of a main trunk, fed from a variety of branches, each running in a valley proportional to its size, and all of them together forming a system of valleys connecting with one another, and having such a nice adjustment of their declivities that none of them join the principal valley at either too high or too low a level; a circumstance which would be infinitely improbable if each of these valleys were not the work of the stream which flows in it. (2)

Some rivers travel in relatively direct routes to their destinations while others develop meanders. **Meanders** are broad, looping bends in a river caused by the natural behavior of flowing water. A profile of a meandering stream shows a series of pools and shallows with deeper sections found downstream of a looping river bend. The inner sides of meanders, or bends, become areas where geologic material is deposited, while outer edges of meanders generally incur erosion. This combination of geologic deposition and erosion creates an asymmetrical channel cross section in a meandering river system. Pools that exist at the outsides of bends will be scoured during high flows but will

FIG. 3.5 The Mississippi River is the third largest drainage basin in the world, exceeded in size only by the watersheds of the Amazon River in South America and the Zaire River in Africa. The Mississippi provides drainage for 31 states and 2 Canadian provinces, and essentially funnels all river flows past downtown New Orleans, Louisiana (shown at right). Water from as far west as Montana and as far east as New York contributes to the flow of the river at this point. Garciliaso de la Vega, a member of DeSoto's Spanish expedition in search of gold in 1543, described the first recorded flood that winter. It began on March 10 and crested 40 days later. The Mississippi River finally returned to its banks in late May, a flood of nearly 80 days. Major floods in 1849, 1850, 1882, 1912, 1913, and 1927 led to increased federal involvement in flood-control efforts in the basin and will be discussed in Chapter 9.

receive deposits of geologic and organic materials during periods of low flow. *River crossings* are locations where flowing water moves from an inner side of a meander to outer edges, called *crossovers*. These crossovers are scoured during low flow but become covered with deposits during high-flow periods. (3) Meanders tend to migrate downstream as a result of erosion and deposition on opposite sides of meander bends.

SIDEBAR

The word *meander* is derived from the old Roman name for the Menderes River in Turkey. In Roman times it was called the Maeander River and had such a winding, circuitous course that it was thought to occasionally flow backward. (4)

A river will often develop meanders if the bank material is erodible. **Oxbow lakes** and wetlands often form in the river channel of an abandoned meander. Meanders may become isolated from the main channel if the stream becomes so sinuous that the narrow neck of land that separates adjacent meanders becomes breached during a flood event. These isolated lakes are very common along the Missouri, Mississippi, and Rio Grande rivers in the United States.

A CLOSER LOOK

The DeSoto National Wildlife Refuge in western Iowa is the site of the buried steamboat *Bertrand*. In 1865, a week before the Civil War ended, the boat was headed up the Missouri River for the goldfields of Montana. It had been heavily loaded in St. Louis, Missouri, with crates of mercury (for mining), clothing, tools, housewares, and food, including olive oil and mustard from France, canned fruits, alcoholic beverages, powdered lemonade in a can, brandied cherries, pickles, and a host of other items. The steamboat hit a snag hidden in the Missouri River and sank 20 miles (32 k) north of Council Bluffs, Iowa. Today an impressive museum has been built overlooking the abandoned oxbow lake where the *Bertrand* was found. The site is located miles from the present-day Missouri River. (5)

Braided rivers, consisting of intertwined channels separated by small, temporary islands, form when excess geologic materials cannot be removed by the flow of a river and the channel simply moves to a location of less resistance. These rivers tend to be very wide and relatively shallow, with coarse bed material. Islands are sometimes barren or can have tree and shrub communities such as willows, cottonwood trees, or other vegetation common to the area. Floods can sometimes remove this growth, but low river flows allow the plants to thrive, changing the local habitat for wildlife. Good examples of braided river systems are the Platte River in Nebraska and the Rakaia River in New Zealand.

Braided rivers are common in glacial regions where a great deal of sand and gravel are available and washed away into broad, flooded plains. Some braided rivers spread out to a width of more than 0.5 mile (0.8 km) and have a depth of only a few feet (1 m or less). Braided rivers seldom contain many living organisms because the sands and gravels of the riverbed offer little cover. In addition, the constant shifting of the riverbed reduces suitable habitat for invertebrates in the hyporheic zone. However, some braided rivers, such as the Platte River, provide excellent habitat for numerous migratory waterfowl (Figure 3.6).

FIG. 3.6 Braided Platte River channel in central Nebraska. This area of the Platte contains numerous endangered species such as the whooping crane, piping plover, and least tern, and will be discussed in Chapter 12.

TYPES OF RIVERS

Rivers located in dry climates are often **ephemeral streams** that are not fed by any continuous water source and flow only after storm events. A stream that is **intermittent** flows both after storm events and during wet seasons when fed by groundwater. In the Sahara Desert of northern Africa, the beds of ephemeral rivers are called *wadies*. These dry riverbeds are so smoothed by floods that they are often used as roads during the dry season. Flash floods can catch caravans by surprise if a sudden thunderstorm in the distance is not detected. Similar dangerous situations often occur in the southwestern United States and northern Mexico where ephemeral rivers flow down *arroyos* (Spanish for "creek" or "gulch"). Campers sometimes sleep in dry *arroyos* since these makeshift campsites have soft beds of sand. However, sudden thunderstorms have swept unwary campers to their deaths. Crayfish *(Paranephrops planifrons)*, flatworms *(Polycelis nigra)*, and fairy shrimp *(Branchinecta packardi)* can survive in the harsh environment of ephemeral streams, some by burrowing a few feet (1 m or less) underground until moist sand or soil is reached. Some species produce eggs that can survive this dry environment until floods recur.

River channels that are located above groundwater systems often discharge some water through percolation. These are called **influent** or **losing rivers** since surface water moves from the stream channel into areas of groundwater storage. An **effluent** or **gaining river** is one that receives baseflow from groundwater and increases discharge (Figure 3.7). An effluent river is usually found in humid or wetter climates. A river can be effluent during parts of the year and influent during other months. Groundwater movement, and the relationship with rivers, will be discussed in detail in Chapter 4.

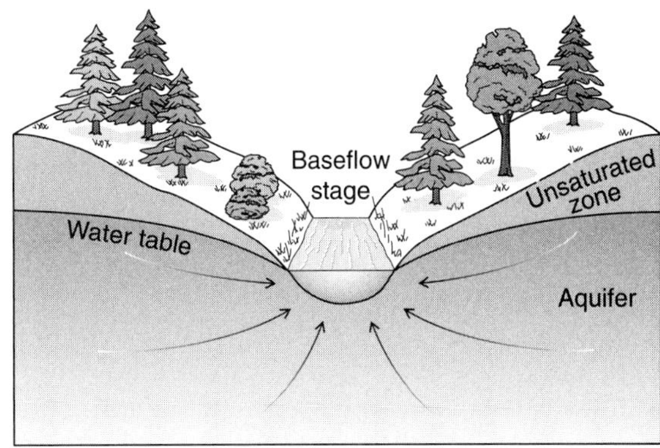

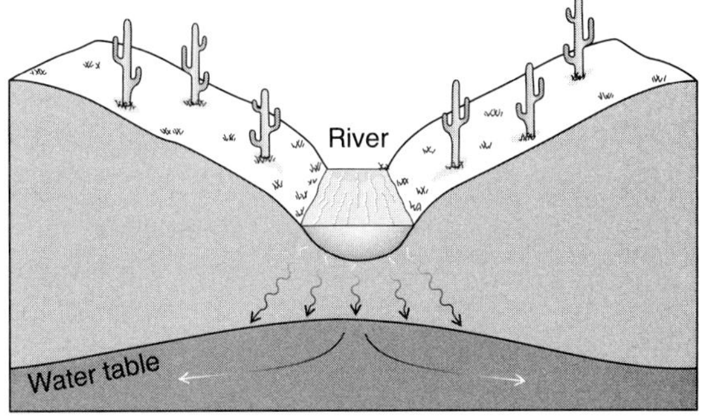

FIG. 3.7 Cross section of a gaining (or "effluent") stream, common in humid regions, and a losing (or "influent") stream, often found in arid or semiarid locations. Effluent streams actually gain water from the local groundwater system, while influent streams allow surface water to replenish groundwater in an area. Influent streams often become dry streambeds during periods of drought or in regions of arid climate.

GRADIENT

Typically, the **gradient** of a river decreases as it continues downstream, resulting in reduced stream velocity. The gradient is the slope or fall of a river and is measured in terms of feet per mile or meters per kilometer. For example, the Little Conemaugh River in Pennsylvania has an average slope of 53 feet per mile (10.0 m per km) over its 29-mile (46.7 km) length, whereas the Ohio River has an average slope of less than 6 inches per mile (9.5 cm per km). These variations in gradient affect stream velocity and material transport (discussed later in this chapter). The Red River in North Dakota and Manitoba averages a gradient of only 5 inches per mile (7.9 cm per km) and as little as 1.5 inches per mile

(2.4 cm per km) in some locations. This nearly flat gradient is located where the Red River flows through the bed of ancient Lake Agassiz, a glacial lake that covered portions of Manitoba, Ontario, Saskatchewan, North Dakota, and Minnesota. The lack of "fall" along the Red River is a major factor in the severity of floods, which can devastate the region, such as the one that occurred in 1997. During such high water, the floodwaters of the Red River essentially form a massive, slow-moving shallow lake and inundate broad areas of the ancient lake bed.

LAKES

A **lake** is any body of water, other than an ocean, that is of reasonable size, impounds water, and has little or no horizontal movement of water. Lakes can be created by glacial activity, volcanic explosions, stream channel abandonment, landslides, and human activity. Lake Superior, located between the United States and Canada, has the largest freshwater surface area in the world at 31,700 square miles (82,103 km^2).

The term *lake* can include shallow bodies of water only a few feet (meters) deep or "ponds" that can be 10 miles (16 km) long and hundreds of feet (over 60 m) deep. In Newfoundland, for example, almost every lake is called a pond, whereas in Wisconsin almost every pond is called a lake. Regardless of size, all lakes are considered **lentic** (Latin for "sluggish") habitats. The science of **limnology** deals with the characteristics and behavior of lakes. Table 3.2 presents the 10 largest lakes (in terms of surface area covered) in the world. (The Caspian Sea is listed since it is a static body of water not directly connected to an ocean.)

TYPES OF LAKES

Many natural lakes in the world were created by glacial activity. The process of scouring, gouging, and scraping by glaciers created depressions in the landscape that hold surface water. Most natural lakes are filled by rain or melting snow, although some are filled by glacial meltwater. If glacial debris blocked the upper reaches of a mountainous valley, a **cirque lake** was created when it filled with water. These are common in the Rocky Mountains of the United States and Canada, particularly in most ranges of Colorado, Wyoming, Montana, Alberta, and British Columbia. Glacial activity will be discussed in Chapter 4.

TABLE 3.2 Largest Lakes in the World

		Area		Maximum Depth	
Name	Location	(mi^2)	(km^2)	(ft)	(m)
Caspian Sea (salt water)	Azerbaijan, Russia, Kazakhstan, Turkmenistan, Iran	146,100	378,399	3363	1025
Lake Superior	Canada and United States	32,162	83,300	1332	406
Lake Victoria	Tanzania and Uganda, Africa	26,988	69,899	302	92
Lake Huron	Canada and United States	23,089	59,800	751	229
Lake Michigan	Canada and United States	22,400	50,016	935	285
Lake Tanganyika	Tanzania and Congo, Africa	13,127	33,999	4823	1470
Great Bear Lake	Northwest Territories, Canada	12,275	31,792	1460	445
Lake Baikal	Russia	12,162	31,500	5712	1741
Great Slave Lake	Northwest Territories, Canada	11,031	28,570	2014	614
Lake Erie	Canada and United States	9930	25,719	210	64

Source: Environment Canada, "World's Largest Lakes," http://www.ec.gc.ca/water/descrip/nature/prop/a2f4e.htm, June 2003.

Pluvial lakes were created in dry climates when favorable changes in precipitation and evaporation occurred during quaternary climate changes. These lakes have long since disappeared through evaporation. Examples of pluvial lakes are Pyramid Lake near Reno, Nevada, and Utah's ancient Lake Bonneville, which remains today as Great Salt Lake.

Although glaciers created many lakes in Canada, Lake Chubb in northern Quebec was created by extraterrestrial forces. It is an eerily shaped perfect circle, two miles (3.2 km) across, and occupies a meteorite impact crater estimated to be 1.4 million years old. Meteor Crater near Flagstaff, Arizona, is another meteorite impact crater formed about 50,000 years ago that once contained a similarly shaped, but much smaller, pluvial lake.

Kettle lakes are depressions created by blocks of stranded, buried glacial ice that gradually melted during the Pleistocene epoch. The melting of the ice caused the overlying land surface to collapse and create a hole. Where the depression is deep enough to reach groundwater, a lake is created. Kettle lakes are generally found in Ohio, Minnesota, North Dakota, Wisconsin, Michigan, Alaska, Colorado, Idaho, Pennsylvania, British Columbia, Manitoba, Ontario, Saskatchewan, Quebec, and central and northern Europe.

Lakes can be young, middle-aged, or old. Young, or **oligotrophic lakes,** have bottoms that are too clean (i.e., they lack enough organic material as food sources) to provide appropriate habitat to produce plants and freshwater organisms. However, earthen particles and other organic materials from decaying plants and animals can eventually build layers of materials for aquatic plant and animal habitat.

Middle-aged lakes that allow for aquatic growth are called **mesotrophic.** As a lake ages, excessive organic material and mineral deposits can actually inhibit or stop the growth of aquatic plants and animals. Such an old lake is called **eutrophic** and is typically filling in with excessive organic and mineral materials.

A **reservoir** is a human-made feature, also called a lake, created by construction of a dam or dikes. (Dams will be discussed in Chapter 7, and dikes—sometimes called levees—are explained in Chapter 10.)

ECOLOGICAL ZONES

Lakes have three ecological zones inhabited by varying organisms. The **littoral zone** is in the shallows, generally near the lake shore, where adequate sunlight penetrates the water surface to promote growth of shallow-rooted plants. This is generally a rich zone of plant and animal life. The **limnetic zone** is located further out in the open water that extends toward the center of the lake and to a depth where sunlight cannot penetrate. The limnetic zone is home to drifting and swimming organisms. The **profundal zone** is found at the lake bottom where organic material has drifted down to become lake bottom material. This bottom zone, too dark to support rooted plants, is inhabited by bacteria, worms, mollusks, insects, and algae.

THERMAL CYCLES

Lakes have stratification zones in temperate regions created by changes in water temperature (and salt content if the lake is saline). Warm air in the spring and summer heats the surface of a lake, causing the water to become less dense. (Warm water causes water molecules to move faster and further apart, similar to the atmospheric heating process described in Chapter 2.) This warmer water becomes less dense and tends to be pushed up above cooler, and more dense, water masses.

A surface layer of warm water, called the **epilimnion,** will form above a cooler layer of lake water, called the **hypolimnion.** The thickness of the epilimnion layer depends on latitude, season, and temperature. The **thermocline** is a transition zone between the epilimnion and the hypolimnion in which there is a rapid change in water temperature with depth.

The heating process of a lake is opposite to the process of atmospheric heating. In the atmosphere, air is warmed by the land surface. Cooler, denser air that sinks below the warmer air near the land surface pushes it up to mix with colder air at higher elevations. This upward movement of warm air, together with the settling of cooler air, creates a continual mixing cycle within the atmosphere in temperate climates. In a lake, some water mixing does occur at the surface owing to wind and wave action. However, during the summer months little water movement occurs between the cold hypolimnion layer and the warmer epilimnion since the cold, deeper water is much denser than the warmer water above. Later, as air temperatures cool in the autumn, surface water temperatures decrease. The water at the surface cools and settles, and replaces the now warmer water at the bottom of a lake. This is called **lake turnover** (see Figure 3.8).

Cold winter temperatures and wind often equalize lake temperatures between autumn and spring, and there is little water mixing during that time. Some lakes can have normal thermal stratification in the summer and an inverted one (warm water on the bottom) in the winter. Water at 4°C (39°F) may be at the bottom of the lake in the grips of winter, with colder water above it until the coldest water is found in a layer of ice at the lake's surface. Water has the greatest density at 4°C (39°F).

The epilimnion is an active area of surface water mixing, high oxygen concentration, and plant and animal life. The hypolimnion, by contrast, is a cold, dark, and still area. Organic material for this region is limited to drifting from above. If thermal stratification is permanent, the hypolimnion will not have oxygen levels replenished and the area will have little or no life. However, if a lake is located in a climate with hot and cold seasons, it will periodically turn over and mix water between the epilimnion and hypolimnion zones. This will bring dissolved materials to the top of the lake and oxygen to the bottom region. Lake turnover generally occurs in the autumn when the epilimnion cools to 4°C (39°F) and in the spring when the cold surface water warms to 4°C (39°F). This mixing of lake water is very important to the ecology of the lake but can produce unfavorable taste and bad odor for consumers if a lake is used as a source of drinking water. (6)

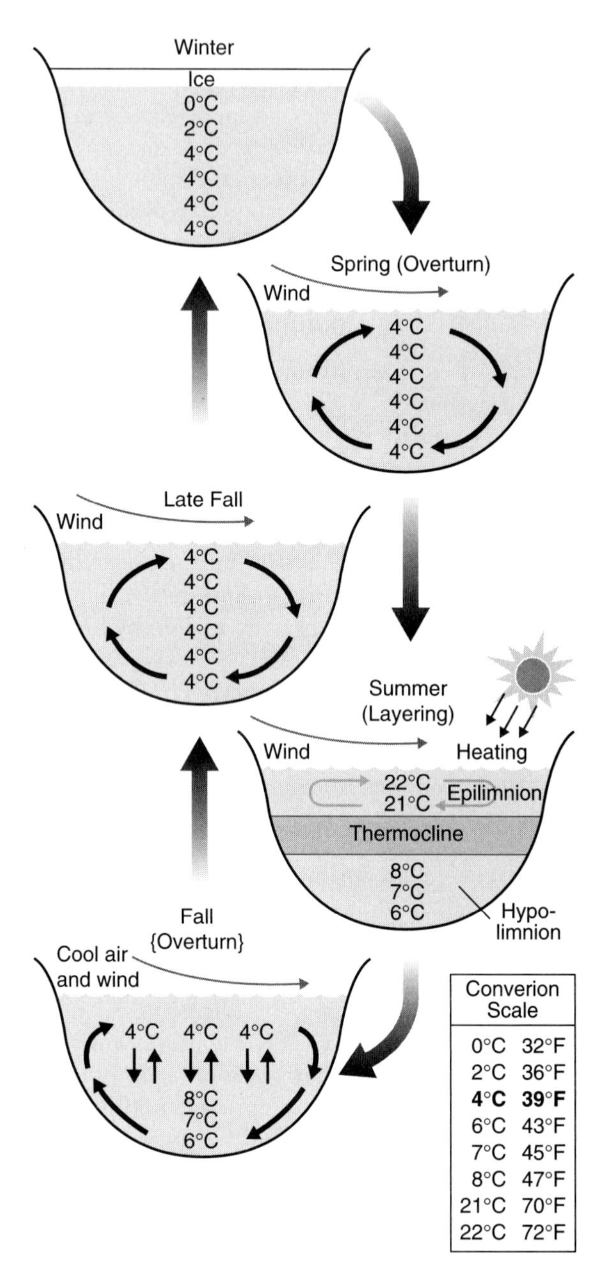

FIG. 3.8 Seasonal layering and turnover in a lake at midlatitudes.

SEICHES

Water levels in lakes routinely oscillate as a result of wind or sudden changes in atmospheric pressure. The highest recorded difference in water elevations at the Great Lakes in North America was during January 1942 when Lake Erie was 13 feet (4 m) higher at the shore of Buffalo, New York, than it was at Toledo, Ohio (a distance of approximately 300 mi, or 483 km). This difference of water-level elevation in a lake, called a **seiche** (pronounced **saysh**), is common around the world. The word was first used to describe the peculiar oscillating waves creating this phenomenon in Lake Geneva in Switzerland in the early 1800s. Seiches continue to be common at Lake Geneva and are about 3 feet (0.9 m) in variation. At Lake Erie, seiches routinely raise and lower water levels in harbors and have caused vessels to run aground. Freighters often schedule docking and departing times to avoid seiches on the Great Lakes. (7)

A CLOSER LOOK

Although seiches are usually harmless, a monster wave occurred on Lake Michigan on June 26, 1954. The lake was calm in the early morning, but a strong thunderstorm with 50 mph (80 km/h) winds blew in from the northwest around 8:00 A.M. At Michigan City, Indiana, 100 people were fishing on a pier but had to run to avoid a 2-foot-high (0.6 m) wall of water created by the wind on the lake surface. This was an *incident wave* created by a seiche pushed by wind and atmospheric pressure. No one was injured, but the *incident wave* rebounded against the shore at Michigan City and returned toward the northwest. This created a *reflective wave* that rapidly moved along the bottom of Lake Michigan toward Chicago, Illinois, some 50 miles (80 km) away. The compressed water at the lake bottom created a 10-foot-high (3 m) wall of water. People on jetties and piers along the Chicago waterfront were completely surprised by the sudden seiche, and dozens were washed into Lake Michigan. Eight people drowned. (8)

TRANSPORT AND DEPOSITION

It was the river which had laid down the new land; it was the river which took it away. The endless cycle of building up, tearing down and rebuilding, using the same material over and over, was contributed to by the river. It was the brawling, undisciplined, violent artery of life and would always be. (9)

From James A. Michener, *Centennial*

Throughout history, rivers have been given names that reflect their appearance. The Yellow River (or Huang He) in China, the Red River in North Dakota and Manitoba, and the "Muddy Mississippi" of the central United States generate images of color and content. The Platte River in Nebraska was described by pioneers in the 1800s as "too thin to plow but too thick to drink." These images were created and enhanced by earthen materials being transported and deposited by moving water.

Transport and deposition are the movement and settling of materials such as rock, gravel, sand, soil, and other geologic materials (called **sediment**) in moving water. Moving water, which creates overland flow and swiftly flowing rivers, provides one of two primary means of transporting sediment (wind is the other main factor). Sediment settles out of moving water as the rate of flow is reduced. This process is called **sedimentation.** Certain sections of rivers in flooded lowlands and lake bottoms are common locations for sediment deposits.

Natural levees are broad, low embankments adjacent to riverbanks created from sediment deposition during flood events. **Floodplains** are relatively flat areas adjacent to rivers created by sedimentary deposits as a meandering channel migrates laterally across a valley floor, and from flood events. These landforms can be as narrow as a few hundred feet (60 m) or as wide as over 25 miles (40 km). During a flood, water often leaves the main river channel and inundates a floodplain. Water velocity decreases across the floodplain owing to lower flows and increased friction from the land surface. As this occurs, sediment rapidly falls out of the floodwater and is deposited. Floodplains can include wetlands, oxbow lakes, natural levees, and rich soils for cultivation (see Figure 3.9). Occasionally, a **yazoo stream**, one that runs parallel to a larger

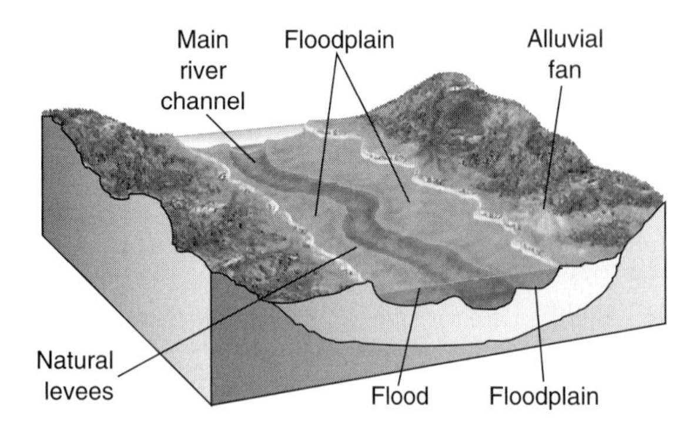

FIG. 3.9 Rivers have a profound effect on landforms. This illustration shows common landforms created by river deposits, including alluvial fans, natural levees, and floodplains.

river and helps drain a floodplain, will form. Yazoo streams are common along the Mississippi River and other large drainage systems with broad floodplains.

Sediment deposited by flowing water is called **fluvial material** or **alluvial deposits.** An **alluvial fan** is created when a river deposits material on the land surface, generally at the base of a mountain valley. A **delta** is created when sediment is deposited at the mouth of a river system, such as the Mississippi Delta located downstream of New Orleans or the Mekong Delta in Vietnam.

OUR ENVIRONMENT

Why is sediment transport important in the study of lakes and rivers? Sediments can be considered either a detriment or a benefit to a river system. Sediments deposited after flood events can benefit the health of farmland by providing nutrient-rich soils for crops. However, sediment deposits in lakes and reservoirs can destroy aquatic habitat, reduce water storage capacity, and eliminate deep water areas necessary for boating. High sediment loads in surface water can reduce the oxygen-carrying capacity of water and harm aquatic wildlife.

VELOCITY

Sedimentation is directly affected by the speed, or **velocity,** of flowing water. As velocity decreases, sediment sorting takes place. In addition, a kind of sorting occurs as finer particles are moved more frequently and farther during peaks of fluctuating discharge. **Sorting** is the process whereby particles of a similar size settle out of moving water caused by changes in velocity and fluctuating discharge. As water velocity decreases, heavier materials first sort, or fall out. This causes larger rocks to settle out, followed by gravels, sand, and finally silt and clay (particles smaller than grains of sand) as water velocity decreases. The process of sorting creates deposits of somewhat uniform size along rivers that can be utilized for mining. This accounts for the location of gravel mines (pits) near mountain foothills and sand pits farther out downstream in broad valleys. It also explains the deposition of rocks and boulders in or near mountain canyons.

A CLOSER LOOK

A meandering river causes water velocity to vary. Water flowing along the outside of a meander has greater velocity than water flowing along the inside of a meander. If you canoe, you may have noticed that deeper, faster water flows on the outside of a meander, while slower water is found on the inside of the curving river. Anglers often fish the inside of a meander because larger fish, such as trout, wait for food to come to them in slower-moving water. The lower water velocity at this location means a fish expends less energy waiting for food to arrive.

SEDIMENT LOAD

A river carries most of its sediment load in suspension, called **suspended load.** These suspended materials consist primarily of silt, clay, and some fine sand, although larger particles can be carried during flood events when water volumes and velocities are greater. A river can transport suspended materials for hundreds or thousands of miles (or km). As water velocity slows, sediments settle to the riverbed, on to adjacent natural levees, or near the mouth of a river to form a delta.

All rivers carry a **dissolved load** consisting of dissolved materials that remain in solution. Additional water flows will dilute these solutions but may not totally eliminate such dissolved materials unless the water chemistry changes.

Rivers that receive groundwater inflow generally have higher dissolved loads than rivers composed only of surface water runoff due to dissolved minerals available from underground geologic formations.

Coarse geologic material, called **bed load,** is pushed along the bottom of a river by moving water. These larger particles of sand and gravel may account for less than 10 percent of total river sediments but are important to the health of a river ecosystem. These particles roll and slide along a riverbed or can move by **saltation.** The saltation process occurs when a particle is pushed up due to a drop in pressure above the grain of sand. It is similar to the processes that affect the wings of a plane. Faster currents above the grain cause a drop in pressure above it. The relatively higher pressure below the grain lifts it up into the current, at which time the water current pushes it down the stream. Often, the settling particle collides with other material at the bottom of a river and is hurled upward again by the water current. (The word "saltate" literally means to bounce along the bottom.)

Sediment yield—the total amount of erosional material carried from a drainage basin—is generally measured in terms of weight (tons or metric tons) per day or year. A sampling device can be used to gather a known unit volume of flowing water. The water sample is then allowed to evaporate so that only dry matter remains. The difference between the weight of the water and the weight of the sediment provides the proportion of sediment to water per unit volume. Multiplication of the concentration of sediment by the rate of discharge will provide the weight of sediment per acre-foot, gallon, or cubic meter.

Sediment load is extremely important in the health of aquatic ecosystems. High sediment loads cause rivers to be smothered and leads to the loss of habitat for fish. Turbid water (water high in sediments) may be warmer and stressful for a variety of aquatic species. In addition, sediments often carry pollutants that are fixed (attached) to the sediment particles. See Chapter 5 for more information on sediment load and water quality.

A CLOSER LOOK

The Maumee River watershed drains land in Ohio, Michigan, and Indiana, and delivers surface water to Lake Erie at Toledo, Ohio. The river has an average slope of 1.3 feet per mile (0.25 m/km) and carries over 10 million tons (9.1 million metric tons) of suspended sediment to Lake Erie annually. Every year, the U.S. Army Corps of Engineers dredges approximately 850,000 cubic yards (649,910 m^3) of sedimentary material from Toledo Harbor to keep shipping viable along the Maumee. The State of Ohio, as well as many groups, are working to reduce soil erosion and sediment loading of the Maumee River through improved land-use practices upstream. (10) The dredging activities of the U.S. Army Corps of Engineers will be discussed in Chapter 9.

WATER MEASUREMENT

Surface water measurement can be fairly simple in some situations and very complex in others. Depth of water, terrain, and velocity are important when determining flow rates and volumes. Three types of surface water measurement are overland flow, discharge of rivers, and water storage in lakes and reservoirs.

OVERLAND FLOW

Overland flow, also called *sheet flow,* as defined earlier in this chapter, is surface water runoff that is moving within a watershed toward a river. Overland flow does not usually continue for more than a short distance because it combines with other moving water to collect in natural or artificial waterways. Overland flow can be mathematically calculated using the **Rational Formula:**

$$Q = KiA$$

where Q = peak rate of runoff in cubic feet or cubic meters per second
K = runoff coefficient (see Table 3.3)
i = intensity of rainfall in inches or centimeters per hour
A = watershed area in acres or hectares

The Rational Formula was proposed in 1889 by Emil Kuichling and continues to be widely used

TABLE 3.3 Runoff Coefficients for the Rational Formula

Runoff Area	Value of K
Business	
Downtown	0.70–0.95
Neighborhood	0.50–0.70
Residential	
Single-family	0.30–0.50
Apartments	0.50–0.70
Industrial	
Light	0.50–0.80
Heavy	0.60–0.90
Parks, cemeteries	0.10–0.25
Playgrounds	0.20–0.35

Source: American Society of Civil Engineers, "Design and Construction of Sanitary and Storm Sewers," *Manuals and Reports of Engineering Practice* No. 37, 1970. Reproduced by permission of the publisher, ASCE.

today around the world. It is used to design storm drains, culverts, and other structures that control runoff, primarily in urban areas.

RIVER DISCHARGE

Overland flows usually combine and eventually form a river. The amount of water carried in a river at any one time is called the river's **discharge,** which is defined as the volume of water flowing past a given point during a given period of time. It is measured in **cubic feet per second** (cfs) or gallons per minute (gpm) in the United States and as **cubic meters per second** (cms) or liters per minute in most other parts of the world (see Figure 3.10).

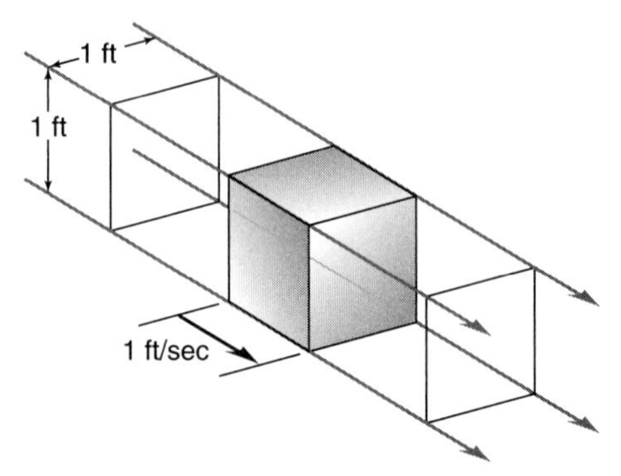

FIG. 3.10 One cubic foot per second, or cfs (or one cubic meter per second, or cms) is equivalent to one cubic foot (or cubic meter) of water flowing past a given point in a one-second time interval.

A CLOSER LOOK

A cubic foot is a volume with dimensions of 1 foot (0.3 m) on each side. A basketball would fit nicely in a box this size. It takes approximately 7.5 gallons (28.4 l) of water to fill 1 cubic foot of space. A cubic foot per second (cfs) is a volume of water equal to 1 cubic foot moving past a given point every second (1 cubic ft/sec; see Figure 3.10). Double that flow rate and a volume of water equal to 2 cubic feet will move past a given point every second, or 2 cubic feet per second (sometimes called 2 second-feet). One cfs of water is equal to approximately 7.5 gallons moving past a given point every second (or 450 gal per minute).[1]

Water managers outside the United States generally use cubic meter per second (cms) to express water flow. The concept is the same as cubic feet per second, except that a cubic meter is a volume with dimensions of 1 meter (3.28 ft) on each side and equals 1000 liters (264 gal). A cubic meter would be the approximate size of a washing machine or dryer.

Discharge in rivers can vary from a few cubic feet per second (0.06 cms) in a small stream to seasonal variations from 1.5 million to over 12 million cfs (42,480 to 339,840 cms) in the Mississippi River at New Orleans. The Colorado River in the Grand Canyon of Arizona usually has a discharge in the range of 4000 to 90,000 cfs (110 to 2550 cms), but a flood in 1921 produced a discharge of more than 200,000 cfs (5660 cms). The highest recorded river discharge in the world was 52.5 million cfs (1.49 million cms) in the Amazon River of Brazil in South America.

River discharge can be calculated by first measuring the depth of a river at a particular

[1] An excellent interactive resource on river discharge can be found on the Internet at http://vcourseware.sonoma.edu/VirtualRiver. This website was developed in part with grants from the National Science Foundation and the California State University System.

cross section. Next, water velocity is measured at several points and depths along the same cross section of the river. The results are then placed in the following calculation:

$$Q = AV$$

where Q = discharge
A = cross-sectional area of a channel
V = average water velocity

This equation provides the mean (average) velocity of flowing water in a channel waterway. The cross-sectional area is determined by the width times the depth of a channel. Width is determined with a measuring tape or laser device. The depth of a river channel can be measured with a pole, if shallow, or with sounding cables (wires with weights attached at the end) if wading is not possible. Some river channels have continually shifting riverbeds (generally consisting of sand and gravel) owing to saltation and rapidly changing bed loads. These locations are often measured weekly to provide more accurate data on the size and shape of the channel bottom. Shifting sand and gravel can drastically change the cross-sectional area of a river channel in a short time, especially during periods of high discharge (see Figure 3.11).

FIG. 3.11 Steep gradients and high water velocity are great combinations for moving boulders, sediment, and kayakers. This photo was taken along the White Salmon River in Washington, a tributary of the Columbia River, near Hood River, Oregon.

OUR ENVIRONMENT

Agencies around the world collect river discharge data. In the United States, the U.S. Geological Survey (USGS) is the primary agency with responsibility for river-flow data. The USGS has real-time river discharge records of nearly every river in the United States. This data is available on the Internet at http://waterdata.usgs.gov/nwis.rt. Real-time data are recorded at 15- to 60-minute intervals but may be more frequent during critical events. The data are transmitted by telephone, radio, or satellite and are available for viewing on the Internet within three minutes of arrival. Data on streamflow in Canada can be found at http://www.ec.gc.ca/water_e.html at which one selects "Monitoring."

Water velocity can be measured with an electronic device called a **flow meter.** Flowing water turns a small propeller at the end of a shaft on the meter. Propeller rotations are electronically recorded over a set period of time and allow a hydrographer to calculate average water velocity. Since flow velocity in rivers often pulsates, observations are usually taken between 40- and 70-second intervals.

Flow meter measurements are made by dividing a river into cross-channel segments of not greater than 10 percent of the total river width. Therefore, 10 to 20 vertical-section measurements are typical. Water velocity often varies from near zero at the bottom of a river to near maximum at the surface. The average of the velocities at 20 percent and 80 percent depth provides the average velocity for a given vertical section.

The **stage** of a river or lake is the height of the water surface above a set reference elevation. If the elevation of a streambed is known and the water surface stage is subtracted, the result is the depth of water in the river. **Staff gages** (usually a metal ruler attached to a permanent fixture) can be used for this purpose (Figure 3.12). The elevation of water on the staff gage is called the **gage height.** Discharge is sometimes determined directly from the river stage using a **rating curve.** Such a curve is generated by making discharge

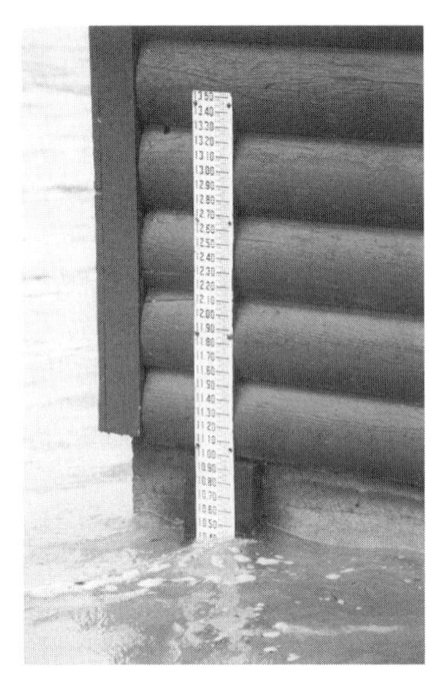

FIG. 3.12 A staff gage is nothing more than a long ruler placed in a stream or lake to monitor water depth. This staff gage is used to measure the stage of the Snake River within the Grand Teton National Park near Moran, Wyoming.

measurements during various times of the year, usually monthly, and plotting these against gage height. Once constructed, a rating curve eliminates the need for velocity and depth measurements unless the channel cross section changes.

A river **hydrograph** is a graph of discharge over time and can be plotted daily, weekly, monthly, or annually. Seasonal variations are evident on an annual hydrograph, whereas flood events are displayed on hourly, daily, or weekly hydrographs (Figure 3.13).

A **recorder** is a device used to record the elevation of water flowing through an open channel or river over time. A recorder is usually located in a vertical stilling well on a riverbank. The stilling well is connected to flowing water through a small horizontal intake, such as a 1-inch-diameter (2.5 cm) pipe, at the base of the stilling well. This opening allows river-stage changes to occur inside the stilling well but reduces fluctuation and turbulence. A cable and pulley system is attached from a float to the recorder, with a float at one end of the cable and a counterweight at the other end. The float rests on the surface of the water at the bottom of the stilling well. Changes in the stage of the river also occur inside the stilling well and are reflected on a paper chart or electronic data logger in the recorder.

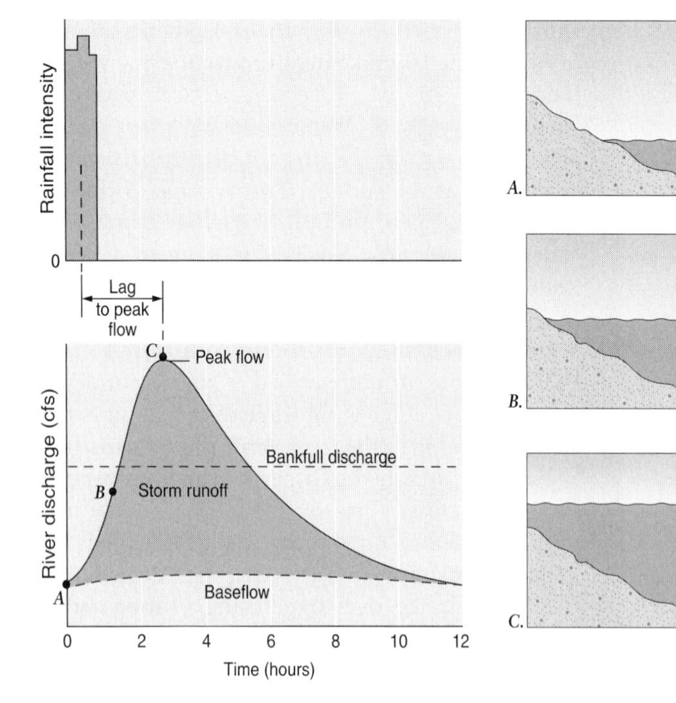

FIG. 3.13 The hydrograph of a river can look similar to this example after a brief but intense rainfall event. A heavy rainstorm can cause a rapid increase in stream discharge. As runoff collects and moves downstream in a watershed, the volume of water in a river channel will increase and flooding may occur. After a few hours, the stream discharge may decrease to near baseflow conditions.

FIG. 3.14 This Nilometer was carved into rock on the shore of the Nile River near Aswan, Egypt. During ancient times, Nilometers were used to determine taxes as well as the proper time to break dikes to begin irrigation of fields.

Measuring water is not a new concept. **Nilometers** have been used for over 5000 years and were developed in Egypt to measure the stage of the Nile River (Figure 3.14). The first Nilometers were simple marks etched on riverbanks. Later, elaborate towers protected stones etched with measurement scales to determine flood stages. A high-river stage meant the deposition of more sediments on floodplains for crops, while a low-stage meant a shortage of irrigation water and poor crops.

Taxes were based on the discharge of the Nile River—good crops meant higher taxes, whereas poor crops produced tax relief for the peasants. Nilometers were located at many locations along the Nile. As Egypt's boundaries extended further south, the Pharaoh ordered installation of additional gages upstream to provide an early warning system of future water discharges. (11)

A CLOSER LOOK

Modern-day water managers closely monitor water discharge in rivers around the world. In some locations, satellite monitoring systems are operated to provide up-to-the-minute water discharge and stage information. For example, the U.S. Geological Survey operates 18 satellite data transmitters in northeastern New Jersey as part of the Passaic Flood Warning System, one of the most flood-prone basins in the United States. River-discharge monitoring networks are also used along the Passaic River to protect water requirements for aquatic species below water supply reservoirs and pumping stations. In Arizona and Colorado, satellite monitoring systems are used to allocate scarce irrigation water supplies throughout the state during dry summer months. Irrigators in New Mexico and Colorado can access streamflow information, gathered by a satellite network, over the telephone or on the Internet.

WATER STORAGE IN LAKES AND RESERVOIRS

Reservoir storage volumes are expressed in terms of million gallons or billion gallons in the eastern United States, while acre-feet is commonly used in most western states. Million cubic meters (mcm) is used in most other countries. An **acre-foot** of water is the amount needed to cover 1 acre (0.4 ha) of land to a depth of 1 foot (0.3 m). One acre contains 43,560 square feet (4047 m^2) of surface area. Therefore, 1 acre-foot of water contains 43,560 cubic feet or 325,851 gallons (43,560 cu ft × 7.48 gal per cubic foot =

SIDEBAR

Here are some estimates of size in acres and hectares of various items:

Item	Approximate Size Acres	Approximate Size Hectares
Average house and yard	0.25–0.3	0.10–0.12
Soccer field	1.5–2.5	0.6–1.0
Coal barge on the Mississippi River	6	2.4
Regional shopping mall with parking lots	100–300	40–120
One square mile	640	259
Central Park in New York City	843	341
Average farm in Iowa	320	130
Average ranch in Wyoming	5,000–15,000	2023–6070

325,851 gal). This is also equivalent to 1235 cubic meters, or 1.2 million liters.

An acre-foot is a volume and does not include the dimension of time. Water stored in a reservoir is measured in acre-feet, million or billion gallons, or cubic meters. Standing water is not measured in cubic feet per second, gallons per day, or cubic meters per second because there is no function of time.

How is the storage capacity of a reservoir measured? First, the storage volume is computed by taking measurements of the depth and area of the inundated area. Depth can be determined with electronic sounding equipment or cables. Area is obtained by developing topographic contour lines at various depth intervals (elevations) of the reservoir and then calculating the area at various water levels. A geographic information system (GIS) can define volumes of a site through use of various software programs.

After the storage capacity is determined, how are water quantities determined for various reservoir stages? A **stage-capacity curve** or **rating curve** is calculated to show reservoir storage volumes for various water depths. An example is shown in Table 3.4 for Loch Lomond Reservoir near San Francisco, California. In some cases, a staff gage can be installed to determine reservoir water levels. The reading on the staff gage can be correlated to the known storage volume calculations for various elevations of the water surface. Determination of changes in reservoir storage must take into account evaporation, groundwater infiltration, seepage out of the reservoir, and

TABLE 3.4 Storage Capacity for Water-Surface Elevations[a] at 2-foot Intervals, Loch Lomond Reservoir, Santa Cruz County, California, November 1998

Water-Surface Elevation	Storage Capacity	Water-Surface Elevation	Storage Capacity	Water-Surface Elevation	Storage Capacity	Water-Surface Elevation	Storage Capacity
577.5[b]	8991	542	4120	504	1340	466	148
577	8900	540	3920	502	1250	464	120
576	8730	538	3730	500	1160	462	99
574	8400	536	3550	498	1070	460	81
572	8070	534	3370	496	991	458	68
570	7750	532	3200	494	912	456	56
568	7440	530	3030	492	836	454	46
566	7140	528	2870	490	763	452	39
564	6840	526	2710	488	694	450	32
562	6550	524	2560	486	628	448	27
560	6270	522	2420	484	565	446	22
558	6000	520	2280	482	506	444	18
556	5730	518	2150	480	449	442	14
554	5470	516	2020	478	397	440	10
552	5230	514	1890	476	348	438	7
550	4990	512	1770	474	301	436	5
548	4760	510	1660	472	258	434	3
546	4530	508	1550	470	217	432	1
544	4320	506	1440	468	181	430	0

[a]Water-surface elevation is in feet above sea level. Storage capacity is in acre-feet.

[b]Maximum capacity of Loch Lomond Reservoir.

Source: Kelly R. McPherson and Jerry G. Harmon, "Storage Capacity and Sedimentation of Loch Lomond Reservoir, Santa Cruz County, California, 1998," U.S. Department of Interior, U.S. Geological Survey, Water-Resources Investigations Report 00-4016, Sacramento, California, 2000. Also found at http://www.wr.usgs.gov/rep/wrir004016, September 2001.

inflow and outflow of surface water. The following equation is generally used:

$$\text{Reservoir Storage} = Qi + Gi - S - E - Qo$$

where Qi = surface water inflow
Gi = groundwater infiltration
S = seepage
E = evaporation
Qo = surface water outflow

Reservoir storage includes the static body of water in the lake behind a dam, as well as water held as bank storage in the walls of the reservoir. The amount of bank storage depends on the geology of the reservoir site and can be great or small (similar to riverbank storage discussed earlier). Some of this water will return to the reservoir if water levels decline, while, under the force of gravity, some bank storage will percolate downward and move toward groundwater. At Hoover Dam and Lake Mead in Nevada and Arizona, reservoir operators use a bank storage rate of 6.5 percent of surface water storage and a 100 percent recovery rate when reservoir levels drop. These calculations are necessary because of the water-holding capacity of sandstone bedrock in the area. (12)

FLOOD EVENTS

Floods occur when precipitation and runoff exceed the capacity of a river channel to carry the increased discharge. Nature has provided river channels and floodplains to carry runoff water from the land surface to the oceans. A floodplain is simply an extended channel that carries high-water volumes. The amount of water carried by a river changes daily but often receives little attention in humid areas unless water overflows its banks. Floods are an inevitable component of any river system, since runoff volumes vary with time.

FLOOD FREQUENCY

The frequency of flooding is based on history—how often have floods occurred in a region, what are the historic extremes of high precipitation, and what land-use changes and development have occurred in a watershed. In some locations, flood frequency has been reduced as a result of flood-control dams and strict floodplain development regulations. However, predicting floods is still a function of determining the odds of high precipitation.

The laws of probability state that the chance of an event occurring are equal to the number of times it has occurred in the past. Floods follow the same rules of probability. Hydrologists refer to a "recurrence interval" of various flood flows, such as a **100-year flood (Q100)** (Figure 3.15). The laws of probability assign a value of 1 in 100 for a Q100 flood, or 0.01. This means that a 100-year flood has a 1 in 100 chance of a given discharge occurring (based on a river hydrograph) and flooding a given area. However, such a flood could occur once in 100 years or twice within the same month. Similarly, the laws of probability for a **500-year flood (Q500)** assign a

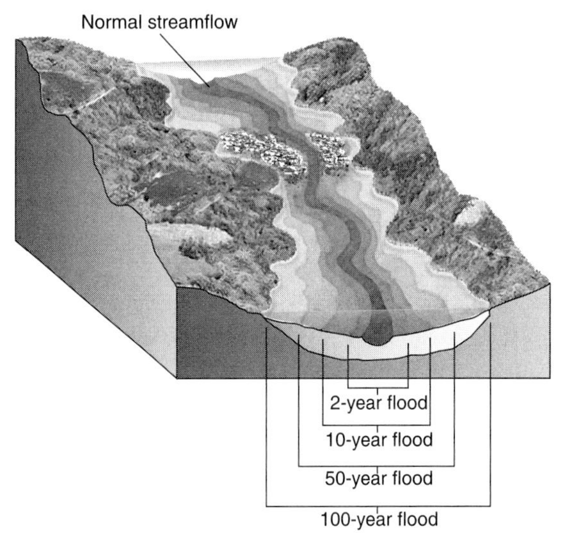

FIG. 3.15 Flood damage can be predicted based on the intensity of a storm and the topography of a region. A flood with a recurrence interval of 50 years means that it has a 1 in 50 chance of occurring in any given year (a 50-year flood, or Q50). A 100-year flood, or Q100, has a 1 in 100 chance of occurring in any given year.

frequency of a 1 in 500 chance, or a probability of 0.002 in any one year, of a given discharge (based on a river hydrograph) of flooding a given area. (13)

A 100-year flood is most commonly used for floodplain management, flood insurance regulations, and engineering designs of dam spillways, road culverts, bridge abutments, and other structures to meet estimated maximum surface water flow volumes. The term *100-year flood* is commonly misunderstood by the public and leads to confusion that can have important consequences regarding flood frequency and the hazards associated with purchasing a home or other property in a flood zone. This subject will be discussed further in Chapter 9.

PROBABLE MAXIMUM PRECIPITATION

Precipitation amounts have been recorded for thousands of years. As improved measurement methods are developed, such as with Doppler Radar, precipitation records are continually broken. The question that then arises is: Will precipitation extremes continue to break records, or is there some finite limit on the atmosphere's ability to produce rain at any given location due to climate, topography, and atmospheric moisture limits?

The answer is, yes, there is a limit. The concept of a finite limit for precipitation from a single storm event is called **Probable Maximum Precipitation (PMP).** A PMP is the maximum depth (amount) of precipitation that is reasonably possible during a single storm event, and it is based on previous storm records, accepted meteorological knowledge, probability, and statistics. As expected, PMP estimates are higher in hot, humid equatorial regions and much lower in colder climates of the midlatitudes where the atmosphere cannot hold as much moisture.

PROBABLE MAXIMUM FLOOD

Flood events also have maximum extremes and are called a **Probable Maximum Flood (PMF).** A PMF is the maximum surface water flow in a drainage area that would be expected from a Probable Maximum Precipitation event. The concept of PMF is useful for engineering purposes because it permits computation of a maximum water inflow for structures such as dams, culverts, or other hydraulic structures. High-hazard dams (dams whose failure would result in loss of lives and widespread property damage) are required, by modern standards, to contain 100 percent of a PMF without water overtopping (spilling over) the dam. A dam overtopped would quickly erode and eventually breach (fail). PMP and PMF rates can be estimated through statistical analysis or through estimates of past meteorological events in a region. (14)

Not all watershed areas with the same PMP have the same magnitude of a PMF, primarily because drainages may differ in slope, vegetation, size, and shape. Rainfall will run off much more quickly on a steep slope than on a level area. In sandy areas, more precipitation will percolate into the soil than in urban areas covered with concrete or rooftops. All of these factors will affect runoff patterns and the discharge of floodwaters in a watershed. Excellent interactive online flood exercises are available at http://vcourseware.sonoma.edu/VirtualRiver.

GUEST ESSAY

GIS and Flooding

Jake Freier with daughter Maizie.

by **Jake Freier**

Jake Freier formerly served as the GIS coordinator for the Iowa Emergency Management Division, Des Moines, Iowa. He earned his Master's degree in Geography from the University of Iowa and his Bachelor's degree in Geography and Environmental Studies from Gustavus Adolphus College in Minnesota. Jake is now with the Evans, Colorado, Planning and Zoning Division. He lives in Greeley, Colorado, with his wife April, daughter Maizie, and dog Tucker.

FIG. 3.16 Coast Guard flight mechanic Trent Ganz views the flooded John O'Donnell Stadium near downtown Davenport, Iowa, during the Mississippi River flood of April 2001.

During April of 2001, near-record-level floods ravaged the towns of the upper Mississippi River (Figure 3.16). River towns on the eastern border of Iowa were particularly hard hit, and Davenport, a city of more than 100,000 people, saw flood levels that had been surpassed only twice before in recorded history. In response, the State of Iowa Emergency Management Division opened its Emergency Operations Center to help coordinate the flood-fighting efforts among local, state, and federal government agencies. The computer-based mapping tool called GIS aided in predicting the spatial extent of flooding and determining what critical facilities might be affected by floodwaters.

GIS, or geographic information systems, are computer-based systems used to store, retrieve, map and analyze geographic, or spatial information. Spatial information links real-world location data, such as latitude and longitude coordinate pairs, with descriptive attribute data. With such spatial information at its core, GIS is a much more powerful tool than a traditional paper map. As an illustration, we will use the example of mapping schools within a town. Both paper maps and digital GIS maps can show the location of each school in town and perhaps provide the school's name. However, with the click of a mouse button, a GIS user can gather additional information about a particular school (Figure 3.17). For example, a GIS user may be able to learn the school's address and phone number, the name of the school's principal, the school's total enrollment, the year the school was built, and more. A digital image of the school may also be available for viewing. The ability of a GIS to ask questions of a spatial data set has earned it the nickname "smart maps."

Within a GIS, similar geographic features are grouped together in a single data set and are shown in a map display as a single layer. These layers are basically transparent maps that, when stacked on top of each other, allow the user to determine whether any relationships exist between different data sets. In the preceding example, each school and road is a separate GIS layer. These layers may be turned on and off, and added or removed from display. Let's add the following layers to our map: streams, lakes, and medical facilities (see Figure 3.18). As illustrated here, GIS maps are far easier to create and customize than traditional paper maps, which may take months to produce.

Now let's do some geographic analysis using our GIS. As a town official, we are looking to

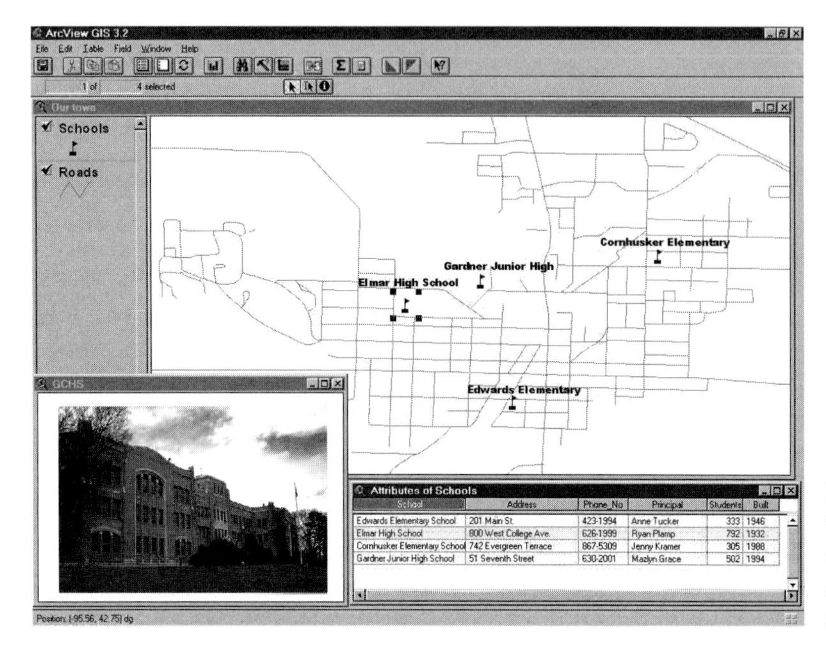

FIG. 3.17 A GIS map display showing the schools in a town. Notice that a query was made of Elmar High School, and the corresponding attribute data and digital image are shown.

analyze the hazards associated with a major flooding event so that we can take measures to ensure the safety of our citizens. With the correct GIS data, we can predict whether a 100-year flood (Q100) will affect important community facilities, such as schools. When we turn on the streams layer, we can see that two schools (Gardner Junior High and Edwards Elementary) are located near the two streams that flow through town (Figure 3.18). At this point,

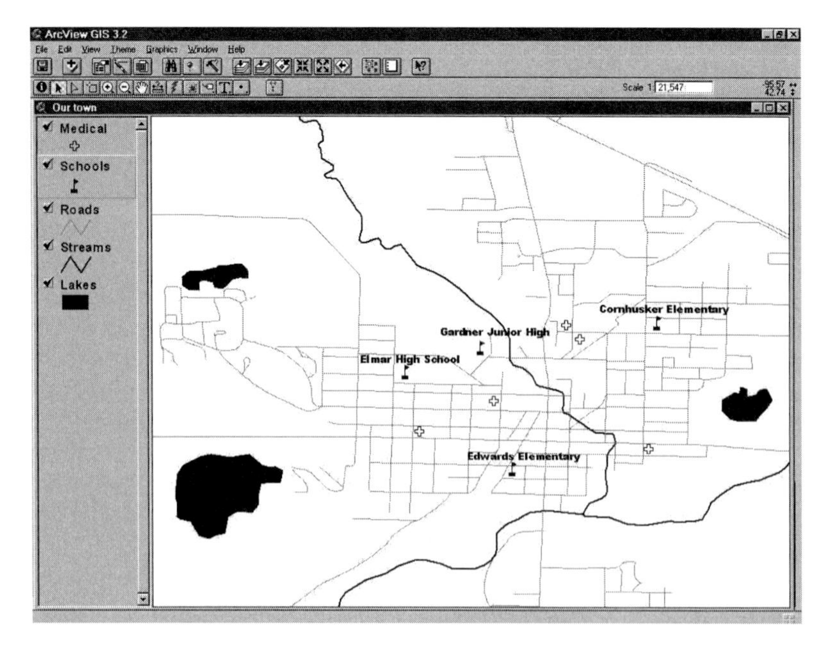

FIG. 3.18 Streams, lakes, and medical facilities are added to the original map display.

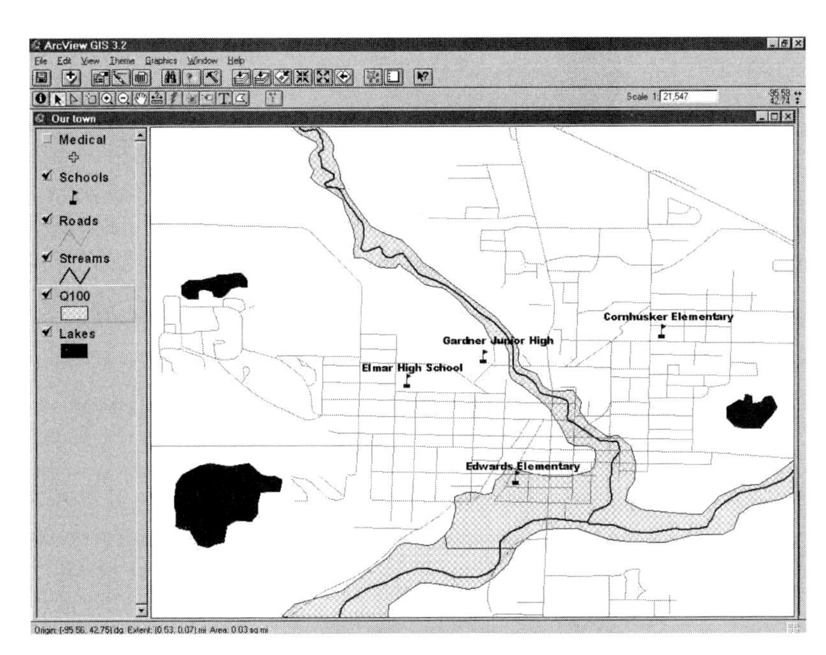

FIG. 3.19 With the Q100 layer added to the map display, it is easy to see that Edwards Elementary is the only school in the 100-year floodplain.

we can only guess as to whether or not these schools will be affected by Q100 floodwaters. To be certain, we must add a Q100 flood layer. By overlaying the Q100 floodplain data on the map (see Figure 3.19), we can clearly see that only one school, Edwards Elementary, lies within the Q100 floodplain. This knowledge allows us to plan accordingly for such an event.

This is a relatively simple example of how GIS can be used to plan for a flood event. During the Mississippi River flooding of 2001, much more complex geographic analyses were performed. State and federal agencies used GIS technology to predict the extent of a flooded area based on a reported sandbag levee breach; to model Mississippi River levels based on upstream tributary flows and recent weather events; and to determine the demographics of citizens whose property was damaged by the flood.

GIS is also used for other areas of water resources work including modeling sediment yields and discharge during extreme weather events, and predicting the effects of land-use planning on downstream water quality. However, the use of GIS is not limited to the field of water resources. Professionals in urban planning, agriculture, forestry, emergency vehicle routing, epidemiology, and others have also developed applications for use in GIS.

Questions

1. Put yourself in the role of a local official looking at the hazards associated with a 100-year flood. Aside from schools, what other features would be important to include in your GIS when doing a geographic analysis?
2. What are some of the other possible advantages of a GIS over a paper map?

POLICY ISSUE

Computer models are often used to predict water discharge for various hydrologic conditions and locations. These mathematical models can predict precipitation runoff on various slopes, river stage and discharge, interaction between surface water and groundwater, sediment transport, water quality changes, and many other scenarios.

The Storm Water Management Model (SWMM) is a popular flood prediction model developed by a consortium of engineers for the U.S. Environmental Protection Agency (USEPA) in the 1970s. Its purpose is to estimate runoff from storm events by utilizing data on rainfall, watershed characteristics, and routes of overland flow and channelized flow patterns. The model then predicts the duration and discharge of runoff to receiving waters downstream. This provides useful data for construction projects, floodplain protection, land-use patterns, and other human activities in flood-prone locations. (15)

CHAPTER SUMMARY

The natural processes of rivers and lakes can have significant effects on humans and the surrounding environment. Surface water runoff patterns, sediment transport and deposition, and lake cycles greatly affect the use and management of water resources around the world.

The measurement of water has developed into an elaborate science that relies on meticulous field measurements, engineering equations, and the use of satellites to closely monitor river and lake stages. Predictions are now being calculated on probable maximum precipitation and probable maximum floods in order to help planners and engineers design various public projects such as culverts, bridges, and other developments that could be impacted by flooding. Finally, GIS is being used to develop urban planning tools to prevent property damage from floods.

QUESTIONS FOR DISCUSSION

1. Describe the watershed that provides water supplies for your home region. Describe the location of water sources, and explain how these supplies were developed.
2. How could you delineate your local watershed?
3. During a storm event, how does interflow differ from overland flow?
4. Describe the process of bank storage and discuss why this process is important when measuring river or lake storage.
5. What was the purpose of a Nilometer? How is that same hydrologic concept used today?
6. Why is it important to know the volume of water stored in a lake or flowing in a river?
7. What insight can a stream hydrograph provide to a hydrologist regarding the climate of a region?
8. What role does water velocity play in the formation of floodplains?
9. Discuss the relationship between meanders and wetlands.
10. Describe how water management and use is dependent on a river gaging network.
11. Go to the USGS streamflow site at http://waterdata.usgs.gov/nwis.rt and determine the current flow in a river in or near your community.
12. Explain the difference between a 100-year (Q100) and a 500-year (Q500) flood.
13. What types of assumptions would be necessary when establishing a PMP (Probable Maximum Precipitation) value for an area? a PMF (Probable Maximum Flood)?
14. Discuss the differences between a gaining and a losing river.
15. Explain how a lake can turnover during autumn months. What problems can this create?

KEY WORDS TO REMEMBER

100-year flood (Q100) p. 75
500-year flood (Q500) p. 75
acre-foot p. 73
alluvial fan p. 68
aspect p. 59
bank storage p. 60
bed load p. 69
braided rivers p. 62
cirque lake p. 64
confluence p. 60
contour lines p. 58
cubic feet per second (cfs) p. 70
cubic meters per second (cms) p. 70
delta p. 68
discharge p. 70
dissolved load p. 68
downstream p. 60
effluent (gaining) river p. 63
ephemeral streams p. 63
epilimnion p. 65
eutrophic lake p. 65
floodplain p. 67
flow meter p. 71
fluvial material (alluvial deposits) p. 68
gage height p. 71
gradient p. 63
headwaters (source) p. 60
hydrograph p. 72
hypolimnion p. 65
hyporheic zone p. 60
influent (losing) rivers p. 63
interflow p. 59
intermittent stream p. 63
kettle lakes p. 65
lake p. 64
lake turnover p. 66
lentic p. 64
limnetic zone p. 65
limnology p. 64
littoral zone p. 65
meanders p. 61
mesotrophic lake p. 65
natural levees p. 67
Nilometer p. 73
oligotrophic lakes p. 65
orientation p. 59
overland flow p. 59
oxbow lakes p. 62
percolate p. 59
pluvial lakes p. 65
Probable Maximum Precipitation (PMP) p. 76
Probable Maximum Flood (PMF) p. 76
profundal zone p. 65
rating curve p. 71
Rational Formula p. 69
recorder p. 72
reservoir p. 65
saltation p. 69
scale p. 58
sediment p. 67
sediment yield p. 69
sedimentation p. 67
seiche p. 67
sorting p. 68
staff gages p. 71
stage p. 71
stage-capacity curve (rating curve) p. 74
surface water hydrology p. 56
suspended load p. 68
thalweg p. 60
thermocline p. 65
topographic (topo) maps p. 58
tributaries p. 60
upstream p. 60
velocity p. 68
watershed p. 56
yazoo stream p. 67

SUGGESTED RESOURCES FOR FURTHER STUDY

READINGS

Brater, Ernest F., Horace W. King, James E. Lindell, and C.Y. Wei. *Handbook of Hydraulics*. 7th ed. New York: McGraw-Hill, 1996.

Dennis, Jerry, and Glenn Wolff. *The Bird and the Waterfall: A Natural History of Oceans, Rivers, and Lakes*. New York: HarperCollins, 1996.

Dunne, Thomas, and Luna B. Leopold, *Water in Environmental Planning*. New York: W.H. Freeman, 1978.

Linsley, Ray K., Jr., Max A Kohler, and Joseph L.H. Paulhus. *Hydrology for Engineers*. 4th ed. New York: McGraw-Hill, 1991.

Manning, John C. *Applied Principles of Hydrology*. 3rd ed. Upper Saddle River, NJ: Prentice Hall, 1996.

Morisawa, Marie. *Streams: Their Dynamics and Morphology*. New York: McGraw-Hill, 1968.

Skinner, Brian J., Stephen C. Porter, and Daniel B. Botkin. *The Blue Planet*. New York: John Wiley & Sons, 1999.

WEBSITES

Environment Canada, "Freshwater Website," July 2003, http://www.ec.gc.ca/water/

Land and Water of Australia, Homepage, July 2003, http://www.lwrrdc.gov.au/

U.S. Department of Interior, U.S. Fish & Wildlife Service, "Wildlife Refuges," July 2003. http://midwest.fws.gov/desoto/

U.S. Department of Interior, U.S. Geological Survey, "Homepage," July 2003. http://pubs.usgs.gov

U.S. Geological Survey, "Water Resources of the United States," July 2003, http://www.water.usgs.gov/

REFERENCES

1. Dennis, Jerry, and Glenn Wolff. *The Bird and the Waterfall: A Natural History of Oceans, Rivers, and Lakes.* New York: HarperCollins, 1996, 109–110.

2. Ibid., 95.

3. Morisawa, Marie. *Streams: Their Dynamics and Morphology.* New York: McGraw-Hill, 1968, 137–139.

4. Dennis and Wolff, 84.

5. U.S. Department of Interior, U.S. Fish & Wildlife Service, "Steamboat Bertrand Collection," http://www.bluegoose.arw.r9.fws.gov/NWRSFiles/CulturalResources/Bertrand/Bertrand.html, July 2001.

6. Dennis and Wolff, 173–174.

7. Ibid., 190–191.

8. Ibid., 191.

9. Michener, James A. *Centennial.* New York: Random House, 1974, 57–58.

10. "River Basin Characteristics—Maumee River, Ohio/Indiana." http://www.glc.org/projects/sediment/maumee.html, July 2001.

11. Casson, Lionel. *Ancient Egypt.* Alexandria, VA: Time-Life Books, Time, Inc., 1965, 29–31.

12. Manning, John C. *Applied Principles of Hydrology.* 1st ed. Columbus, Ohio: Merrill Publishing Co., 1987, 233.

13. Black, Peter E. *Watershed Hydrology.* Englewood Cliffs, NJ: Prentice-Hall, 1991, 216.

14. Shaw, Elizabeth M. *Hydrology in Practice.* 3rd ed. London: Chapman & Hall, 1994, 235–237.

15. Black, 469.

CHAPTER 4

GROUNDWATER HYDROLOGY

For two decades scientists have debated whether liquid water might have existed on the surface of Mars as recently as a few billion years ago. With today's discovery, we're no longer talking about a distant time. The debate has moved to present-day Mars. The presence of liquid water on Mars has profound implications for the question of life not only in the past, but perhaps even today. If life ever did develop there, and if it survives to the present time, then these landforms would be great places to look. (1)

Dr. Ed Weiler, Jet Propulsion Laboratory,
National Aeronautics and Space Administration,
Pasadena, California, June 22, 2000

Interplanetary explorers may soon prove that life exists on Mars, and groundwater could provide the vital clue to such a discovery. The National Aeronautics and Space Administration's (NASA) Mars Global Surveyor spacecraft entered Martian orbit in 1999 and has gathered photographic evidence that liquid water may exist on, or just beneath, the surface of the Red Planet. Analysis of these photos has led NASA scientists to believe that gullies and other channels on the Martian surface were formed by running water. The images also show fluvial landforms that could indicate the presence of water beneath the surface (Figure 4.1).

In the 1970s, photos from *Mariner 9* also showed ancient geologic features shaped by flowing water on Mars. However, the more detailed Global Surveyor images display fairly recent flows. If surface water actually did exist on Mars billions of years ago, where did the water go? The best hypothesis provided by scientists is that it percolated below the Martian surface and could still exist there today. NASA's 2004 Rover missions and the European Space Agency's Mars Express orbitor will provide more conclusive information.

On Earth, groundwater accounts for only 0.06 percent of the total water supply and yet represents 98 percent of all freshwater readily available to humans. Abundant freshwater is tied up in glaciers and polar ice, making it essentially unavailable. Other water sources, such as rivers, lakes, and reservoirs, have local importance but are much less significant on a global scale.

In the United States, groundwater provides drinking water to approximately one half of the

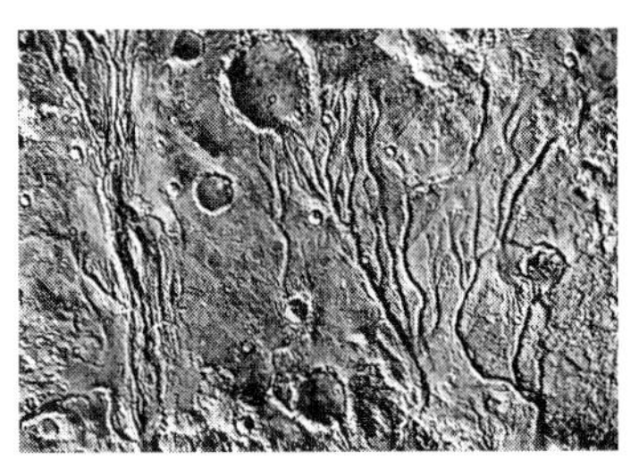

FIG. 4.1. Mars appears very dry today, but images of gullies like these have led scientists to suspect the presence of water in the past on the Red Planet. Such channels may have formed as a result of the sudden release of frozen groundwater to the planet's surface. In this NASA photo, note how some of the channels cut through craters, which indicates that water flowed after the craters were formed.

population; in Canada, the figure is approximately 30 percent. These figures are high because most groundwater is readily accessible and can often be used without any treatment. Irrigators in the United States obtain 37 percent of all irrigation water from groundwater sources. In some states, such as Nebraska, Mississippi, Nevada, Oklahoma, and Texas, the figures are in the 69 to 94 percent range.

Our extensive reliance on groundwater can be tenuous at times owing to its vulnerability to pollution. Just one gallon of gasoline (3.8 l) can contaminate one million gallons (3.8 million l) of groundwater for drinking water purposes. Hazardous waste spills, percolation of fertilizer, pesticides, and other chemicals can pollute vast quantities of groundwater supplies. For more details on groundwater pollution, see Chapter 5.

WHAT IS GROUNDWATER?

Groundwater is water found within the pore spaces of geologic material beneath the surface of the Earth. It exists in saturated layers of sands and gravels, in certain types of clay material, and in cracks within crystalline rock. Moving water, wind, ice, and tectonic forces create opportunities for surface water to seep into underground material. Groundwater moves through porous geologic materials under the force of gravity or sometimes by the sheer weight of atmospheric pressure. This movement continues downward until an impervious layer of rock, shale, clay, or other water-tight formation is encountered. These geologic barriers can cause a localized area to remain completely saturated with groundwater, sometimes up to the land surface. Such a wet, underground geologic setting can create a wetland, pond, or baseflow for a river at the land surface.

The source of all groundwater is precipitation. As discussed in Chapters 2 and 3, groundwater can be replenished from surface water runoff, or through the beds or banks of rivers, lakes, ponds, or wetlands. Groundwater is sometimes found within a few feet (or meters) of the Earth's surface and can be hydraulically connected to rivers, lakes, and wetlands. In other locations, well drillers must core hundreds of feet (100 m or more) to reach usable supplies. Large quantities of groundwater are not generally found below 10,000 feet (3048 m). This is due to the tremendous pressures at such depths that cause any small openings between geologic material to be tightly pressed together and closed.

SIDEBAR

The word *groundwater* is often written as two words *(ground water)* to correspond to the term *surface water*. However, many scientists prefer to write *groundwater* as a single word to represent a technical term. Either form is appropriate, although we will use the single-word form in our discussions.

Groundwater is a small but integral part of the hydrologic cycle. According to the U.S. Geological Survey, there are an estimated 1 million cubic miles (4.2 million km^3) of groundwater within one-half mile (0.8 km) of the Earth's surface. This compares to only 30,000 cubic miles (125,045 km^3) of water in freshwater lakes and a mere 300 cubic miles (1250 km^3) in streams. (2)

Groundwater is the largest source of freshwater on Earth and provides drinking water to 53 percent of the people in the United States. (3) Cities such as San Antonio and El Paso, Texas; Albuquerque, New Mexico; Dayton, Ohio; and Lincoln, Nebraska, rely almost exclusively on groundwater. In many regions of the world, however, groundwater cannot be obtained in sufficient quantities to economically justify installation of a single well (sometimes called a *bore hole* outside the United States). The spatial distribution of groundwater can be quite irregular and variable in supply (see Figure 4.2).

The United Nations, during its observance of World Day for Water in 1998, described groundwater as the "Invisible Resource." Though unseen, groundwater is playing a decisive role in global economies, political conflicts, and personal

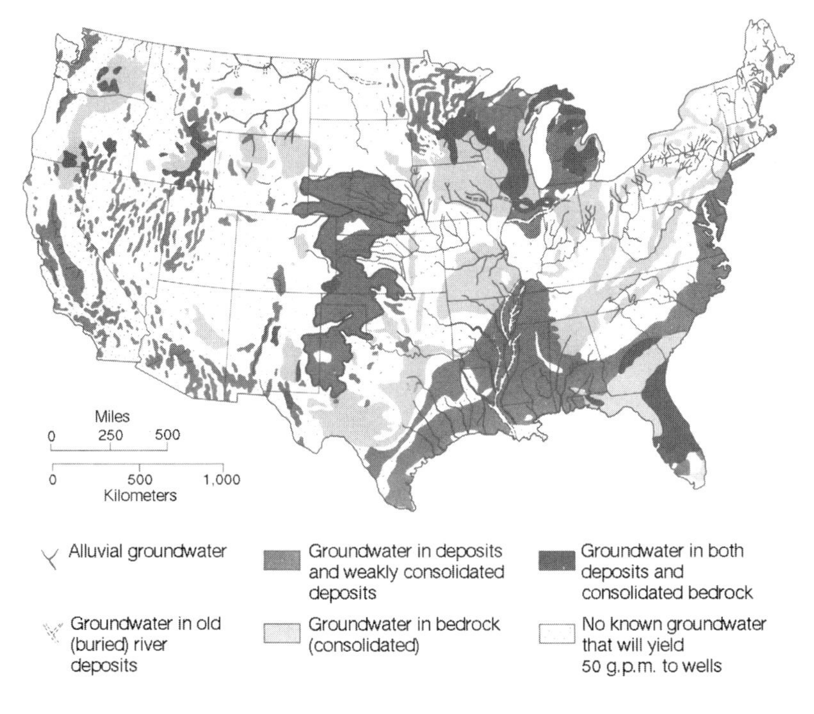

FIG. 4.2. This map of major aquifers in the United States shows an interesting distribution of groundwater formations. Do any regional surface features coincide with the geographic distribution of groundwater? Which region appears to have the greatest concentration of groundwater? Can any correlations be made between climate and the distribution of groundwater in the United States? The various types of groundwater shown on this map will be explained later in this chapter.

health around the world. This focus on groundwater is inevitable as population increases and as demand for water supplies accordingly escalates.

The study of groundwater requires an understanding of many physical processes. Knowledge of these attributes and methods is critical in determining groundwater availability, trends, and characteristics to meet the needs of our natural and developed environment. A number of questions surround these physical processes:

How is groundwater formed?

What role does geology play in the movement of surface water into groundwater settings?

How does groundwater interact with surface water?

What role does geology play in the movement of groundwater?

What methods are used to measure the movement of groundwater?

How are groundwater quantities quantified?

WHAT IS GROUNDWATER HYDROLOGY?

Groundwater hydrology is the study of the characteristics, movement, and occurrence of water found beneath the surface of the Earth. Groundwater professionals have expertise in engineering, geology, earth science, or other scientific fields that require knowledge of Earth systems, mathematics, and chemistry. Groundwater hydrologists provide water managers, planners, and others involved in water resource management with invaluable information regarding groundwater attributes.

The hydrology of groundwater has some characteristics similar to those of surface water. In Chapter 3, we discussed how watersheds are a basic geographic unit for the study of surface water. Groundwater and aquifers are somewhat analogous to surface water and watersheds. We've seen how the movement of surface water is affected by the gradient of the land surface. Groundwater movement, in turn, is affected by subsurface geologic material and the gradient of water found at such depths. Surface water discharge can be measured by using a flow meter, gaging station, or mathematical equations such as the Rational Formula. Groundwater levels (elevations below the land surface) can be measured by inserting a tape measure or electronic device down a well or by using mathematical formulas developed in the 1800s.

The study of groundwater existed even during ancient times. Although the level of scientific knowledge was then limited, the development of *qanats* in Africa, the Middle East, China, and South America provides an impressive record of hydrologic insight. Hydrologists and engineers of the Roman Empire utilized groundwater from springs (which will be discussed later in this chapter) as water sources for aqueducts to their cities.

These impressive ancient groundwater supply systems notwithstanding, the ancients also propagated certain bizarre theories regarding the origin of groundwater, some of which continue today. Early Greek philosophers, such as Homer (c. 800 B.C.), Thales (636–546 B.C.), and Plato (427–347 B.C.), believed that groundwater originated as sea water. These learned scholars deduced that water moved through subterranean channels beneath mountains and was then purified as it moved to the land surface. The Roman architect and water manager Marcus Vitruvius Pollio (c. 70–25 B.C.) corrected this misconception and hypothesized that precipitation and surface water percolation served as the source of all groundwater. In the seventeenth century, Pierre Perrault (1611–1680) measured precipitation and its relationship to runoff in the upper Seine River watershed of France and discovered that regional precipitation exceeded the discharge of the Seine River by 600 percent. His research supported the theory of Vitruvius developed in Rome nearly 1700 years earlier. (4)

THE GEOLOGY OF GROUNDWATER

Groundwater is found in a variety of geologic settings constrained by lithology, stratigraphy, and the structure of geologic deposits and formations. *Lithology* is the study of the physical characteristics of rocks, including mineralogy, composition, grain size, and density of geologic materials; *stratigraphy* describes the composition and age of deposit beds (such as sediments), lenses, and other formations; and *structure* refers to cracks, folds, and other deformations of geologic systems. In some locations, wind and erosion created opportunities for surface water to fill underground layers of sand, gravel, and other sediments. In other regions, deep geologic faults allowed surface water to migrate (travel) to depths of hundreds and even thousands of feet or meters. Understanding regional geology is the road map to finding groundwater.

SEDIMENTARY ROCKS

Sedimentary rocks are made up of particles created through weathering and erosion of igneous and metamorphic rock. Such materials can form underground layers of **conglomerate** (boulders, gravels, pebbles, cobbles), **sandstone** (sand), **siltstone** (silt), or **shale** (clay). These layers, or formations, can have a thickness of just a few inches (or centimeters) or as much as hundreds of feet (over 100 m).

Conglomerate is often rice- to pea-sized and may have irregular or smooth surfaces. An irregular surface means that the conglomerate was not transported far from its origin and was not exposed to the smoothing action of water. Sandstone is generally of uniform size and may have irregular or smooth surfaces, depending on the method of transport and exposure to wind and water. Siltstone consists of very small particles, often called **fines,** which are smaller than grains of sand. Shale contains an abundance of tightly bound, adhesive clay materials but may also include numerous fines dispersed throughout such a formation.

Sediments can be transported by wind, gravity, ice, and water. Such materials are deposited when the carrying capacity of the transportation method is exceeded (as discussed in Chapter 3). Common locations of sediment deposition are in river valleys (deposits of gravels, sands, and fines), lake shores and bottoms (well-sorted sands), and glaciated regions (random-sized materials ranging from clay particles to boulders). Sediments can also be carried by the wind to create large deposits of sand and smaller grained materials called *eolian deposits* (named after Aeolus, the Greek god of wind). The Great Sand Dunes National Park in the San Luis Valley of southern Colorado, and the Sandhills of north-central Nebraska, are excellent examples of eolian deposits.

Limestone formations are generally composed of consolidated (hardened) lime mud, marine algae, and sand. Tremendous pressures from overburden compressed these materials into hardened deposits. Limestone tends to dissolve and can create large openings for groundwater movement. Some dissolved limestone areas can be so extensive that people frequently walk through such formations, such as at Carlsbad Caverns in New Mexico, which has one dissolved chamber more than one-half mile (0.8 km) long, 650 feet (198 m) wide, and almost 330 feet (100 m) high. The Mammoth Cave system in Kentucky is another large dissolved limestone formation.

Karst (a German word meaning "bare, stony ground," named for the Karst region in Slovenia) is terrain where solution, or the dissolving of limestone, dolomite, gypsum, or marble, creates highly erodible areas on the land surface and underground. These regions can resemble Swiss cheese and contain numerous caves, sinkholes, and rivers that disappear underground. Karst terrain is commonly found in Florida, Texas, Kentucky, China, Slovenia, and Turkey.

Sinkholes are smaller features where geologic formations such as limestone, carbonate rock, or salt beds are naturally dissolved by water. Although deterioration is very slow, formation of a sinkhole can be sudden and dramatic. Vehicles and houses have fallen into sinkholes with little or no warning. In the United States, most sinkholes occur in Florida, Texas, Alabama, Missouri, Kentucky, Tennessee, and Pennsylvania. Sinkholes are also found in the Shan Plateau of China, Nullarbor Region of western Australia, Atlas Mountains of north Africa, Belo Horizonte of Brazil, and the Carpathian Basin of southern Europe.

GLACIATED TERRAIN

Glaciers have carved the surface of the Earth. However, these ancient ice sheets also left deposits of material that today often contain groundwater. During the Pleistocene epoch, much of Canada, the northern United States, Scandinavia, Russia, and Siberia, as well as portions of southern Africa and South America, were covered by massive sheets of ice. These ice layers were as much as 1 to 2 miles (1.6 to 3.2 km) thick. Glaciers increased in size during periods of global cooling and then receded as global warming occurred. During this same period, much smaller mountain glaciers also formed in adjacent regions and had traits similar to the continental ice sheets.

More glaciers existed in the Northern Hemisphere than other parts of the world probably because the larger landmasses cooled more quickly than the large expanses of ocean that existed in the Southern Hemisphere. The ice sheets of the Northern Hemisphere could move along landmasses to the south, while the glacial ice in the Southern Hemisphere tended to break off along the coast of Antarctica and formed icebergs before reaching nearby southern continents. Scandinavian ice sheets moved southward to present-day Germany, the Ukraine, and Kazakhstan, while glaciers in North America extended as far south as the midwestern states of Indiana, Iowa, and Illinois (Figure 4.3). Surface water drainage patterns changed, valleys were created, and huge sedimentary deposits of glacial materials were left behind.

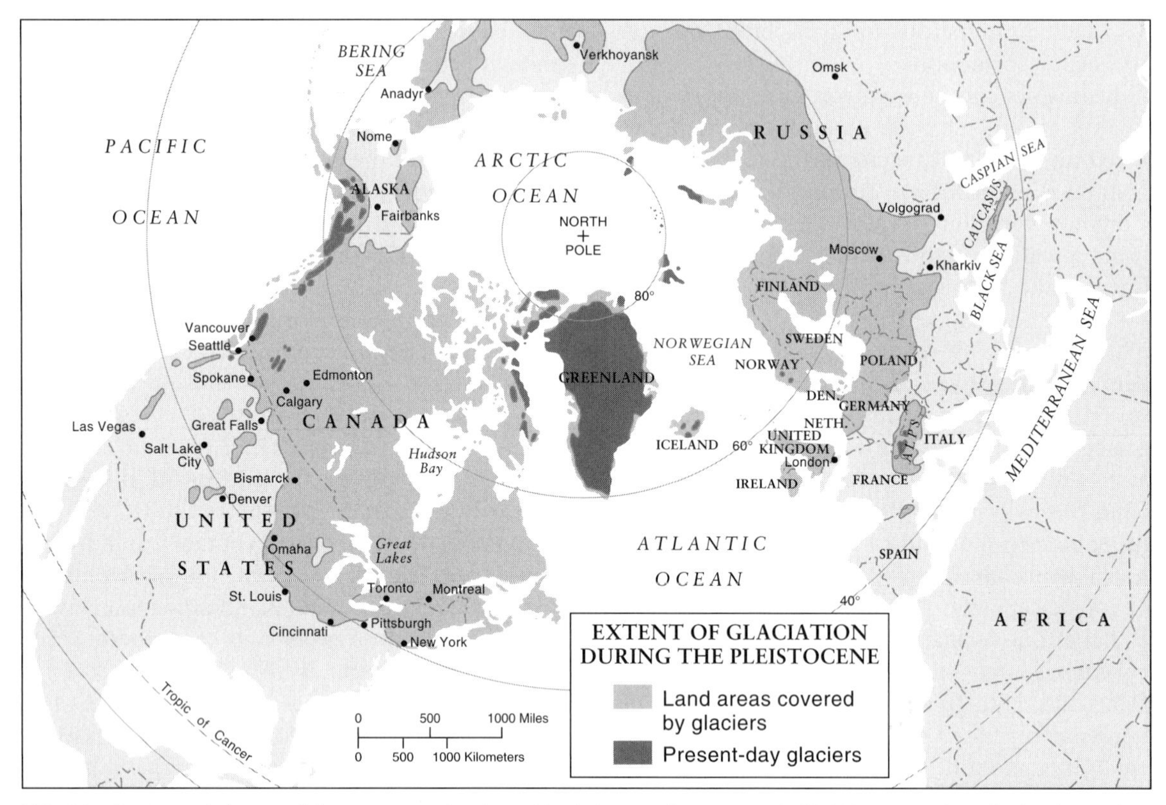

FIG. 4.3. Continental glaciers of the most recent Ice Age in North America (approximately 20,000 years ago) reached as far south as the Ohio and Missouri River valleys. Most of Canada and the northern United States were covered with ice sheets between 500 and 2500 feet (152 to 762 m) deep. Much of the world's water supply was frozen at that time in glaciers, and sea level fell about 300 feet (91 m). *Geography: Realms, Regions, and Concepts,* 10th Edition, by Harm de Blij and Peter O. Muller. Copyright © 2002 H. J. de Blij and John Wiley & Sons, Inc. This map was originally produced in color. Adapted and reprinted by permission of John Wiley & Sons, Inc.

Several glacial epochs have occurred throughout Earth's history. The most recent period of deglaciation began about 15,000 years ago, following global warming, and continues today.

During periods of glacial activity, underlying terrain was scraped, carved, and reshaped through the plowing and grinding action of moving glacial ice, rocks, and boulders. Rocks and boulders were carried at the base of the ice sheet and were deposited as a glacier melted. In North America, materials were carried from the Canadian Precambrian Shield regions of Ontario and Quebec to the northern United States (Figure 4.4). In Europe, rocks from Scandinavia were dragged into Germany, Poland, and the former Soviet Union. Boulders from Antarctica can be found in South America and the southern tip of Africa. Mountain glaciers in the Rockies, Alps, and Andes altered local landscapes in much the same way, though on a much smaller scale.

A CLOSER LOOK

Sir Charles Lyell (1797–1875) was an esteemed British geologist who announced the theory of the Pleistocene epoch in 1839. His study of fossil mollusks, as well as his profound understanding of geology and geologic processes, led to his explanation of the natural glacial forces that caused erosion and other terrestrial changes during the most recent Ice Age. Others of that same era developed the Glacial Theory through research of the odd and

FIG. 4.4. A piece of Canadian rock, called an *erratic* (so-called because of its "erratic" occurrence) sits in central Illinois. Farmers in regions where erratics are found were forced to clear fields of rock obstructions before plowing or other cultivation. Some erratics were used to build fences and foundations, while others were simply piled out of the way in the corner of fields.

erratic occurrence of boulders in regions far from native bedrock. Scientists of the eighteenth and early nineteenth centuries had argued that a great flood deposited these freak boulders around the world. Later, another theory proposed that these boulders were first frozen within drifting icebergs and then deposited in wild patterns as melting occurred during a great flood. These out-of-place boulders were called *drift*, a term still used today even though the theory of drifting boulders was discredited over 150 years ago. (5)

Rock debris that is transported by glaciers and then deposited is called **glacial till.** Glacial deposits often contain boulders intermixed with sand and silt (material between the size of fine sand and clay particles), gravel, and large rocks. Some buried valleys of glacial material are not apparent from surface topography, and coring (drilling) is necessary to locate these potential sources of groundwater. Groundwater found in glacial till may not be as plentiful as in sedimentary formations because of the highly irregular size, distribution, and location of geologic materials.

As glaciers melted, some boulders, rocks, pebbles, gravel, sand, and fines were carried by floodwaters and deposited as **glacial outwash.** These materials settled in valleys, moraines (ridges of glacial material), and beyond the terminus (end point) of glaciers. Glacial deposits of sand and gravel can sometimes be found in buried valleys tens of miles (over 16 km) long and several miles (over 3.2 km) wide.

ALLUVIAL VALLEYS

Flowing rivers deposit sediments called **alluvium** (also called alluvial or fluvial material). In Chapter 3 we discussed the process of sediment transport and deposition during flood events. During a flood, gravels and larger materials are typically deposited near or adjacent to a riverbed. Smaller gravels and sands are generally transported greater distances and deposited across a floodplain during high-water events. As geologic time passed, alluvial material could build to a thickness of hundreds of feet (over 100 m).

Rivers that flow through an alluvial valley are often hydraulically linked to groundwater. This physical connection creates opportunities for surface water in a river to recharge groundwater or for groundwater to replenish flows in a river as baseflow. The direction of water movement between groundwater and surface water is dependent on gradients, climatic conditions, and water volume as discussed in Chapter 3.

TECTONIC FORMATIONS

Tectonic activity, the movement of rock formations, can create fissures and fractures that hold groundwater. A **fracture** is the separation of a rock surface which creates a hairline crack in the rock (Figure 4.5). A **fissure** is a location where the walls of a fracture have become separated and moved apart. Some fissures are filled with materials (veins) such as rocks, sediments, minerals, or water.

Many homeowners in mountainous regions rely on groundwater found in fissures (called *L'eau des roches* or *rock water* in France) for water supplies. Variability is a problem, however. A well driller may find groundwater at one

FIG. 4.5. Ms. Cech inspects rock fractures along the Big Thompson River near Estes Park, Colorado.

location but be unsuccessful only a few hundred feet (100 m) away. Fissures in igneous rocks are generally not extensively interconnected and often provide limited sources of groundwater because of the limited open space within fissures.

GROUNDWATER RECHARGE

The hydrologic cycle has a major impact on groundwater storage. Precipitation and surface water slowly move below ground until they are intercepted by plant roots or stopped by an impervious layer of material such as clay or shale. This naturally occurring process of downward water migration is called **groundwater recharge** or **percolation.**

Groundwater recharge rates depend on climate, terrain, geology, and vegetative ground cover. Percolation occurs slowly if geologic materials are tight (somewhat impervious) and limit movement of percolating water. It has been estimated that groundwater in some regions of the High Plains in the central United States and Canada could take centuries to recharge if depleted. This is due to the slow rate of groundwater recharge. By contrast, a small, shallow body of groundwater, located in geologic material with high recharge capacity, could refill after one significant rainstorm.

Groundwater recharge is greatly reduced in urban areas. Paved roads, rooftops, and other impermeable surfaces prevent surface water from percolating to groundwater. Reduced recharge rates in such developed locations can cause downstream flooding problems as a result of increased surface water runoff. Preservation of wetlands along streams, stormwater detention ponds, and open space such as parks, golf courses, and wildlife areas can help preserve groundwater recharge zones.

High-precipitation events do not always lead to increased percolation rates. Arid locations, in particular, can have low percolation rates due to hard, sun-baked land surfaces. A heavy thunderstorm in a desert may generate significant volumes of surface water runoff that rapidly collect in an *arroyo* (a creek or gulch). The result is usually a flash flood but negligible groundwater recharge.

When precipitation or surface water begins to recharge, it enters an area just below the land surface called the **vadose** (from the Latin word *vadosus* meaning "shallow") **zone.** The vadose zone (also referred to as the **unsaturated zone**) extends vertically from the land surface down to the area completely saturated with groundwater (called the **saturated zone**). The top of the saturated zone is known as the **groundwater table.** The slope of the groundwater table generally (but not always) follows the topography of the land surface, though usually in less detail (Figure 4.6).

Soil is a combination of inorganic weathered geologic material, decomposed organic material, bacteria and other living organisms, air, and water. Water found in the small openings between soil particles is called **soil moisture** (also called *soil water*). Soil moisture resides between soil particles in quantities that vary with precipitation and evapotranspiration. Plant roots

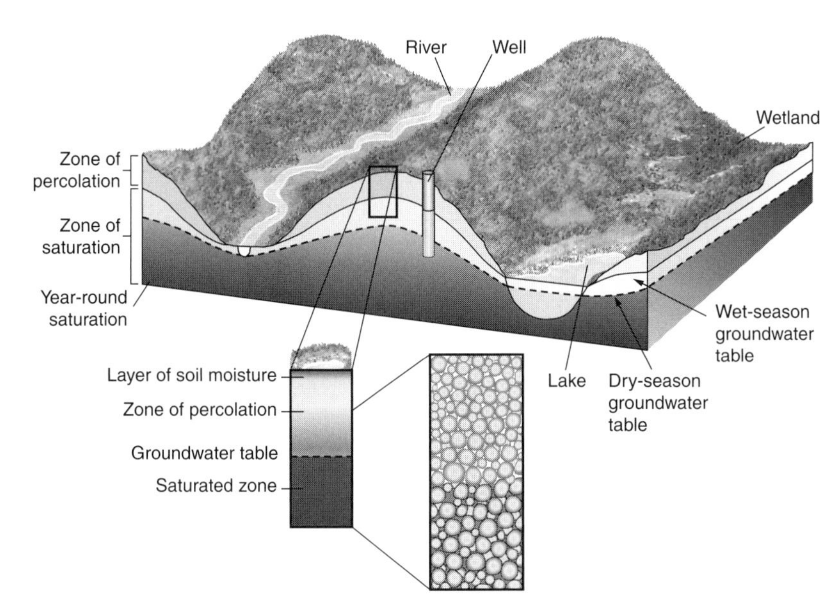

FIG. 4.6. Lakes and wetland complexes often exist in areas with shallow groundwater elevations that intercept the land surface. Variations in precipitation, between seasons and during wet or dry cycles, can greatly affect the elevation of shallow groundwater.

capture soil water through **capillary action.** This process draws water into a root system (called the **root zone**) and moves it upward to all parts of the plant. Surface tension within the root causes water molecules to be attracted upward into the plant. Surplus soil water percolates downward past the root zone until it reaches a saturated zone. These processes are of particular interest to soil scientists, agronomists, and botanists in determining plant water requirements and groundwater recharge rates.

POLICY ISSUE

Groundwater recharge is an important natural process for replenishing groundwater supplies. In some areas of the world, however, drought and overuse of groundwater for urban and rural uses have led to alarming declines. Coupled with these conditions is ongoing urban sprawl, which effectively seals potential recharge zones with paved streets, sidewalks, and rooftops.

If you were a groundwater hydrologist, how would you collect and develop data in an urban area to prove the need to preserve open space for groundwater recharge? What information would be necessary to convince a city council, development community, or homeowners of the importance of groundwater recharge? Suppose your community relied solely on surface water for its water supplies. How would this change your argument regarding the need to protect groundwater recharge zones?

AQUIFERS

An **aquifer** is a water-bearing geologic formation that can store and yield usable amounts of water. The word "aquifer" comes from the Latin words *aqua,* meaning "water," and *ferre,* meaning "to bear or carry." Aquifer materials include sand, gravel, sandstone, limestone, and fractured rock such as granite, which has sizable fissures. Aquifers are analogous to surface watersheds in that both are basic units of water management.

An aquifer is identified by characteristics such as type, areal extent, depth from the land surface, thickness, yield, and direction of groundwater movement. Some jurisdictions manage an aquifer by implementing rules (laws) that can regulate pumping rates, as is done in Nebraska, for example, and by spacing (spatial separation) of groundwater wells, as is required in Kansas. In Colorado, some groundwater users are required to provide alternative water supplies to surface water users in locations where a hydraulic connection exists between surface water and groundwater. (Groundwater allocation laws will be discussed in Chapter 8.)

AQUIFER TYPES

Aquifers are classified as consolidated or unconsolidated rock. **Consolidated rock** includes sandstone, limestone, granite, or other rock. Some are very low water-yielding formations since the material is almost impervious and does not allow groundwater to move easily through the geologic material. Limestone aquifers, however, can yield large amounts of groundwater because of extensive porous space created by solution. The Floridian Aquifer in Florida is an excellent example of a high water-yielding limestone aquifer. **Unconsolidated rock** consists of granular material such as sand and gravel and generally yields larger amounts of groundwater.

Aquifers can range in size from very small formations of a few feet (1 m) thick that extend less than 1 mile (1.6 km) to massive systems that extend hundreds of miles (hundreds of kilometers) across multiple state, provincial, or international borders. Aquifers can vary greatly in depth from the land surface. In some locations, the top of an aquifer may extend to the land surface and then tilt gradually downward for hundreds of feet (over 100 m). Again, regional geology provides the setting for groundwater and aquifers.

The **saturated thickness** of an aquifer is the total water-bearing thickness of a geologic formation. An aquifer may be a few feet (1 m) thick or hundreds of feet (over 100 m) thick. The saturated thickness of an aquifer significantly affects its potential water yield.

Numerous types of aquifers exist around the world. A **perched aquifer**, for example, is often found in formations of glacial outwash where clay layers (sometimes called *lenses*) form impermeable layers above a primary aquifer. This upper or perched groundwater usually covers a small area but allows groundwater to exist above the saturated zone of a lower aquifer system. A perched aquifer is often located relatively close to the land surface, and in some cases, the upper limit of a perched aquifer (called a *perched water table*) can provide baseflow for wetlands or streams. The process of drilling a well can actually puncture a perched aquifer and allow it to drain into lower geologic formations.

A **fractured aquifer** is found in rocks, such as granite and basalt, that contain usable amounts of groundwater in cracks, fissures, or joints. Limestone formations are sometimes found in fractured aquifers but often contain cracks or other openings enlarged by solution (dissolving of rock). Large channels or caverns can be created in this type of geologic setting, such as the limestone caves of Kentucky discussed earlier. An **aquiclude** (from the Latin word *claudere* meaning "to shut down") is a formation that contains groundwater but cannot transmit it at significant rates to supply a well or spring.

Groundwater exists in an aquifer under two different conditions: **confined** (also called *artesian*) or **unconfined** (sometimes called *water table*). An unconfined aquifer is generally located near the land surface. It is often composed of highly permeable and uniform materials, such as sand, gravel, and other sedimentary rock, and it is recharged directly by surface water. Alluvial aquifers are excellent examples of unconfined aquifers. Recharge can occur from the downward seepage of surface water through the unsaturated zone or from lateral movement or upward seepage of groundwater from underlying geologic strata.

Confined, or artesian, conditions occur when an inclined water-bearing formation is located at depth below an impermeable layer of geologic material such as rock, clay, or shale. This geologic barrier "confines" groundwater and causes it to be under pressure. If the pressure is

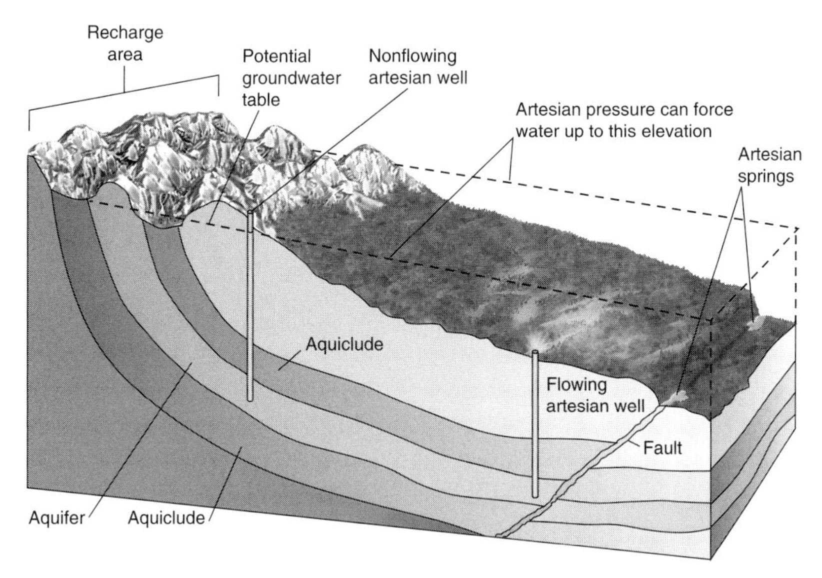

FIG. 4.7. Two conditions are necessary to create an artesian groundwater system: a confined aquifer and sufficient pressure in the aquifer to force groundwater in a well or other opening to rise above the static water level (or groundwater table) of the aquifer. If groundwater pressure is great enough to force water to the same elevation as the recharge zone (shown by the dashed line), water at the surface will flow out of the ground naturally and create artesian springs.

great enough, groundwater can emerge at the land surface as an **artesian spring.** A spring can also occur if groundwater in an unconfined aquifer moves from a higher to a lower elevation and emerges on the land surface (Figure 4.7).

A CLOSER LOOK

A confined aquifer is similar to a volume of water in a pressurized container. The sides of the container act as an impermeable barrier that confines the water within a given space. If a small opening is made, water will flow out to release pressure built up inside the confined space of the container. A confined aquifer has similar properties. Water pressure caused by gravity will cause confined groundwater to find exit points anywhere in the geologic system. Occasionally, the path of least resistance is upward to the land surface. If enough pressure exists in the aquifer, a spring may form.

If pressure is great enough, confined groundwater will flow up a well to the land surface, called a *flowing artesian well* (Figure 4.8). If the pressure in a confined aquifer is less, groundwater may travel only partially up a rock fissure or well but not reach the surface. Confined aquifers generally have small recharge zones and yield small amounts of water. However, an exception is the Dakota Sandstone Aquifer of South Dakota. This formation is recharged by surface

FIG. 4.8. The citizens of Bad Oeynhausen, Germany, owe a large portion of their prosperity to a local farmer and his pigs who discovered the first flowing artesian spring in the area. *Bad,* German for *spa,* was added to the town name, and the community was soon discovered by thousands seeking therapeutic relief from the minerals contained in the hot waters, called *Heilwasser* or healing waters by Germans. The grateful residents erected this *Schweinebrunnen* (pig fountain) in honor of their discovery.

FIG. 4.9. Groundwater can sometimes be seen at unique geologic locations. Thermal heating far beneath the Earth's crust has created hundreds of geysers, mud pools, fumaroles, and boiling chloride pools here at the Whakarewarewa Thermal Area on the North Island of New Zealand. Pohutu Geyser, shown above, is the largest in New Zealand. It usually erupts 20 times a day to a height of 59 feet (18 m) and at times exceeds 100 feet (30.5 m).

water from across the Black Hills in western South Dakota and from adjacent formations, and supplies groundwater to much of the state. Pressure in this confined aquifer was as high as 130 pounds per square inch (9.1 kg/cm^2) in the early 1900s. The first wells were drilled into the Dakota Sandstone Aquifer in 1882 and required coring through overlying layers of shale that varied in thickness from 985 feet (300 m) to 1640 feet (500 m). These shale layers acted as the confining beds of the aquifer. Artesian pressure was so great at the time that groundwater jetted over 100 feet (30 m) into the air in some locations. However, as more wells were drilled in the late 1800s, artesian pressure decreased. By 1915, 10,000 artesian wells had been drilled into the formation, with a corresponding head reduction (drop in groundwater levels) of approximately 13 feet (4 m) per year between 1902 and 1915. (6)

In the mid-1900s, many artesian wells in the San Luis Valley of south-central Colorado (elevation 8000 feet, or 2438 m) flowed as "fountains" above ground due to the high artesian pressure of groundwater in the area. As the weather cooled in the fall, the artesian water froze, creating beautiful ice sculptures. Valley farmers sometimes placed food coloring in the nearly frozen water to create colored fountains—a unique tourist attraction throughout the Valley. Today, artesian pressures have declined significantly owing to increased groundwater pumping in the area. Groundwater recharge from the nearby Sangre de Cristo Mountains no longer provides enough pressure to maintain historic artesian conditions. Today, the rainbow artesian fountains of the San Luis Valley are only memories.

A CLOSER LOOK

The word *artesian* comes from the province of Artois in northwestern France where the first artesian well was drilled by Carthusian monks in A.D. 1126. The monks used a percussion method to drill their well (they hit a sharp metal rod with a heavy hammer). The bore hole was only a few inches (or centimeters) in diameter, but the confined groundwater was under enough pressure that it flowed out of the ground.

The bottling of such artesian spring water has become a huge business worldwide. According to UNICEF (the United Nations Children's Fund), the consumption of bottled water was almost nonexistent in the 1950s but grew to 843 million gallons (3.2 billion l) in 1984. It reached a staggering 2.95 billion gallons (11.2 billion l) in 1997. Bottled water may seem like a luxury in the United States and other more-industrialized countries, but in less-industrialized regions, where water pipelines may not extend to poorer neighborhoods, bottled water can be a necessity.

Thermal springs discharge groundwater that has a higher temperature than normal ambient (native) groundwater (an average of approximately 50°F, or 10°C, in the United States). Thermal spring water is found only in locations where groundwater has been heated by the Earth's hot interior. Steam pressure then forces superheated groundwater back to the surface through fault zones. Warm Springs in Georgia, Hot Springs in Arkansas, Yellowstone Park in Wyoming, Glenwood Hot Springs in Colorado, and Bath, England, are famous resorts where groundwater is heated by thermal activity. It's been estimated that some thermal spring water may circulate as deep as 2.5 miles (4 km) before returning to the Earth's surface with temperatures as high as 115°F (46°C). (7)

Mud pots, such as those found in Yellowstone National Park, are similar to thermal springs. The addition of clay and other undissolved particles in suspension creates the mud effect. A *geyser* is a thermal spring that intermittently builds up pressure from thermal expansion and steam to force heated water out of its crater and into the air. Geyser water is forced up narrow plumbing systems of fissures, vents, and shafts until it reaches the surface in a wild, steamy explosion (see Figure 4.9).

The Ogallala Aquifer, an unconfined aquifer located in the central United States, is the largest groundwater aquifer in North America (Figure 4.10). Also known as the High Plains Aquifer, it stretches from South Dakota to the Texas

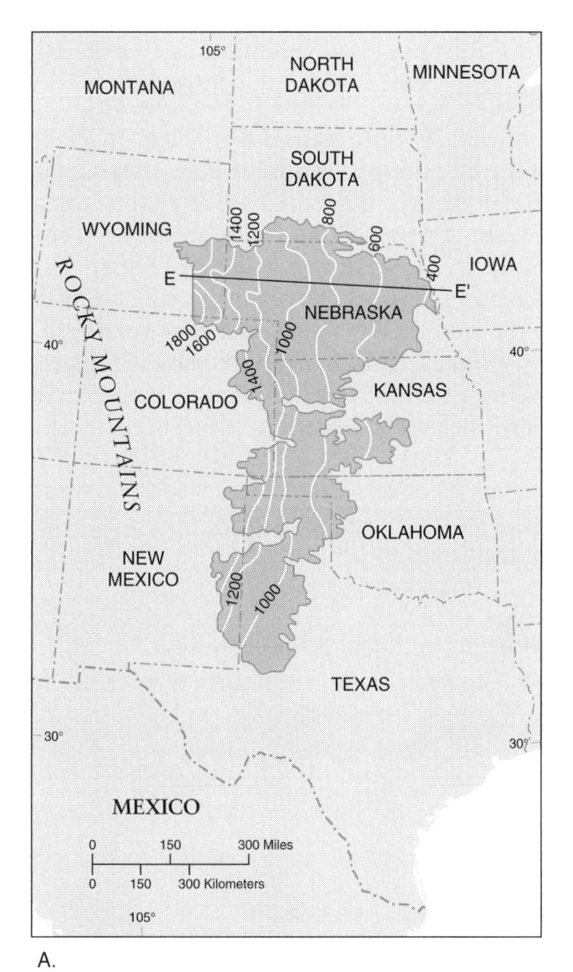

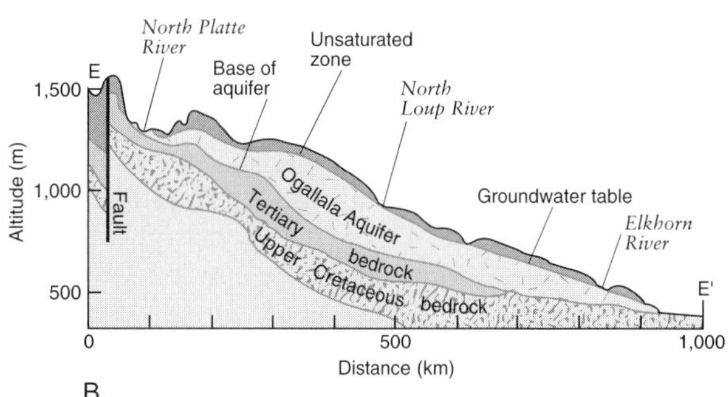

FIG. 4.10. The Ogallala Aquifer provides water to irrigators, cities, and other groundwater users in parts of South Dakota, Nebraska, Wyoming, Colorado, Kansas, Oklahoma, Texas, and New Mexico. The thickness of the aquifer varies generally from 20 to 1000 feet (6 to 305 m) while the elevation of the land surface in the region ranges from approximately 5900 feet (1800 m) in Wyoming to 1300 feet (400 m) in eastern Nebraska. Note the land surface elevation contour lines represented in A., and the corresponding profile view across Wyoming and Nebraska in B. above. The edges of the Ogallala Aquifer, particularly in north Texas, western Kansas, and eastern New Mexico and Colorado, tend to have significantly less saturated thickness than areas at the center of the aquifer in the Sandhills of central Nebraska.

Panhandle. It is composed of buried remnants of the ancient Rocky Mountains that were carried by wind and water to be deposited over millions of years. The aquifer covers nearly 175,000 square miles (453,250 km^2) and contains enough groundwater to fill Lake Huron.

The thickness of the Ogallala varies from less than a foot (0.3 m) at its edges to over 1300 feet (396 m) in central Nebraska. In the sandhills of Nebraska, the aquifer has springs that actually bubble, and some even form geysers several feet (meters) in height. In other locations, the saturated zone of the aquifer can be seen along eroded bluffs at the base of small spring-fed ponds.

Most water use from the Ogallala Aquifer is for irrigation. **Groundwater mining** has occurred in some areas where more groundwater is withdrawn than is recharged. Declines of over 100 feet (30 m) are common in many locations of north Texas, while smaller declines have occurred in South Dakota, Nebraska, Kansas, Colorado, and Oklahoma. Some irrigation wells have been abandoned because the expense of pumping groundwater from such depths exceeds the economic benefits from crop production.

PROPERTIES OF AQUIFERS

The amount of groundwater that is contained within aquifers varies tremendously with the pore spaces created between geologic materials. **Pore spaces,** or **voids,** are the open spaces found between geologic material and provide opportunities for groundwater to reside or move through under the force of gravity. Pores range in size from microscopic openings between very fine material of chalk formations (found in Great Britain) to dissolved limestone caverns (found in Kentucky). Interconnected pores provide residence sites for groundwater, and in enough quantity, they combine to form an aquifer.

Porosity is the percentage of the total volume of pore spaces within a geologic formation that can fill with water. An aquifer that contains a relatively high percentage of void space is considered to be porous or to possess a high porosity. Porosity is defined as:

$$n = V_v/T_t$$

where n = porosity
V_v = volume of void space in a unit volume
T_t = total unit volume of earth material within a geologic formation

In principle, porosity can be determined in a laboratory by taking a known volume of geologic material T_t, drying the material in an oven until all water is removed, and then submerging the dried matter in a known volume of water. The difference between the original water volume and the amount remaining after removal of geologic material is the volume of the void space. Dividing void space by the known volume gives the porosity of the material.

EXPERT ANALYSIS

Calculate the porosity of a sample of sand given the following information:

Total volume of sand = 12 cubic inches (197 cm^3)

Initial water level in a graduated cylinder = 26 cubic inches (426 cm^3)

Displaced water level after dried sand is placed in graduated cylinder = 34 cubic inches (557 cm^3)

Using
$n = V_v/T_t$
$n = (12 \text{ cu in.} - 8 \text{ cu in.})/12 \text{ cu in.}$
$n = 4 \text{ cu in.}/12 \text{ cu in.}$
$n = 0.33$ or 33%

Grains of sand of uniform size are considered to be well sorted and provide many open spaces for water to accumulate. However, if grains are poorly sorted, smaller particles of sand can fill in the pore spaces between larger grains. This intermixing results in fewer voids available for water

TABLE 4.1 Grain-Size Classification

Material	Size (inches)	Size (mm)	Example
Boulder	>12	>300	Basketball
Cobbles	3–12	75–300	Grapefruit
Coarse gravel	0.7–3	18–75	Grape
Fine gravel	0.2–0.7	5–18	Pea
Coarse sand	0.08–0.2	2–5	Water softener salt
Medium sand	0.02–0.08	0.5–2	Table salt
Fine sand	0.003–0.02	0.075–0.5	Powdered sugar
Fines	<0.003	<0.075	Talcum powder

storage. A porosity of 30 percent (often found in sand and gravel formations) provides significant storage space for groundwater. A porosity of 15 percent, found in a formation of small sands packed with finer-grained clay materials, is fairly tight and restrictive. Materials such as igneous rock or tight sedimentary material like clay or shale may have a porosity of less than 1 percent (basically a water-tight barrier). See Figure 4.11 and Table 4.1.

Clay and shale formations contain numerous porous openings, but the voids are often too small to allow movement of water. Therefore, impervious formations of clay and shale act as barriers to groundwater percolation and can greatly influence groundwater location and movement.

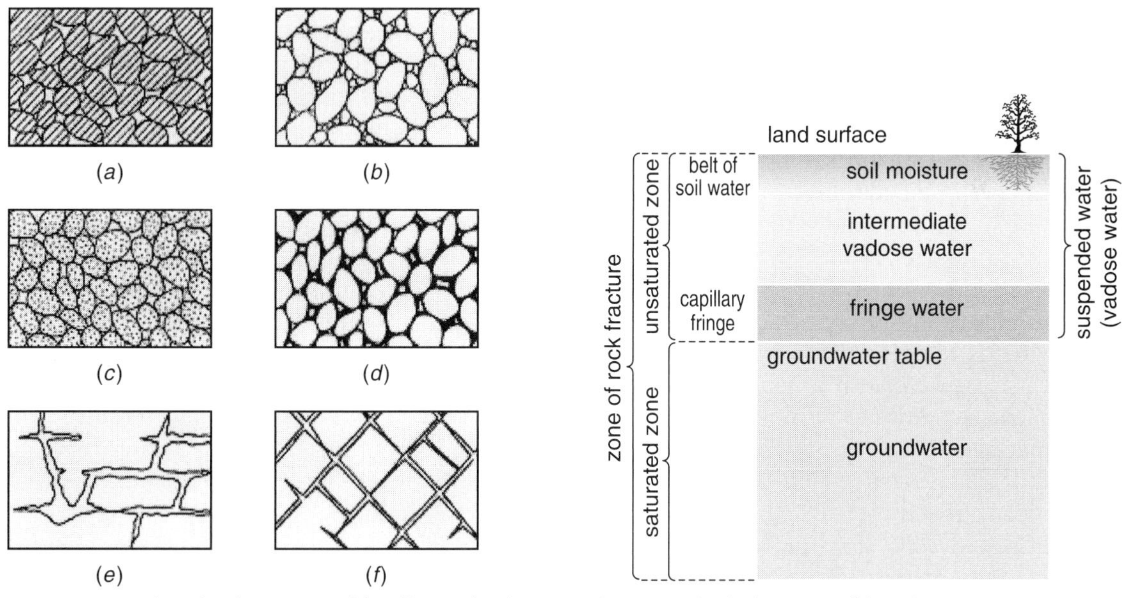

FIG. 4.11. Examples of rock interstices. *(a)* Well-sorted sedimentary deposits with a high porosity. *(b)* Poorly sorted sedimentary deposits with a much lower porosity. *(c)* Well-sorted material containing porous pebbles. *(d)* Well-sorted material that has a low porosity due to nonporous sediments found between pore spaces. *(e)* Porous rock due to solution. *(f)* Porous rock due to fracturing. These types of rock interstices form the zone of rock fracture illustrated to the right.

GROUNDWATER MOVEMENT

Through the force of gravity, groundwater naturally moves to lower elevations. However, the direction and rate of movement are determined by the lithology, stratigraphy, and structure of geologic deposits. For example, in the Great Plains of the United States and Canada, several formations of Cenozoic and Paleozoic sandstones warp up along the foothills of the Rocky Mountains. This produces opportunities for artesian conditions as surface water from higher elevations is recharged into lower sandstone formations. In other locations, groundwater movement may generally follow the topography of the land surface, and slowly move toward areas of lower elevation. A hydrologist may find conditions where the groundwater table is somewhat parallel to the land surface.

Within the intermountain basins of the western United States and Canada, alluvial valleys underlain with clay lenses and silt deposits provide excellent aquifers composed of sand and gravel. Groundwater levels within these aquifers are often closely linked to surface water levels in adjacent rivers. The direction of groundwater movement in these alluvial settings is typically in a relatively perpendicular direction toward local rivers. Why would a groundwater hydrologist be interested in understanding the direction of groundwater movement within such an aquifer?

Permeability is the ability of porous materials to allow fluids to move through it. Formations with low permeability (such as tight sands or clay) do not allow groundwater to move as rapidly as gravels that have a high permeability. Grain size, shape, and arrangement will have an effect on the ability of groundwater to move through an aquifer. Groundwater may move only a few inches (or centimeters) per *year* in clay, while it can move several feet or meters per *day* in gravel. Surface water in a river, on the other hand, travels many miles (or kilometers) in a single day.

Aquifers that contain large openings, such as the dissolved limestone formations discussed earlier, may have low porosity but high permeability. Why? Because the large openings in these aquifers (such as are found in karst formations) allow groundwater to move at a high velocity even though the formation itself is impervious. The term *underground rivers* could actually be used to describe groundwater movement in these unique geologic settings.

Hydraulic conductivity is the actual measurement of the rate of flow of a fluid through porous material. For example, the permeability of sand remains the same whether water or maple syrup is present. However, the hydraulic conductivity (or permeability coefficient) would be much slower for syrup than it would be for groundwater. Permeability is expressed as a coefficient, while hydraulic conductivity is shown as a rate of discharge in feet or meters per day (similar to the measurement of surface water movement).

Tracer tests can be used to determine hydraulic conductivity by placing a dye in a monitoring well and then measuring the time necessary for the dye to move to the next monitoring station. Several test holes generally need to be installed in fairly close proximity to intercept the slowly migrating dye if exact groundwater movement patterns and directions are uncertain.

Hydraulic head (denoted as h in hydrology formulas) is the driving force that moves groundwater. The hydraulic head combines fluid pressure and gradient, and is the height of a column of water that can be supported by water pressure at the point of measurement. (It can also be thought of as the height that groundwater will rise inside a well.) Generally, groundwater elevations are the same as the hydraulic head in a well. Groundwater always moves from an area of higher hydraulic head to an area of lower hydraulic head. Therefore, groundwater not only moves downward under the force of gravity, but it can also move laterally and upward. The actual direction of groundwater movement is dependent on local conditions.

Hydraulic gradient—the slope of the top of the groundwater table—is a function of the hydraulic head within an aquifer. The gradient indicates the direction of groundwater movement.

Hydraulic gradient is expressed as:

$$i = dh/dl$$

where i = hydraulic gradient
dh = change in head (elevation) between two points at the top of the groundwater table
dl = distance between the two points

Generally, the groundwater table will slope toward low spots on the land surface, often toward a river or lake. If the groundwater table is flat (a hydraulic gradient of 0), there will be no groundwater movement unless it is withdrawn by wells or consumed by deep-rooted plants. If groundwater has no gradient, or if an inclined groundwater table has reached an equilibrium between recharge and discharge, then an aquifer has reached a constant or steady state. This is analogous to a volume of water in a container filled with gravel and placed on a level surface. The water surface will have a hydraulic gradient of 0, and there is no water movement. If the container of gravel is tipped so that water flows over the edge, the water level will momentarily have a very slight incline or slope.

The direction of groundwater movement is also dependent on porosity and the connectivity of geologic voids. An aquifer of sand and gravel has elaborate networks of pores throughout the material. This allows unconfined groundwater to percolate downgradient (to a lower elevation) along the path of least resistance. If the aquifer is in a confined geologic setting, groundwater under pressure will seek a geologic pathway through voids in any direction. In a dissolved limestone cave, the direction of groundwater movement will be along the course of the bottom of the cave. Groundwater found at higher elevations will naturally seek exit points to land surfaces at lower elevations.

OUR ENVIRONMENT

Pollutants from industrial, urban, and agricultural sources can percolate into aquifers and then migrate (travel) great distances. The direction, speed, and extent of groundwater contamination can be predicted by groundwater hydrologists through the use of mathematical formulas. This type of analysis is vital to protect downgradient (lower elevation) wells used for drinking water and other purposes, or to determine the appropriate locations for well installation to remove contaminated water for cleanup (to be discussed in Chapter 5).

Depth of groundwater often changes with climatic conditions. For example, a prolonged drought may cause groundwater table declines owing to reduced recharge and increased groundwater use by cities, irrigators, and industry. A wet cycle could produce higher groundwater table levels because of increased recharge. Variations in shallow aquifers, as well as in hydraulically connected wetlands, ponds, rivers, and lakes, are often directly related to climatic patterns.

Transmissivity is the rate at which groundwater moves laterally through the saturated thickness of an aquifer with a hydraulic gradient of 0. Transmissivity is equal to the hydraulic conductivity of an aquifer multiplied by its saturated thickness. It is expressed as square feet (or square meters) per day.

$$T = Kb$$

where T = transmissivity (square feet or square meters per day)
K = hydraulic conductivity (feet or meters per day)
b = saturated thickness of an aquifer (feet or meters)

EXPERT ANALYSIS

Calculate the transmissivity of a confined aquifer with a hydraulic conductivity of 5.1 feet (1.6 m) per day and a saturated thickness of 196 feet (59.7 m).

Use $T = Kb$

where T = 5.1 feet per day × 196 feet
= 999.6 square feet per day,
or 92.9 m^2 per day

In 1855–1856, the French engineer Henry Darcy (1803–1858) conducted experiments (Table 4.2) showing that the water discharge through a uniform bed of sand could be expressed mathematically. His discovery was the beginning of the science of groundwater hydrology and is still in use today. **Darcy's Law** is known to groundwater hydrologists and hydraulic engineers as:

$$q = Ki$$

where q = discharge per unit area
K = permeability of the medium (or hydraulic conductivity)
i = hydraulic gradient

In 1855, Darcy designed an apparatus and used the plumbing system in a Dijon, France, hospital to test his theory. He made a tubular device 8.2 feet (2.5 m) in height, with a diameter of 13.8 inches (35.1 cm). It was filled with sand from the Saone River, and then a hose was attached between the test apparatus and a water faucet in the hospital. The water tap allowed him to regulate the flow of water during the experiment. The bottom of the tube had a small pipe attached to a pressure gage. Darcy and his assistant packed the tube tightly with sand and then filled it with water to remove all air from the voids. Next, the height of the sand column was measured, and the faucet was turned on.

Darcy immediately encountered a small problem as every pipe in the hospital began vibrating violently when the water pressure was increased. Fortunately, the cause of the noise was determined to be simply the loose fittings in the water faucets of the facility. Deciding there were no negative effects on his experiment (or the hospital pipes), Darcy continued.

Measurements were recorded every minute after the flow of water through the packed sand became constant. The experiment was then altered by varying times, water pressures, and volumes and types of sand. Darcy discovered that the ratio between water pressure and the volume of water forced through his sand-filled device remained almost constant even as water pressure changed. This implied that groundwater movement through aquifers of uniform sands would have the same characteristics.

TABLE 4.2 Results of Darcy's Experiments, Dijon, France, October 29, 30, and November 2, 1855

Experiment Number	Duration (minutes)	Mean Flow (liters/minute)	Mean Pressure (meters)	Ratio of Volumes and Pressure	Observations
1	25	3.60	1.11	3.25	Sand not washed
2	20	7.65	2.36	3.24	
3	15	12.00	4.00	3.00	Weak movements
4	18	14.28	4.90	2.91	
5	17	15.20	5.02	3.03	
6	17	21.80	7.63	2.86	
7	11	23.41	8.13	2.88	
8	15	24.50	8.58	2.85	
9	13	27.80	9.86	2.82	Very strong oscillations
10	10	29.40	10.89	2.70	

Source: "The Public Fountains of the City of Dijon," Report to the City of Dijon, Henry Darcy, Inspector General of Bridges and Roads, 1856.

A CLOSER LOOK

Henry Philibert Gaspard Darcy (Figure 4.12) was born in Dijon, France, in 1803 and attended L'Ecole Polytechnique (Polytechnic School) and L'Ecole des Ponts et Chaussee's (School of Bridges and Roads) in Paris where he was an outstanding student. Soon after graduation, he began working on the water supply system for his hometown of Dijon.

Water systems in nineteenth-century France were abhorrent. In Paris, the River Seine was a public sewer, and conditions during the summer months were unbearable. Conditions in Dijon weren't much better. The city relied on groundwater, but the wells often went dry or sewage wastes contaminated the aquifer. In addition, the drinking water gave off terrible odors.

In 1844, Henry Darcy completed a new water delivery system for his hometown. It started with water from a 2000 gallon per minute (8 m^3 per minute) spring at Rosori. From there it flowed through a 7-mile (11 km) underground aqueduct that delivered spring water to a covered 1.5 million gallon (5678 m^3) reservoir. Buried distribution lines of over 17 miles (27 km) provided clean water to public fountains, hospitals, and major buildings throughout the city. A total of 142 public street fountains were also installed 300 feet (91 m) apart. The fountains provided the first dependable water supplies for the citizens of Dijon.

The townspeople were thrilled. Darcy received commendations from the Municipal Council and a bouquet of flowers from his workmen. Later, the City of Dijon provided him with free water for life. Unfortunately, Darcy died just a few years after his famous experiment in the hospital at Dijon. (8)

FIG. 4.12. Henry Philibert Gaspard Darcy—the discoverer of Darcy's Law for flow in a porous medium.

EXPERT ANALYSIS

How do groundwater hydrologists use Darcy's Law? One of the most frequent and basic methods is to determine the natural movement of groundwater through an aquifer. Suppose an aquifer of uniform material has a permeability *(K)* of 180 feet per day and a hydraulic gradient *(i)* = 10 feet/1000 feet = 0.01. What is the specific discharge per unit area *(q)* in this aquifer? Using Darcy's Law:

$$q = Ki$$

$$q = 180 \text{ feet/day} \times 0.01 = 1.8 \text{ feet/day } (0.5 \text{ m/day})$$

This is the specific discharge per unit area of the aquifer. We then must divide this result by the porosity of the aquifer since not all of the cross-sectional area in the aquifer is open to water movement. Thus, the actual groundwater velocity (v) will be $v = q/n$ where n is porosity. If we use 33% for the porosity, $v = (1.8$ feet/day$) \div 0.33 = 5.5$ feet/day (1.7 m/day). Note that the actual groundwater velocity is much higher than the specific discharge because water can only move through pore spaces and not the entire cross-sectional area of an aquifer.

In one year, groundwater in this formation would move 2008 feet (5.5 ft/day × 365 days, or 612 m).

Specific yield of an unconfined aquifer is the ratio of the water that will drain freely from the geologic material to the total volume of the formation.

$$Y = V/T$$

where Y = specific yield
V = volume of water released
T = total volume of aquifer

Specific yield is always less than porosity because it is impossible to remove every drop of groundwater from an aquifer. The relationship of specific yield to porosity depends on the size of particles in a formation. The specific yield of a fine-grained aquifer will be small, while coarse grains will yield greater amounts of water (see Table 4.3).

TABLE 4.3 Specific Yield

Material	Maximum %	Minimum %	Average %
Clay	5	0	2
Sandy clay	12	3	7
Silt	19	3	18
Fine sand	28	10	21
Medium sand	32	15	26
Coarse sand	35	20	27
Gravelly sand	35	20	25
Fine gravel	35	21	25
Medium gravel	26	13	23
Coarse gravel	26	12	22

Source: A. I. Johnson, "Specific Yield—Compilation of Specific Yields for Various Materials," U.S. Geological Survey Water-Supply Paper 1662-D, Washington, DC, 1967.

POLICY ISSUE

Both the direction and speed of groundwater movement are extremely important in many facets of groundwater hydrology. For example, the states of Colorado and Nebraska share surface water supplies from the South Platte River as it flows from the plains of northeast Colorado into the panhandle of Nebraska. However, groundwater movement from the South Platte Alluvial Aquifer in Colorado, under the state line into Nebraska, was not addressed in an agreement between the two states (called an interstate compact) in 1923. As a result, in recent years disagreements have arisen between water officials in the two states over water delivery requirements for the endangered whooping crane in central Nebraska (to be discussed in Chapter 12).

AGE OF GROUNDWATER

Groundwater can remain underground for a few days, years, centuries, and up to many thousands of years depending on geologic conditions and pumping by wells. By contrast, water in a surface stream may completely replace itself within just a few weeks. The period of time that groundwater remains in an aquifer is called its **residence time.** You'll note in Table 4.4 that the residence time of groundwater can vary from weeks to thousands of years. Refer back to Table 2.1 in Chapter 2, and compare the residence times listed in Table 4.4 to the various water storage locations shown in Table 2.1. In Chapter 5, we'll discuss the implications of residence time, location of water storage, and water quality.

SIDEBAR

Age dating has determined groundwater beneath the Sahara Desert to be between 20,000 and 30,000 years old. (9) This groundwater was probably recharged during the more humid Pleistocene epoch in the region. This explains the availability of groundwater in desert regions for *qanats* as discussed in Chapter 1.

Tritium (^3H), a radioactive isotope of hydrogen with a half-life of 12.4 years, can be used to determine the age of water that has been underground since 1953. That year hydrogen bomb explosions filled the atmosphere with tritium, a byproduct of the testing program. Prior to the explosions, naturally occurring tritium resulted in precipitation with only 2 to 4 tritium units (TU). After the testing began, levels greater than 10 to 20 TU were common.

TABLE 4.4 Estimated Residence Time of the World's Water Supply

Water Type	Residence Time
Oceans and seas	4000 years (approx.)
Lakes and reservoirs	10 years (approx.)
Swamps	1–10 years (approx.)
Rivers	2 weeks
Soil moisture	2 weeks–1 year
Groundwater	2 weeks–10,000 years
Icecaps and glaciers	10–1000 years
Atmospheric water	10 days

Source: Adapted from R. Allen Freeze and John A. Cherry, *Groundwater* (Englewood Cliffs, NJ: Prentice-Hall, 1979), 5.

The approximate age of groundwater can be determined by measuring levels of tritium in water samples. Levels above 10 to 20 TU indicate that the water was exposed to the atmosphere after 1952. Lower levels identify groundwater that was not exposed to tritium and was underground prior to the start of hydrogen bomb explosions. (10)

Radiocarbon analysis can also be used to date groundwater. Carbon activity can be measured to determine the time when surface water percolated below ground. Since the half-life of Carbon 14 is 5730 years, a water sample with one-fourth the original carbon activity has an elapsed time of two half-lives, or 11,460 years underground. Adjustments must be made if carbonate or other organic materials are found in the groundwater to be tested since it can alter results.

LOCATING AND MAPPING GROUNDWATER

Geologic history provides clues to the sources of groundwater. If a region has been glaciated, aquifers probably contain boulders, gravel, sand, and fines. The depth and quantity of groundwater may vary greatly, and some groundwater might be found beneath hills or within valley fill materials. A region with alluvial formations of sedimentary material can provide obvious locations to find groundwater. Such a formation would be composed of materials of uniform size, and the groundwater table would probably be somewhat horizontal to the water surface of an adjacent river. By contrast, a fractured rock region in mountainous terrain would be a much more difficult place to find groundwater. Numerous drilling attempts may be required before adequate supplies of groundwater are found in fissures.

Information regarding the elevation of a groundwater table (also called the **potentiometric surface** or *piezometric surface* of an aquifer) can be determined by measuring the depth to groundwater in surrounding domestic water wells, irrigation wells, or groundwater monitoring wells. Monitoring wells are typically installed in areas that have no water supply wells available or in locations where additional groundwater data are needed. Occasionally, monitoring wells are "nested" for water quality studies. Nesting occurs when several wells are installed in close proximity but are drilled to different depths. For example, monitoring well A may have a depth of 10 feet (3 m), monitoring well B is at a depth of 15 feet (4.6 m), monitoring well C at 20 feet (6 m), and so on. This allows a groundwater researcher to collect groundwater quality samples at various depths through the saturated thickness of an aquifer.

Once groundwater data are collected, a **potentiometric map** of an aquifer can be produced to indicate the direction of groundwater movement within an aquifer (Figure 4.13). Well locations are plotted on a base map along with surface water features such as ponds, lakes, wetlands, and rivers. Land elevations are noted at each well location with data from either a global positioning unit (GPS) or a topographic map. Groundwater table elevations are also listed next to each well shown on the map. Contours can then be drawn between measurement points to give a general description of groundwater table elevations and hydraulic gradients. Readings will generally correspond to topographic features if the wells are within an unconfined aquifer.

Areas with a shallow gradient will have groundwater contours spaced far apart. A steep hydraulic gradient will be reflected in contours that are close together (similar to a topographic map of a land surface). A groundwater table map for a confined aquifer will generally not correspond with surface features since confined groundwater is under pressure and has different properties than unconfined groundwater. It is not unusual for changes in atmospheric pressure to produce large fluctuations in wells drilled into confined aquifers. Therefore, accurate measurements can be difficult to obtain on a day-to-day basis in confined systems.

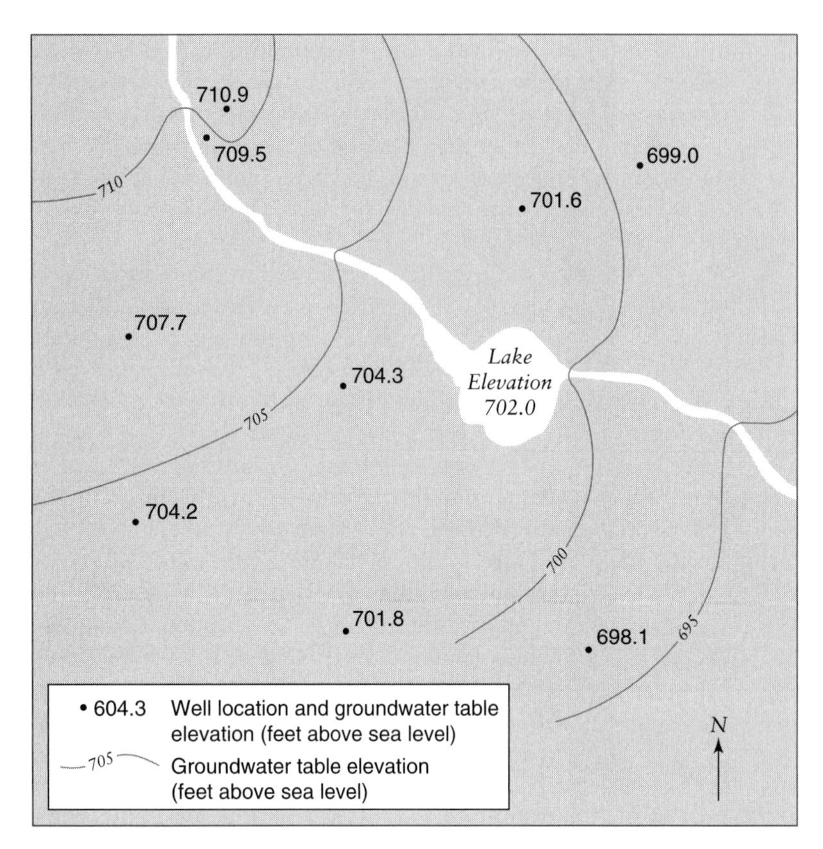

FIG. 4.13. Surface water bodies can be an important feature of a potentiometric map if they are hydraulically connected to the groundwater table. Note that groundwater contours cross the stream in this example, forming a "V" that points upstream. This indicates a gaining or "effluent" river. If the "V" on the contours pointed downstream, the river would be a losing or "influent" system.

Computer software programs are used to model (generate maps based on mathematical equations) in order to show changes in groundwater elevations over time. This information can be extremely valuable in determining total groundwater in storage and groundwater use trends. The field of groundwater modeling is a highly specialized area of study generally performed by groundwater engineers, hydrologists, or computer scientists.

A CLOSER LOOK

Finding groundwater can be both an art and a science. Around 30 B.C., the Roman author Vitruvius wrote extensively on prospecting for groundwater. He recommended digging in locations where the mist rose from the ground in the early morning. He also noted that the quality and quantity of groundwater could be predicted based on surface topography and local geology.

Water dowsing is a technique that has been used for centuries to search for groundwater. Even today, many claim they can locate groundwater by holding a forked stick, metal rods, or a swaying weight at the end of a string. A particular movement, or pattern of movements, in the instrument signals the presence of groundwater to the dowser. During the 1977 drought in California, for example, a suburban water well was reportedly located using a bent coat hanger. (11)

In the late 1950s, an anthropologist and a psychologist conducted a well-documented study to assess the extent and credibility of water dowsing. The researchers surveyed county agricultural extension agents in the United States and found an average of 181 water dowsers per one million population. More were located in rural areas, particularly in regions where groundwater was difficult to locate. Although the authors of the study found no convincing scientific evidence that water dowsing worked, they concluded that it provided a process and belief system that relieved anxiety over groundwater availability and shortages. (12)

DRILLING A GROUNDWATER WELL

Groundwater wells are a vital component of many irrigation systems (in addition to providing drinking water, livestock watering, and urban uses). A well can be used to pump water into farm ditches, gated-pipe, lateral sprinkler systems, or center pivots (discussed in Chapter 6).

There are four main steps in constructing a well: (1) drilling; (2) casing; (3) developing; and (4) pump installation. The most common drilling methods are the auger, fluid, and percussion methods. A drilling rig mounted on a truck is used for each of these methods and usually requires a team of two or three technical people to complete the drilling operation (see Figure 4.14).

The **auger drilling** method uses a bit attached to the end of a rotating column of pipe to drill and grind through geologic material. A drill bit has cutting "teeth" generally made of steel; for very hard geologic material, the teeth may be carbon, titanium, or diamond tipped. A hollow opening in the center of the bit allows lubrication with drilling mud. A well drilling crew uses a large gasoline or diesel engine, mounted on the drill rig, to supply energy to rotate and lower the bit attached to the end of drilling pipe. As the drill bit rotates and moves downward creating a bore hole, the drilling crew adds more lengths of 20-foot (6-m) pipe. This "string" of pipes is increased in length until the desired well depth is reached, often hundreds of feet (over 100 m) below ground. Typically, a well is drilled to the base of the saturated thickness of a water-bearing formation.

Drilling mud is forced inside the pipe stem, down to the drill bit, and back up to the land surface in a constant cycle. This recirculation of drilling mud cools the drill bit and forces tailings (geologic material) from the bore hole through the *annulus* (the space between the drilling pipe and the outside of the bore hole) up to the land surface. Drilling mud also exerts pressure on the walls of the bore hole and prevents it from caving in during the drilling process.

FIG. 4.14. Well drilling rigs like this one are used around the world to tap precious groundwater supplies.

One member of the drilling team, called the *mud logger,* monitors tailings produced during the drilling process. This record (or log) gives a relatively accurate cross section of the formation being drilled and the elevation of the groundwater table. The landowner generally decides how deep to drill below the groundwater table since the cost of a well is based on the number of feet drilled. The ideal situation is to drill down to shale or bedrock, but this is not always economical if the depth to the bottom of an aquifer is too great. Cost of drilling a well varies, but $80 to $100 per foot ($262 to $328 per m) is common for a completed, 8-inch-diameter (20 cm) irrigation well to depths of 100 to 250 feet (30 to 76 m).

Fluid drilling is similar to the auger method except that a high-pressure system of air or water, rather than a drill bit, is used to cut through geologic material. The **percussion drilling** method uses hydraulic pressure to hit a sharp rod against material, somewhat like a jackhammer breaking concrete. The method is simply an advanced method of the procedure used by the Carthusian monks of Artesia (discussed earlier in this chapter) to drill their artesian well in France in A.D. 1126.

After the desired depth is reached, well casing is installed. **Casing** is a plastic or steel pipe installed inside the bore hole to prevent the sides of the well from caving in. The bore hole is generally 2 to 3 inches (5 to 8 cm) larger than the casing to be installed. This extra space allows the well driller to place grout (a type of cement impervious to water) around the top section of the well, thereby preventing surface water contamination from percolating around the outer wall of the well casing and polluting groundwater. Many states require that wells be cased to specified depths for water quality purposes.

The bottom section of the casing, called the **well screen** (see Figure 4.15), is perforated with numerous narrow slots (openings). This allows groundwater to move into the well casing but prevents most sand and gravel from entering the casing when the well is pumped. A well screen also prevents excessive unconsolidated material from being removed around the casing, which could cause the surrounding formation to subside (cave in). This could lead to collapse of the well or at least could damage the moving parts of a pump. The bottom 10 to 20 feet (3 to 6 m) of a well casing is usually screened, but 50 to 100 feet of screen (15 to 30 m) is not uncommon. The length of screening generally depends on the saturated thickness of the aquifer. Screen slots vary in size based on the grain size of aquifer material found during the drilling process. A 0.50-slot screen, for example, has openings of fifty thousandths (0.05) of an inch (0.127 cm).

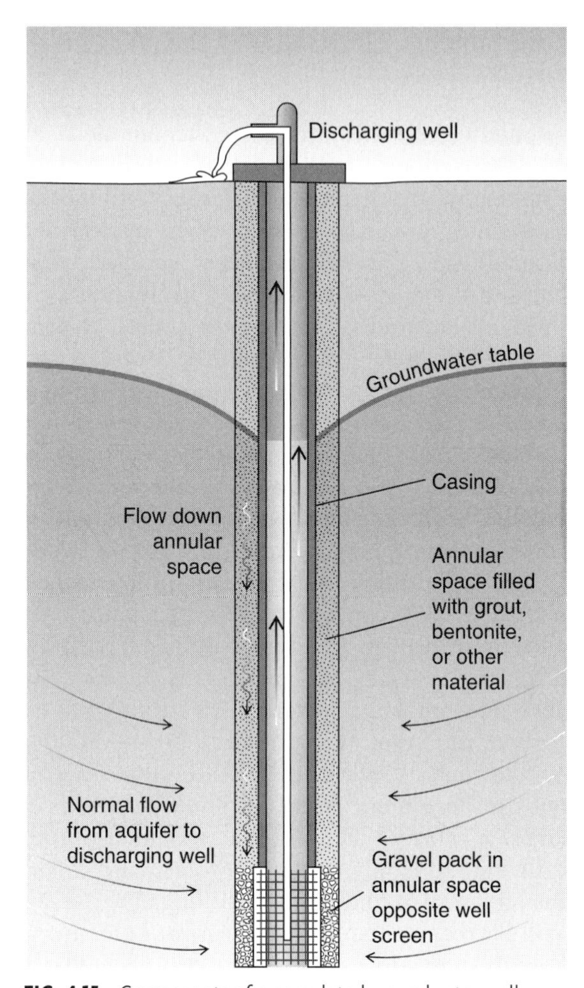

FIG. 4.15. Components of a completed groundwater well.

The outside of the casing is often packed with large gravel. This prevents excessive pumping of sand and small gravel into the well and helps keep screen openings clear of sand buildup that could clog and plug the well screen slots.

After drilling is completed, the next step is to develop the well. This is accomplished by pumping or bailing (lifting groundwater out of the bore hole with a special bucket) to clean sediments, tailings, and other geologic material left inside the well. This process increases the transmissivity of the aquifer in the immediate proximity of the well screen by clearing pore spaces of the aquifer.

The final step in constructing a well is pump installation. The pump assembly generally consists of a power plant (pump), which is installed above the well casing on a concrete pad at the land surface. The pump rotates a shaft that turns impellers located near the base of the well inside the well casing. The impellers are similar in appearance to the propeller on a boat and are housed inside pump bowls (large cast iron cases). The pump turns a shaft that spins the impellers at a high rate of speed to lift water from the bottom of the well to the land surface. The power plant is usually operated by gasoline or diesel fuel engines, or electric motors. The cost of a complete pump assembly can run approximately $10,000, depending on the horsepower of the power plant selected.

Pumping rates of groundwater can range from less than 15 gallons per minute (less than 57 l per minute) to over 3000 gallons per minute (11,356 l per minute). Pumping rates depend on the size of the well and power plant, depth to groundwater, and transmissivity of the aquifer. The cost of pumping groundwater is based on the amount of energy needed to "lift" groundwater to the surface. **Lift** can be defined as the total distance between the land surface and the depth to groundwater, within the well casing of a well being pumped, after the area of groundwater being pumped has reached equilibrium. Horsepower is the common power unit used when discussing the energy required to lift water.

SIDEBAR

It takes 1 horsepower unit of energy to lift 2 cubic feet of water a vertical distance of 1 foot per second if the pumping plant is 100 percent efficient. If the pumping plant is only 50 percent efficient, it will require 2 horsepower units to life 2 cubic feet of water a vertical distance of 1 foot per second.

Pump efficiencies vary based on friction and heat, similar to a car engine that runs less efficiently if it needs a tuneup or the tires are low on air. Pumping plants never run at 100 percent efficiency due to friction, engine wear, temperature, and other normal factors that affect mechanical engines. The efficiency of moving water will also be affected by friction created within the well casing, bowls, and impellers. Other factors that affect pumping efficiency include plugged well screens, worn impellers, and reduced aquifer transmissivity. Pumping plant efficiencies of 70 percent are very good, while 20 percent efficiency is considered very poor.

Pumping plants with greater horsepower needs require more energy to operate. Pumping groundwater from a depth of 80 feet (24 m) will require double the energy costs of a well operating at a depth of 40 feet (12 m). (See Table 4.5.) Therefore, static water levels (the depth to the groundwater table) and drawdown are extremely important factors to pump operators.

TABLE 4.5 Horsepower Required to Lift Different Quantities of Water to Elevations of 10 to 80 Feet[a]

Gallons per Minute	Cubic Feet per Second	10 ft.	30 ft.	50 ft.	80 ft.
100	0.22	0.5	1.5	2.5	4.0
200	0.45	1.0	3.0	5.0	8.1
300	0.67	1.5	4.6	7.6	12.1
400	0.89	2.0	6.1	10.1	16.2
500	1.11	2.5	6.7	12.6	20.2
600	1.34	3.0	9.1	15.2	24.2
700	1.56	3.5	10.6	17.7	28.3
800	1.78	4.0	12.1	20.2	32.3
900	2.01	4.6	13.6	22.7	36.4
1000	2.23	5.0	15.2	25.2	40.4
1250	2.78	6.3	18.9	31.6	50.5
1500	3.34	7.6	22.7	37.9	60.6

[a]All figures are for a pumping plant efficiency of 50 percent.

Source: A. S. Curry, *New Mexico Agricultural Experiment Station Bulletin 237,* New Mexico State University, Las Cruces, New Mexico, 1937. Reprinted with permission.

EXPERT ANALYSIS

Determine the cost of pumping an irrigation well for 60 days if the static groundwater level is 65 feet (20 m), drawdown is 15 feet (5 m), the pumping rate is 1500 gallons per minute (5678 l per min), pumping plant efficiency is at 50 percent, and the cost of a single horsepower unit is \$0.50 per day. First, determine the lift: 65 feet + 15 feet = 80 feet. (The static groundwater level was 65 feet below the land surface, and well pumping created a cone of depression that lowered the groundwater level within the well casing an additional 15 feet.)

Next, use an appropriate table to determine the horsepower requirement for a pumping rate of 1500 gallons per minute, with a 50 percent pumping plant efficiency from a depth of 80 feet (24 m). Table 4.5 provides a requirement of 60.6 horsepower.

Finally, to determine the cost, take 60.6 horsepower × \$0.50 per horsepower unit per day × 60 days = 60.6 × \$0.50 = \$30.30 per day

$$\$30.30 \times 60 \text{ days} = \$1818$$

In this scenario, it will cost the irrigator \$1818 to pump this well for 60 days. The actual cost per horsepower unit per day will vary greatly based on the type of energy used (such as diesel or gasoline fuel, or electricity) to operate the pump, and on current energy rates.

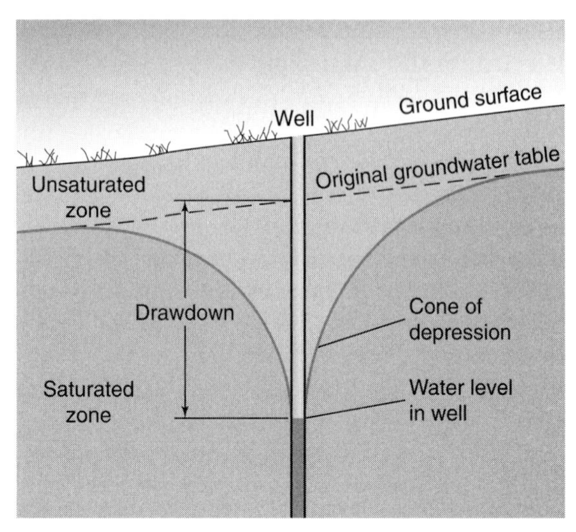

FIG. 4.16. A cone of depression is caused when a groundwater table is depressed due to pumping. Wells that are adjacent to one another can impair pumping rates due to well-to-well interference.

A CLOSER LOOK

Water in a well will reach a steady state (constant head) after the well is developed. However, once groundwater pumping begins, the water level in the well will decline. The difference between the original groundwater level and the reduced groundwater level caused by pumping is called **drawdown** (Figure 4.16). The pumping capacity of a well, and the hydraulics of the groundwater aquifer, will determine the discharge of a well. Pumping from an alluvial aquifer, for example, will create aquifer responses different from those associated with pumping from a fractured aquifer. A small drawdown means that plentiful supplies of groundwater exist in material of high transmissivity. A large drawdown can mean low transmissivity in an aquifer.

The yield (or pumping capacity) of a well divided by the drawdown is called the **specific capacity** of a well and is usually expressed as cubic feet/day/foot of drawdown or cubic meters/day/meter of drawdown. Specific capacity varies over several orders of magnitude for different geologic formations and provides an easily obtainable, useful measure of aquifer or well performance.

Groundwater pumping causes the hydraulic gradient to decline in the vicinity of the well casing and is called a **cone of depression.** This area forms in a radial pattern around the intake point of the well screen. If wells are located in close proximity, well-to-well interference can occur if pumped simultaneously. Most states have well-spacing restrictions that regulate the distance required between wells.

EXPERT ANALYSIS

Well pumping can have a direct effect on surface water if the groundwater system is hydraulically connected to a river, lake, or wetland system. **Stream depletion** is the reduction of flowing water in a river caused by groundwater pumping of a well. A **stream depletion factor** is defined as the effect of well pumping on streamflow at any given time. It is expressed as:

$$SDF = a - S/T$$

where SDF = stream depletion factor
a = distance of the well from the stream in feet (or meters)
S = specific yield of the aquifer
T = aquifer transmissivity

When a groundwater well is pumped continuously and the volume of stream depletion caused by the pumping reaches 28 percent of the volume pumped, the pumping time will be approximately equal to one SDF at the well. This analysis can be used to generate maps with contour lines (similar to topographic maps) that show the depletive effects of well pumping, located at various distances from a river, on flows in the river.

OUR ENVIRONMENT

Stream depletion factors are used in a variety of situations to assess the effects of well pumping on streamflow. In Nebraska, the U.S. Geological Survey, with the assistance of water officials in Wyoming and Colorado, are assessing the effects of groundwater pumping on streamflow in the Platte River between North Platte and Grand Island. This work is being conducted to determine the impacts of reduced streamflows on endangered species such as the whooping crane, least tern, and piping plover along the Platte River in central Nebraska.

CHAPTER SUMMARY

Groundwater can be found in a variety of geologic settings. Sedimentary rocks, alluvial landforms, glacial till, and tectonic formations can all provide suitable locations for groundwater. The hydrologic cycle provides the source of all groundwater through precipitation. Percolation from surface water runoff migrates downward as groundwater recharge. This migration eventually reaches a saturated zone or impervious geologic layer.

Groundwater may move short or long distances annually depending on geologic material underground. This movement can be predicted with a variety of mathematical formulas that were developed over one hundred years ago. Knowledge of the direction and speed of groundwater movement is critical in surface and groundwater management.

QUESTIONS FOR DISCUSSION

1. Why is it important to understand the relationship between surface water and groundwater?
2. What is groundwater hydrology?
3. What role has glacial activity played in the formation of aquifers?
4. What is the geologic area called where water collects underground, and how can we find it?
5. How did Darcy develop his theory of groundwater movement?
6. How can Darcy's Law be used to determine groundwater movement?
7. What data are necessary to study potential overpumping of groundwater?
8. Explain the differences between an unconfined and a confined aquifer.
9. How can changes in atmospheric pressure cause the water level in a well to rise?
10. Discuss how the age of groundwater is determined.
11. What is groundwater mining?
12. Discuss the process of drilling a well to obtain groundwater.

KEY WORDS TO REMEMBER

alluvium p. 89
aquiclude p. 92
aquifer p. 91
artesian spring p. 93
auger drilling p. 105
capillary action p. 91
casing p. 106
cone of depression p. 108
confined aquifer p. 92
conglomerate p. 86
consolidated rock p. 92
Darcy's Law p. 100
drawdown p. 108
fines p. 86
fissure p. 89
fluid drilling p. 106
fracture p. 89
fractured aquifer p. 92
glacial outwash p. 89
glacial till p. 89
groundwater p. 84
groundwater hydrology p. 85
groundwater mining p. 96
groundwater recharge p. 90
groundwater table p. 90
hydraulic conductivity p. 98
hydraulic gradient p. 98
hydraulic head p. 98
karst p. 87
lift p. 107
perched aquifer p. 92
percolation p. 90
percussion drilling p. 106
permeability p. 98
pore space (voids) p. 96
porosity p. 96
potentiometric map p. 103
potentiometric surface p. 103
residence time p. 102
root zone p. 91
sandstone p. 86
saturated thickness p. 92
saturated zone p. 90
sedimentary rock p. 86
shale p. 86
siltstone p. 86
sinkholes p. 87
soil p. 90
soil moisture p. 90
specific capacity p. 108
specific yield p. 101
stream depletion p. 108
stream depletion factor p. 108
thermal spring p. 95
transmissivity p. 99
unconfined aquifer p. 92
unconsolidated rock p. 92
vadose zone (unsaturated zone) p. 90
well screen p. 106

SUGGESTED RESOURCES FOR FURTHER STUDY

READINGS

Fetter, C.W. *Applied Hydrogeology.* 4th ed. Upper Saddle River, NJ: Prentice Hall, 2001.

Freeze, R. Allan, and John A. Cherry. *Groundwater.* Upper Saddle River, NJ: Prentice Hall, 1979.

McWhorter, David B., and Daniel K. Sunada. *Ground-Water Hydrology and Hydraulics.* Littleton, CO: Water Resources Publications, 1977.

Murck, Barbara W., and Brian J. Skinner. *Geology Today—Understanding Our Planet.* New York: John Wiley & Sons, 1999.

Price, Michael. *Introducing Groundwater.* 2nd ed. London: Chapman & Hall, 1996.

Rushton, K. R., and S. C. Redshaw. *Seepage and Groundwater Flow.* New York: John Wiley & Sons, 1979.

Skinner, Brian J., Stephen C. Porter, and Daniel B. Botkin. *The Blue Planet.* 2nd ed. New York: John Wiley & Sons, 1999.

Todd, David Keith. *Groundwater Hydrology.* 2nd ed. New York: John Wiley & Sons, 1980.

U.S. Department of the Interior, Geological Survey. "Ground Water." Washington, DC: U.S. Government Printing Office, 1986, 491–402/04.

WEBSITES

Fiji Mineral Resources Department, Ministry of Lands and Mineral Resources, "Fiji Groundwater," July 2003. http://www.mrd.gov.fj/gfiji/geology/educate/grndwatr.html

Environment Canada, Home Page, June 2003. http://www.ec.gc.ca/water

The Groundwater Foundation, "Groundwater Basics," June 2003. http://www.groundwater.org/GWBasics/gwbasics.htm

"Ground Water Atlas of the United States," U.S. Geological Survey, July 2003. http://capp.water.usgs.gov/gwa/

Government of British Columbia, Ministry of Water, Land and Air Protection, "Groundwater Resources of British Columbia," June 2003. http://wlapwww.gov.bc.ca/wat/gws/gwbc/C02_origin.html

UNICEF, "Groundwater, The Invisible and Endangered Resource" July 2003. http://www.unicef.org/wwd98/

U.S. Geological Survey, June 2003. http://va.water.usgs.gov/GLOBAL/AWWALAST.htm

REFERENCES

1. "New Images Suggest Present-Day Sources of Liquid Water on Mars," Jet Propulsion Laboratory, National Aeronautics and Space Administration, http://www.jpl.nasa.gov/releases/2000/marswater.html, June 22, 2000.

2. "Ground Water," U.S. Department of the Interior, U.S. Geological Survey, Washington, DC: U.S. Government Printing Office, 1993.

3. "Groundwater Basics," Groundwater Foundation, http://www.groundwater.org/GWBasics/depend.htm, September 20, 2000.

4. Todd, David Keith. *Groundwater Hydrology.* 2nd ed. New York: John Wiley & Sons, 1980, 5.

5. Flint, Richard Foster. *Glacial and Quaternary Geology.* New York: John Wiley & Sons, 1971.

6. Price, Michael. Introducing Groundwater. London: Chapman & Hall, 1985, 68–70.

7. Ibid., 167.

8. Fancher, G. "Henry Darcy—Engineer and Benefactor of Mankind," *Journal of Petroleum Technology* 8, October 1956.

9. Todd, 25.

10. Fetter, C. W. *Applied Hydrogeology.* 3rd ed. Upper Saddle River, NJ: Prentice Hall, 1994, 419–420.

11. Ibid., 426.

12. Vogt, E. Z., and R. Hyman. *Water Witching U.S.A.* Chicago: University of Chicago Press, 1959.

CHAPTER 5

WATER QUALITY

Each component of the hydrologic cycle—precipitation, surface water runoff, surface water and groundwater storage, and evaporation—changes the quality of a water body. For example, precipitation in the form of rain or snow can carry airborne pollutants to the Earth's surface; surface water runoff can cause erosion and transport sediments; groundwater recharge can leach chemicals into aquifers; and evaporation can elevate concentrations of pollutants in bodies of water by reducing the total volume of stored water. Each natural component of the hydrologic cycle can have a negative effect on surface and groundwater quality.

Humans also have a tremendous effect on water quality. All of us contribute waste to the environment through the consumption of resources such as food, clothing, housing, and fuel for transportation. The rapidly growing world population is contributing to the deterioration of our existing water quality and is creating significant challenges for water managers, industry, and fish and wildlife agencies. This chapter will explore problems and solutions that are currently addressed throughout the world.

WATER POLLUTION

Pollution (from the Latin word *pollutus* meaning "to soil or defile") can occur either naturally or through human activity. Water is considered to be polluted if it is unusable for a particular purpose. Natural processes such as chemical reactions between rocks and water, erosion and sedimentation caused by flowing water, percolation of surface water into groundwater aquifers, and the residence time of water stored in rivers, lakes, wetlands, and aquifers, can all create or compound pollution. In some locations, water is naturally of such poor quality that plants and animals cannot survive.

Unfortunately, humans have caused incredible levels of water pollution. According to the *National Water Quality Inventory: 2000 Report to Congress,* as reported by the U.S. Environmental Protection Agency, only 61 percent of the streams, lakes, and estuaries that were assessed (19 percent of all rivers and streams, and 43 percent of all lakes, ponds, and reservoirs in the United States) met the water quality standards evaluated. Leading pollutants in these impaired waters included sediments, bacteria, nutrients, and metals (primarily mercury). Runoff from urban areas and agricultural lands were the primary sources of these pollutants. (1)

The *National Water Quality Inventory* showed that 78 percent of the state-assessed shoreline miles of the Great Lakes were impaired. This classification was given, in part, because of the high level of pollutants found in fish tissue that could be harmful to human health if

Summary of the National Water Quality Inventory, Office of Water,
U.S. Environmental Protection Agency, 1998

Waterbody Type	Total Size	Amount Assessed* (% of Total)	Good (% of Assessed)	Good but Threatened (% of Assessed)	Polluted (% of Assessed)
Rivers (miles)	3,662,255	842,426 (23%)	463,441 (55%)	85,544 (10%)	291,264 (35%)
Lakes (acres)	41,593,748	17,390,370 (42%)	7,927,486 (46%)	1,565,175 (9%)	7,897,110 (45%)
Estuaries (sq. miles)	90,465	28,687 (32%)	13,439 (47%)	2,766 (10%)	12,482 (44%)

*Includes waterbodies assessed as not attainable for one or more uses.
Note: Percentages may not add up to 100% due to rounding.

FIG. 5.1. Summary of the *National Water Quality Inventory,* Office of Water, U.S. Environmental Protection Agency. "Good" means the water quality will support all designated uses. "Good but Threatened" indicates that data show a declining trend in water quality and will be impaired in the future unless action is taken to prevent further degradation. "Polluted" is a water body that does not support one or more designated uses.

eaten. The Inventory made no report on groundwater, but in 1998, it reported to Congress that groundwater quality was generally "good" in the United States (see Figure 5.1), meaning that the water quality sampled would support all designated uses. However, measurable negative impacts have been detected from leaking sources such as underground storage tanks, septic systems, and landfills. The *National Water Quality Inventory* can be found at http://www.epa.gov/305b.

In spite of continuing problems with water quality, the nation's waters have improved significantly since passage of the Clean Water Act Amendments of 1972. At that time, only a third of U.S. waters were safe for fishing and swimming. By the end of the 1990s, the figure had increased to two-thirds. Agricultural runoff has been reduced, and modern wastewater treatment facilities served 173 million people in the United States at the end of the twentieth century. (2)

Generally, water quality is classified in four or five categories—A through D—with A being the highest quality. In New York State, for example, the five-letter classification system, reflecting the actual use or intended best use of a water body, is as follows:

Class	Best Water Use
AA and A	drinking and all other uses
B	swimming/recreation
C	boating, fish propagation, and fishing
D	fishing

The purpose of these classifications is to alert the public to appropriate and safe water use activities, based on water quality, in local rivers, streams, and lakes.

POINT SOURCE AND NONPOINT SOURCE POLLUTION

Where does pollution come from, and how is it transported to rivers, lakes, wetlands, and estuaries? Pollution sources are divided into two categories: point source and nonpoint sources.

Point source pollution is generally defined as contamination discharged through a pipe or other discrete, identifiable location. Pollution from a point source is relatively easy to quantify, and impacts can be directly evaluated. **Nonpoint source pollution** is generated from broad, diffuse sources that can be very difficult to identify and quantify. Nonpoint source pollutants enter rivers, lakes, and other water bodies through surface and groundwater movement, and even from the atmosphere through precipitation.

Point source and nonpoint source pollution are caused by human activities. It is important to separate these activities from natural water quality degradation, sometimes called background pollution or natural contamination. As will be discussed, naturally degraded water quality can be caused by chemical reactions between water and metals and minerals, natural erosion, forest litter, natural migration of salts, and other normal processes of the hydrologic cycle.

Point Source Pollution Historically, concern about pollution has focused primarily on point sources of contamination, such as those described below. Since these locations are easy to identify, it is logical that water quality programs focused on these sources of pollution first. Only in the past few decades have state and federal water quality protection programs shifted emphasis to nonpoint source pollutants. Generally, in the United States, Canada, Australia, Western Europe, Japan, and other, more industrialized areas of the world, most point source problems have been identified and remediated. Currently in these areas, most aquatic pollution comes from nonpoint sources.

Factories and Wastewater Treatment Plants Common sources of point source pollutants are factories and wastewater treatment plants. Mechanization during the Industrial Age (the second half of the nineteenth century) required large labor forces and led to rapid urban growth near major manufacturing centers. Increased population created and concentrated more human wastes in a region since sewage treatment methods remained crude or nonexistent. Mechanization of factories also led to increased point source discharges of pollutants into local watersheds from manufacturing processes. Many local and regional water supplies were exploited and soon became widely polluted.

Factories were often located near waterways to dispose of wastes. In 1900, 40 percent of the pollution load of U.S. rivers was from industrial waste. By 1968, that figure had doubled to 80 percent. (3) The total urban population (cities exceeding 2500) in the United States was just over six million people in 1860 but exploded to over 200 million by 2000 (see Table 5.1). The Passaic River, for example, was used as a water supply source by the City of Newark, New Jersey, and was also a major recreation area during the 1800s. However, the Passaic became badly polluted by sewage and industrial waste from population and industrial growth in the region. Odors from the river became so bad during the summer that homes were actually abandoned. During hot weather, the stench from the Passaic River even forced many polluting factories along its banks to close. (4)

Landfills Old landfills are a common point source of groundwater contamination. Prior to the 1970s, most landfills in the United States were open sites where trash was burned without controls. During open burning, ash regularly washed offsite during storm events. Later, in the 1970s, when the federal government banned burning at landfills, refuse was simply buried beneath soil with little or no regard for groundwater contamination.

As pollutants dissolved, the solution beneath a landfill became known as leachate, a kind of

TABLE 5.1 Urban Population in the United States, 1790–2000[a]

Year	Urban Population	Percent of Total Population
1790	201,656	5
1860	6,216,518	20
1920	54,263,282	51
1990	187,053,487	75
2000	222,353,453	79

[a]Cities with populations exceeding 2500.

Source: U.S. Census Office, Washington, DC, http://www.census.gov/population/censusdata/table-4.pdf, June 2003.

garbage "tea." Leachate often contains toxic metals and other hazardous wastes from improperly disposed paints, solvents, household cleaners, oil, and other illegally disposed items. Precipitation carried the leachate downward into groundwater sources and often created an underground zone of contamination called a **plume.** Such contamination plumes often slowly migrate (move) with groundwater in local aquifers, depending on its residence time (discussed in Chapter 4). This movement of groundwater can be as fast as a few feet (meters) per *day* or as slow as a few feet (meters) per *year.*

New federal regulations require that no dumping can occur in landfills unless disposal sites meet rigid federal standards to prevent surface and groundwater contamination. Many abandoned landfills that continue to leach contaminants are in the process of being cleaned up, while new ones are rigorously being monitored to minimize the likelihood of contaminant release and to comply with various government regulations. (5)

Abandoned Mines Mine tailings and abandoned mines are significant sources of point source pollution. Mining operations, such as silver, gold, and other ore extraction operations, often leave waste piles or heaps of mined rock adjacent to mines or smelters. These natural materials can dissolve through normal processes of the hydrologic cycle and can leach into groundwater or be transported by runoff into nearby streams.

Acid mine drainage (water pollution caused by water percolation through mines or mine tailings) can have lethal effects on aquatic plants and fish of receiving streams. (6) Water flowing from an abandoned coal mine, for example, can have a pH of 2, which is very acidic and would kill fish and wildlife downstream. If we used the logarithmic scale of pH, water in this example would be 100,000 times more acidic than neutral water with a reading of 7. (7)

Underground and Above-Ground Storage Tanks Underground storage tanks (USTs) are used to store fuel at gasoline stations, and solvents and other industrial liquids at factories. Storage tanks are often made of steel and can corrode after many years of operation, thus becoming leaking underground storage tanks (LUSTs). Thousands of LUSTs and noncompliant tanks have been replaced over the past 20 years at a high cost to businesses but of great benefit to human health and the environment. The USEPA implements extensive monitoring programs to regulate the integrity of underground and above-ground storage tanks. (8)

Nonpoint Source Pollution Nonpoint sources of pollution enter waterways as overland flow, surface water percolation into groundwater, and other diffuse sources. These sources of pollution often occur as a result of chemical use on lawns, gardens, and golf courses, improper agricultural practices, street refuse, construction activities, and dredging of rivers and reservoirs. This pollution can be very difficult to identify, quantify, and regulate since nonpoint sources are caused by such broad, general uses.

Lawns, Gardens, and Golf Courses Lawns, gardens, and golf courses are major sources of nonpoint source pollution. Excessive or improper use of fertilizers and pesticides, inadequate removal of pet refuse, and other pollutants can migrate into aquifers or be carried away by street gutters to streams, lakes, and reservoirs.

Agricultural Practices Agriculture is another potential major contributor of nonpoint source pollutants to a watershed. Examples include soil erosion, fertilizer and pesticide leaching into groundwater or adjacent rivers, and leaching of salts from soils from irrigation. Animal feedlots can be sources of pollution from runoff of animal manure or from excessive application of wastes on cropland.

Street Refuse Organic pollution from oil and gasoline, rubber from tires, asphalt compounds, pesticides, fertilizers, lawn clippings, heavy metals, and bacterial loadings from dust can all be generated from street refuse. Highway salting, though reduced in some areas, often migrates to rivers or into groundwater after snow and ice begin to melt. (9)

Construction Activities Soil erosion at construction sites can cause increased sedimentation in local water bodies. Improper storage of construction materials, chemicals, and fuels can also become sources of pollution due to construction site runoff. Proper construction site management can greatly reduce nonpoint source pollution runoff into local water bodies.

Stormwater Runoff Surface water runoff from precipitation events flush large amounts of pollutants—including chemicals, fertilizer, trash, and other waste—from streets, construction sites, agricultural fields, golf courses, factories, and sediment from erosion. All of these pollutants enter water bodies at a variety of locations—sewer systems, overland flow, drainage ways, and so on.

Dredging Activities Dredging activities in rivers and lakes can also cause water quality problems. Pollutants often become sequestered, or bonded, to bottom sediments and can reside there for decades. Dredging can stir up these materials and often causes them to be transported downstream or to be ingested by fish. Deposition of suspended sediments in reservoirs and temperature changes of deeper waters created by dredging can all affect water quality. (10)

BASIC PARAMETERS OF WATER

The basic chemical and biochemical processes that affect water quality are the result of nature. Long before humans settled along the banks of rivers such as the Yangtze in China and the Nile in Egypt, sediment-laden floods carried metals and minerals that contributed to poor quality of water. Ancient floods of the Mississippi River filled adjacent oxbow lakes and marshes with organic materials such as decaying plants and animals. The aridity of the Colorado River watershed caused salt from alkaline soils to enter the river for thousands of years before human cultivation began in Mexico and the United States. Groundwater in certain regions, or at great depths, contained dissolved minerals that rendered it unfit for human consumption. These natural processes greatly affected water quality around the world long before the negative influences of humans.

TEMPERATURE

Many physical, biological, and chemical characteristics of surface water are dependent on temperature. Excessive temperature changes can accelerate chemical processes and can be detrimental to aquatic plants and wildlife. Increased heat in water can reduce its ability to hold dissolved oxygen, while sudden temperature "shocks" (often caused by heated industrial water released into a lake or stream) can be deadly to many aquatic species. Removal of shade trees and shrubs along a shoreline can also affect the temperature of a water body, particularly during warmer seasons of the year. Fish respond to water temperature variations and often move to new locations when temperature changes vary by little more than 1 to 4°F (1 to 2°C). (11)

Water temperature is greatly affected by depth. Surface water is generally much colder at greater depths than shallow water, since it requires more time to absorb heat. Such temperature variations can cause lakes to "turn over" in the spring and fall (see Chapter 3), creating variable water quality characteristics. By contrast, groundwater at depths less than 300 feet (91 m) generally maintains a constant temperature of approximately 50°F (10°C) in the United States, while surface water in lakes can range between a frozen state to 70°F–80°F (21°C–27°C) and higher during the summer.

DISSOLVED OXYGEN

Oxygen comprises about 21 percent of the atmosphere but only a fraction of 1 percent of water. Where atmosphere and water meet, the great difference in proportions causes oxygen to become dissolved in water. **Dissolved oxygen (DO,** pronounced **dee-oh**) is comprised of microscopic bubbles of oxygen gas (O_2) in water and is critical for the support of aquatic plants and wildlife. DO is produced by diffusion from the atmosphere, aeration of water as it passes over falls and rapids, and as a waste product of photosynthesis. It is affected by temperature, salinity, atmospheric pressure, and oxygen demand from aquatic plants and animals. Dissolved oxygen is measured in parts per million (ppm), milligrams per liter (mg/L) or percent saturation.

Most aquatic plants and animals need dissolved oxygen in water to survive. Species such as rainbow trout *(Salmo gairdneri)* require medium to high levels of DO, while warm-water fish such as blue catfish *(Ictalurus furcatus)* or carp *(Cyprinus carpio)* require lower concentrations. High levels of dissolved oxygen allow a variety of aquatic organisms to thrive. In most areas, dissolved oxygen levels are often the single most important measure of habitat quality. When DO concentrations drop below 5 mg/L, sensitive organisms such as coldwater fish can become stressed, especially if these conditions prevail for long periods of time. Bottom-dwelling organisms, such as worms, are usually more tolerant, and some species can survive at levels as low as 1 mg/L in some cases. However, many visible living organisms cannot survive in waters with DO levels of less than 1 mg/L for more than a few hours.

Elevated levels of dissolved oxygen also make drinking water taste better but can be corrosive to metal water pipes.

pH

pH, a unit of measure that describes the degree of acidity or alkalinity of a solution, is one of several primary indicators of water quality. The pH scale (p for power and H for the chemical symbol for hydrogen) refers to the power or concentration of hydrogen ions (atoms) in water. pH values generally range from 0 (very acidic with a high concentration of positive hydrogen atoms, H^+) to 14 (very alkaline, or basic, with a very high concentration of negative hydroxyl ions, OH^-). However, pH values can range into negative numbers and above 14 in extreme situations. A pH of 7.0 represents exact neutrality of water at 46°F (8°C) where the positive hydrogen atoms and negative hydroxyl ions are in equilibrium.

pH is reported in logarithmic units, similar to the Richter Scale used for earthquakes. Each number represents a tenfold change in acidity or basicness of a solution. Water with a pH of 4 is 10 times more acidic than water with a pH of 5 (Figure 5.2). pH can be measured in a variety of ways, including the use of pH paper, a pH pen, and pH meters. pH paper changes color with varying degrees of acidity; a pH pen is a simple electrode that measures the electrical potential of hydrogen ions in the water sample; and a pH meter measures voltage and contains a reference electrode that provides a constant electric potential. A pH meter also includes a temperature compensation device since pH is temperature dependent.

HEALTH CONCERNS

Normal ranges of pH in drinking water do not have a direct effect on human health. Carbonated beverages such as soft drinks have an expected pH value of 2.0 to 4.0. Foods such as apples (2.9 to 3.3) are also on the acid side of the pH scale. The maximum and minimum allowable ranges of pH for drinking water have been set at 6.5 to 8.5 by the World Health Organization and the U.S. Safe Drinking Water Act.

Water treatment plant operators alter pH in drinking water by adding slightly acidic or slightly alkaline (basic) chemicals. A pH between 7.0 and 7.8 is generally targeted to avoid corrosion problems in distribution pipes throughout the delivery system. Corrosion can cause stripping of metal pipes, which introduces contaminants into drinking water even if mildly acidic water is present. Lead pipes are of particular concern because lead is a toxic metal that can cause severe health problems in humans.

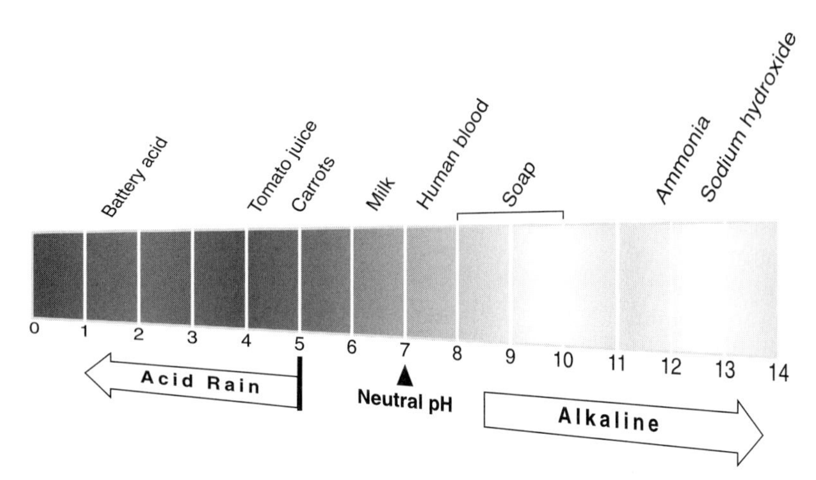

FIG. 5.2. pH is a measure of the amount of hydrogen ions in solution and is equal to the negative logarithm of the hydrogen ion concentration. The term *pH* is from the French *pouvoir hydrogene,* meaning "power of hydrogen." The common pH scale extends from 0 to 14 and covers the range of pH of most common solutions. A few extremely acidic solutions, such as concentrated battery acid, can have a negative pH, and a few highly caustic (basic) solutions have pH values of 15 or higher.

Raw water found in rivers and lakes generally has a pH between 4 and 9, while pure distilled water is at 7. Fish have a narrow range of pH preference that varies by species. pH levels outside this narrow range can cause severe health problems for a species. For example, the Japanese koi carp, or *Nishikigoi,* prefer a pH range between 7 and 8.5. Water outside the normal pH range for a particular species of fish can cause physical damage to skin, gills, and eyes, and, in severe cases, can be fatal.

TURBIDITY

Turbidity is the relative measure of clarity and is the result of suspended matter in water that reduces the transmission of light. It can be caused by silt, very small organic particles, salt, plankton, or decaying vegetation. The presence of these constituents can result in a "cloudy" appearance of water.

Turbidity in water is measured in units called NTUs, or Nephelometric Turbidity Units. A nephelometer, which electronically measures light scatter in water, can be used to determine turbidity. However, a simple Secchi disk is a very popular method to determine the depth of water that causes the visibility of the disk to disappear (Figure 5.3). A Secchi disk is inexpensive, simple to use, and provides accurate measurements of turbidity. A Secchi disk is 20 centimeters (8 in.) in diameter and is divided into alternating black and white quadrants to enhance visibility. The disk is tied to the end of a white nylon rope, which is marked in black every tenth of a meter (0.33 ft) and in red every meter (3.3 ft). Water with turbidity levels greater than 5 NTU is not safe for recreational use or human consumption. Levels less than 25 NTU cannot sustain aquatic life. As of January 1, 2002, the U.S.

FIG. 5.3. A Secchi disk provides valuable data on turbidity. Markings on the rope indicate depth to the scientist.

Environmental Protection Agency has set a turbidity standard for drinking water of not more than 1 NTU and not exceeding 0.3 NTU in 95 percent of the daily samples taken in any month.

Turbidity can create water quality problems because toxic chemicals can attach to suspended particles. In addition, drinking water treatment can be hindered if turbid surface water sources are used. Treatment for turbidity can remove some chemicals and waterborne diseases that bond to the fine suspended matter. (These processes will be discussed further in Chapter 11.)

HARDNESS

Hardness is the amount of dissolved calcium, magnesium, and iron present in water. "Hard" water is sometimes described as an inability to create soap suds or lather when washing. Hardness is not a contaminant but rather a characteristic of water that varies greatly around the world. Hard water is common in the United States in Florida, New Mexico, Arizona, Utah, Wyoming, Nebraska, South Dakota, Iowa, Wisconsin, Indiana, and in the Prairie Provinces and Ontario in Canada. (12)

Hard water often creates a buildup of scale on hot water heaters, showers, and porcelain surfaces. Scale is caused by calcium and magnesium, which form a solid precipitate. This buildup of minerals can clog hot water pipes, water heaters, and boilers, but for the most part is purely an economic problem. Excessive hardness can be removed through softening processes such as filtering hard water through layers of salt. However, water softeners are used less in drinkable cold water supplies because the excessive salt they contain can cause health problems. (13)

Soft water can be difficult to use for washing since soap is not easily removed from skin or other surfaces. Surface water is softer than groundwater because it has less contact with soil minerals and originated more recently as precipitation.

The degree of hardness can be described in milligrams per liter of calcium carbonate ($CaCO_3$):

0–75 mg/L = soft
75–150 mg/L = moderately hard
150–300 mg/L = hard
300 mg/L and higher = very hard

Hardness of water can be a factor in degenerative cardiovascular disease such as heart disease, hypertension, and stroke. According to the National Academy of Sciences, mortality rates increase by 15 to 20 percent for people who use very soft drinking water as compared to populations that use hard water. The exact biological reasons are not well understood, but it could be that the corrosive aspects of soft water contribute to the human ingestion of heavy metals from pipes. (14)

INORGANIC CHEMICALS

All substances are poisons: there is none which is not a poison. The right dose differentiates a poison and a remedy.

Dr. Theophrastus Hohenheim (Paracelsus) (1493–1541), Swiss physician and chemist

Chemicals that do not contain carbon are called **inorganic chemicals.** Metals and minerals, and nonmetals that do not contain carbon, fall into this category.

METALS

Metals are elements found naturally in the mineral ores of the Earth's crust. Some metals, such as calcium, zinc, and iron, can be healthful in proper quantities, but copper, lead, cadmium, mercury, arsenic, and chromium can be toxic (poisonous)—see Figure 5.4. Mining, excavation, and other construction activities can stir and expose these metals to the natural forces of the hydrologic cycle. Surface water runoff, groundwater percolation, and groundwater pumping can lead to natural chemical reactions that can be harmful to humans and wildlife. Activities that disperse ("stir") bottom sediments of

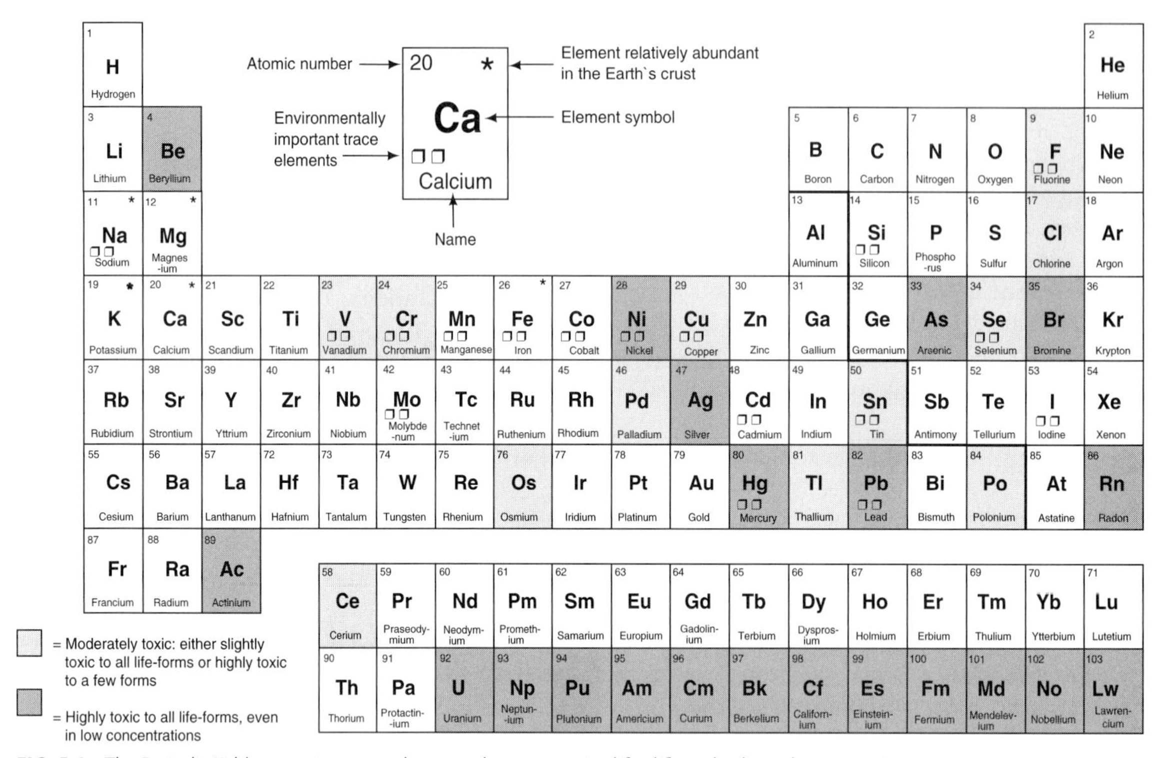

FIG. 5.4. The Periodic Table contains many elements that are required for life and others that are toxic to living organisms.

lakes or streams can cause metals to remain in suspension for long periods of time.

The production processes of factories can generate heavy metals, such as lead, copper, and zinc, and can create serious problems for human health and the environment. Since the beginning of the Industrial Age, the production of these three metals has increased tremendously, with a tenfold increase between 1850 and 1990. (15)

Naturally occurring metals in drinking water are common. Copper is present in tap water if blue or green stains are present on porcelain fixtures such as sinks or bathtubs; red or brown stains on porcelain indicate iron (rust); and black stains are probably caused by manganese.

LEAD

Lead (Pb) is a heavy, very soft metallic element used to conduct electricity and was formerly used in water pipes in homes, businesses, and industrial plants. Lead is also sometimes a byproduct of mining processes. High levels of lead in the human body (called lead poisoning) can lead to toxic reactions, brain damage, and death. Children are particularly susceptible to lead poisoning from lead-based paint and lead plumbing found in older homes.

SIDEBAR

Marcus Vitruvius Pollio (first century B.C.) was a Roman architect and engineer who authored the famous treatise *De architectura,* a 10-volume work regarding city planning and architecture. Vitruvius wrote about building materials, temple construction, clocks, and hydraulics among other topics. With regard to water pipes, he strongly believed that water was much more wholesome if stored in earthenware than if it was delivered through lead pipes. He was definitely ahead of his time regarding public health and municipal water supplies.

In the United States and around the world, lead pipes were commonly used to distribute water because of its low cost and high durability. It has been estimated that over 50 percent of U.S. cities still use lead pipes. (16) The 1986 Amendments to the Safe Drinking Water Act required the use of lead-free pipes in new or replacement construction, but existing lead pipes were not affected by the ban, owing to the high cost of replacement.

ARSENIC

Arsenic (As), a highly toxic metal found naturally in the environment and as an industrial byproduct, can be found in raw water, particularly groundwater, as a result of natural geologic and erosional processes. Arsenic is also used in wood preservatives, paints, dyes, semiconductors, pesticides, and smelting operations. (17) Arsenic can in addition be released into the air from volcanic action and forest fires. Long-term exposure to the metal can lead to cancer of the bladder, lungs, skin, kidneys, nasal passages, liver, and prostate. It can also have adverse effects on cardiovascular, pulmonary, immunological, neurological, and endocrine systems. (18)

In 1975, the USEPA set a drinking water standard of 50 ppb (parts per billion) based on a 1942 Public Health Service standard. However, a 1999 report by the National Academy of Sciences concluded that the old standard was too high to protect public health. In response to the study, in the final days of the Clinton Administration, the USEPA proposed lowering the drinking water standard to 10 ppb. Under the Safe Drinking Water Act Amendments of 1996, USEPA was required to issue a final rule on this new standard. The Bush Administration later rescinded the new standard in early 2001 and requested additional study of the economic costs and health benefits of a new federal arsenic standard for drinking water. The standard of 10 ppb was later upheld in a federal appeals court in June 2003 and will go into effect on January 23, 2006.

SIDEBAR

What's a ppm and a ppb? A ppm is parts per million and is the measure of a concentration equivalent to three drops in 42 gallons (159 l) of water. A ppb is parts per billion and is the measure of a concentration equivalent to approximately one drop in 14,000 gallons (52,990 l) of water.

MINERALS

Minerals are naturally occurring, crystalline materials that have a distinct inorganic chemical composition. Some minerals are used in their natural state (such as gold and diamonds), whereas others are refined to extract desired metals (iron, copper, and zinc).

All surface water and groundwater contain minerals. Minerals are consumed through food and water, and are needed by the human body to regulate ("turn on" or "turn off") chemical reactions of enzymes. Important minerals in the human diet include calcium, phosphorus, potassium, magnesium, sulfur, sodium, and chloride. Trace elements required by the body include iron, zinc, selenium, manganese, and chromium. Minerals may compete with each other in the body for absorption. For example, a diet rich in manganese may decrease the amount of iron absorbed by the body.

SALT

Salt is composed of sodium and chloride, and occurs naturally in soil and water. Surface water runoff and groundwater percolation from precipitation and irrigation can cause salts to **leach** (dissolve from the soil) and contaminate surface and groundwater supplies.

Salinity (the presence of excess salts in water) can be harmful to certain plants, aquatic species, and humans. High levels of salt in drinking water can lead to high blood pressure and other health concerns for humans. Saline soils can harm plants by pulling moisture out of roots and by reducing the uptake of water and fertilizer. However, plant tolerance for salt varies greatly by vegetation type. For example, Kentucky Bluegrass *(Poa pratensis)* and alfalfa *(Medicago sativa)* have a relatively low tolerance for salt in water or soils. By contrast, Tamarisk *(Tamarix gallica)* and Fourwing saltbush *(Atriplex canescens)* thrive in saline environments that would kill most other plants. (19)

POLICY ISSUE

The presence of salts in the Colorado River is a major problem for residents of the United States and Mexico. High salinity levels have caused an estimated $750 million in annual damages in the United States and approximately $100 million in Mexico due to lost vegetable and fruit production, damage to water pipelines, and destruction of fish and wildlife. High salinity is caused by surface water runoff on alkali soils and by irrigation.

In 1974, the U.S. Congress passed the **Colorado River Basin Salinity Control Act** (the Salinity Act), which authorized desalting and salinity control projects to improve the quality of Colorado River water entering Mexico (see Figure 5.5). The 1974 agreement requires that the United States not allow more than 115 ppm of average annual salinity in

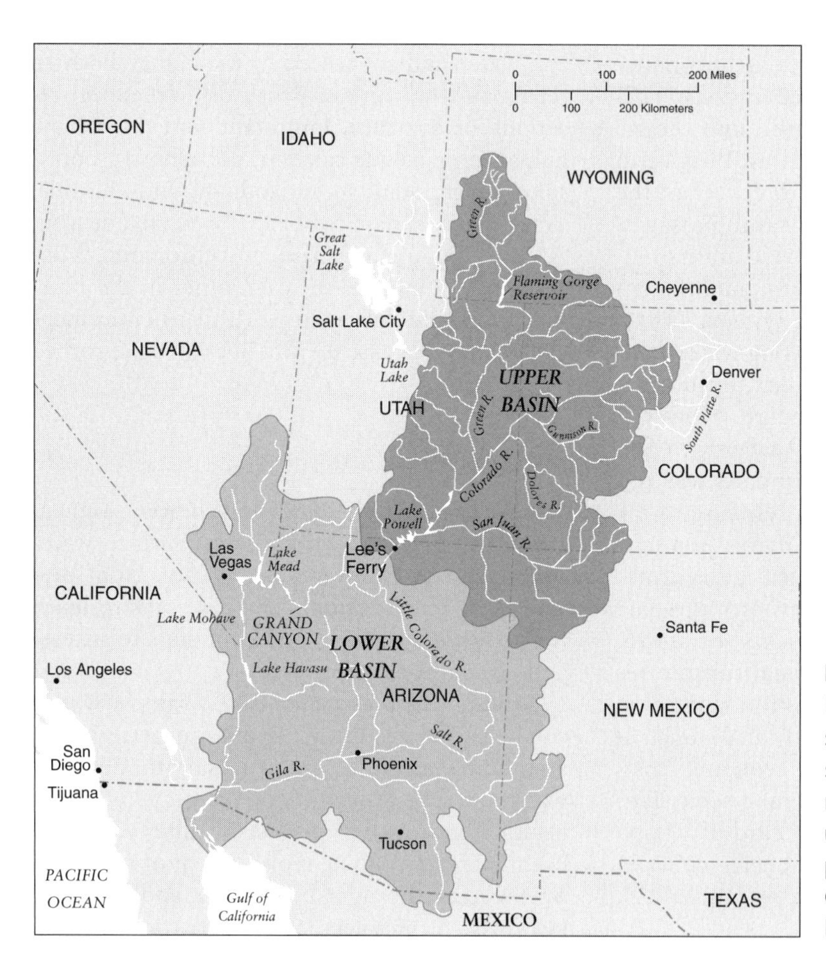

FIG. 5.5. The Colorado River Basin is located in a region of aridity and saline soils. Natural and human-caused surface water runoff, from irrigation return flows, stormwater runoff, and urban drainage, all compound the problem of high-salinity levels in the Colorado River, particularly in the lower reaches of the river basin.

the Colorado River as it crosses the border into Mexico. The Salinity Act provided funds to the Bureau of Reclamation, through the U.S. Department of Interior, for construction of a desalination complex near Yuma, Arizona, installation of groundwater wells near San Luis Rio Colorado, Sonora, Mexico, and concrete-lining of the Coachella Irrigation Canal in southern California. In 1987, Congress amended the Salinity Act to direct the U.S. Department of Agriculture to develop salinity control measures with upstream irrigators in the Colorado River Basin. In 1994, the Salinity Act was amended yet again to direct the U.S. Bureau of Land Management (BLM) to develop a comprehensive program that will minimize nonpoint source pollution of salts from BLM lands.

One issue of the original Salinity Act was the groundwater wellfield (called the Protective and Regulatory Pumping Unit) along the U.S./Mexico border in southwestern Arizona. A battery of groundwater wells are located within a 13-mile-long (21 km) and 5-mile-wide (8 km) strip of land, intercepting groundwater that is slowly moving beneath the border to Mexico. The wells along the U.S./Mexico border are spaced 0.5 mile (0.8 km) apart and are drilled to a depth of 600 feet (183 m), with the lower 300 feet (91 m) of casing screened. The pumped groundwater is less saline than water found in the nearby Colorado River and helps dilute salt concentrations present in the surface water. The Bureau of Reclamation has estimated that the benefit of salinity control is $340 per ton (in 1994 dollars), while the cost of salinity control ranges from $20 to $100 per ton. (20)

The pumped groundwater also helps the United States meet its requirements under the Mexican Water Treaty of 1944. This international treaty commits the United States to deliver 1.5 million acre-feet (488.8 billion gal, or 1.9 billion m^3) of Colorado River water to Mexico annually. (No water quality requirements were imposed when the treaty was signed.) U.S. groundwater pumping is limited to 160,000 acre-feet (52.1 billion gal, or 197.4 million m^3) per year.

If you were appointed to administer the Colorado River Salinity Program, would you focus on improving upstream irrigation practices to reduce salinity, or would you seek federal funds to construct more desalination plants? Would you encourage or discourage construction of dams on smaller tributaries to capture snowmelt and rain from thunderstorms that could be used to dilute salinity levels during critical periods? Would you carry out an education program to alert water users of the basin to the problems of high salinity? How would you coordinate the numerous federal agencies involved in this project?

What if you were a water official with the Mexican government and you were directed to reduce quantities of salt flowing across the border in the Colorado River? Although excess salts are damaging vegetable and fruit crops in Mexico, the U.S. representative argues that they are meeting the average annual salinity obligations under the 1974 Salinity Act. How would you convince U.S. officials to adopt lower daily salinity standards at the border to protect the local Mexican economy?

FLUORIDE

Fluoride (F), another natural mineral found in surface and groundwater, is also present in foods such as tea and fish. (21) Around 1950, many municipal water systems began adding fluoride to drinking water to improve dental health. Approximately two-thirds of U.S. public drinking water systems are fluoridated today, although some agencies are not convinced that adding fluoride is completely safe. Dental fluorosis (excessive fluoride exposure which causes white, yellow, or brown flecks or spots on teeth) is low when fluoride in drinking water is between 1 and 2 mg/L. The U.S. Environmental Protection Agency has determined that fluoride does not cause cancer and recommends an **MCL (maximum contaminant level)** of 4.0 mg/L. (22)

SIDEBAR

What's an MCL? An MCL is the maximum contaminant level set by the USEPA for public water supplies. Public water supply systems are defined as those serving 25 persons and/or 15 house connections. Although MCLs are not enforceable for individual water supply systems, the criteria of what constitutes safe drinking water quality apply equally to all people. (23) MCLs are set at a level where the health risks are considered acceptable, in contrast to the economic costs of requiring all public water supply providers to treat drinking water to a higher level.

ORGANIC CHEMICALS

NATURAL ORGANIC CHEMICALS

Organic chemicals contain carbon and can be classified as natural or synthetic, whereas **natural organic chemicals** occur in nature upon the decomposition of plants and animals. Prehistoric forests, grasslands, and wetlands provide complex carbon-based compounds found naturally in water and soil. These organic chemicals can develop into naturally occurring nitrates, nitrites, and ammonia. High levels can cause health problems in both animals and humans.

SYNTHETIC ORGANIC CHEMICALS

Synthetic organic chemicals (SOCs) are not found naturally and are generally developed in laboratories for mass production by industry. Synthetic organic chemicals can persist in the environment for very long time periods because natural decomposing processes are unable to degrade these complex compounds. Many SOCs are **carcinogens**—substances or agents that stimulate the formation of cancer. Petroleum-based industries are a major source of SOCs in water supplies. Synthetic organic chemicals include industrial solvents such as benzene, carbon tetrachloride, polychlorinated biphenyls, and pesticides:

Benzene (C_6H_6) A carcinogen used as a commercial solvent in petroleum refining and coal processing.

Carbon Tetrachloride (CCl_4) A carcinogen used in the manufacture of fire extinguishers, solvents, and cleaning agents.

Polychlorinated Biphenyls (PCBs) A carcinogen formerly used in fluids of electrical transformers and capacitors in the electronics industry. Its use was discontinued in 1976 when it was found to cause ecological damage.

Synthetic organic chemicals can be divided into volatile and nonvolatile categories. **Volatile organic chemicals (VOCs)** are lightweight chemicals that can be dispersed into the air through aeration processes. Examples of VOCs are MTBE (Methyl tert-butyl ether), a gasoline additive used to oxygenate gasoline to decrease carbon monoxide emissions; PCE (Perchloroethylene), a solvent used in dry cleaning processes; and trichloroethylene (TCE), which has had extensive industrial use as a degreasing agent and in gasoline products like benzene.

Nonvolatile organic chemicals (NVOCs) are usually heavier than VOCs and will settle to the bottom of rivers and lakes into sediments. These chemicals include polychlorinated biphenyls (PCBs) and dichlorodiphenyltrichloroethane **(DDT)**, a banned pesticide that was used extensively in the United States prior to 1973. Synthetic chemicals can have a wide variety of negative health effects, including carcinogenic problems. Skin absorption by simple contact with contaminated water, as well as breathing in VOCs that can be dispersed into the air during a hot shower, are all significant health issues.

OUR ENVIRONMENT

The U.S. Environmental Protection Agency (USEPA) and the General Electric Company (GE) are involved in the cleanup of more than 100,000 pounds (45,360 kg) of polychlorinated biphenyls (PCBs) from a 40-mile (64-km) stretch of the Hudson River in upstate New York. The $460 million cleanup, paid for by GE, is an effort to remove point source pollution generated from GE factories between 1947 and 1977. During that period, the USEPA contends that GE released over 1.3 million pounds (589,680 kg) of PCBs directly into the Hudson River as a byproduct of industrial waste from its facilities at Hudson Falls and Fort Edward, New York. Over 2.6 million cubic yards (2.0 million m^3) of contaminated sediments must be dredged and transported to licensed landfills along the East Coast. (24)

The USEPA also contends that the massive dredging operation is necessary to "restore the environmental health" of the Hudson River. (25) Officials from GE argue that the dumping was legal at the time, that dredging would stir up these materials and cause them to be transported downstream and ingested by fish, and that the USEPA is punishing the company for actions several decades old (called *past liabilities* under the Comprehensive Environmental Response, Compensation and Liability Act of 1980, or CERCLA, also known as Superfund).

Debate continues over the ability of microbes to clean these pollutants and over the safety of "stirring" bottom sediments that will occur during dredging of the Hudson River. The USEPA maintains that the health of fish, other aquatic species, and humans is at risk if the pollutants are not removed immediately. The USEPA and others are concerned that since PCBs bioaccumulate (build up) in the environment, increased concentrations of the pollutant will occur in the food chain. Contamination levels in fish have exceeded food safety standards by a hundredfold in most contaminated areas.

In 2000, GE filed a lawsuit in federal court claiming that CERCLA is unconstitutional. (26) The federal government could use provisions of CERCLA to impose fines and the threat of jail to enforce compliance with federal water quality regulations. However, in 2002, USEPA Administrator Christie Whitman signed a "record of decision" for the Hudson River Superfund site after GE agreed to pay up to $28 million in partial reimbursement of EPA's past and future cleanup costs. Dredging is expected to begin in 2006. See http://www.epa.gov/hudson/ for updated information.

PESTICIDES

In 1962, Rachel Carson wrote a groundbreaking book called *Silent Spring* in which she alerted the world to the potential long-term negative effects of pesticides. Carson was concerned about the global use of chemicals, particularly insecticides, to control pests. She argued that sprays, dusts, and aerosols were too widely dispersed into the air, onto the land, and inadvertently into waterways, and caused the "inadvertent" killing of insects, song birds, fish, and other plant and animal life. (27) The concerns were quite valid, and her book is still widely read today.

Pesticides are synthetic organic chemicals (SOCs) used to eliminate unwanted pests such as insects, mice, and other animals, unwanted plants (weeds), fungi, and microorganisms like bacteria and viruses. SOCs include **insecticides,** which kill insects and other arthropods, **herbicides** which kill plants and weeds, **fungicides** which kill fungus, **nematacides** which kill nematodes, and **rodenticides** which kill rodents such as rats and mice. Pesticides are used to increase crop yields, improve public health, and enhance the appearance of landscaped areas.

Approximately 50 percent of pesticides are used for nonagricultural purposes such as at golf courses, lawns, parks, schools, roadways, railways, utility rights-of-way, and public buildings. Lawn and garden use of pesticides is so great that more pesticides are used on lawns and gardens than on farmland. In addition, application rates have been found to be over 500 percent greater on lawns and gardens than on cropland. (28)

Pesticides are designed to remain in the application area to control target pests and then degrade into harmless products. However, some degradation rates (measured as half-life) allow contaminants to reach groundwater or surface water sources before breaking down. The half-life of many common pesticides (the time it takes for half of the active ingredient to break down) varies from 3 to 150 days. (29)

In 1972, the **Federal Insecticide, Fungicide and Rodenticide Act (FIFRA)** was passed to regulate the use of pesticides. The USEPA measured the risks of chemical use against the amount of benefits received. Specifically, health risks, such as damage to the central nervous system, reproductive problems, and other genetic problems, were compared with the costs and benefits obtained by regulating pesticides. FIFRA authorized the USEPA to require all chemical manufacturers to print information on the chemical container label regarding potential risks of pesticide use and proper handling procedures.

A CLOSER LOOK

DDT was one of the very first pesticides used to kill pests. Although invented in 1873, DDT was not used until 1939 when Paul Herman Müller (1899–1965) of Geigy Pharmaceutical in Switzerland discovered its effectiveness as an insecticide. He was awarded the Nobel Prize in medicine and physiology in 1948 for his work. (30)

During World War II, military use of DDT was widespread for the control of malaria (spread by mosquitos) and typhus (lice). The World Health Organization estimates that approximately 25 million lives were saved during its period of use. After the war, DDT was widely applied to control insect pests in crops and on forest lands, around homes and gardens, and for industrial and commercial purposes since it was relatively inexpensive and highly effective. Users often called it a "miracle" pesticide, and over 80 million pounds (36.3 million kg) were used in 1959, its peak year of use. Later, declining populations of birds, particularly bald eagles, were attributed to DDT use. The USEPA finally banned it in the United States in 1972 in an effort to protect the environment and public health. However, DDT is still widely used today in some foreign countries.

The USEPA has prohibited the use of many other insecticides and herbicides since DDT was outlawed in 1972. These chemical bans have improved the overall health of the environment. (31) At the same time, these restrictions on chemical use have allowed some unwanted pests to thrive, reducing crop yields and allowing the spread of certain dreaded diseases, such as West Nile Virus in the United States and malaria in many less-industrialized countries around the world. In 2000, the World Health Organization called for the continued use of DDT to control the spread of malaria in less-industrialized nations. (32)

The issue of environmental health versus economic opportunity and human health has been debated for decades. Would you support a ban on certain chemicals if their use caused any environmental harm? What role should economics or human health play in this consideration? An estimated 25 million lives were saved during the period of use of DDT, and the inventor was awarded the Nobel Prize. However, its widespread use produced great environmental destruction. How should the value of human life be weighed against the value of environmental protection?

Integrated pest management (IPM) is the use of a combination of economical, common-sense practices to manage pest damage with the least possible hazard to people, property, and the environment. (33) The IPM approach can be used in both agricultural and nonagricultural settings, such as in the home, garden, or workplace.

Appropriate pest management options can include the selective use of pesticides, natural predators, or no action if pest thresholds (the point at which pest populations become a problem) are not exceeded. Biological controls, such as lady beetles, predatory mites, minute pirate bugs, big-eyed bugs, and predatory thrips can be used to control spider mites (Family: *Tetranychidae*) on corn, a serious problem in eastern Colorado. Spider mites are common plant pests that injure leaves and can cause plant loss. Often, spider mites become invasive since insecticides also kill their natural predators. (34)

In the United States, the silverleaf whitefly, *Bemisia argentifolii*, a 0.04-inch-long (0.1 cm) insect has caused billions of dollars of damage to agriculture. It first appeared on poinsettias in Florida greenhouses in 1986 and by 1990 had spread to dozens of crops throughout Florida, Texas, California, Arizona, and northern Mexico. Losses have probably reached several billions of dollars since then. (35) Responses to the silverleaf whitefly have included the use of insecticides, crop breeding improvements to increase resistance, and introduction of the *Eretmocerus* wasp. The female wasp deposits eggs under a silverleaf whitefly larva, and upon hatching, the immature wasp feeds on the whitefly, killing it. In 1995, the USEPA approved the commercial use of Mycotrol, a whitefly-killing fungus *(Beauveria bassiana)*. (36) The use of these various controls presents a broad spectrum of pest control typical of IPM programs.

NUTRIENTS

Plants, animals, microorganisms, and even single-cell bacteria must extract substances from the environment for energy and growth. These substances are called **nutrients** and include nitrogen, phosphorus, magnesium, calcium, and iron. Nitrogen, phosphorus, and associated compounds are particularly important in the study of water quality.

Nitrogen **Nitrogen (N)** is important as a plant nutrient for crop production, lawns, landscaping, golf courses, forest growth, and other vegetation. It is most abundant in its atmospheric form (N_2), or nitrogen gas. **Nitrogen gas (N_2),** or dinitrogen, comprises 78.1 percent of the Earth's atmosphere, by volume. (37) This naturally occurring gas was first discovered by chemist and physician Daniel Rutherford (1749–1819) in 1772.

Through an experiment in his laboratory, Rutherford was able to remove oxygen and carbon dioxide from an atmospheric sample and proved that the remaining residual gas was a newly discovered element (nitrogen). In 1775, the French chemist Antoine Laurent Lavoisier (1743–1794) was appointed commissioner of the Royal Gunpowder and Saltpeter Administration and continued Rutherford's work. Lavoisier created an excellent laboratory in the Paris Arsenal in which he conducted scientific investigations on weaponry. This led to chemical experiments on gases, particularly nitrogen, since it could be used in explosives. He later renamed the nitrogen element Rutherford had discovered as *azote,* meaning without life. (38)

Nitrate (NO_3^-) is created by bacterial action on ammonia, by lightning, or through artificial processes that include extreme heat and pressure. Nitrate is found in soluble form in both surface and groundwater. It is not bound by soil particles, it is consumed by plants, and it converts into gaseous forms by microbial action. Nitrates can pollute groundwater aquifers by leaching through soils, or they can move laterally with surface water or subsurface flow to contaminate surface waters. In proper amounts, nitrates are very beneficial. However, excessive concentrations in water can cause health problems if consumed by humans. (39)

The MCL established by the U.S. Environmental Protection Agency for nitrate (NO_3^-) is 45 ppm, which equals 10 ppm nitrate-nitrogen (NO_3^-N). This USEPA Safe Drinking Water Standard means that a maximum of 10 units of nitrate-nitrogen can be present for every million units of drinking water. Boiling water does not help reduce nitrate levels and will actually increase concentrations due to water lost through evaporation.

Nitrite (NO_2^-), also found in soluble form in water, is an intermediate form created by bacterial action on ammonium (NH_4^+) or nitrate (NO_3^-). Ammonium is converted to the nitrite and nitrate forms rather quickly by nitrifying bacteria. These add oxygen to the ammonium ion and convert it to nitrate. Ammonia toxicity is a problem for aquatic life, whereas nitrite toxicity is a problem for infants. Nitrates turn to nitrites in an infant's stomach and react directly with the blood to produce methemoglobin. Methemoglobin destroys the ability of red blood cells to carry oxygen and can cause a condition called *methemoglobinemia,* or "blue baby" syndrome, in infants primarily under three months of age. Water with nitrite levels exceeding 1.0 mg/L should not be used for feeding infants. (40) Older children and adults are not as susceptible to methemoglobinemia because their stronger stomach acids kill the bacteria required for this process to occur.

Nitrite toxicity in fish is greater in water with low DO (dissolved oxygen) levels because nitrite reduces the ability of blood to carry oxygen. Nitrites can produce "brown blood disease" in fish and occur mostly in farm or commercial fish ponds when sediments are disturbed. Disturbance typically occurs during the spring and fall turnovers or through mechanical agitation to increase dissolved oxygen levels. Although fish tolerance for nitrite (NO_2^-) is low (often less than 0.15 mg/L), their tolerance for nitrate (NO_3^-) is high, typically greater than 1000 mg/L. (41) Fish with acute nitrite toxicity may become very dark in color and extremely sluggish.

Ammonia (NH_3) and **ammonium (NH_4^+)** are commonly found in surface water, in the soil, and as a byproduct of decaying plant tissue and decomposition of animal waste. Ammonia and ammonium are rich in nitrogen and excellent fertilizers. Ammonia levels at 0.1 mg/L usually indicate polluted surface waters, whereas readings above 0.2 mg/L can be toxic for many aquatic species. (42) High levels of ammonia are often found downstream of wastewater treatment plants and near ponds that have large populations of waterfowl, such as ducks and geese, which produce waste. If ammonia levels in water are too high, fish can experience ammonia toxicity and become hyperactive, appearing to skim along the water's surface.

Phosphorus In contrast to nitrogen, **phosphorus (P)** does not exist in a gaseous state but occurs naturally as a salt in the mineral apatite, which is found in igneous, metamorphic, and sedimentary rocks. Phosphorus is a common nutrient found in soil and water, and is quickly bound to soil particles or is consumed by plants. Phosphorus can originate from dissolved leachate from rocks, from decomposing organisms, animal waste, manufacturing processes, effluent from wastewater treatment plants, and as artificial fertilizers applied to crops, lawns, gardens, and golf courses. Much of the phosphorus found in sewage effluent is from synthetic detergents (dish soap). Street runoff also contributes phosphorus to waterways from grass clippings and fertilizer runoff.

Phosphorus by itself does not have any notable health effects on humans. However, phosphorus levels above 1.0 mg/L may interfere with coagulation processes at water treatment plants. This can hinder the removal of microorganisms bound to sediments and other particles from drinking water (see further discussion in Chapter 11).

Nonpoint source pollution of eroding sediments in runoff is the primary mechanism whereby phosphorus enters surface water. In many cases, point sources, such as wastewater treatment plants, are also very important contributors. Lake and reservoir sediments serve as phosphorus sinks (holding areas) and can cause excessive growth of algae and phytoplankton. Such growth often occurs when summer warming conditions of the hypolimnion (the normally cooler water at the bottom of a lake) stimulate the release of phosphorus from the benthos (sediments at the bottom of the water body).

THE NITROGEN CYCLE

The **nitrogen cycle** (Figure 5.6) is driven by nitrogen, oxygen, and bacteria. It is the natural process of converting the reservoir of nitrogen gas from the atmosphere into usable forms of nutrients for plants and animals. The nitrogen cycle includes complex interactions with various forms of nitrogen, notably:

Atmospheric nitrogen (N_2)
Organic nitrogen (N)
Nitrite (NO_2^-)
Nitrate (NO_3^-)
Ammonia (NH_3)
Ammonium (NH_4^+)

Each form of nitrogen affects plant utilization and can have negative impacts on water quality. The nitrogen cycle includes four main components:

Nitrogen fixation
Ammonification
Nitrification
Denitrification

Nitrogen Fixation Animals, including humans, cannot utilize nitrogen gas (N_2) from the atmosphere or from inorganic compounds. Instead, nitrogen must first be converted into an organic form (nitrogen combined with carbon) through a process called **nitrogen fixation.** This process requires substantial amounts of energy to break apart the nitrogen molecule, since it has a triple

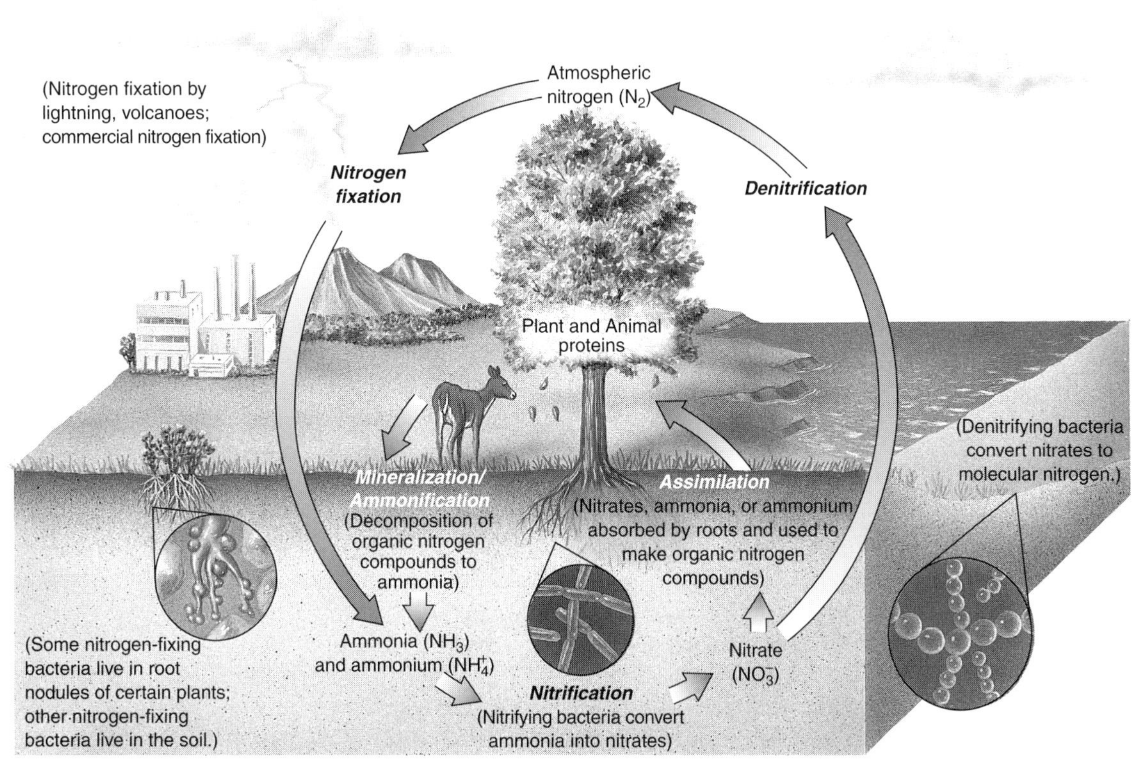

FIG. 5.6. The Nitrogen Cycle.

bond between the two nitrogen atoms, making the molecule almost inert. Nitrogen gas (N_2) will react with oxygen only in the presence of high temperatures and pressures, or through bacterial activity, to create organic nitrogen (N), nitrate (NO_3^-) or ammonia (NH_3). This process can be caused by atmospheric fixation by lightning, biological fixation by bacteria and algae, and industrial fixation caused by combustion reactions in power plants, chemical processes to make fertilizers, or inside internal combustion engines. Industrial fixation requires great pressure and temperatures.

Biological nitrogen fixation accounts for about 70 percent of the total conversion of nitrogen into biologically useful forms of nitrate. The *Rhizobium* bacteria in root nodules of legume crops such as clover, alfalfa pinto beans, and soybeans, in surface water environments such as wetlands complexes, and in the soil can convert nitrogen gas in the atmosphere into biological matter. Approximately 20 percent of all nitrogen fixation—from nitrogen gas (N_2) to ammonia (NH_3)—occurs through industrial processes. Atmospheric nitrogen fixation from lightning accounts for less than 5 percent of total fixed nitrogen conversion from atmospheric nitrogen (N_2) to nitrate (NO_3^-). Agriculture may now be responsible for approximately 35 percent of all nitrogen fixation on Earth through the use of fertilizers produced by industrial fixation and biological fixation caused by the production of legume crops. (43) See Table 5.2.

Mineralization/Ammonification The process of organic matter, such as dead plants and animal waste, decomposing in the presence of oxygen is called **mineralization**, or decay, and includes

TABLE 5.2 Nitrogen Fixation

Type of Fixation	N_2 Fixed (10^6 metric tons per year)
Nonbiological	
Industrial	about 50
Combustion	about 20
Lightning	about 10
Total nonbiological	about 80
Biological	
Agricultural land	about 90
Forest and nonforest land	about 50
Sea	about 35
Total biological	about 175

Source: D. F. Bezdicek and A. C. Kennedy, *Microorganisms in Action*, J. M. Lynch and J. E. Hobbie (Eds.) (Oxford: Blackwell Science Ltd., Osney Mead, 1998). Used with permission.

ammonification. The principal storehouse for nutrients found in the soil is within organic matter, such as decaying plants or animal waste. Organic matter can hold more than 96 percent of all soil nitrogen. (44)

Nitrification The process of organic nitrogen (N) changing into nitrate (NO_3^-) is called **nitrification**—a rapid anaerobic process that requires bacterial action. The common soil bacterium *Nitrosomonas* oxidizes ammonia (NH_3), or organic matter, into nitrite (NO_2^-). *Nitrobacter,* another common bacterium, relatively quickly oxidizes nitrite (NO_2^-) into nitrate (NO_3^-).

Oxidation within the nitrification phase of the nitrogen cycle is very important. When oxygen (O_2) is readily available, the system is said to be in an *oxidized state,* or **aerobic,** and the O_2 is ready to accept negatively charged electrons. If O_2 is not present, or is very low as in marshes, swamps, or saturated soils, the system is said to be in a *reducing state,* or **anaerobic.** Aerobic conditions are required to promote nitrification, whereas anaerobic conditions favor denitrification.

Denitrification **Denitrification** is the change from nitrate (NO_3^-) to atmospheric nitrogen (N_2), or other gaseous forms such as NO, NO_2^-, or N_2O, caused by bacterial action. Denitrification is a rapid process and can cause the loss of up to 60 percent of nitrogen-based fertilizer added to a crop. (45)

Plants use nitrate as fertilizer and convert it into proteins and amino acids for growth and energy. In turn, decaying plants convert nitrates into organic nitrogen (N) or ammonia (NH_3).

The process of denitrification continues the nitrogen cycle. A side product of this reaction is the production of nitrous oxide, N_2O, also known as "laughing gas." Nitrous oxide is also a greenhouse gas that contributes to global warming. Temperature, pH, and dissolved oxygen concentrations are very important parameters in denitrification rates. Increased temperatures accelerate denitrification processes, while pH ranges between 8.0 and 9.0 provide the best conditions. (46)

THE PHOSPHORUS CYCLE

Phosphorus occurs in organic and inorganic forms through microbial activity and converts in a relatively simple process. Processes of the hydrologic cycle add phosphorus to the soil where it can be consumed by plants. Decaying plants and animal waste decompose and return phosphorus to organic form in the soil, where the phosphorus cycle continues.

EXPERT ANALYSIS

Commercial fertilizers are made of a combination of nitrogen, phosphorus, and potassium, and are used to enhance plant growth. Too often, consumers over apply fertilizers to lawns and gardens, and cause problems with nitrate leaching and runoff in street gutters and ultimately to local lakes and rivers. According to a study by the Cooperative Extension Service of Oklahoma State University, some lawns receive 10 or more pesticide application rates per season and two or three times as much nitrogen as required by a typical field crop. (47)

In April 2002, Minnesota's governor Jesse Ventura signed SF 1555, the Phosphorus Lawn Fertilizer Bill, into state law. This law prohibits the use of phosphate in lawn fertilizer in a seven-county area of Minneapolis and St. Paul, and limits to 3 percent the amount of phosphate that can be applied to lawns in all other counties in the state unless a soil or tissue test shows a phosphorus need. The purpose of this law, the first and only one of its kind in the United States, is to reduce nutrient loading in area ponds, lakes, and wetlands. A 1996 study estimated that 2 to 4 million pounds (0.9 to 1.8 million kg) of phosphorus were applied to turf in the Twin Cities area of Minneapolis and St. Paul, with up to 35 percent of the total phosphorus in urban runoff caused by fertilizer that fell on driveways and sidewalks. (48) The Phosphorus Lawn Fertilizer Bill took full effect in January 2004. See http://www.extension.umn.edu/extensionnews/2003/Phosphe.html for additional information.

Contact your local county Cooperative Extension Office and determine the recommended application rate for lawn fertilizer. Determine the number of pounds that would be applied per acre if the directions were followed, and multiply your result by the number of applications applied per year (usually three or four). Determine the amount for 200 and 300 percent of the recommended application rate, based on the Oklahoma State University study results. Then contact a local planning agency to determine the approximate number of acres within the boundaries of your selected town or city. Generally, 25 percent of a developed municipal land area is utilized as lawns around homes, businesses, parks, golf courses, and so on. How many tons of fertilizer are being applied to green areas in your community?

EUTROPHICATION

Eutrophic waters are characterized by the presence of a lower number of aquatic species due to the dominance of algae. An eutrophic lake will contain a large amount of algae but little else. This state is caused by excessive nutrients in water that provide food for algae and other excessive aquatic plant growth. **Eutrophication** is the process of a water body moving from oligotrophic to mesotrophic and finally to eutrophic conditions. Oligotrophic waters contain low levels of essential nutrients such as nitrogen, phosphorus, and iron, and support only small volumes of aquatic life. Lake Superior between the United States and Canada, Lake Tahoe between California and Nevada, and Crater Lake in Oregon are examples of oligotrophic lakes. Mesotrophic waters have an abundance and diversity of aquatic life due to adequate levels of essential nutrients.

Mesotrophic water bodies are commonly found around the world. Eutrophic waters place a high demand on dissolved oxygen. Organic materials, particularly nutrients such as phosphorus and nitrates, cause the microbes decomposing them to consume oxygen from the water, thereby creating anaerobic conditions. Eutrophication, though a natural process, is accelerated by human activities that introduce excessive nutrients into a water body. A major goal of water management is to reduce or mitigate eutrophication.

The measurement of oxygen demand by organic materials in a water body is called **biological oxygen demand (BOD).** It provides an excellent indicator of the amount of oxygen stress, caused by organic pollutants, on living aquatic organisms. Excess decaying plant material (as well as living plants) cause increased organic requirements for dissolved oxygen, followed by high BOD requirements. Major sources of nutrients are municipal effluent, fertilizers, animal feeding operations, and septic systems. Oxygen depletion is more common in standing bodies of water such as ponds, lakes, and wetlands since moving water in rivers is constantly oxygenated through wind action and turbulence.

Most wastewater treatment plants release treated effluent that contains nitrogen in the form of nitrate, a nutrient. If adequate water supplies are available, either in the treated wastewater or in the receiving water body, dilution can prevent the accumulation of nitrates, and the conversion to excessive concentrations of nitrite and ammonia, in waterways. However, if nitrate-laden wastewater is loaded into a river, lake, or estuary, BOD can cause the waterway to become anaerobic. In addition, benthic (bottom-dwelling) organisms such as clams, scallops, and other small invertebrates can be destroyed by algae growth that absorb sunlight for photosynthesis. Eventually, the algae will die, decompose, and form a mat (a layer of organic material) and will promote more dissolved oxygen demand. Algae and dead organic material became so thick along the bottom of Chesapeake Bay in the late 1960s that they obstructed shipping lanes.

WATERBORNE DISEASES

HISTORICAL PROBLEMS

Early concerns regarding water pollution were motivated by waterborne diseases. Infectious hepatitis, cholera, bacterial dysentery, and giardiasis were common prior to the twentieth century throughout the world, and health impacts were staggering. Plagues in the eleventh and twelfth centuries in Europe wiped out entire villages. Over 53,000 people died during a cholera epidemic in London in 1848–1849 alone. (49)

It was not until a now famous study was completed in 1854 by Dr. John Snow (1813–1858) of England that a scientific connection was made between drinking water, water pollution, and disease. Dr. Snow's Broad Street Pump Study in London found that cholera cases were clustered in an area served by a community water pump at Broad Street (in the Soho region of present-day London). Dr. Snow isolated the Broad Street pump as the source of the contamination after extensive mapping of cholera cases and drinking water supplies in the area.

A CLOSER LOOK

Dr. John Snow's 1854 Broad Street Pump Study is a landmark in the field of epidemiology (the study of infectious diseases). Dr. Snow was a London physician who practiced as an obstetrician (Figure 5.7). During the 1830s and 1840s, when several severe cholera epidemics struck London, Snow became interested in the cause and transmission of the disease. In 1849, he published a brief pamphlet, *On the Mode of Communication of Cholera*, and suggested the disease was transmitted by contaminated drinking water. The pamphlet contained just one of many theories on the cause and spread of the disease which were being circulated at the time, and his theory was not widely considered.

FIG. 5.7.
Dr. John Snow.

In 1854, however, Snow painstakingly plotted the locations of illness during a subsequent spread of the disease and compared it to the subscribers of the city's two private water companies. His research showed that the incidence of cholera appeared more frequently with one company, the Southwark and Vauxhall. This company obtained its water supply from the lower Thames River, which was contaminated with London sewage, while the other company obtained less polluted water from the upper Thames River.

Dr. Snow's maps showed a strong correlation between reported cases of cholera and proximity to the intersection of Cambridge and Broad streets. The concentration was so great that 500 deaths occurred in only 10 days in that area. The conclusion was obvious. The cause of the cholera outbreak was the water taken from the Broad Street pump, a community source of water.

On the evening of September 7, 1854, Dr. Snow met with the Board of Guardians of St. James's Parish, the local governmental agency. The board was responsible for safety, order, and the public health. The members often relied on advice from professionals in the field of health, such as Dr. Snow, before making community decisions.

Snow's good friend, Dr. Benjamin Richardson, wrote later of the historic meeting:

When the Vestry [board members of the Parish] men were in solemn deliberation they were called to consider a new suggestion. A stranger had asked in a modest speech for a brief hearing. Dr. Snow, the stranger, was admitted, and in a few words explained his view of the "head and front of the offering." He had fixed his attention on the Broad Street pump as the source and center of the calamity. He advised the removal of the pump handle as the grand prescription. The Vestry was incredulous but had the good sense to carry out the advice. The pump handle was removed and the plague was stayed. There arose, hereupon, much discussion among the learned . . . but it makes little for the plague was stayed. (50)

The pump handle was immediately removed to prevent any further use of the contaminated water, and Dr. John Snow became a legend. Today the site of the Broad Street pump is located near the Underground (the "Tube") Station near Piccadilly Circus. A walk along the intersection of Broadwick Street (the "wick" was added to "Broad" in 1936) and Lexington Street (called Cambridge Street on maps of the 1800s) will place you at a historic spot in the annals of epidemiology. A replica of the famous pump (its handle removed) sits at the intersection with a plaque that reads:

It marks a pioneering example of medical research in the service of public health.

The John Snow Pub is located nearby, and includes portraits, photos, and narrative frames regarding Dr. Snow and his famous study. The original well is located near the back wall of the pub by a red granite curbstone and is capped (permanently sealed).

The pump handle remains a symbol of effective epidemiology. Today, the original handle of the Broad Street pump is displayed at the John Snow Pub and serves as a tribute to Dr. Snow's seminal work in the study of infectious diseases. Go to the UCLA Department of Epidemiology website at http://www.ph.ucla.edu/epi/snow.html for an excellent resource of Dr. John Snow information.

MICROORGANISMS

The term **microorganism** comes from the Greek word *micro* meaning "small" (see Figure 5.8). All water, including freshwater, salt water, lakes, ponds, rivers, wetlands, and groundwater, contain microorganisms. Microbes range in size from small viruses between 0.004 mm and 0.085 mm (microns or micrometers) in diameter to large protozoans (5 mm to 15 mm in diameter). Most microorganisms are smaller than 100 mm (100 microns are slightly larger than the diameter of a human hair), and a microscope is necessary to make them visible.

The existence of microorganisms was not known to science until the nineteenth century. In

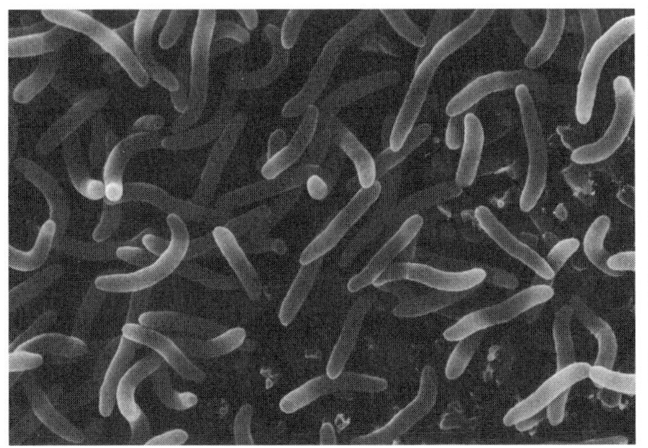

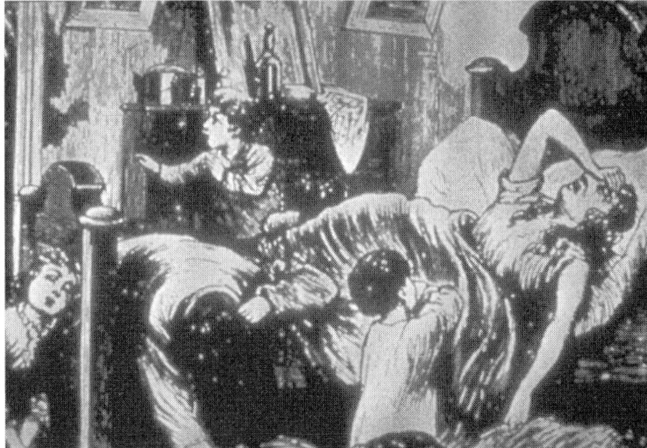

FIG. 5.8. The cholera bacteria, magnified 1390X under an electron micrograph, was the cause of a major cholera outbreak in Chicago in 1885. On August 2, 1885, a torrential rainstorm dumped 6.8 inches (17.3 cm) on the city and flushed raw sewage into Lake Michigan. The lake was the primary source of drinking water for residents and became polluted with raw sewage at drinking water intakes. This cartoon from the *Chicago Tribune* of 1885 depicts scenes of death and despair caused by the plague that killed over 80,000 people.

1861, Louis Pasteur (1822–1895) performed experiments that proved microbes existed in the air, on solids, and in liquids. He also proved that microbial life could be destroyed by heat and that nutrient processes required air. (51)

Indicator Organisms Because of the wide range of infectious microorganisms found in water, it is difficult and expensive to conduct a complete analysis of every type. To save time and money, **total coliform** is used as an **indicator organism** to determine the presence of pathogenic bacteria in water. Coliforms are bacterial microorganisms found in human and warm-blooded animal wastes, and normally live in the intestinal tracts of animals and humans. This group of microorganisms is a strong indicator of the presence of unsanitary or possible disease-producing conditions in water.

Total coliform is used as an indicator organism to indicate pollution because it is easy and inexpensive to test. However, some forms of total coliform bacteria are harmless to humans since fish and even insects produce the microorganism. Therefore, additional tests may be warranted if coliform is present in water supplies. A sample of water is placed on an incubation plate for several hours to allow coliform colonies time to "grow" in a warm environment. After a set incubation time is met, coliform colonies are counted, and total colonies per 100 milliliters of water are recorded. The USEPA Safe Drinking Water Act standards allow no more that 5 percent of the samples collected in one month to test positive for total coliforms. (For systems that collect fewer than 40 routine samples per month, only one result can test positive.) All samples that test positive for total coliforms must be retested for fecal coliform.

Other Waterborne Organisms **Fecal coliform** is a member of the total coliform group that gives a strong indication of recent contamination from sewage or waste from warm-blooded animals. ***Escherichia coli*** (also called ***E. coli,*** pronounced **ee kole**-eye) is a type of fecal coliform that indicates the presence of pathogenic organisms. There are hundreds of strains of *E. coli* that help to break down and ferment food in the intestine and are necessary to promote digestion in humans. However, ingestion of the *E. coli* O157:H7 microorganism in food or water is particularly dangerous because it can cause health problems such as diarrhea and occasionally kidney failure. It is estimated that 73,000 cases of infection and 61 deaths occur each year in the United States from this infection. (52)

OUR ENVIRONMENT

In 2000, 2300 people became ill, and seven deaths occurred, in Walkerton, Ontario, after the community's drinking water supply became contaminated with the *E. coli* 0157:H7 microorganism. Walkerton is a community of 5000 located approximately 125 miles (200 km) northwest of Toronto. The water crisis hit in May 2000, and nearly half of the community became ill with gastrointestinal problems, with almost 1000 requiring medical treatment. It was the worst *E. coli* breakout in Canadian history.

Three groundwater wells were used to supply drinking water to residents of Walkerton when the epidemic occurred. The town's water distribution system was privatized (turned over to a private contractor) in 1996 and consists of approximately 26 miles (41.5 km) of water mains (water pipelines) and 1900 individual service connections. It appears that the majority of illnesses occurred from May 11 to June 6, 2000. A Boil Water Advisory was issued on May 21, 2000 by the local Health Unit agency.

Potential contamination sources, as identified by the Ontario Ministry of Environment, included new watermain construction, fire events, storage structures, cross connections (where water mains and sewer collection pipes were in adjacent buried trenches), and flooding (which occurred on May 12, 2000, in the area). The Ministry concluded that the probable cause was cattle manure which entered wells from nearby feeding operations. Surface water runoff from the May 12 rainfall event probably transported the fecal contaminants into one of the town's wells. (53)

Genetic fingerprinting could be used to determine if, and which local livestock operation carried an identical bacteria. This would be done by identifying the *E. coli* DNA taken from Walkerton victims and then tracing it back to livestock feeding operations in the area. Large-scale feedlot operations have been introduced into rural towns across North America. These intensive animal feeding facilities provide strong economic benefits to small communties, but if not constructed and operated properly, they can create water quality problems such as high concentrations of nitrates and fecal matter in local surface and groundwater systems. Wellhead Protection Programs can prevent disasters like the one in Walkerton from occurring and will be discussed later in this chapter.

Giardia (gee-**are**-dee-uh) was first identified in 1681 but was first reported officially in the United States in Aspen, Colorado, in 1965. (54) ***Giardia lamblia*** is a one-celled microscopic parasite that lives in the intestines of humans and other animals (see Figure 5.9). The parasite exists in animal waste and can survive in the form of a cyst for months at a time. When ingested, the cyst can move into a reproductive stage and cause infection of the gastrointestinal system. However, infection does not always produce illness, and humans can carry the disease with no symptoms of illness. Symptoms generally begin one to two weeks after infection and include diarrhea, stomach cramps, and upset stomach.

Cryptosporidium parvum (**krip**-toe-spo-**rid**-dee-um **par**-vum), a microscopic parasite that infects humans and other animals, causes the disease known as **Cryptosporidiosis** (**krip**-toe-spo-**rid**-ee-**oh**-sis). The parasite is found in raw sewage, treated wastewater, and effluent from cattle slaughterhouses. Infections are routinely found in cattle, dogs, cats, deer, rabbits, and other animals. Children are most susceptible, with symptoms including diarrhea, abdominal cramps, nausea, vomiting, and low fever. There is no special drug or treatment for cure. Prevention includes protecting watersheds from human and animal wastes. The hard outer shell of the *Cryptosporidium parvum* parasite makes it very resistant to chlorine disinfection, although the use of ozone for drinking water treatment has proven to be effective.

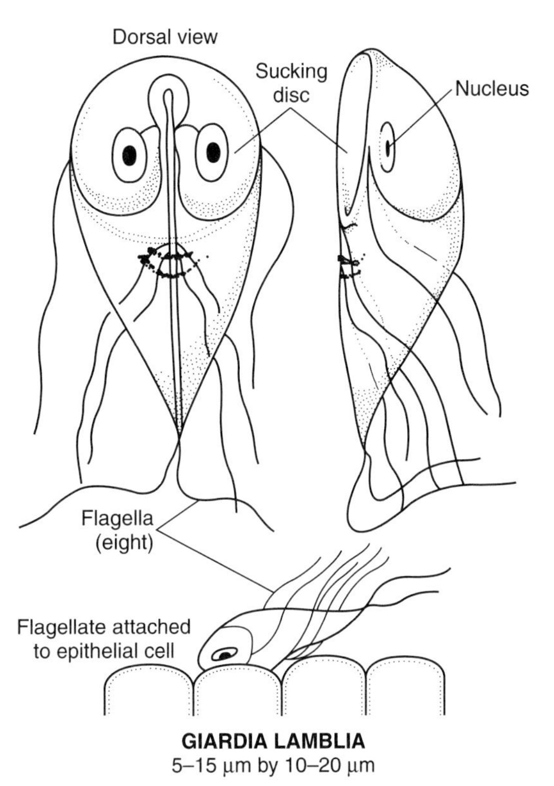

FIG. 5.9. *Giardia lamblia* is a single-celled animal (a protozoa) that moves with the aid of flagella. In Europe, it is sometimes referred to as *Lamblia intestinalis.*

Giardiasis is most commonly associated with drinking contaminated water and is more prevalent in children than adults because many individuals become immune after infection.

Unfortunately, outbreaks of *Crypto* have become widespread in humans worldwide. In April 1993, 400,000 people became infected in Milwaukee, Wisconsin, with over 100 deaths registered. Outbreaks of the disease have also been associated with swimming pools, amusement park wave pools, and water slides. (55)

HEALTH CONCERNS

Hiking, camping, and *Giardia* too often go together if safe drinking water precautions are not taken. Since most streams are also used by wild animals, *Giardia* can easily be transmitted to humans (and dogs) if water is consumed without boiling, filtering, or chemical treatment.

The safest method of raw water purification is boiling. Dispute continues over the length of time necessary to boil water before such water is safe to drink. Many authorities believe that a rolling boil of one minute is adequate to decontaminate water. (High altitude decreases the temperature needed to boil water and may increase the required boiling time to three minutes.) The Centers for Disease Control and Prevention in Atlanta, Georgia, recommends three minutes of a rolling boil at lower altitudes to ensure that all bacteria and viruses are destroyed. If the treated water tastes like wood smoke, consider adding a few tea bags, or pour the treated water back and forth between water containers to allow aeration to rid the water of odor.

Iodine and chlorine are popular water treatment methods used by hikers since they are lightweight and simple to use. Iodine and chlorine can be used in liquid or tablet form; however, the correct amount required will depend on the pH, temperature, turbidity, and quantity of water treated. Treatment times of 30 minutes to over an hour may be required. If treatment time is inadequate, disinfection will not be complete. Some experts do not believe this type of chemical treatment will kill *Cryptosporidium* even if proper procedures are used.

Filters are gaining popularity with hikers since fire or chemicals alter the taste of water. Various sizes of filters are available, but an "absolute" pore size of 1 micron or less should be used in most situations. Since filters do not remove all microorganisms, the Centers for Disease Control does not currently recommend the use of filters for water purification. Many hikers use a combination of filters and chemical treatment to increase the likelihood of complete disinfection. Given these conflicting safety recommendations, boiling water for an adequate period of time or bringing water from home are the two safest methods of water use in the backcountry. (56)

WATER QUALITY MANAGEMENT

FATE AND TRANSPORT

Fate and transport are the movement and ultimate disposition (and residence) of pollutants. Determining the fate and transport of pollutants is extremely important in predicting potential sites of surface and groundwater contamination. Fate and transport in surface water depend on stream velocity, mixing, temperature, concentration of contaminants, variability of stream cross sections, sedimentation, and residence time of water. For example:

- Heavy metals often become embedded in stream sediments.
- Nitrates can migrate in groundwater until removed by groundwater pumping.
- Volatile organic compounds can float on the surface of a groundwater aquifer and slowly migrate with the movement of groundwater.

Fate and Transport in Surface Water Pollutants in surface water may be in a dissolved form or may occur as particulates (very small particles). Dissolved compounds have the same movement patterns as flowing surface water, whereas particulates move similar to sediments of equal size and weight that can move by saltation (discussed in Chapter 3). Turbulent drag forces on the bottom of a stream must be considered when determining the transport and eventual fate of particulate contamination. Mixing zones of turbulent surface water, or additional flows from tributaries, effluent streams, or tidal action will also affect the dispersion of pollutants. (57)

Fate and Transport in Groundwater Pollutants in groundwater generally move in the form of a plume of contamination. Depending on the weight of the pollutant, the rate of movement may be faster or slower than groundwater located in the aquifer. If the contaminant is a metal and heavier than water, pollution movement may exceed the movement of groundwater. If the contaminant is lighter and not water soluble, such as hydrocarbons or other petroleum-based products that float near the surface of an aquifer, the contaminated plume may move slower than groundwater.

Pollution in groundwater moves by the force of gravity. A plume of pollution may move only a few *feet per year*, or may travel several *miles within a single decade*. Predicting the movement rate, direction, and extent of a plume is very difficult. For analysis and testing purposes, groundwater monitoring wells are usually drilled, based on geology, direction and rate of groundwater movement, and samples taken to determine the extent of contamination. The edges, or fringe, of a contamination plume generally have lower concentrations of pollutants than locations nearest the source of contamination. The location of a plume is generally determined with monitoring wells (see Figure 5.10) and the use of groundwater computer models. Researchers often inject dye into a groundwater aquifer to track the movement of a plume of contamination.

Pollutants in aquifers degrade at variable rates under natural conditions but generally change very slowly owing to the lack of sunlight, oxygen, and microbial activity. Recent field demonstrations of technologies used to remediate contaminated groundwater show, however, that biodegradation of contaminants can be enhanced significantly through injection of specialized microbes, oxygen gas, or other amendments in both aerobic and anaerobic environments.

Currently, researchers within government, private industry, and academia are studying and debating the feasibility of employing **monitored natural attenuation (MNA)** as a remedy for cleaning up contaminated aquifers. MNA involves the long-term monitoring of groundwater conditions to ensure that contamination degradation is occurring under natural conditions (albeit slowly), without efforts to treat a contaminated aquifer. This approach is most often considered in areas where complete groundwater cleanup costs are prohibitive, health and environmental risks are very low, or cleanup goals may not be achievable for many decades. (58)

A CLOSER LOOK

The USEPA's Superfund Program was established in 1980 to locate, investigate, and clean up hazardous waste sites around the United States. It was authorized under CERCLA in response to deadly spills, improper storage, and abandoned toxic chemical factories. Thousands of sites have been located since passage of the law, and cleanup costs have been in the billions of dollars. Currently, there are more than 1200 sites across the nation, of which approximately 80 percent involve contaminated groundwater. (59)

An example of a Superfund site, in the remediation (cleanup) stage, is the former Nebraska Ordnance Plant near Mead, Nebraska. This World War II munitions factory assembled bombs, shells, and rockets on a 17,250-acre (6981 ha) site northeast of Lincoln. Extensive groundwater and soil contamination was

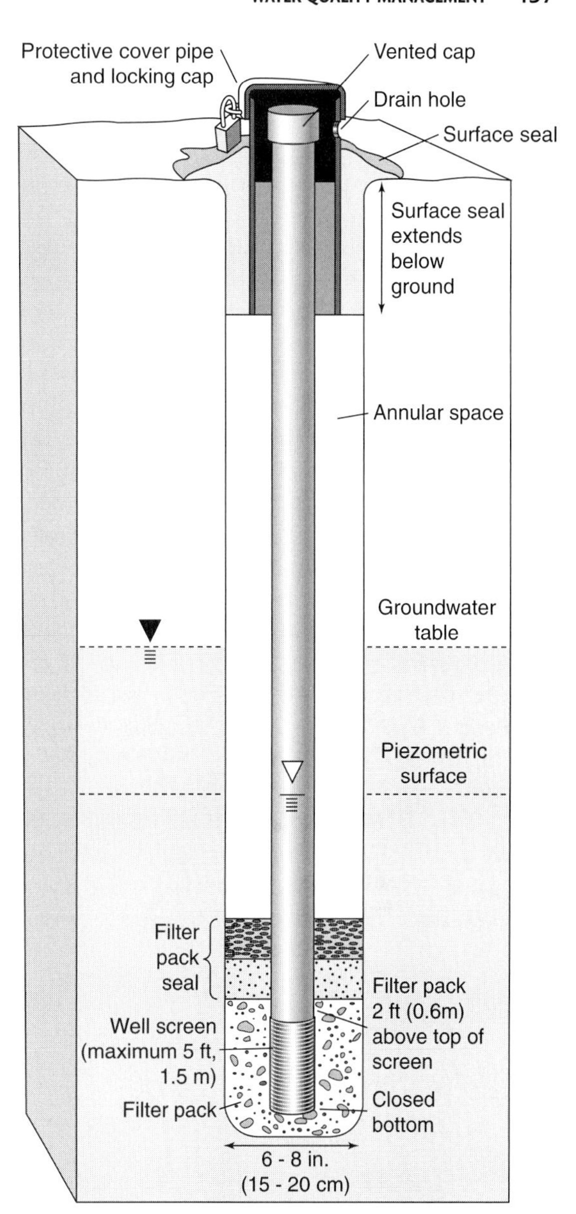

FIG. 5.10. Typical groundwater monitoring well. Note the piezometric (also called potentiometric) surface located below the groundwater table. This is the level to which water rises in a tightly cased well in a confined aquifer. In an unconfined aquifer, the piezometric surface and the groundwater table would be the same.

caused by improper disposal of the solvent trichloroethylene (TCE) and the explosive compound RDX (Royal Demolition explosive, chemical name 1,3,5-trinitro-1,3,5-triazine). During production, it was a regular practice to wash out the interiors of the production buildings to prevent a dangerous buildup of explosive dust residue. Unfortunately, the pollution this practice created was not understood or deemed significant at the time. Contaminated wash water was flushed from the buildings into ditches and eventually into the local groundwater aquifer. Many downstream and downgradient water users obtained drinking and irrigation water supplies from these contaminated sources.

It's estimated that 69,000 acre-feet (22.5 billion gal, or 85.1 million m^3) of contaminated groundwater underlie 6000 acres (2428 ha) or are migrating from the former munitions factory. Remediation activities include bottled water and other sources for residents in the area, incineration of contaminated soils, and groundwater cleanup by recirculation through an extensive filtering system at a rate of 2550 gallons per minute (9653 l per minute). (60)

The contamination of groundwater supplies around the former Nebraska Ordnance Plant is a tragedy for residents in the area. However, Frank Gwin, Jr., of Tchula, Mississippi (Figures 5.11 and 5.12), provided a different perspective while on a tour of groundwater experts to the site in May of 2000. Gwin served in the U.S. Army during World War II but was captured by the Germans and held prisoner for nearly four years. The prison was located near Potsdam, Germany, and conditions were beyond description. Lack of food, torture, and disease were so common that the outlook for survival was bleak for all the prisoners. Then U.S. Air Force planes started flying overhead toward factory targets nearby. Explosions lit the night sky, and the sounds of war continued for months. Finally, after day upon day of bombings, American troops arrived at the Nazi prison gates to free their G.I. comrades. The war was soon over, and Frank Gwin went home.

After the Mead groundwater tour 55 years later, Gwin addressed the busload of participants:

I feel so sorry for the residents around this munitions plant, for having to deal with the deadly chemicals in their groundwater. But I wish I could thank everyone of them personally, to shake their hands, and thank them for the sacrifice they are making today. There are many of us who wouldn't be alive today if it wasn't for the bombing of those German factories near our prison camp. I don't know how much longer I would have lasted if the war hadn't ended when it did. This munitions plant saved my life.

What types of handling procedures should have been implemented in the 1940s to prevent the tragic groundwater pollution of the regional drinking water supply? What role did the federal government's concern about winning World War II probably have in allowing lax environmental protection in the United States during the war? Does a hazardous waste site exist in your state, and was it caused by industrial activities during World War II? What restrictions would be placed on such a production facility today to prevent nonpoint source pollution?

FIG. 5.11. POW Frank Gwin, Jr., at the time of his capture in Germany during World War II.

FIG. 5.12. Frank Gwin, Jr., today.

GUEST ESSAY

Dr. Curt Elmore

Managing Data for a Groundwater Restoration Project

by **Curt Elmore, Ph.D., P.E.**
University of Missouri—Rolla

Curt Elmore participated in the groundwater restoration activities at the former Nebraska Ordnance Plant over a 10-year period beginning in early 1992. His professional expertise includes the design of groundwater remediation systems, decision analysis and stochastic methods applied to groundwater restoration, and the design of groundwater circulation wells for groundwater restoration and conservation. Currently, Dr. Elmore is on the faculty of the University of Missouri—Rolla, in the Department of Geological and Petroleum Engineering.

The former Nebraska Ordnance Plant is located in the east-central portion of the state in a primarily agricultural area. The facility was used during World War II and the Korean conflict to assemble bombs and shells, and explosives compounds were released to the environment during those activities. In the late 1950s and early 1960s, the Strategic Air Command used portions of the facility as an intercontinental ballistic missile site and for activities supporting other missile sites. Apparently, the common solvent called TCE was released to the environment at that time.

Today most of the site is used for the University of Nebraska Agricultural Research and Development Center. The groundwater underlying approximately 6000 acres (2428 ha) of the Superfund site has unacceptable concentrations of TCE and the explosive compound RDX. The U.S. Army Corps of Engineers and the U.S. Environmental Protection Agency, with the support of the Nebraska Department of Environmental Quality, are currently implementing groundwater restoration activities.

Collection and analysis of data, development of remedial alternatives, and design and construction of the groundwater remedy have taken more than 10 years. During that period, more than 160 monitoring wells were installed and routinely sampled. Soil gas investigations were conducted, soil samples were collected, and domestic and irrigation water wells were routinely monitored. All of the data were collected in a systematic manner in order to support specific data needs and requirements.

The project has several data users. The engineers and scientists who have developed the various groundwater restoration remedies have used the data to define the nature and extent of groundwater contamination, to track the changes in the distribution of the contamination over time, to define the physical parameters of the aquifer and the soil above the aquifer, and to evaluate the performance of the groundwater restoration systems. The data provided necessary information to the project decision makers who selected the course of action that would be followed at the site. The data were also made available to the representatives of the local community who were formally involved in restoration activities at the site through the Technical Review Committee and the Restoration Advisory Board.

Hundreds of thousands of individual data records for chemical concentrations in soil and groundwater, aquifer water levels, land survey coordinates of monitoring points, and other parameters were collected for the project. The management of the data evolved to meet the needs of the projects. Initially, the data were stored in a proprietary database specially designed for the management of environmental data. Later, the data were transferred to a widely available commercial database to accommodate additional data users and expanded data use. Today, a large portion of the data have been included in a commercially available geographic information system. One of the primary benefits of the GIS system is that it allows the U.S. Army Corps of Engineers to rapidly access project data from a graphical interface.

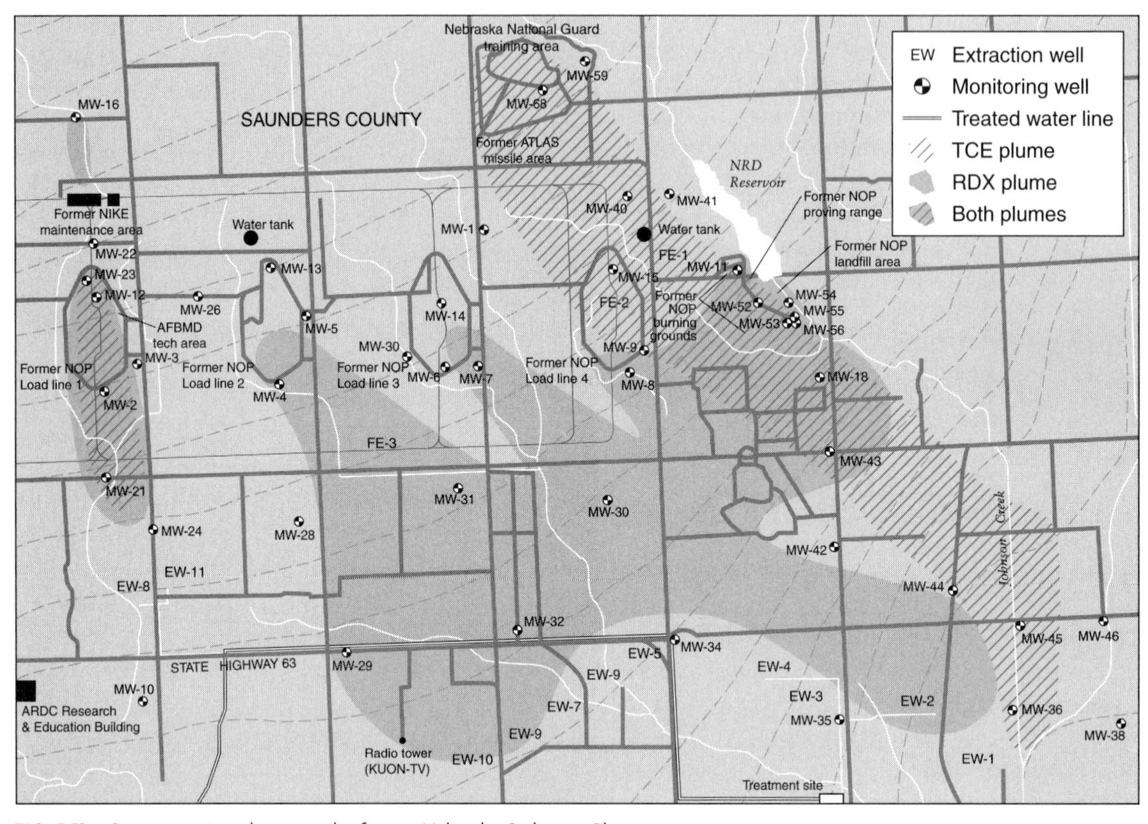

FIG. 5.13. Contamination plumes at the former Nebraska Ordnance Plant.

This feature is valuable when assessing the performance of the restoration systems and when presenting the data to the public. Figure 5.13, developed using ArcView, shows several features of the project.

Two high-capacity extraction wells are currently creating hydraulic capture zones to keep portions of the contamination plumes from migrating to uncontaminated aquifer areas. The extracted groundwater is piped to a facility where it is treated using granular-activated carbon prior to being discharged to a nearby stream. The construction of nine additional high-capacity wells is complete, and an expanded treatment plant will soon be treating all of the groundwater extracted for the purpose of hydraulic containment. The infrastructure that is in place will allow local farmers to use some of the treated groundwater for agricultural purposes, and treated groundwater will also be available for rural firefighters during wildfire season.

Two groundwater circulation well systems are currently removing contamination from groundwater hotspots; 12 more systems have been designed and will be constructed to remediate even more hotspots. All of these remedial action systems include instrumentation for the collection of performance monitoring data, and given the size of the groundwater contamination problem, it is estimated that the systems will operate for decades. Activity and interest at the site will continue throughout the life of the project. The project engineers, the regulatory agencies, and the public will continue to have an active interest in knowing what is happening at the site. The efficient management of both the existing and yet uncollected data will remain a priority for the project.

WELLHEAD PROTECTION PROGRAMS

Congress established the **Wellhead Protection Program (WHPP)** in 1986 through amendments to the Safe Drinking Water Act. The WHPP program is a proactive citizen, local, state, and federal process designed to protect **source water** (untreated groundwater used for public drinking water supplies). Protection methods can include land-use controls through zoning, restrictions on storage of hazardous materials, surface water runoff regulations, and other criteria. Individual communities generally develop Wellhead Protection Programs under the guidance of state and USEPA officials. (61)

There are five steps in establishing a Wellhead Protection Program: (1) organization; (2) delineation; (3) contamination source inventory; (4) source management; and (5) contingency planning.

Organization *Organization* of a WHPP (sometimes called a *whip*) involves several different steps, including establishing a local team to develop the program, contacting the state to announce intent to develop a program, applying for state or federal grants to assist with development, and organizing local informational meetings or other activities to begin the planning process. This first step is often the most difficult component since it requires a long-term commitment from local sponsors.

Delineation Rigorous technical requirements are used to establish flow lines, cones of depression, and contribution areas in order to determine the surface land area that will be managed under the program (see Figure 5.14).

The USEPA allows the use of three methods of delineation: fixed radius, analytical model, and numerical model.

- *Fixed radius* simply involves drawing a circle around a wellhead based on the relative pumping rate of the well. This technique is very crude and does not take into account flow lines and well capture zones. (Wells rarely have a concentric well capture zone.)
- The *analytical model* involves groundwater flow equations such as Darcy's Law to determine the areas of contribution to the well. Factors such as aquifer slope, transmissivity, and the rate and direction of groundwater flow are taken into account to determine the protection area of the well.
- *Numerical modeling* is similar to analytical models except additional data are used to obtain more realistic results of aquifer characteristics. The study area is divided into cells of 1 mile (1.6 km) square or smaller, and attributes are assigned to aquifer characteristics in each cell. This fine level of detail provides a more accurate picture of groundwater characteristics in an area but can be quite expensive owing to the cost of additional monitoring wells, computer and engineering analysis, and interpretation.

Vulnerability mapping of soils should be considered when delineating a wellhead protection area. The susceptibility of groundwater pollution from land-use practices can be mapped and developed into classifications or indices to show leaching potentials. This mapping would include soil thickness, permeability, slope, pH of soil, and classification of soil types such as gravel, sand, or loess. Other factors would include ground cover, depth to groundwater, and hydraulic gradient of the groundwater table. (62)

Contamination Source Inventory Land uses within the delineated area are inventoried, using county planning maps, site inspection, or aerial photography in order to identify potential sources of contamination. Factories, farming operations, chemical storage locations, landfills, and similar facilities or land uses are identified.

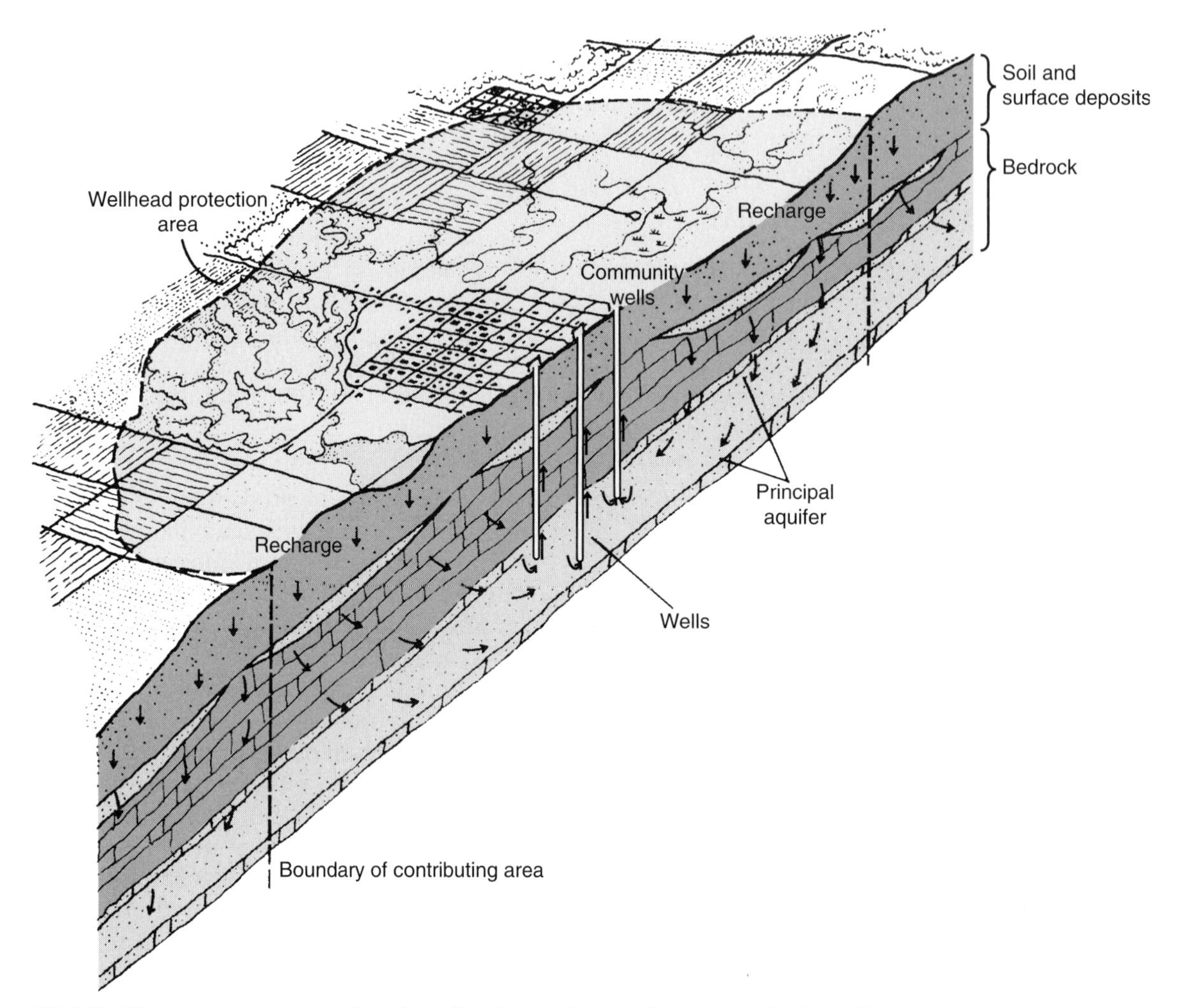

FIG. 5.14. This conceptual representation of a wellhead protection zone for a community shows the complex nature of groundwater recharge, groundwater pumping, land use, and geology. All are taken into consideration when delineating a WHPP.

Source Management After the WHPP is established, delineated, and inventoried, a community can then begin regulating land uses that could contaminate groundwater supplies. Nonregulatory options include education, land acquisition, and groundwater monitoring. Regulatory options include zoning restrictions that restrict or prohibit certain types of future land uses such as landfills, hazardous waste disposal, storage of hazardous wastes, feedlots, golf courses, or other uses that could pollute the aquifer. Existing land uses are generally grandfathered (allowed to continue) unless appropriate compensation is given for the lost use of private property.

Contingency Planning If the prior actions do not provide safe drinking water supplies for a community, what actions will be taken? What alternative sources of water are available? How will residents be alerted to unsafe water supplies? These and other emergency measures must be established.

WATERSHED PROTECTION PROGRAMS

Many public and private organizations are working together to form partnerships to improve the quality of surface and groundwater in watersheds. This process, which the USEPA calls a **Watershed Protection Approach (WPA),** is providing significant results across the country. Senior managers at the USEPA endorsed this method of local, state, regional, and tribal cooperation in 1991 and continue to support comprehensive approaches to watershed protection.

The Watershed Protection Approach involves a variety of elements, but generally follows these guiding principles: (1) partnerships; (2) geographic focus; and (3) sound management techniques based on strong science and data.

Partnerships require the inclusion of a broad spectrum of people most affected by management decisions made under a watershed protection program. This ensures the integration of economic stability as well as the consideration of local social and cultural goals.

The delineation of the *geographic focus* is extremely important so that reasonable geographic boundaries are selected for the program. The area may include an entire watershed, a tributary of a larger watershed, or the drainage area into a lake or other water body. The geographic focus may include the area that overlays a local groundwater aquifer or a combination of surface and groundwater protection zones. The focus area may be further divided into management units that delineate smaller watersheds such as source water for a community well, special wetlands protection areas, or riparian zones along streams.

Sound management techniques must be based on strong science, as well as relevant and accurate data. Generally, the **ambient** (existing) **conditions** of water quality in the study area will be assessed, water demands by local communities and other water users will be quantified, goals and objectives will be identified, problems will be prioritized, specific management options and action plans will be considered, and finally, implementation and evaluation of programs will begin. Ideally, management techniques will include improvements in the coordination of existing programs rather than implementation of new regulatory processes and procedures.

Since stakeholders must work together, all information is shared, concerns are addressed openly, and pollution prevention techniques are adopted when possible. Establishment of attainable goals and targets is vital to the success of the watershed program. Partnership development is extremely important between representatives of local, regional, state, tribal, and federal agencies, public interest groups, industry, academia, private landowners, concerned citizens, and others. (63)

Data acquisition may include an inventory of existing information in the study area, including groundwater data, sources of drinking water, wetlands and riparian areas, unique wildlife habitat locations, and restoration sites. Additional data may be collected to fill in data gaps. Data are stored in a readily accessible format to allow use by all parties. A geographic information system can be used to map information for ease of visual observation of data.

Management goals can include establishment of groundwater protection areas, wetlands protection plans, estuary protection programs, consultation on NPDES permits for dischargers, and establishment of Total Maximum Daily Loads (TMDLs).

Total Maximum Daily Loads The Federal Water Pollution Control Act (Clean Water Act) of 1972 implemented many water quality programs (which will be discussed in Chapters 9 and 11). However, a current program receiving much attention by state and local agencies around the United States is the **Total Maximum Daily Load (TMDL) Program.** The purpose of TMDL standards is to improve water quality conditions in rivers, lakes, and along shorelines by calculating the maximum amount of a pollutant that a water body can receive and still meet state and federal water quality standards. A TMDL is, therefore, the sum of all pollutants from point sources and nonpoint sources into a water body. The calculation for a TMDL must consider seasonal variations such as temperature, flow rates, and surface water runoff, and must include a margin of safety for designated uses of the water.

The USEPA estimates that 300,000 miles (482,790 km) of impaired rivers and shorelines exist in the United States today. Contamination is caused primarily by sediments, excess nutrients, and microorganisms. It is also estimated that most of the U.S. population lives within 10 miles (16 km) of these impaired waters. (64) Section 303(d) of the Clean Water Act requires that local regulators (agencies of states, territories, and authorized tribes) develop a list of impaired waters within their jurisdictions. Next, these regulatory agencies must establish a priority ranking for impaired waters in their area and then develop TMDLs to protect or improve existing water quality. (65) Federal policymakers believe that TMDLs are an appropriate method to improve water quality.

POLICY ISSUE

TMDLs have been required by Section 303 of the Clean Water Act since 1972 and are set by states, territories, and authorized tribes. The USEPA must by law approve or disapprove 303(d) lists of impaired waters and established TMDLs. The agency also readjusts inadequate 303(d) lists and TMDLs.

Extensive litigation has stalled implementation. On July 13, 2000, the USEPA proposed new rules for TMDLs, but Congress prohibited the USEPA from spending any funds to implement those new rules in fiscal year (FY) 2000 and FY 2001. (66) On March 13, 2003, USEPA Administrator Christie Whitman withdrew the proposed July 2000 Rule after receiving more than 34,000 comments (90 percent of which supported its withdrawal). (67)

Are TMDLs being administered in your area? How were these standards developed? What state, territorial, or tribal agency is responsible for taking water samples, monitoring data, and enforcing the TMDLs? How does the TMDL program differ from previous federal water quality programs? Do you agree that the USEPA should be able to change the 303(d) list or TMDLs selected by local regulators?

WATER SAMPLING

Many water agencies collect surface and groundwater samples to assess the quality of water resources in an area. Water testing can involve the collection of "grab" samples that are simply "grabbed" in an appropriate container from a river, lake, or discharge pipe of a well or distribution line. The sample is then placed in an ice cooler to remain at a constant temperature and taken to a laboratory for analysis. Additional samples may be obtained from groundwater monitoring wells.

Some agencies have developed in-house laboratories to conduct analysis of basic water parameters such as pH, hardness, nitrate, nitrite, and dissolved oxygen. A rigorous quality control program must be in place to ensure accurate results. A QA:QC (quality assurance:quality control) document is usually prepared to set standards for sample collection, transport, chain of custody, and number of samples.

In accordance with applicable regulations, the sample collection protocol involves criteria such as minimum volume of water to collect, type of collection container to use, transportation, temperature, and recordkeeping. Chain of custody provides signed documentation by all individuals in control (custody) of each water sample. It will show the sampling location, date, and time of sampling events, as well as other relevant information such as field conditions and water discharge.

The number of samples will be scheduled by week, month, and time of day, and may include "splits," "spikes," "duplicates," and "blanks" that help ensure the accuracy of analytical laboratory results. A split sample comprises a single sample that is split into two containers while in the field, without the laboratory's knowledge, so that analytical results can be compared. The results should be identical for each of the splits if laboratory procedures are being followed correctly. Spike samples are created by adding a known concentration of pollutant to a sample, while duplicate samples (dupes) are replicates collected under identical field conditions. Blank samples are made by filling a sampling container with nothing but distilled water.

SIDEBAR

Standard Methods for the Examination of Water and Wastewater, 20th edition, is published by the American Water Works Association, American Public Health Association, and the Water Environment Federation. This definitive work describes individual analytical procedures in detail for water sampling and analysis. Consult this publication for further information on water sampling.

GUEST ESSAY

Sandra Novotny

Environmental Consulting as a Career

by **Sandra Novotny, Environmental Consultant, Kensington, Maryland**

Sandra Novotny, president of Nova Environmental Services in Kensington, Maryland, specializes in environmental policy and technical analysis concerning hazardous waste site remediation. As a consultant for the past 20 years, she has assisted local, state, and federal agencies such as the U.S. Environmental Protection Agency (USEPA), the U.S. Department of Energy, and the U.S. Department of Defense in the implementation of environmental protection and restoration programs. She received a B.A. in Biology and Psychology from the University of Colorado in 1975 and an M.S. in Earth Science from Adelphi University in 1983. Over the years, she has worked for major environmental consulting firms such as SAIC, ABB Environmental Services, and DynCorp.

My interest in water quality likely began in the early 1960s during childhood walks along Maple Creek, which ran across my family's farm in Nebraska. During these hikes, I learned to ignore the unrepairable farm equipment, household trash, and used oil and chemicals that were routinely dumped along the banks of the creek as it made its way downstream to the Platte River. At that time, no one generally talked about the adverse impacts of practices such as this, and the government had little involvement in the relationship between natural resources and private industries or consumers.

All of that changed dramatically during the 1970s and 1980s, however, with passage of the Clean Water Act (CWA), Clean Air Act (CAA), Resource Conservation and Recovery Act (RCRA), and Comprehensive Environmental Response, Compensation, and Liability Act (CERCLA, or Superfund), and creation of the USEPA. I left college to join one of the many environmental consulting firms that were forming to help private industry comply with the increasing number of rules and regulations issued by federal and state agencies. While helping clients in the mining industry to develop permitting and compliance plans, I found the network of interwoven regulations to be fascinating and came to prefer such paper studies over traditional field work.

To enhance my credibility in the consulting business, which is extremely competitive, I obtained a graduate degree in earth science with a concentration in environmental management. My interest in regulatory analysis blossomed when I began assisting the USEPA through its Superfund Program. My work began with "reportable quantities" (RQs), the concentrations and volumes of hazardous substances that must be reported to authorities when accidentally released. It broadened over time to address the many activities involved with the distinct phases of Superfund site cleanup: identification, characterization, preliminary assessment, detailed field investigation, remedial option planning, remediation, and technology development. My projects have included technical support for several RQ rulemakings, development of an emergency response notification information system, hazard ranking system (HRS) evaluations determining whether a site should be listed on the National Priorities List (NPL), outreach programs promoting the use of innovative cleanup technologies, and development of the agency's annual reports to Congress.

The tools and methods used to provide regulatory and analytical support for agencies such as the USEPA are important and vary in accordance with the needs of the end-product "deliverable." Large numbers of chemical- and site-specific technical reports, guidance materials, and databases have been developed by the USEPA and other federal agencies such as the Centers for Disease Control and Prevention (CDC) to help regulatory agencies and regulated communities. Knowing where to obtain background and up-to-date source materials, how to best analyze and interpret raw data and technical or policy updates, and how to package a document for effective use by the intended audience are key elements in the environmental consulting business.

One of my most interesting experiences occurred early in my work with the USEPA, perhaps in part due to the excitement of close contact with the "movers and shakers" in Washington, D.C., but primarily due to the project urgency. My consulting office had been housed for a few months at USEPA's headquarters location, within the Office of Emergency and Remedial Response, which administers the Superfund Program and Oil Program. On January 2, 1988, the offices and corridors suddenly were buzzing with the bad news that a 400-million-gallon (1228 acre-foot, or 1.5 million m^3) aboveground storage tank had ruptured at the Ashland Oil facility in Floreffe, Pennsylvania, and had released one million gallons (3.1 acre-feet or 3785 m^3) of fuel that was making its way to the adjacent Monongahela and Ohio rivers. As a result of the spill, 15 water intakes were shut down, water supplies to over 2.7 million residents were disrupted, schools and businesses were closed, and thousands of fish and waterfowl were destroyed.

Soon after the spill, the USEPA formed an interagency Oil Spill Prevention, Control, and Countermeasures (SPCC) Program Task Force. The Task Force was charged with exploring alternatives to ensure that similar spills would not occur in the future, or that if such a spill were to take place the negative impacts would be minimal. Over the following 90 days, we quickly organized Task Force activities, analyzed comparable industry standards and regulatory requirements, compiled a lengthy and high-profile technical report, negotiated and integrated diverse (and sometimes divergent) comments, and produced a final report for review by Congress and regulatory agencies. It was found that the violating facility had neither an up-to-date SPCC Plan, as required under Section 311 of the CWA, nor an adequate contingency plan.

Pursuant to a major Task Force recommendation, the USEPA began developing new regulations that would require more stringent controls for storage of hazardous chemicals and oil. It was estimated that SPCC planning was required at more than 650,000 nontransportation-related facilities that could potentially discharge oil into or upon navigable waters of the United States or adjoining shorelines. At the time of the Ashland Oil spill, the SPCC regulations (40 CFR 109-114) did not require facilities to meet minimum standards for design and operation.

Following issuance of the Task Force report in May 1988, my support to the agency was converted to assistance with the complex and intensive rulemaking process, which involved drafting of regulatory language for a revision of the original 1973 SPCC regulation, analysis of public comments on the agency's proposed revision, additional technical and policy research on tank-related issues, preparation of a comment/response document summarizing the agency's response to comments, and revising text for the final rule. In early 2001 (13 years later), the rule revision finally was signed by the USEPA administrator.

In 1997, I established an independent consulting practice to provide analytical and management expertise in the environmental sector of Washington, D.C. My work on issues surrounding hazardous waste site cleanup continues, with a focus on emerging remediation methods, technologies, and programs. As cleanup begins to near completion at some of the priority sites across the nation, more attention now may be given to "brownfields" where contamination is less extensive, or simply the perception of contamination has prevented productive use.

CHAPTER SUMMARY

This chapter has attempted to show that water quality involves a wide range of parameters, including temperature, minerals, metals, dissolved oxygen, and nutrients. Pollution can occur from point sources (discharge from pipes) as well as from nonpoint sources such as chemical runoff from lawns, golf courses, and fields. Pollution can also originate from natural sources but is too often caused by human activities. Although severe water pollution has occurred in the United States for centuries, federal water quality programs, such as the Clean Water Act, Watershed Protection Programs, and TMDLs are improving the quality of the nation's rivers and lakes. The public policy implications of these programs are tremendous, with the health of rivers, lakes, wetlands, and groundwater hanging in the balance. See Table 5.3 for a listing of major water quality events discussed in this chapter.

TABLE 5.3 Selected History of Water Quality Events

Date	Event
Antiquity	Floods around the world create sedimentation and salinity pollution. The hydrologic cycle causes minerals and metals to dissolve and enter surface and groundwater sources. Decaying animals and vegetation cause eutrophication of swamps and wetlands.
100 B.C.[a]	Marcus Vitruvius Pollio promotes use of lead pipes to deliver drinking water in Rome.
A.D. 1000–1200	Plagues wipe out entire villages in Europe.
1681	*Giardia* first identified.
1790	5 percent of U.S. population lives in urban areas.
1848–1849	Cholera epidemic in London kills over 53,000 people.
1854	Dr. John Snow convinces local government officials to remove handle from the Broad Street pump in London to prevent use of contaminated water.
1860	20 percent of U.S. population lives in urban areas.
1861	Louis Pasteur performs experiments regarding control of microorganisms.
1873	DDT invented.
1885	Over 80,000 people die from contaminated drinking water in Chicago, Illinois.
1899	Rivers and Harbors Act passed by the U.S. Congress.
1900	40 percent of pollution load in U.S. rivers derives from industrial waste.
1920	51 percent of U.S. population lives in urban areas.
1939	Paul Herman Müller of Switzerland awarded Nobel Prize for work with DDT.
1950[a]	Fluoride added to a public drinking water supply in the United States.
1959	80 million pounds (36.3 million kg) of the pesticide DDT used in the United States.
1960s	Shipping lanes of Chesapeake Bay obstructed due to algae growth caused by eutrophication.

(continued)

TABLE 5.3 Selected History of Water Quality Events *(continued)*

Date	Event
1962	*Silent Spring,* landmark environmental book regarding the dangers of pesticides by Rachel Carson, published and becomes best seller.
1965	First case of *Giardia* in the United States officially reported in Aspen, Colorado.
1968	80 percent of the pollution load on U.S. rivers derives from industrial waste.
1972	Water Pollution Control Act (Clean Water Act) passed by the U.S. Congress.
1972	Federal Insecticide, Fungicide and Rodenticide Act (FIFRA) passed by Congress to regulate the use of pesticides.
1972	Use of DDT banned in the United States.
1974	Safe Drinking Water Act passed by the U.S. Congress.
1974	Colorado River Basin Salinity Control Act passed by U.S. Congress.
1980	Comprehensive Environmental Response Compensation and Liability Act (CERCLA, or Superfund Program) established by U.S. Congress.
1986	Wellhead Protection Program (WHPP) authorized by Congress and implemented by the USEPA.
1986	Amendments to the Safe Drinking Water Act prohibit the use of lead water distribution pipes in new or replacement construction.
1990	75 percent of U.S. population lives in urban areas.
1991	USEPA endorses Watershed Protection Approach (WPA) to develop comprehensive watershed protection programs.
1993	400,000 people infected with *Cryptosporidium parvum* in Milwaukee, Wisconsin, with over 100 deaths registered.
1999	National Academy of Sciences determines that drinking water standards for arsenic, set by the USEPA, are too high to protect public health and safety.
2000	World Health Organization calls for continued limited use of DDT to prevent malaria.
2000	2300 people become ill and 7 die in Walkerton, Ontario, from *E. coli* contaminated drinking water.
2002	Communities in Minnesota ban the use of fertilizers that contain phosphorus.

[a]Approximate date.

QUESTIONS FOR DISCUSSION

1. Explain the difference between point source and nonpoint source pollution.

2. Describe potential sources of point source and nonpoint source pollutants.

3. Describe several basic parameters of water.

4. Explain the relationship between irrigation and salinity. Why are Mexico and the United States working on this problem along the Colorado River?

5. Explain the nitrogen cycle.

6. How are some areas working to reduce eutrophication in ponds, lakes, and wetlands?

7. Why would a water resources manager be concerned with biological oxygen demand in a waterway?

8. Discuss the relationships between safe drinking water and Dr. John Snow.

9. What steps are necessary to establish a Wellhead Protection Program?

10. TMDLs will be a major water resource management factor in the future. Why?

11. Discuss the role of phosphorus in the pollution of waterways in Minnesota. What actions have been taken to reduce this problem?

12. Discuss some governmental actions that could have prevented the tragedy that occurred in Walkerton, Ontario, in May 2000.

KEY WORDS TO REMEMBER

aerobic p. 130
ambient conditions p. 143
ammonia (NH_3) p. 128
ammonium (NH_4^+) p. 128
anaerobic p. 130
arsenic p. 121
biological oxygen demand (BOD) p. 131
carcinogens p. 124
Colorado River Basin Salinity Control Act (1974) p. 122
Cryptosporidiosis p. 135
Cryptosporidium parvum p. 135
DDT p. 124
denitrification p. 130
dissolved oxygen (DO) p. 117
Escherichia coli (E. coli) p. 134
eutrophic p. 131
eutrophication p. 131
fate and transport p. 136
fecal coliform p. 134
Federal Insecticide, Fungicide, and Rodenticide Act (FIFRA) of 1972 p. 125
fluoride (F) p. 123
fungicides p. 125
Giardia p. 135
Giardia lamblia p. 135
hardness p. 119
herbicides p. 125
indicator organism p. 134
inorganic chemicals p. 119
insecticides p. 125
Integrated Pest Management (IPM) p. 126
leach p. 121
lead p. 120
maximum contaminant level (MCL) p. 124
metals p. 119
microorganism p. 133
mineralization/ammonification p. 129
minerals p. 121
monitored natural attenuation (MNA) p. 137
natural organic chemicals (NOCs) p. 124
nematacides p. 125
nitrate (NO_3^-) p. 127
nitrification p. 130
nitrite (NO_2^-) p. 127
nitrogen (N) p. 127
nitrogen cycle p. 128
nitrogen fixation p. 128
nitrogen gas (N_2) p. 127
nonpoint source pollution p. 113
nonvolatile organic chemicals (NVOCs) p. 124
nutrients p. 127
organic chemicals p. 124
pesticides p. 125
pH p. 117
phosphorus (P) p. 128
plume p. 115
point source pollution p. 113
pollution p. 112
rodenticides p. 125
salinity p. 121
salt p. 121
source water p. 141
synthetic organic chemicals (SOCs) p. 124
total coliform p. 134
Total Maximum Daily Load (TMDL) Program p. 143
turbidity p. 118
volatile organic chemicals (VOCs) p. 124
Watershed Protection Approach (WPA) p. 143
Wellhead Protection Program (WHPP) p. 141

SUGGESTED RESOURCES FOR FURTHER STUDY

READINGS

Barzilay, Joshua I., Winkler G. Weinberg, and J. William Eley. *The Water We Drink—Water Quality and Its Effects on Health.* New Brunswick, NJ: Rutgers University Press, 1999.

Canter, Larry W. *Nitrates in Groundwater.* Boca Raton, FL: Lewis Publishers, 1997.

Carson, Rachel. *Silent Spring.* New York: Houghton Mifflin Co., 1962.

Csuros, Maria, and Csaba Csuros. *Microbiological Examination of Water and Wastewater.* Boca Raton, FL: Lewis Publishers, 1999.

De Zuane, John. *Handbook of Drinking Water Quality.* 2nd ed. New York: Van Nostrand Reinhold, 1997.

Donahue, Roy L., Raymond W. Miller, and John C. Shickluna. *Soils: An Introduction to Soils and Plant Growth.* 5th ed. Englewood Cliffs, NJ: Prentice Hall, 1983.

Krenkel, Peter A., and Vladimir Novotny. *Water Quality Management.* Orlando, FL: Academic Press, 1980.

Stewart, John Cary. *Drinking Water Hazards.* Hiram, Ohio: Envirographics, 1990.

Viessman, Warren Jr., and Mark J. Hammer. *Water Supply and Pollution Control.* 6th ed. New York: Harper & Row, 1998.

WEBSITES

Hach Company, "Important Water Quality Factors," March 2001. http://www.hach.com/h2ou/h2wtrqual.htm

Public Broadcasting System (PBS), Bill Moyers Reports, "Earth on Edge," March 2002. http://www.pbs.org/earthonedge

U.S. Environmental Protection Agency, Office of Water, Water Quality Conditions in the United States: A Profile from the 1998 *National Water Quality Inventory Report to Congress,* EPA-841-F-00-006 (4503F), Washington, DC, June 2000, September 2001. http://www.epa.gov/305b/98report/98summary.html

U.S. Environmental Protection Agency, Office of Water, "Clean Water Act: A Brief History," Washington, DC, September 2001. http://www.epa.gov/owow/cwa/history.htm

U.S. Environmental Protection Agency, Office of Water, "Current Drinking Water Standards," Washington, DC, September 2001. http://www.epa.gov/safewater/mcl.html

U.S. Environmental Protection Agency, "Hudson River PCBs," September 2001. http://www.epa.gov/region02/superfnd/hudson

U.S. Environmental Protection Agency, "Superfund," September 2001. http://www.epa.gov/superfund/about.htm

U.S. Environmental Protection Agency, History Office, "New DDT Reports Confirms Data Supporting 1972 Ban, Finds Situation Improving," Washington, DC, September 2001. http://www.epa.gov/history/topics/ddt/03.htm

U.S. Environmental Protection Agency, "Summary of State Biennial Reports of Wellhead Protection Program Progress," September 2001. http://www.epa.gov/safewater/protect/gwr/biennial.html

University of California at Los Angeles, School of Public Health, Department of Epidemiology, Los Angeles, California, September 2001. http://www.ph/ucla.edu/epi/snow/removal.html

World Resources Institute, "Industrialization: Heavy Metals and Health," Washington, DC, September 2001. http://www.wri.org/wr-98-99/metals2.htm

VIDEOS

A Civil Action. Directed by Steven Zaillian. Touchstone Pictures/Paramount Studios, 1998. 118 min.

Erin Brockovich. Directed by Steven Soderbergh, Universal Studios, 2000. 126 min.

REFERENCES

1. U.S. Environmental Protection Agency, Office of Water, *Water Quality Conditions in the United States: A Profile from the 1998 National Water Quality Inventory Report to Congress,* EPA-841-F-00-006, (4503F), Washington, DC, June 2000, http://www.epa.gov/305b/98report/98summary.html, September 2001.

2. U.S. Environmental Protection Agency, Office of Water, "Clean Water Act: A Brief History," Washington, DC, http://www.epa.gov/owow/cwa/history.htm, September 2001.

3. American Public Works Association. *History of Public Works in the United States: 1776–1976.* Chicago: American Public Works Association, 1976, 410.

4. Galishoff, Stuart. *Safeguarding the Public Health: Newark, 1895–1918.* Westport, CT: Greenwood Press, 1975, 54–55.

5. Personal communication with Sandra Novotny, President, Nova Environmental Services, Kensington, Maryland, September 10, 2001.

6. Krenkel, Peter A., and Vladimir Novotny. *Water Quality Management.* Orlando, FL: Academic Press, 1980, 235–236.

7. U.S. Geological Survey, Washington, DC, "Water Science for Schools," http://wwwga.usgs.gov/edu/characteristics.html, September 2001.

8. Personal communication with Sandra Novotny, September 10, 2001.

9. Krenkel and Novotny, 209–212.

10. Ibid., 236–239.

11. Hach Company, Ames, Iowa, "Important Water Quality Factors," http://www.hach.com/h2ou/h2wtrqual.htm, September 2001.

12. U.S. Geological Survey, http://wwwga.usgs.gov/edu/characteristics.html, September 2001.

13. Stewart, John Cary. *Drinking Water Hazards.* Hiram, Ohio: Envirographics, 1990, 51–53.

14. Ibid., 50; and personal communication with Dr. Irina Cech, Professor of Environmental Studies and International Health, School of Public Health, University of Texas at Houston, Houston, Texas, August 23, 2001.

15. World Resources Institute, "Industrialization: Heavy Metals and Health," Washington, DC, http://www.wri.org/wr-98-99/metals2.htm, September 2001.

16. Stewart, 56.

17. De Zuane, John. *Handbook of Drinking Water Quality.* 2nd ed. New York: Van Nostrand Reinhold, 1997, 57.

18. Bergeson, Lynn L. "The Arsenic Rule: The Debate Continues," *Pollution Engineering,* September 2001, 36–37.

19. Colorado State University, Cooperative Extension, "Salt Tolerance of Various Temperate Zone Ornamental Plants," Fort Collins, Colorado, wysiwyg://6/http://www.colostate.edu/Depts/CoopExt/TRA/PLANTS/stable.html, September 2001.

20. U.S. Department of Interior, Bureau of Reclamation, "Colorado River Basin Salinity Control Program—Overview," http://dataweb.usbr.gov/html/crwq.html, March 2001.

21. De Zuane, 76.

22. U.S. Environmental Protection Agency, Office of Water, "Current Drinking Water Standards," Washington, DC, http://www.epa.gov/safewater/mcl.html, September 2001.

23. Personal communication with Dr. Irina Cech, August 23, 2001.

24. U.S. Environmental Protection Agency, "Hudson River PCBs," Washington, DC, http://www.epa.gov/region02/superfnd/hudson, September 2001.

25. Statement by Carol Browner, Director, U.S. Environmental Protection Agency, December 6, 2000.

26. U.S. Environmental Protection Agency, http://www.epa.gov/region02/superfnd/hudson, September 2001.

27. Carson, Rachel. *Silent Spring.* New York: Houghton Mifflin Co., 1962, 7–8.

28. Stewart, 151–153.

29. Criswell, Jim, et al., "Pesticides in Residential Areas—Protecting the Environment," Cooperative Extension Service, Division of Agricultural Sciences and Natural Resources, Oklahoma State University, Stillwater, Oklahoma, http://hermes.ecn.purdue.edu:8001/cgi/convertwq?6813, September 2001.

30. "Pops Producers," Greenpeace Toxic Site, http://www.greenpeace.org/,toxics/tbg/tbg1.html, September 2001.

31. U.S. Environmental Protection Agency, History Office, "New DDT Reports Confirms Data Supporting 1972 Ban, Finds Situation Improving," Washington, DC, http://www.epa.gov/history/topics/ddt/03.htm, September 2001.

32. U.S. Department of State, International Information Programs. "World Health Organization on DDT and Malaria Control," Washington, DC, http://usinfo.state.gov/topical/global/environ/latest/00120406.htm, September 2001.

33. U.S. Environmental Protection Agency, Office of Pesticide Programs, "Integrated Pest Management and Food Production," Washington, DC, http://www.epa.gov/pesticides/citizens/ipm.htm, September 2001.

34. Cranshaw, W. S., and D. C. Sclar, "Spider Mites," Colorado State University Cooperative Extension, Fort Collins, Colorado, http://www.ext.colostate.edu/pubs/insect/05507.html, September 2001.

35. U.S. Department of Agriculture, "The Whitefly Plan: A Five-Year Update," *ARS News & Information,* Agricultural Research Service, Washington, DC, http://www.ars.usda.gov/is/pr/1997/fly10297.htm, September 2001.

36. Ibid.

37. Los Alamos National Laboratory, "Nitrogen," Los Alamos, New Mexico, http://pearl1.lanl.gov/periodic/nutshell.html, September 2001.

38. Ibid.

39. Personal communication with Dr. Jim Loftis, Department of Civil Engineering, Colorado State University, Fort Collins, Colorado, September 14, 2001.

40. Canter, Larry W. *Nitrates in Groundwater.* Boca Raton, FL: Lewis Publishers, 1997, 15.

41. "Overview of the Hydrologic Cycle in Livestock Farming Systems," *Integrated Animal Waste Management.* Task Force Report No. 128, Council for Agricultural Science and Technology, Ames, Iowa, 1996.

42. Hach Company Homepage, http://www.hach.com, September 2001.

43. Bezdicek, D. F., and A. C. Kennedy. "Microorganisms in Action," J.M. Lynch and J.E. Hobbie (Eds.). *Blackwell Scientific Publications,* 1998, 243.

44. Donahue, Roy L., Raymond W. Miller, and John C. Shickluna. *Soils: An Introduction to Soils and Plant Growth.* 5th ed. Englewood Cliffs, NJ: Prentice-Hall, 1983, 210.

45. Ibid., 218.

46. Viessman, Warren, Jr., and Mark J. Hammer. *Water Supply and Pollution Control.* 4th ed. New York: Harper & Row, 1985, 696–697.

47. Criswell, et al., http://hermes.ecn.purdue.edu:8001/cgi/convertwq?6813, September 2001.

48. *City of Burnsville Bulletin,* Burnsville, Minnesota, Volume 11, No. 2, March/April 2001.

49. Stevens, Leonard A. *Clean Water—Nature's Way to Stop Pollution.* New York: Dutton & Co., 1974, 25.

50. Department of Epidemiology, School of Public Health, University of California at Los Angeles, http://www.ph/ucla.edu/epi/snow/removal.html, September 2001. Reprinted with permission.

51. Csuros, Maria, and Csaba Csuros. *Microbiological Examination of Water and Wastewater.* Boca Raton, FL: Lewis Publishers, 1999, 1–5.

52. Centers for Disease Control and Prevention, Atlanta, Georgia, http://www.cdc.gov, September 2001.

53. Ontario Ministry of the Environment, "Executive Summary," October 20, 2000, http://www.ene.gov.on.ca/news/0072sum.htm, August 2003.

54. Stewart, 88.

55. Centers for Disease Control and Prevention, "Cryptosporidiosis: Sources of Infection and Guidelines for Prevention," Atlanta, Georgia, wysiwyg://193/http://www.cdc.gov/ncidod/diseases/crypto/sources.htm, March 2001.

56. Great Outdoors Recreation Pages (GORP), "Treating Water," wysiwyg://21/htp://www.gorp.com/gorp/activity/hiking/skills/watertrt.htm, March 2001.

57. Krenkel and Novotny, 299–300.

58. Personal communication with Sandra Novotny, September 10, 2001.

59. U.S. Environmental Protection Agency, "Superfund," http://www.epa.gov/superfund/about.htm, March 2001; and U.S. Environmental Protection Agency, Office of Emergency and Remedial Response, http://oaspub.epa.gov/oerrpage, September 2001.

60. U.S. Army Corps of Engineers, "Project Fact Sheet—Former Nebraska Ordnance Plant," Kansas City District, Kansas City, Missouri, Summer 2000.

61. U.S. Environmental Protection Agency, "Summary of State Biennial Reports of Wellhead Protection Program Progress," http://www.epa.gov/safewater/protect/gwr/biennial.html, March 2001.

62. Cantor, 74–77.

63. U.S. Environmental Protection Agency, Office of Water, "Watershed Approach Framework," http://www.epa.gov/OWOW/watershed/framework.html, September 2001.

64. Bergeson, Lynn L. "One Hot Issue: TMDL Litigation." *Pollution Engineering,* February/March 2001, 17.

65. Ibid., 17–18.

66. U.S. Environmental Protection Agency, Office of Water, "Total Maximum Daily Load (TMDL) Program," Washington, DC, wysiwyg://203/http://www.epa.gov/owow/tmdl/overviewfs.html, September 2001.

67. U.S. Environmental Protection Agency, Headquarters Press Release, "Environmental News," http://yosemite.epa.gov/opa/admpress.nsf, June 2003.

CHAPTER 6

MUNICIPAL AND IRRIGATION WATER DEVELOPMENT

MUNICIPAL WATER SYSTEMS

IRRIGATION

Sir Ebenezer Howard (1850–1928) was a farmer, urban planner, and nontraditionalist whose radical ideas on the growth of cities are still debated today. His 1898 book, *Tomorrow: A Peaceful Path to Real Reform,* and the 1902 revised edition, *Garden Cities of To-morrow,* promoted a radical concept of community development around the world. (1)

Ebenezer Howard was born in London and was the son of a shopkeeper. He became a bookkeeper, and at the age of 21 traveled to America with some friends to become a farmer. He settled on 160 acres (65 ha) of native prairie land in Howard County, Nebraska, and attempted to grow crops such as corn, potatoes, cucumbers, and watermelons. Erratic rainfall, harsh winters, and Howard's inexperience limited his success. He left after a few years, and after some time in Chicago returned to England.

Through his travels, Ebenezer Howard developed concepts designed to improve the way cities are allowed to evolve. He visited the vast developing agricultural lands of the United States, as well as the great urban centers of Chicago, New York City, and London. Everywhere he traveled, he saw unplanned, chaotic growth. Howard felt strongly that urban development should be orderly, bucolic (agriculturally oriented), and controlled. One of his ideas was to limit the size of centralized cities to no more than 58,000 residents and to surround these urban centers with smaller satellite cities that had populations no greater than 32,000. Broad boulevards would connect these communities surrounded by large fields to grow food. Also, a large, irrigated garden would be located at the center of each city (Figure 6.1).

Howard's theory of a strong rural/urban connection was intended to produce a healthier, more productive citizenry distributed efficiently across the countryside. The chaotic urban development in Howard's day was abhorrent, with

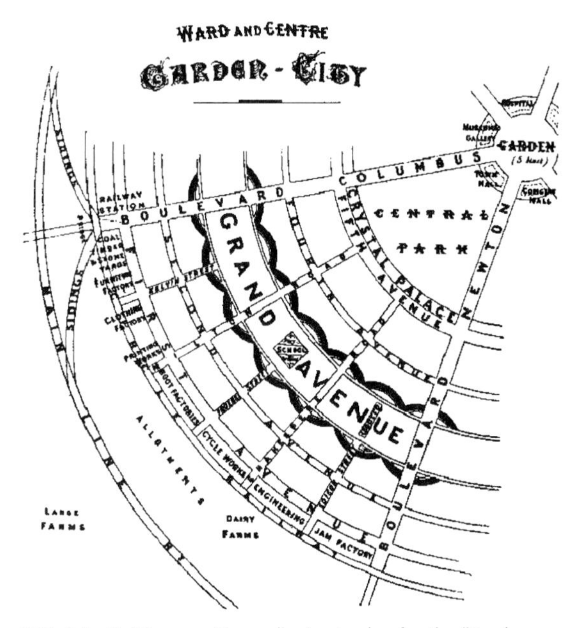

FIG. 6.1. Sir Ebenezer Howard's classic plan for the "Garden City" included broad boulevards radiating out from the city center garden.

disease, slum dwellings, and incredible poverty running rampant. Howard envisioned communities that used walking paths to get to work and a municipal irrigation system to provide water to the large central garden, as well as garden plots near each house. He was also a proponent of social order, organization, and public ownership of land. Controlled rents would pay for civic improvements, health care, and pensions.

Although the Garden City Movement was only a small success, it provided a vision for an "improved" society through orderly growth in a physically and mentally healthy environment. These concepts of community order and health correspond with the philosophy behind the construction of public water supply systems. Since ancient times, organized societies have engineered and constructed water projects to provide healthier, more reliable supplies of water for growing communities. In addition, municipal water systems were developed to promote economic growth, to meet the needs of in-migration of new residents, and to provide reliable water supplies during drought.

What has been the result of such a paradigm of municipal water development? On the one hand, it has been extremely successful. Almost all settled areas of the United States, Canada, Europe, Australia, New Zealand, and Japan have excellent public drinking water supplies. Centuries of work by engineers, hydrologists, politicians, and local business leaders have promoted reliable, ever-expanding supplies of water to meet the needs of urban and rural residents, at least in more-industrialized countries.

On the other hand, the model has had certain negative impacts. Water development does not occur without costs to the environment or to society as a whole. Too often, these costs (such as loss of plant and wildlife species, degraded water quality, and loss of habitat) go undetected until too late. The challenge of water managers and water users is to accomplish both goals: to develop clean, safe, reliable water supplies for increasing populations and at the same time to protect the fragile environmental systems of our natural world. This is an extremely difficult challenge that our society continues to grapple with today.

MUNICIPAL WATER SYSTEMS

Modern municipal water delivery systems are not significantly different from the waterworks developed for Roman cities thousands of years ago. Although today's water distribution networks are more elaborate and use modern technology, the basic principles of water delivery remain the same. Gravity is still used whenever possible to transport water, reserves are stored in surface reservoirs or underground storage vessels, water is delivered through pipes that run beneath city streets, and treated wastewater (sewage) is generally returned to rivers downstream of water supply areas.

The developers of ancient water systems were concerned primarily with moving water from point A to point B, with little or no regard for the impacts of such diversions. Today, competition for water supplies exists between urban, rural, and environmental interests in our finite hydrologic system. As population increases, whether it be at the local or the worldwide level, water demands by humans also increase. Conservation can reduce the growth rate of these demands, but ultimately, increased numbers of people require greater quantities of water for municipal water supplies and food production.

The three case studies that follow present a historical perspective of municipal water development in Los Angeles, California; Lincoln, Nebraska; and New York City, New York. As you read these histories, note the similarities and differences in the development and expansion of these water systems, the importance of water conservation, and the potential environmental effects of water development activities. What role did economic growth play in water development decisions? What were the probable impacts to the rural communities that lost their water supplies to Los Angeles or New York City? What types of water conservation alternatives could have been considered to meet a portion of the water needs of these growing population centers?

CASE STUDY

Los Angeles Department of Water and Power Los Angeles, California

http://www.ladwp.com/home.htm

Los Angeles

Average annual precipitation: 15.1 inches (38.4 cm)
Population: 3.9 million (city), 10.0 million (county)

Overview

The Los Angeles Department of Water and Power (DWP) was established in 1902 and is the largest municipal utility in the United States. A staff of 7300 provides water and electricity to nearly four million residents. A five-member Board of Commissioners establishes policy for the Department and is appointed by the mayor of Los Angeles.

Prior to creation of the DWP, water was delivered from the Los Angeles River through a system of small dams, waterwheels, and ditches (called *zanjas*) that were constructed in the late 1700s. Since then, other sources of water have been developed, and the Los Angeles River has been turned into a cement-lined, graffiti-marred flood-control channel that snakes its way through the city. Efforts are underway to beautify the Los Angeles River by creating parks and other green spaces along its course, but its role as a natural watercourse has been lost for over a century.

History

In 1886, William Mulholland became the first superintendent and chief engineer of the Los Angeles Water Department. Under his leadership, the water supply system was greatly enlarged through the construction of aqueducts, reservoirs, and pipelines across hundreds of miles (hundreds of km) of rough terrain to capture new water sources (Figure 6.2). His ultimate goal was to bring water from the eastern side of the Sierra Nevada Mountain Range, some 200 miles

FIG. 6.2. This classic construction photo (circa 1908) shows a 52-member mule team moving 1-inch-thick (2.5 cm) steel pipe to construct the Jawbone Syphon along the Los Angeles Aqueduct. Steel pipe was extremely expensive to use since it was produced on the East Coast of the United States. Because of its size, the pipe had to be transported to California by ship around Chile's Cape Horn.

(322 km) to the north of the Los Angeles Basin. Mulholland initiated the installation of water meters to measure the use of water, which was a new concept in water management. Mulholland Drive, on the northwest side of Los Angeles, is named after him.

The Owens Valley In 1905, Los Angeles voters approved a $1.5 million bond issue to purchase farms and orchards to acquire water in the Owens Valley, a small farming community located in central California. A $23 million bond was approved two years later to build a 233-mile (375 km) aqueduct to transport water from the valley by gravity to the Los Angeles area (Figure 6.3). The water acquisition program was very controversial and has left the Owens Valley without any substantial water resources other than limited precipitation.

Water is stored in eight major reservoirs along the Los Angeles Aqueduct and in five large reservoirs within Los Angeles. Storage tanks of treated water range in size from 10,000 gallons (0.03 acre-ft, or 37.9 m^3) to 10 million gallons (30.7 acre-ft, or 37,854 m^3). Total storage capacity in the system, including reservoirs, is approximately 115.7 billion gallons (355,070 acre-ft, or 437.9 million m^3). (2) Since water demands change constantly with the time of day or shift in weather, water supplies must be stored to meet peak demands and to maintain constant water pressure throughout the system. The major aqueducts that supply Los Angeles and southern California are shown in Figure 6.4.

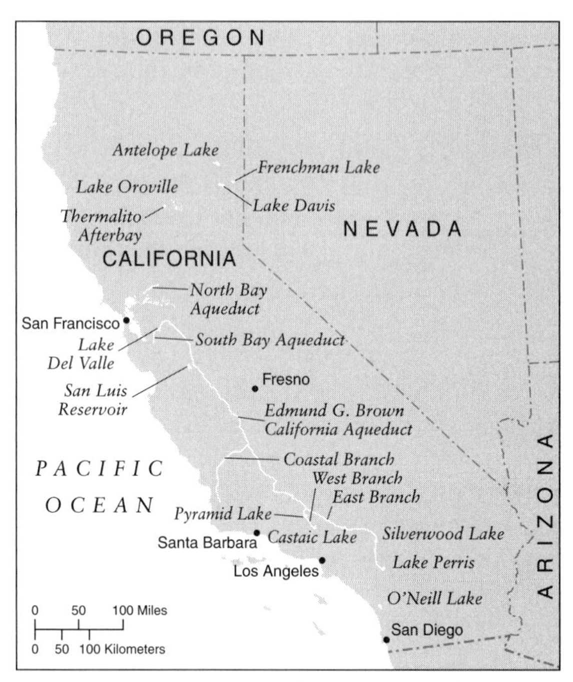

FIG. 6.4. Major aqueducts supplying water to southern California.

FIG. 6.3. A portion of the completed Los Angeles Aqueduct is shown here, which greatly expanded the supply of water to Los Angeles. Without the project, the city would have been limited to a population of approximately 500,000 since no additional water supplies existed in the local area. Notice the extremely dry terrain in the photograph.

The Metropolitan Water District In 1928, a group of 13 southern California cities united to build the Colorado River Aqueduct Project under the umbrella of the Metropolitan Water District (MWD) of southern California. The MWD acts as a wholesale provider of water to communities to save the expense of each city working to develop individual water storage and delivery projects. Today, the MWD of southern California is made up of a consortium of 26 cities and water districts that provide drinking water to 18 million people in parts of Los Angeles, Orange, San Diego, Riverside, San Bernardino, and Ventura counties.

Most MWD water must be pumped because there is no pipeline route that can rely solely on gravity for water delivery. The MWD now provides approximately 30 percent of the total water supply for Los Angeles, although during a drought that figure can increase to 60 percent. MWD has five water treatment plants and delivers an average of 1.8 billion gallons (5524 acre-ft, or 6.8 million m^3) of water a day and up to 2 billion gallons (6138 acre-ft, or 7.6 million m^3) on a hot day. (3)

EXPERT ANALYSIS

Contact your local water provider and inquire about local variations ("peaks") for water use on a daily and seasonal basis. Daily peak demands generally occur from 6:00 A.M. to 8:00 A.M. and from 5:00 P.M. to 7:00 P.M. as people prepare meals, perform personal hygiene, and wash dishes. Seasonal peak demands occur during the summer months when lawn irrigation, car washing, and other intense water uses increase.

POLICY ISSUE

The creation of the Metropolitan Water District in southern California was a logical and yet radical concept in the historic development of urban water supplies. Too often, municipal water departments tend to develop parochial viewpoints toward water development and distribution. It's common for such agencies to focus only on individual system water needs, to develop water supply projects for local residents, and to resist efforts to join with other water providers, even though such efforts could provide economies of scale. The result is often isolated and uncoordinated water supply plans for individual cities.

The Metropolitan Water District has provided a more coordinated approach to water development and supply in southern California. However, if each of the 27 original water providers had continued to work separately on water development projects, what would the implications be? What are some of the issues that would have surfaced as individual cities developed individual water systems? What additional costs would have been incurred? Would southern California's growth rate been slowed if the MWD had not been created?

Los Angeles Water Delivery System

Today, the City of Los Angeles has over 7000 miles (11,265 km) of water mains (pipes) ranging in size from 4 to 10 inches (10 to 25 cm) in diameter. Water pipes are usually buried several feet below city streets. Since the water service area in Los Angeles ranges from mean sea level to an elevation of 2400 feet (732 m), the city is divided into pressure zones. Over 85 booster stations (pumps) are needed to lift water to the higher elevations above the gravity flow system. A computerized map shows the size and location of every pipe, valve, and fire hydrant in the city. (4)

SIDEBAR

The Los Angeles River had a rich natural and cultural history as a provider of intermittent water flows across the arid southern California landscape. Centuries ago, the river provided water for wildlife and native, desert vegetation. Later, Spanish settlers relied on it as a source of irrigation and drinking water.

Today, the culture of the 50-mile (80 km) Los Angeles River is quite different with its cement-lined banks throughout the greater Los Angeles area. A local politician recently campaigned on the promise that, if elected, he would order the cement banks and bed of the Los Angeles painted blue to make it look more like a real river. Shopping carts, barbed wire, discarded sofas, and trash litter the waterway, and some maps don't even bother to show its course through the city. Its primary purposes today are to transport floodwaters quickly out of the area and to serve as a backdrop in many Hollywood movies. The modern Los Angeles River is an excellent example of human development encroaching, overwhelming, and substantially destroying a river that once served the water needs of a local population. (5)

TABLE 6.1 Monthly Residential Water Rates, Los Angeles, California

Rate	High-Use Season (June–October)	Low-Use Season (November–May)
First tier	$1.90 per hcf	$1.83 per hcf
Second tier	$2.98 per hcf	$2.33 per hcf

Source: Los Angeles Department of Water & Power, http://www.ladwp.com/water/rates, January 2001, and personal communication with Eric Tharp, Office of Public Affairs, Los Angeles Department of Water & Power, Los Angeles, California, December 27, 2000.

Los Angeles Water Rates

Residential water rates are based on the use of one hundred cubic feet (hcf) units, or 750 gallons (2839 l), which is a standard rate unit. Water pricing varies significantly based on lot size, temperature zones, household size, time of year, zip code, and amount of water used. The prices shown in Table 6.1 are for residential use that employ a two-tiered rate structure.

For example, to compute the cost of using 35 units of water in a Los Angeles residence in one month during the high-use season, multiply the number of hcf units used by the cost per unit. Note that the first 28 hcf units used in this example are priced at $1.90/hcf (the first-tier rate), and the additional 7 hcf units cost $2.98/hcf (the second-tier rate):

$$[28 \text{ hcf} \times \$1.90/\text{hcf}] + [(35 \text{ hcf} - 28 \text{ hcf}) \times \$2.98/\text{hcf}]$$
$$= \$53.20 + (7 \text{ hcf} \times \$2.98/\text{hcf})$$
$$= \$53.20 + \$20.86$$
$$= \$74.06$$

Notice, too, that the cost of using water under the second-tier rate is over 50 percent more than the cost of water under the first-tier rate ($2.98/hcf for the second-tier versus $1.90/hcf for the first-tier during the high-use season).

The cost of using 35 units of water for that same residential customer during one month of the low-use season would be:

$$[28 \text{ hcf} \times \$1.83/\text{hcf}] + [(35 \text{ hcf} - 28 \text{ hcf}) \times \$2.33/\text{hcf}]$$
$$= \$51.24 + (7 \text{ hcf} \times \$2.33/\text{hcf})$$
$$= \$51.24 + \$16.31$$
$$= \$67.55$$

In Los Angeles, residential and apartment dwellers are charged on the two-tiered billing rate, while commercial, industrial, and governmental customers pay a slightly lower rate than residential customers based on the size of the water meter. (6) Water meters are the same size as the water line serving a building and vary in size based on the amount of water needed. A 3/4-inch (1.9 cm) water pipe is generally adequate for a house, whereas a skyscraper may have several 4- to 6-inch (10 to 15 cm) water lines supplying the building. Why would billing rates be based on the size of a building's water meter? (Consider the differences in water demands and usage between a building with a 3/4-inch meter and one requiring a 6-inch meter.)

CASE STUDY

Lincoln Water System
Lincoln, Nebraska

http://www.ci.lincoln.ne.us/city/pworks/water

Lincoln, Nebraska

Average annual precipitation: 26.9 inches (68.3 cm)
Population: 232,000

History

Residents in Lincoln, Nebraska, turned to groundwater early in their history because nearby surface water in Salt Creek was too salty to drink. The first city-owned well was drilled in the artesian aquifer of the Salt Creek Basin in 1875. The well produced an adequate volume of water but contained excess salt brine from natural salt deposits in the area. Even though the groundwater could not be used for drinking, people believed the minerals in the water cured health problems, and so many traveled great distances to fill buckets and jars for washing and bathing.

In 1881, the City of Lincoln passed a $10,000 bond issue to drill new wells and install water delivery lines. However, excessive salt in the groundwater continued to cause problems. Later, water from one of the wells caused a typhoid fever breakout, and the well was finally abandoned in 1912. Other wells also had to be abandoned as salt content continued to increase. (7)

SIDEBAR

Salt was a major reason for the early settlement of Lincoln. Fur trappers of the early 1800s learned of the salt deposits in the area from indigenous Native Americans. Caravans of western travelers along the cutoff to the Oregon Trail stopped to purchase salt as a dietary supplement, for seasoning of food, and for curing meats. Salt was a driving force in the settlement of the area and was a major reason Lincoln was later selected as the capital of Nebraska.

Ashland Wellfield

Lincoln's water engineers eventually abandoned the wells in the Salt Creek Basin and established a new wellfield 25 miles (40 km) northeast of Lincoln along the Platte River near Ashland. A $2 million bond issue was approved around 1930 to construct five wells, a pump station, and a 36-inch (91 cm) transmission (delivery) line over hilly terrain to Lincoln. Initially, the Platte River alluvial groundwater had excess iron and manganese, and customers complained. Not long after, city officials built a water treatment plant to remove these mineral contaminants from the drinking water. (8)

By 1984, 44 groundwater wells had been drilled along the Platte River to deliver treated drinking water to 16 storage reservoirs throughout the City of Lincoln. Today the water distribution system contains over 1000 miles (1609 km) of water mains. Water fees helped pay for an additional water treatment plant, new wells, and a new water transmission line to the city. (9)

TABLE 6.2 Monthly Water Rates, Lincoln, Nebraska

Rate 1	Rate 2	Rate 3
0–8 units $0.90/unit	Next 15 units $1.11/unit	All additional units $1.55/unit

Source: Public Works/Utilities Department, City of Lincoln, Lincoln, Nebraska, http://www.ci.lincoln.ne.us/city/pworks/water/history.htm, October 2000, and personal communication with John Miriovsky, Superintendent of Production, Public Works/Utilities Department, City of Lincoln, Lincoln, Nebraska, November 8, 2000.

Lincoln Water Rates

Residential water rates are determined by an increasing block (or progressive) structure in Lincoln (Table 6.2). This means that, as more water is used, a higher rate per unit is charged. Water rates are set on a unit basis of 100 cubic feet (hcf) or 750 gallons (2839 l).

To compute the cost of using 35 units of water in a Lincoln home in one month, you need to determine the amount of water used in each billing rate category:

(Rate 1)	8 units × $0.90/unit	= $ 7.20
(Rate 2)	15 units × $1.11/unit	= $16.65
(Rate 3)	12 units × $1.55/unit	= $18.60
	35 units of water	= $42.45

EXPERT ANALYSIS

How many gallons of water did the Lincoln family in this example use in one month? The answer is 26,250 gallons: (35 units × 750 gallons/unit = 26,250 gallons).

What fraction of an acre-foot does this represent? (One acre-foot = 325,851 gallons, so 26,250 divided by 325,851 = 0.08 acre-foot).

An average family of four in the United States uses almost one acre-foot (325,851 gal, or 1233 m^3) of water per year.

Incentives are built into the Lincoln water supply system to conserve water. Not all communities have a progressive price system, and some continue to price water use at a flat rate. A flat rate means that the price per unit of water delivered is the same whether the use is 10 units or 10,000 units in a single month.

CASE STUDY

New York City Department of Environmental Protection New York City, New York

http://www.nyc.gov/html/dep/home.html

New York City

Average Annual Precipitation: 41.5 inches (105.4 cm)
Population: 8.1 million

History

When Dutch Governor Peter Minuit acquired the use of Manhattan Island from the Lenape Indians in 1626, water supply was not a significant issue. A large alluvial aquifer existed beneath the area in a layer of sand, rocks, and boulders that was left behind by glaciers from the Labrador icecap. The ice reached an estimated thickness of 3000 feet (914 m) in the Catskill Mountains as it pushed down the Hudson River Valley from the north, but it was reduced to about 100 feet (30 m) by the time it reached New York City. The moving sheet of ice deposited glacial till throughout the region and left behind the seemingly out-of-place boulders found in Central Park today.

Early residents of Manhattan obtained water from shallow alluvial groundwater wells that were replenished by seepage from the Hudson and East rivers. Water was raised by buckets tied to the end of a rope or with poles similar to the *shadoufs* used thousands of years earlier in Egypt (discussed in Chapter 1).

By 1776, the population increased to 22,000 residents, and a reservoir was constructed on the east side of Broadway. Groundwater was pumped by hand from wells or by animal power, and it was delivered through hollow logs that were laid along the surface of main streets. Inadequate water supplies (and water pressure) caused problems for firefighters. A huge fire in 1776 destroyed almost one-quarter of the buildings in New York, and fires in the early 1800s again destroyed large portions of the city. Only drastic measures stopped some of these devastating fires, and explosives were sometimes used to blow up buildings in the path of the blaze to deny kindling for the flames. (10)

SIDEBAR

New York City had few restrictions regarding water suppliers at the beginning of the nineteenth century, so in 1801 the vice-president of the United States, Aaron Burr (1756–1836), decided to become a water broker. Burr had a remarkable background and even spent the winter of 1777–1778 as a continental soldier at Valley Forge. A few years later he formed a law firm with Alexander Hamilton. Burr was a banker at heart, but later his former partner, now Secretary of the Treasury Alexander Hamilton, refused to give Burr a charter to open his own bank in New York City. Hamilton and his friends held the only banking charter for the city, the Bank of New York.

Burr was made livid by the rebuff but, undeterred, proceeded to start a water company with his associates. The charter contained language that allowed Burr to use the profits from the sale of water for any purpose, with no restrictions. As soon as the water company became profitable, Burr used the money to open a bank, the Manhattan Bank (today called the JPMorgan Chase Bank). Hamilton had been outwitted and later slandered Aaron Burr at a political dinner in New York. Upon hearing of the comments, an outraged Burr challenged Hamilton to a duel at 10 paces with .56 caliber dueling pistols to save his reputation. The famous Burr-Hamilton Duel was fought on July 11, 1804 in Weehawken, New Jersey. The challenge cost Alexander Hamilton his life. (11)

Sanitation systems in New York City were also lacking in these early years, and it has been estimated that in 1830 one hundred tons (90.7 metric tons) of human waste entered the porous soils of the region daily. The population was just over 200,000, and waste was deposited in shallow pits and cesspools because no wastewater disposal system existed. Inevitably, groundwater in the shallow groundwater wells became contaminated. (12)

The Croton River Basin System Faced with inadequate firefighting and water quality problems, water officials in New York City looked north to the Croton River in Westchester County. In 1842, a reservoir and aqueduct system was constructed to meet growing water demands. The system had a capacity of 90 million gallons per day (276 acre-ft, or 340,687 m^3) and transported water to holding reservoirs throughout the city. The new water delivery system could serve the needs of over 1 million residents, although the population at the time was only 300,000. Planners felt they had met the water needs of New York City forever.

One problem that continued to plague New York water engineers was the issue of head pressure from elevated water supplies. As more water was diverted from higher elevations into the city grid of underground water pipes, head pressure caused water pipes to burst. The technology of the day was unable to construct pipeline connections (joints) that could withstand the enormous pressures generated by the head of millions of gallons of water originating at elevations higher than delivery pipes. In addition, the concrete used in aqueducts in the 1800s was inadequate to remain watertight under the constant presence of water. The problems of water leaks were constant and legendary, and even persist to this day, though on a much smaller scale.

By the turn of the twentieth century, the Old Croton Reservoir and Aqueduct were further enlarged, and by the early 1900s, 12 new reservoirs and 3 controlled lakes were built. Later, between 1915 and 1926, the Catskill System went on-line with additional storage reservoirs. The storage capacity of the New York City water supply system doubled when these projects were completed. (13)

As New York City's population continued to grow, water planners realized that even the new Catskill System could not sustain the predicted growth of the city. By the mid-1920s, the Croton River Aqueduct was being utilized to capacity, and by 1929, construction was underway on a new project to divert water from a tributary of the Delaware River. However, the State of New Jersey objected. Lawyers for the Garden State argued that the diversion of water from the Delaware River Basin would injure the health and economic well-being of the citizens of New Jersey. The case eventually made it to the U.S. Supreme Court, but unfortunately for New Jersey residents, the Court sided with the residents of New York City.

On May 4, 1931, Justice Oliver Wendell Holmes delivered the opinion of the U.S. Supreme Court on the Delaware River lawsuit. He eloquently stated:

A river is more than an amenity, it is a treasure. It offers a necessity of life that must be rationed among those who have power over it. New York has the physical power to cut off all the water within its jurisdiction. But clearly the exercise of such power to the destruction of the interests of lower states could not be tolerated. And on the other hand, equally little could New Jersey be permitted to require New York to give up its power altogether in order that the River might come down undiminished. Both States have real and substantial interests in the River that must be reconciled as best they may be. The effort always is to secure an equitable apportionment without quibbling over formulas. (14)

Construction continued, and by 1965 the city had again doubled its water storage capacity with construction of four new reservoirs and connecting aqueducts.

City Tunnel #3

The largest construction project in the history of New York City, and one of the largest in the world, is ongoing beneath the feet of New Yorkers at this very moment. City Tunnel #3 is a 24-foot-diameter (7.3 m), 64-mile-long (103 km), 800-feet-deep (244 m) water tunnel being drilled, blasted, and scraped through bedrock by tunnel workers of Local 147 (respectfully

FIG. 6.5. Tunnel lining is completed by using this elaborate concrete-forming machine. Note the size of this equipment within New York City Tunnel #3 in contrast to the two workers standing to the right.

called sandhogs). The sandhogs use a tunnel boring machine (TBM), which is a mechanical rotary drill, to grind through rock at a rate of about 40 feet (12.2 m) per day. A tunnel lining machine is used to form concrete along the bored-out bedrock to provide a uniform conduit for water delivery (Figure 6.5). Both pieces of equipment are lowered in pieces into sections of tunnels to be bored and then assembled below ground. Excavated materials are hoisted (lifted) in buckets attached by metal cables to cranes at street level.

Enormous water valve chambers, the length of two underground football fields (600 ft, or 183 m) will control the distribution of 90 percent of the city's water supply after Tunnel #3 is completed. Reservoir water will be redirected by 34 gigantic valves into thousands of miles (thousands of km) of underground pipes, and eventually to faucets and other delivery points throughout the city.

A 13-mile (21 km) stretch of Tunnel #3 was activated in 1998. It begins at Hillview Reservoir in Yonkers and extends south, under Central Park, to about Fifth Avenue and Seventy-eighth Street, then east beneath the East River and Roosevelt Island to Astoria, Queens. The cost of this segment of the tunnel was $1 billion. Water is lifted from Tunnel #3 through 14 supply shafts or "risers" that feed into the City's water distribution system. (This design is strikingly similar to the *qanats* used in the Middle East.) Completion of the first segment of Tunnel #3 was celebrated with a ceremony at Jacqueline Kennedy Onassis Reservoir in Central Park. (15) The remainder of Tunnel #3 will be completed by 2020 at a total projected cost of $6 billion.

GUEST ESSAY

Eileen Schnock

Construction of City Tunnel #3

by **Eileen M. Schnock**
Resident Engineer of Tunnel #3, New York City

Eileen Schnock has always been interested in underground projects. She received her Bachelor's degree in Geology from City University of New York and a Master's degree in Mining Engineering from Columbia University. Eileen has worked on various underground projects in New York City and has been working on City Tunnel #3 for 10 years.

When I was a junior in high school, I was given a career preference test. The test contained questions like: "If you had three hours of free time would you rather (a.) visit a sick friend, (b.) go to a museum, (c.) plant some flowers, or (d.) watch an old movie?" Results indicated that I had an inclination toward becoming a florist or landscape architect. Tunnel engineer was not on their list of career possibilities, but the notion of digging in the earth was relatively accurate.

After obtaining degrees in Geology and Mining Engineering, my relatives and friends wondered how I would ever find work in New York City. I can honestly say, "I got my job through *The New York Times.*" My first position was as an engineering inspector for New York City's Third Water Tunnel. I recall my interview with the Chief of Waterworks Construction. "Would you like to work underground in a tunnel?" With a smile, I said, "Oh yes, certainly," never dreaming of what I was getting into.

On my first day of work, I descended 700 feet (213 m) down a shaft and walked 3.5 miles (5.6 km) below the streets of Harlem to get to the heading. A thousand feet (305 m) back from the face, I put my fingers in my ears and heard "Fire in the hole!!" followed by a boom and an air blast. Tunnel colors ranged from shades of brown to shades of gray. Everywhere the tunnel was muted except for the glaring lights and deafening noise at the heading. There were many colorful characters underground: second- and third-generation sandhogs with accents from Ireland, the Caribbean, and upstate New York. I was treated to many tales of near disasters and death-defying feats. I was made to feel right at home.

There is a sense of dedication and pride in working on New York City's Water Tunnel. This follows the grand tradition of the Board of Water Supply. The five boroughs were consolidated into the City of New York in 1898. Created in 1905, the Board was authorized by the State Legislature to plan, design, and construct a safe and reliable water supply system for New York City. During their 73 years of existence, the list of structures completed is impressive: seven reservoirs with 492 billion gallons (1.5 million acre-ft, or 1.9 billion m^3) of storage capacity, 145 miles (233 km) of pressure tunnel, 55 miles (89 km) of cut and cover tunnel, 32 miles (51 km) of grade tunnel, 6 miles (10 km) of syphon, and 83 shafts. Three watershed areas, totaling almost 2000 square miles (5180 km^2), supply water for New York City.

The City of New York is blessed with hard, competent bedrock that allowed construction of pressure tunnels to deliver water within the city limits. City Tunnel #1, 18 miles (29 km) long, services the west Bronx, Manhattan, and sections of Brooklyn. City Tunnel #2, 20 miles (32 km) long, services Queens and portions of the Bronx and Brooklyn. These two pressure tunnels have been in continuous operation since 1917 and 1936, respectively. Pressure from the elevation of Hillview Reservoir in Yonkers, north of the city, is regulated to deliver water to the height of a six-story building. Engineers recognized the need for a third tunnel as a means of alternate delivery and to allow for maintenance of City Tunnels #1 and #2. In 1970, the Board of Water Supply began construction of City Tunnel #3, Stage 1.

In contrast to former projects, construction of City Tunnel #3 has been hampered by several factors, including legal battles, funding difficulties, and problems with real estate acquisition. Shortly after construction began, a conflict arose regarding rock conditions, leading to default by the contractor and termination of construction in 1975. This event, along with the fiscal crisis of the City of New York, led to the dissolution of the Board of Water Supply. Work on City Tunnel #3 resumed under the authority of the Department of Environmental Protection, an agency of the City of New York.

Stage 1 was completed under 13 additional contracts and was finally placed into service in August 1998. This stage consists of 13.7 miles (22.0 km) of tunnel with 14 supply shafts and three major valve chambers that can be used to shut off portions of the tunnel. Stage 2 of City

Tunnel #3 consists of the Brooklyn section and the lower Manhattan section. Construction on Stage 2 began in 1987 and is scheduled for completion in 2010. Completion of Stage 2 will allow engineers to take the older tunnels off-line for inspection and rehabilitation.

After my introduction to underground construction with the Water Tunnel, I worked for a contractor on several other underground projects in the New York City area. These included sewer tunnels, rail stations, and tunnel rehabilitation. After several years in the private sector, I returned to civil service as the Resident Engineer for the Brooklyn portion of City Tunnel #3, Stage 2. The scope of work consisted of the excavation and concrete lining of 5.5 miles (8.9 km) of high-pressure water tunnel 600 feet (183 m) beneath the streets of Brooklyn, a mammoth job in a highly congested area. I knew it was going to be a challenge. The project seemed like something I had been training for my entire professional life. I wanted to be involved with the construction of the Brooklyn Tunnel, whether on the owner's side or the contractor's side. It was the first tunnel excavated by a tunnel boring machine in the very hard gneissic rock of Brooklyn.

Tunnel engineering turned out to be very similar to mining engineering, with the exception that no one wanted to keep the rock that came out of the ground. The same engineering principles applied, such as those involving ventilation, material handling, rock support, and such. With underground work, one must always be prepared for the unexpected and combine it with a healthy dose of Murphy's Law. Groundwater inflows, fault zones, and unstable rock formed the underlying theme during excavation of the Brooklyn Tunnel.

Operating design pressure for the Brooklyn Tunnel is 1000 feet (305 m) of water head (440 lb/in.^{2}, or 30.9 kg/cm^2). The in situ (in place) rock is actually the load-bearing unit containing the internal water pressure and preventing tunnel failure during operation. After excavation, concrete was placed to form an impervious liner. Extensive grouting operations were performed: to fill voids between the concrete and the rock excavation, and to fill all open joints in the rock. In areas where severely decayed rock was encountered along a major fault zone, a 2-inch-thick (5 cm) full circle steel liner was placed. This liner was capable of withstanding the design pressure without support from the rock.

Although problems of a complex nature were always occurring underground, it seemed that the surface issues took up a disproportionate amount of time. The work site for the Brooklyn Tunnel was tiny: 0.8 of an acre (0.3 ha) and hemmed in by a school, residential structures, an overhead highway, and other critical structures.

There were difficulties in obtaining real estate for the Water Tunnel in densely populated Brooklyn (Figure 6.6). Shaft sites typically required some sort of environmental remediation and were located adjacent to structures requiring underpinning or careful monitoring. This was the case with the main work site for the Brooklyn Tunnel. A pedestrian bridge was underpinned and jacked several times during construction. The soil underlying piers and columns supporting the overhead Brooklyn-Queens Expressway required constant monitoring for movement. The site was too small for any substantial amount of muck storage; therefore, rapid movement of broken rock off the site was of prime importance. Today there are stricter regulations on construction, especially regarding safety and community concerns. Noise, dust, and truck traffic were the main factors that impacted the local community. They required continual oversight to be kept within acceptable levels.

Similar to the work on Stage 1, construction of Stage 2 is proceeding on a piecemeal basis, with contracts covering a portion of the work. In part, this is due to shifts in funding priorities in the city. A heavy emphasis on water conservation has led to substantial reductions in water consumption. Federal and state laws have required

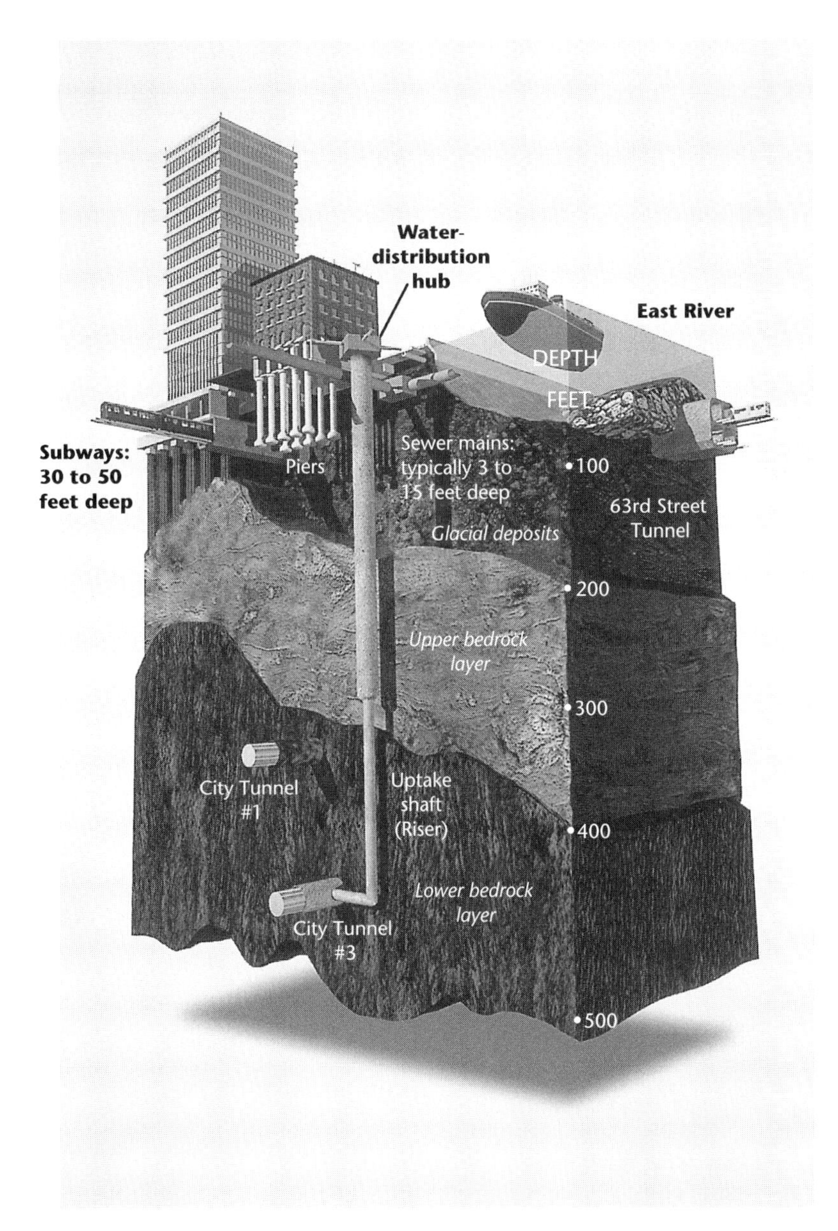

FIG. 6.6. New York City Tunnel #3, identified at the bottom of this *National Geographic* illustration, is part of an intricate web of tunnels, pipes, and other municipal infrastructure located beneath the city. Notice the congested area near the land surface.

diversion of funds to sewage treatment plant upgrades, water filtration, watershed management, and other projects. However, I have found that being a tunnel engineer is a perfect career in a place like New York City. Tunnels seem to be in fashion; everyone wants to put unsightly highways and other structures underground. I look forward to continuing with construction of the Water Tunnel or any other underground project that New York decides to build.

New York City Water Rates

The City of New York has both a metered and a flat rate water pricing structure. The flat rate is being phased out, however, and all buildings are expected to be metered by July 1, 2004. (16) Approximately 95 percent of New York City's water supply is transported by gravity. New Yorkers use 1.2 billion gallons (3684 acre-ft, or 4.54 million m^3) daily, with another 125 million gallons (384 acre-ft, or 473,176 m^3) consumed by residents in other counties that use water from the city's system. The New York City water supply system serves a total of eight million residents, one million upstaters, plus millions of tourists and commuters who visit throughout the year. Approximately 98 percent of the city's water supply is obtained from surface water. (17)

New York City bases its pricing structure on hcf units and charges \$1.31/hcf. The cost to a New York City homeowner for use of 35 units in one month would be:

$$35 \text{ hcf} \times \$1.31 = \$45.85$$

EXPERT ANALYSIS

Table 6.3 shows a comparison of residential water rates for the three cities previously discussed. Why would Los Angeles have a water billing system that charges more during certain months of the year, while New York City has historically used a flat rate? Why are Lincoln's water rates the lowest of the three cities? Why are rates highest in Los Angeles? Consider climate, topography, and availability of water supplies in your analysis.

Learn about the water system of your community. Is water use billed at a flat or a progressive (increasing) rate? Does your community obtain most of its water supply from surface water or groundwater? What types of delivery systems are used, and when were they built?

A CLOSER LOOK

The three municipal water systems just discussed (Los Angeles, Lincoln, and New York City) were developed under the logical premise that safe, reliable water supplies were necessary to maintain and enhance local population and economic growth. Los Angeles, for example, looked to the Owens Valley and then the Colorado River to acquire greater supplies of water. Both Los Angeles and New York City sustained population growth at the expense of taking water from other locations through expansive pipeline systems. Lincoln obtained reliable water supplies by constructing pipelines to alluvial aquifer wells.

The City of Boulder, Colorado, has used a completely different paradigm in dealing with population growth. Simply stated, the city has greatly restricted growth by limiting the expansion of its water supply system. Boulder is a unique community with a population of approximately 100,000 and is located at the foothills of the Rocky Mountains just northwest of Denver. It has long been known as a progressive and innovative city of well-educated, upper income residents with equally progressive elected officials.

In 1959, the city made a radical decision to control urban growth by establishing a "Blue Line" along an elevation line across the close upthrust of foothills. This boundary marked the area above which no water or sewer services would be extended. Driving into Boulder today, we see that the demarcation line is obvious and abrupt—all development ends at the base of the mountains, and open space (undeveloped land) stretches above that point.

TABLE 6.3 Comparison of Residential Water Rates in Los Angeles, Lincoln, and New York City

Cost	Los Angeles	Lincoln	New York City
Cost per 35 hcf	\$67.55–\$74.06	\$42.45	\$45.85
Total accounts	640,000	70,000	826,000
Average annual cost per residential customer	\$534	\$249	\$455

Sources: Los Angeles Department of Water & Power, Lincoln Water System, and the New York City Department of Environmental Protection.

In 1967, Boulder became the first city in the country to enact a special sales tax for the purchase of open space. The tax began at 0.4 cent per dollar and eventually increased to 0.77 cent per dollar. These tax revenues have allowed over 25,000 acres (10,118 ha) of land, above the Blue Line and in other locations, to be purchased for open space. In 1972, the City Council of Boulder, with the support of local residents, approved a growth-management plan that limited residential growth to only 2 percent a year. (The growth rate was later capped at 1 percent per year.) This was accomplished by limiting the number of new residential building permits issued annually by the city's Planning & Zoning Department.

What has been the result of these growth controls? We can make three dramatic observations about the city limits of Boulder. First, the Blue Line has prevented development in the foothills above Boulder, which today provide a recreation mecca for local residents and visitors. Readily accessible hiking trails and bike paths above Boulder receive more visits per year than Yellowstone National Park.

A second observation is that rental and purchase prices for residential properties in Boulder are very high for the area—$900/month to rent a one-bedroom apartment and over $300,000 to purchase a 1500-square-foot (or 139 m^2) house are common and increase constantly. Third, very few vacant lots exist within the city since developers are required to infill these areas with new homes and apartments in existing neighborhoods and not in new, sprawling suburban locations.

Outside its city limits, Boulder's growth controls have led to expansive urban sprawl in neighboring bedroom communities. High housing prices have forced many who work in Boulder to live in and commute from outlying areas. This has created daily traffic jams and has increased energy consumption and pollution of Front Range air from exhaust fumes.

The growth controls of Boulder, Colorado, and the Garden City Movement developed by Ebenezer Howard have many similarities. Both utilize an outer boundary for urban growth, restrict population size or growth rates, and promote an orderly, systematic plan for growth. How would you contrast the Los Angeles or New York City paradigms of water development to these two growth models? Would the implementation of growth controls in Los Angeles have simply shifted population migration to bedroom communities—as happened in Boulder? Is it appropriate to use water service as a method to regulate urban growth?

IRRIGATION

Between 1970 and 2000, the Earth's population increased by 2.3 billion people. Fortunately, on average, food production and distribution have kept pace with demand owing to improved crop varieties that produce higher yields, although extreme food shortages do occur in areas where distribution systems are inefficient and inadequate. In addition, more land is continually being placed under irrigation, particularly in less-industrialized countries around the world. However, as the previously rapid growth rate of food production slows, food shortages will surely occur in the future. Cereal grain production is already being outstripped by population growth in Africa and other regions. Land degradation is forcing the abandonment of agricultural lands, and competition for limited water supplies is increasing between irrigators and burgeoning urban centers. (18)

HISTORICAL PERSPECTIVE

The **100th Meridian** has historically been a hydrologic boundary between the humid climates (annual precipitation of over 30 in., or 76 cm) of the eastern United States, Canada, and Mexico, and the arid regions of the West. This virtual boundary runs north and south across western Manitoba, central North and South Dakota, Nebraska, Kansas, Oklahoma, and Texas, and eastern Mexico. **Irrigation** is the artificial application of water for crop production and is required west of the 100th Meridian as well as in other arid regions of the world to grow most crops.

EXPERT ANALYSIS

Refer back to Chapter 2. What climatic features found in North America account for the disparity of precipitation between lands located east and west of the 100th Meridian?

TABLE 6.4 Irrigation Water Use by State, 1995

State	Irrigated Land (thousand acres)	Surface Water Withdrawals			Groundwater Withdrawals			Total Withdrawals[a]	
		Thousand Acre-Feet per year	*Million Gallons per day*		*Thousand Acre-Feet per year*	*Million Gallons per day*		*Thousand Acre-Feet per year*	*Million Gallons per day*
Alabama	52	98	88	(63%)	57	51	(37%)	155	139
Alaska	1.4	0.1	0.1	(14%)	0.6	0.5	(86%)	0.6	0.6
Arizona	1,090	3,970	3,540	(62%)	2,390	2,130	(38%)	6,360	5,670
Arkansas	3,510	1,130	1,010	(17%)	5,520	4,930	(83%)	6,650	5,940
California	9,480	20,300	18,100	(63%)	12,100	10,800	(37%)	32,400	28,900
Colorado	3,310	12,000	10,700	(84%)	2,260	2,020	(16%)	14,300	12,700
Connecticut	19	13	12	(42%)	18	16	(58%)	31	28
Delaware	66	17	15	(30%)	38	34	(70%)	54	48
Florida	2,130	2,010	1,800	(52%)	1,880	1,670	(48%)	3,890	3,470
Georgia	1,150	273	243	(34%)	537	479	(66%)	810	722
Hawaii	136	537	479	(73%)	194	173	(27%)	731	662
Idaho	3,010	11,800	10,500	(81%)	2,820	2,520	(19%)	14,600	13,000
Illinois	359	0	0	(0%)	202	180	(100%)	202	180
Indiana	241	61	55	(47%)	69	61	(53%)	130	116
Iowa	158	4	3.6	(9%)	39	35	(91%)	43	39
Kansas	3,090	258	230	(7%)	3,540	3,150	(93%)	3,790	3,380
Kentucky	32	12	11	(96%)	0.5	0.5	(4%)	13	12
Louisiana	810	330	294	(38%)	533	475	(62%)	862	769
Maine	27	27	24	(90%)	2.9	2.6	(10%)	30	27
Maryland	74	29	26	(41%)	41	37	(59%)	70	62
Massachusetts	40	60	54	(66%)	31	28	(34%)	91	82
Michigan	354	142	127	(56%)	113	101	(44%)	255	227
Minnesota	401	41	37	(23%)	135	120	(77%)	176	157
Mississippi	1,370	109	97	(6%)	1,840	1,640	(94%)	1,950	1,740
Missouri	786	37	33	(6%)	599	535	(94%)	636	567
Montana	1,810	9,490	8,460	(99%)	92	82	(1%)	9,580	8,550
Nebraska	7,450	1,990	1,770	(23%)	6,480	5,780	(77%)	8,460	7,550
Nevada	560	1,120	1,000	(61%)	719	641	(39%)	1,840	1,640
New Hampshire	8.6	6.8	6.1	(96%)	0.3	0.3	(4%)	7.1	6.3
New Jersey	99	104	93	(74%)	36	32	(26%)	140	125

(continued)

John Wesley Powell, the famous nineteenth-century explorer, observed the dramatic change in rainfall as he crossed the 100th Meridian in the Great Plains of the central United States. Powell, realizing the potential for irrigation in the West, encouraged development of water storage for irrigation. Tens of thousands of miners flocked to California and Colorado during the gold rush periods of 1849 and 1859, respectively, and later constructed irrigation canals on land previously surveyed by Powell. John Wesley Powell's journey down the Colorado River, as well as his views on irrigation and western settlement, will be discussed in Chapter 9.

THE NEED FOR IRRIGATION

Irrigation projects are developed for several reasons, but primarily to grow crops in regions that lack sufficient precipitation. Most areas west of the 100th Meridian in North America (excluding some mountainous areas) receive less than 20

TABLE 6.4 Irrigation Water Use by State, 1995 *(continued)*

		Surface Water Withdrawals			Groundwater Withdrawals			Total Withdrawals[a]	
State	Irrigated Land (thousand acres)	*Thousand Acre-Feet per year*	*Million Gallons per day*		*Thousand Acre-Feet per year*	*Million Gallons per day*		*Thousand Acre-Feet per year*	*Million Gallons per day*
New Mexico	959	1,920	1,710	(57%)	1,430	1,280	(43%)	3,360	2,990
New York	47	16	14	(48%)	17	16	(52%)	33	30
North Carolina	167	203	181	(76%)	64	57	(24%)	267	239
North Dakota	196	64	57	(50%)	66	59	(50%)	131	117
Ohio	59	17	16	(58%)	13	12	(42%)	31	27
Oklahoma	560	110	98	(11%)	859	766	(89%)	969	864
Oregon	1,840	5,930	5,290	(86%)	985	878	(14%)	6,910	6,170
Pennsylvania	23	8.6	7.7	(95%)	9.2	8.2	(5%)	18	16
Rhode Island	7.1	1.8	1.6	(69%)	0.8	0.7	(31%)	2.6	2.3
South Carolina	23	28	25	(47%)	31	27	(53%)	58	52
South Dakota	301	206	184	(68%)	95	85	(32%)	301	269
Tennessee	63	16	15	(59%)	11	9.9	(41%)	27	24
Texas	6,310	3,280	2,920	(31%)	7,320	6,530	(69%)	10,600	9,450
Utah	1,140	3,520	3,140	(89%)	441	393	(11%)	3,960	3,530
Vermont	3.8	3.9	3.5	(91%)	0.4	0.4	(9%)	4.3	3.9
Virginia	69	27	24	(81%)	6.3	5.6	(19%)	33	30
Washington	2,120	6,330	5,650	(87%)	918	819	(13%)	7,250	6,470
West Virginia	2.8	0	0	(0%)	0	0	(0%)	0	0
Wisconsin	331	1.7	1.5	(1%)	187	167	(99%)	189	169
Wyoming	1,990	7,190	6,410	(97%)	203	181	(3%)	7,390	6,590

[a]Figures may not add to totals due to independent rounding.

Source: Wayne B. Solley, Robert R. Pierce, and Howard A. Perlman, *Estimated Use of Water in the United States in 1995*, U.S. Geological Survey Circular 1200 (Washington, DC: U.S. Department of the Interior, U.S. Geological Survey, U.S. Government Printing Office, 1998).

inches (51 cm) of precipitation per year. Crops such as corn, pinto beans, and alfalfa have an annual consumptive use of 20 to 25 inches (51 to 64 cm) and cannot survive in such a dry region without irrigation. Other crops such as wheat, oats, sorghum, and sunflowers require only 12 to 18 inches (30 to 46 cm) of precipitation per year and can be grown in many locations west of the 100th Meridian with **dryland farming** methods (farming without supplemental water). These crops require less water because of a generally shorter growing season and smaller plant size. However, the economic return from these crops is generally much less than that from corn, beans, or higher value vegetable crops.

A less obvious reason for irrigation development is to increase crop production in regions generally thought to receive adequate moisture for cultivation. The total amount of a crop harvested from an acre (or hectare) of land is called **crop yield.** The application of irrigation water during short-term periods of drought will greatly enhance crop yields in humid locations. Irrigation is now used east of the 100th Meridian in states such as Iowa, Indiana, and Illinois, as well as in surprising locations such as northwest Mississippi, North Carolina, and Georgia (see Table 6.4).

The Delta Region in northwest Mississippi, for example, has an average annual precipitation of approximately 50 inches (127 cm) per year. However, rainfall is variable during critical summer months for staple crops such as cotton and rice. Drought periods as short as two to three weeks can stress crop growth and reduce yields. Since a reduction in yield means less income for the irrigator, supplemental water from irrigation wells can be an important investment.

A CLOSER LOOK

Table 6.5 presents data on the top 10 irrigated states in the United States. Notice the relationship between the amount of irrigated land in a state and the type of irrigation that is most predominant in that state. For example, Nebraska has 7.45 million acres (3.0 million ha) of irrigated lands but has total withdrawals of only 8.46 million acre-feet (2.8 trillion gal., or 10.4 billion m^3). By contrast, Idaho has only 3.01 million acres (1.2 million ha) of irrigated land but 14.6 million acre-feet (4.8 trillion gal., or 18.0 billion m^3) of withdrawals. Why such a contrast between water use in Idaho and Nebraska? The answer most likely lies in the primary source of irrigation water used in these two states. Groundwater and center pivots are common in Nebraska and are used by most irrigators to apply groundwater to crops. In Idaho, gravity irrigation is the most common surface water application practice. Differences in application efficiencies probably account for the wide variation in irrigation water withdrawals in the two states.

IRRIGATION TECHNIQUES

Irrigation water can be obtained from either surface water or groundwater. In mountainous regions, snowpack (mountain snow) accumulates during the winter months and then melts during the spring. Melted snow can be captured in reservoirs or diverted directly from a river for irrigation. Many irrigated areas in Colorado, Wyoming, Idaho, Utah, New Mexico, Arizona, Oregon, Washington, and California receive less than 15 inches (38 cm) of precipitation during the summer growing season and rely primarily on irrigation water supplied from melting snow in the mountains. Similar irrigated regions exist in southcentral and western Canada, Mexico, Australia, India, and other locales around the world.

Groundwater is often used for irrigation if the depth to the groundwater table is not excessive (depths not over a few hundred feet, or 100 m), and aquifer permeability and thickness are adequate to provide sufficient quantities of water. If depth to groundwater is great, pumping costs may be too high to lift groundwater to the land

TABLE 6.5 Top 10 Irrigated States in the United States, 1995

State	Irrigated Land (thousand acres)	Surface Water Withdrawals			Groundwater Withdrawals			Total Withdrawals
		Thousand Acre-Feet per year	*Million Gallons per day*		*Thousand Acre-Feet per year*	*Million Gallons per day*		
California	9,480	20,300	18,100	(63%)	12,100	10,800	(37%)	32,400
Nebraska	7,450	1,990	1,770	(23%)	6,480	5,780	(77%)	8,460
Texas	6,310	3,280	2,920	(31%)	7,320	6,530	(69%)	10,600
Arkansas	3,510	1,130	1,010	(17%)	5,520	4,930	(83%)	6,650
Colorado	3,310	12,000	10,700	(84%)	2,260	2,020	(16%)	14,300
Kansas	3,090	258	230	(7%)	3,540	3,150	(93%)	3,790
Idaho	3,010	11,800	10,500	(81%)	2,820	2,520	(19%)	14,600
Florida	2,130	2,010	1,800	(52%)	1,880	1,670	(48%)	3,890
Washington	2,120	6,330	5,650	(87%)	918	819	(13%)	7,250
Oregon	1,840	5,930	5,290	(86%)	985	878	(14%)	6,910

Source: Wayne B. Solley, Robert R. Pierce, and Howard A. Perlman, *Estimated Use of Water in the United States in 1995*, U.S. Geological Survey Circular 1200 (Washington, DC: U.S. Department of the Interior, U.S. Geological Survey, U.S. Government Printing Office, 1998), p. 35.

surface. Pumping costs will be discussed later in this chapter.

Surface water irrigators use a variety of methods to deliver water to crops. A first step is to divert water from a river or reservoir into a **delivery canal.** Irrigation delivery canals vary greatly in size, ranging from widths and depths of a few feet (1 m) to extensive systems like the All-American Canal in the Imperial Valley of southern California, which is 200 feet (61 m) wide and 20 feet (6 m) deep in some locations.

Water is diverted from a delivery canal to a farm through a farm headgate. A **headgate** is a metal structure with a vertically sliding gate that can be raised or lowered. An open headgate allows irrigation water to flow from the delivery canal into a **farm ditch** (or **field ditch**). This smaller conveyance ditch, usually 2 to 3 feet (0.6 to 0.9 m) wide, carries water to individual farm fields. The amount of water discharged through the headgate is determined by the size of the opening created when the headgate is raised and by the water rights owned or controlled by the irrigator. The allocation of water rights will be discussed in Chapter 8.

Irrigation delivery canals are generally located along the highest elevation possible to maximize the amount of cropland that can be irrigated by gravity. Most delivery canals follow contours of the land (elevation lines on a topographic map—discussed in Chapter 3) to deliver irrigation water from rivers and reservoirs to farms. Farm ditches are located at the high end (or top end) of a field so that water can flow by gravity across the entire field to be irrigated.

Gravity Irrigation Numerous methods of irrigation are practiced on farms around the world today, although **gravity irrigation** is the most common. This method has been used since ancient times since it requires no mechanical energy and relies on gravity to distribute water across fields. The two types of gravity irrigation are furrow and wild flood. The method used is determined by the type of crop grown.

Furrow irrigation is a gravity method that utilizes crop rows to convey irrigation water across a field. Corn, soy and pinto beans, sugar beets, and many vegetable crops are grown in furrows (rows) to allow tractor-mounted equipment to cultivate, fertilize, and spray chemicals for insect and weed control (if organic practices are not used). Furrow widths are usually between 24 and 42 inches (61 to 107 cm) to allow adequate room for a tractor tire to fit between a crop row without crushing plants. The slope of the field must be adequate to allow irrigation water to flow slowly down individual furrows. This irrigation technique does not have a high loss of water to evaporation, but groundwater percolation rates can be excessive.

Wild flood irrigation is used on nonrow crops such as alfalfa and on grains such as oats or wheat. Water is released from a farm ditch into channels or between low ridges that are carved at regular intervals across a field. Irrigation water is allowed to slowly follow its own course across the field. Wild flood irrigation is used in areas where water supply is plentiful and inexpensive since the goal is to "wet" 100 percent of the field surface. The rate of water discharge, slope and length of the field, and percolation rates will determine if some areas are overirrigated and others are underirrigated. Areas near the top end of the field, or in low spots, may receive excess water, while areas at the bottom (or far end) of the field may receive inadequate irrigation water. Uniform distribution of irrigation water is difficult to achieve with wild flood irrigation.

A CLOSER LOOK

Flood irrigation is common in arid settings, but similar techniques can also be found in humid locations such as Massachusetts, Wisconsin, and Nova Scotia for cranberry production. To be productive, cranberry bogs require three essential water features: ample supplies of freshwater, excellent drainage, and relatively nonporous soils to retain flood irrigation water for winter protection and fall harvesting.

Cranberry bogs are leveled in the spring to provide flat areas for production. Some growers utilize sprinkler systems to protect fragile cranberry blossoms in the spring and ripening fruit in the fall from frost damage. The heat of the applied water prevents the plant from freezing, even though air temperatures may dip below the freezing point. (19)

Cranberries are harvested between September and November in North America. Originally, the berries were harvested by hand, but pickers later used wooden scoops to comb through vines to gather berries. Two methods are used today: dry harvesting, which involves mechanical pickers that resemble giant lawnmowers, and wet harvesting. Wet harvesting begins by flooding a cranberry bog with up to 18 inches (46 cm) of water and then stirring the bog with water reels called egg beaters. This process loosens cranberries from the vines and allows them to float to the water surface. A flooded cranberry bog turns red with floating berries during harvest, until they are corraled with floating plastic tubes and loaded into trucks for processing. (20)

Irrigation water is released through a delivery canal headgate, or pumped by a groundwater well, into a farm ditch for delivery to fields for furrow or wild flood irrigation. Water can be diverted from a farm ditch to a field with **syphon tubes**—S-shaped aluminum, galvanized iron, or plastic pipes that look somewhat like the Egyptian symbol for water (~) (see Figure 6.7). Irrigation syphon tubes are approximately 48 inches (122 cm) in length, with diameters between 2 and 6 inches (5 and 15 cm). Syphon tubes use vacuum pressure to deliver water from the higher elevation farm ditch to a lower elevation field. (The difference in elevation may only be a few inches or centimeters.)

The use of syphon tubes requires that a **check dam** first be placed in the farm ditch to pool irrigation water entering from the delivery canal or irrigation well. The check dam creates a small reservoir of water in the farm ditch and allows the syphon tubes to "pull" water out of the temporary reservoir. A tube is "started" by submerging one end into the field ditch to fill it with water and then quickly tossing the other end over the side of the field ditch. If done properly, the vacuum created will syphon water out of the field ditch.

Common gravity-irrigated field lengths are 660 to 1320 feet (201 to 402 m). If a field is too long, excessive percolation will occur at the top end of the field since irrigation water is constantly applied in that area. It usually takes irrigation water several hours to reach (or to be "pushed") to the far end of the field.

FIG. 6.7. Flood irrigation with syphon tubes is labor intensive but provides sorely needed irrigation water in arid and semiarid regions of the world.

Normally, several "sets" are needed to gravity irrigate a field. A **set** is the maximum number of syphon tubes that can divert water from a checked farm ditch. Usually, 30 to 40 2-inch (5 cm) syphon tubes are used to irrigate a 1/4-mile-long (402 m) field. If too many tubes are set simultaneously, water in the checked ditch will be depleted too rapidly and water flow through the tubes will stop. Since there is no automatic shutoff on an irrigation ditch, however, water will continue to flow into the farm ditch even though syphon tubes have stopped pulling water. This will cause the farm ditch to overflow and adjacent lands to flood. Gravity irrigators must closely monitor fields when they are "running water."

Another labor issue associated with syphon tubes is that, after a section of a field is adequately irrigated, the check dam must be moved down the farm ditch and a new set started. This requires each syphon tube to be moved and restarted regardless of the time of day. Generally, a set must be changed every 8 to 12 hours, 24 hours a day, all summer long in arid regions. It's not unusual for an irrigator to get out of bed at 2 A.M. to set water.

Some irrigators use **gated-pipe** to gravity irrigate fields. Gated-pipe is simply that: a 30-foot-long (9 m), 6-inch-diameter (15 cm) plastic or aluminum pipe with small slide-gates every 24 to 42 inches (61 to 107 cm, the same width as crop rows). Each length of gated-pipe can be joined together by a metal clasp to form a water delivery pipeline across the top (or upper) end of a field. Gated-pipe eliminates the need to set and reset syphon tubes. Pipe is placed on the ground so that, as each slide-gate is opened, irrigation water discharges directly into a crop row. Irrigators can adjust the width of the slide-gate opening to regulate discharge (similar to the much larger ditch headgates). If a crop requires less water, alternating slide-gates can be closed to irrigate every other row. Water inside gated-pipe must be under pressure to force it to the end of a long strand of pipes, so one end of the delivery pipeline is typically attached to an irrigation well or booster pump at a small farm reservoir.

A variation of gated-pipe is **flexible plastic pipe.** This is a relatively new, lightweight, disposable method of delivering irrigation water to a field. Flexible plastic pipe is generally hundreds of feet (approximately 100 m) long and has a diameter of 8 inches (20 cm) when filled with water. The thin plastic tube is manufactured as a roll and is simply rolled out along the top end of the field to be irrigated, similar to gated-pipe. Because water must be under pressure, one end of the plastic pipe is typically attached to an irrigation well. Slide-gates are placed in the flexible pipe at desired locations (generally every 24 to 42 in., or 61 to 107 cm), and water is released into crop furrows as needed. An advantage of flexible plastic pipe over gated-pipe is that it's very easy to move, is low cost, and eliminates the need to store 30-foot (9 m) lengths of aluminum gated-pipe during the winter months. A primary disadvantage is leakage and disposal. Flexible plastic pipe can tear easily, and care must be taken while handling or opening slide-gates. Disposed plastic pipe is often burned or hauled to landfills, which contributes to environmental degradation.

A disadvantage of all gravity irrigation techniques is the lack of uniform water application across a field. Irrigation water is delivered through syphon tubes or slide-gates at the top end of a field. From there, it generally takes 6 to 8 hours for irrigation water to make its way across the entire length of a field to the lower end. During this time, irrigation water is percolating as it wets a field. The result is very deep percolation at the top end of the field and relatively little percolation at the bottom end.

Irrigation efficiency rates for flood irrigation range, on average, from 40 percent to 65 percent. This means that the crop consumes only 40 to 65 percent of irrigation water that is applied. The remaining 35 to 60 percent of the irrigation water applied to a crop either percolates below the root zone, flows off the end of the field as **wastewater** (excess surface water runoff), or evaporates as it flows across the field. In regions with alluvial aquifers, this water loss is not a major water quantity problem since unused, percolating water will eventually reenter the groundwater aquifer system as groundwater recharge. However, the seepage of irrigation water into groundwater aquifers can lead to water quality problems (discussed in Chapter 5).

OUR ENVIRONMENT

In Australia, it's called the "white death." In the United States millions of dollars have been spent fighting its spread, while in Canada, a "Risk Index" has been developed to assess the potential for future problems. **Salinity,** the buildup of salts in the soil, has existed for thousands of years, particularly in some areas where intensive irrigation is practiced.

There are two types of salinity: dryland salinity (which occurs naturally on land that is not irrigated) and irrigated land salinity. Irrigated land salinity can occur where irrigation has been practiced for long periods of time in areas that contain high levels of salts in the soil. Most agricultural lands contain some salt, but it is generally held deeper in the soil profile (deeper beneath the land surface) so that plant growth is not impaired. Unfortunately, some areas of the world (the Colorado River Basin in the southwestern United States discussed in Chapter 5, western Australia, and some southern regions of the Prairie Provinces in Canada, for example, are facing environmental threats from increased salinity in soils due to irrigation. High levels of salts in soil inhibit normal plant growth. (21)

In Australia and other countries, remote sensing is being used to collect aerial data from planes and satellites to determine the extent of their growing salinity problem. Remote sensors use cameras that can record electromagnetic radiation from the Earth's surface to detect saline soils. Infrared film can be used to record images of different wave lengths within the infrared band to show vegetation under varying degrees of stress—dark-green vegetation produces a bright red image, light-green foliage gives a pink image, barren saline soil a white image, and salt-stressed vegetation a reddish-brown image. Photographs taken over various time periods present a spatial and temporal pattern of salination in an area. (22)

Why should we be concerned with the effects of irrigation and saline soils? Historically, saline soils have led to the abandonment of vast regions of irrigated lands and the devastation of local economies. The Anasazi Indians of southwestern Colorado (discussed in Chapter 1) may have been forced to abandon their settlements due to high soil salinity exacerbated by ongoing irrigation. Irrigated lands in the Middle East have been abandoned due to high salt levels in soils. In her excellent book *Pillar of Sand: Can the Irrigation Miracle Last?* (published by the Worldwatch Institute, 1999), Sandra Postel discusses the advancements that irrigation has made around the world in the past 6000 years, but also examines the collapse of early civilizations as irrigated soils became saline and crops could no longer survive.

What future does irrigation hold in the modern era? With the world's population now above 6 billion, irrigation is vital to maintain adequate food production, but salinity will inhibit food production in some regions. Is salinity a problem in your area? If so, how extensive is the problem, and what actions are being taken by landowners and local, regional, and federal agencies to mitigate the problem?

Sprinkler Irrigation Another common method of applying irrigation water to crops is **sprinkler irrigation.** In the late 1800s, lawn irrigation technology was developed for urban areas and was later modified for use on orchards, commercial greenhouses, and high-profit vegetable crops. Water was applied through elaborate networks of pipes and sprinklers, but inadequate water pressure and heavy lead pipes were major constraints. Then, after World War II, three improvements were made that led to increased use of sprinkler irrigation across the world:

1. Development of lightweight aluminum sprinkler pipe.
2. Availability of low-cost electricity in rural areas.
3. Improvement of electric motor technology.

Sprinkler irrigation opened the way to irrigate lands that were unsuitable for gravity irrigation technology. Sandy soils, land with excessive slope, and regions where surface water supplies were inadequate or nonexistent but groundwater was plentiful, now became potential sprinkler ground.

Sprinkler systems may be stationary, such as in a fruit orchard, or may involve long towers that move across fields on wheels. Stationary sprinkler systems consist of a network of pipes, permanently located at appropriate water application intervals, with spray nozzle attachments at the end of each pipe. Stationary sprinklers are almost identical to lawn sprinkler systems except that larger pipes and spray nozzles are used.

Nonstationary sprinkler systems include lateral and pivot systems. **Lateral sprinkler systems** are sprinkler machines that move suspended irrigation pipes (similar to gated-pipe mounted on wheels) across a field. Water inside the suspended pipe is under pressure to discharge water through nozzles onto a crop. Groundwater (or surface water from a small reservoir) is pumped from a farm ditch and through the lateral sprinkler system as it moves from one end of the field to the other. Water application rates are determined by the speed of the moving lateral system.

A **center pivot sprinkler** system is similar to a lateral irrigation system except that this machine rotates around a fixed pivot point. Several long sections of 6- to 10-inch-diameter (15 to 25 cm) aluminum pipe are suspended approximately 10 feet (3 m) in the air. The apex of a triangular tower supports each end of the suspended pipe and is attached to two wheels at the base of each tower. Each section of aluminum pipe is approximately 130 to 210 feet (40 to 64 m) long. The total length of a center pivot varies by the size of the field to be irrigated. Lengths of 1/4 mile (402 m) are most common, although some center pivot systems are 1/2 mile (805 m) long. Center pivot wheels, located at the base of each tower, are powered by small electric motors or by water pressure inside the pivot system. Wheels near the center of the pivot travel less distance and must be geared to turn at a much slower rate than wheels located at the far end of the pivot (see Figure 6.8).

It generally takes one to three days for a center pivot to make one complete rotation across a field, depending on the speed of the pivot and the water application rate required by the crop. Application rates between 0.25 and 1.0 inches (0.6 and 2.5 cm) per rotation are common. If the weather is dry, a

FIG. 6.8. This center pivot system was manufactured by the Lindsay Manufacturing Company of Lindsay, Nebraska. The pivot point, drop nozzles, and towers are all visible in this photo. Notice the ground-level pipe that extends toward an irrigation well to the right of the irrigator (out of camera view). The speed of the center pivot is regulated at this digital control panel, and various water application rates can be selected.

center pivot may continually move around a field to irrigate for weeks at a time.

Nozzle calibration is very important when applying water through a center pivot system. Each nozzle must be calibrated (sized) to provide the proper amount of irrigation water without causing land erosion. Nozzles at the center of the field are calibrated to release less water than nozzles located at the outer end of the circle. Why?

Originally, irrigators preferred to operate center pivots with high water pressure. The theory was that smaller water droplets sprayed from nozzles would reduce surface water runoff and land erosion. High energy costs in the 1970s changed this practice, and today most irrigators use low-pressure systems to reduce the energy required from an irrigation pump. Initially, water pressures of 90 to 100 pounds per square inch (6.3 to 7.0 kg/cm^2) were common, and it was not unusual to see irrigation water spraying high into the air above a center pivot. While this prevented erosion on hilly terrain, it led to excessive evaporation on windy days and high energy costs. Low-pressure application methods of 35 to 40 pounds per square inch (2.5 to 2.8 kg/cm^2) significantly reduce power bills. Erosion is limited by improved nozzle design and the use of "drop nozzles" that place water spray closer to the ground. Water application efficiency rates can be as high as 75 to 85 percent with a properly calibrated center pivot sprinkler system.

Operating a center pivot requires little physical labor. Capital costs of a center pivot range from $40,000 or more depending on the total length of the system and the type of nozzles used to apply water. Operation costs range from $20 to $25 per acre-foot of water applied but depend on the amount of energy needed to lift water from a well or farm reservoir.

Center pivot circles can be seen from the air and are common in parts of Nebraska, Colorado, Arizona, Kansas, Oklahoma, the Texas Panhandle, northwest Mississippi, Oregon, Washington, southern Alberta and Saskatchewan, numerous countries in the Middle East, and Australia (see Figure 6.9). Most manufacturers of these automated systems are located in Nebraska.

EXPERT ANALYSIS

The length of a center pivot is based on the size of the field to be irrigated. In the United States, a section of land represents 640 acres (259 ha) or 1 square mile (5280 ft, or 1609 m on each side). A quarter of a

FIG. 6.9. Aerial view of extensive center pivot development along the Columbia River in the Oregon–Washington region. Note the tight spacing of circles in this area and the unirrigated corners. What could this type of intense irrigation do to the local aquifer system?

section equals 160 acres (65 ha) or one-quarter of a square mile (2640 ft, or 805 m, on each side). Since center pivots irrigate a circular pattern and are usually placed on 160-acre (65 ha) square fields in the United States, what is the maximum number of acres that can be irrigated by a center pivot on a 160-acre square-shaped field?

Since 1 mile = 5280 feet, and ½ mile = 2640 feet, a quarter section of ground that contains 160 acres has dimensions of 2640 feet (805 m) on each side. To irrigate this parcel, the pivot point will be placed in the center of the field. What's the maximum total length for the center pivot system?

$$(2640 \text{ ft} \div 2 = 1320 \text{ ft})$$

The equation for the area of a circle is pi (which is approximately 3.14) times the radius squared. So,

$$3.14 \times 1320^2 = 5{,}471{,}136 \text{ square feet}$$

Divide 5,471,136 by 43,560 (the number of square feet in one acre) and you get 125.6 acres (50.8 ha) of irrigated ground under the center pivot. The corners often remain unirrigated and will probably be planted to dryland wheat or some other low-consumptive use crop (see Figure 6.10).

The preceding example can be altered if "end guns" are placed at the end of the last center pivot tower. These high-pressured nozzles release irrigation water only when the pivot reaches the corner of a field. Typically, the center pivot circle contains four unirrigated corners due to the circular area irrigated within a square field. Too often, end guns are not "set" properly, and as a result water is sprayed beyond field corners onto roads, farm driveways, or waste ground not intended for irrigation. State laws generally prohibit spraying irrigation water onto roads or highways and include financial penalties for such water waste.

Drip Irrigation A relatively new technology—**drip irrigation**—is gaining wider acceptance as the cost of irrigation water increases. It is used primarily on fruit and vegetable crops because the economic return is much higher than that from crops such as corn or beans. Rigid plastic pipes are either laid on the surface or buried below ground in each crop row at the root line. Small holes allow water to "drip" from the pipe directly to each plant root system. The amount of water applied is regulated by the hole size in the distribution tubes and the water pressure in the system. Although installation of drip systems is very expensive, evaporation rates are greatly reduced and efficiencies can reach over 90 percent. Israel has been a leader in drip irrigation technology, although California and other areas of the western United States are rapidly adopting this very efficient irrigation practice.

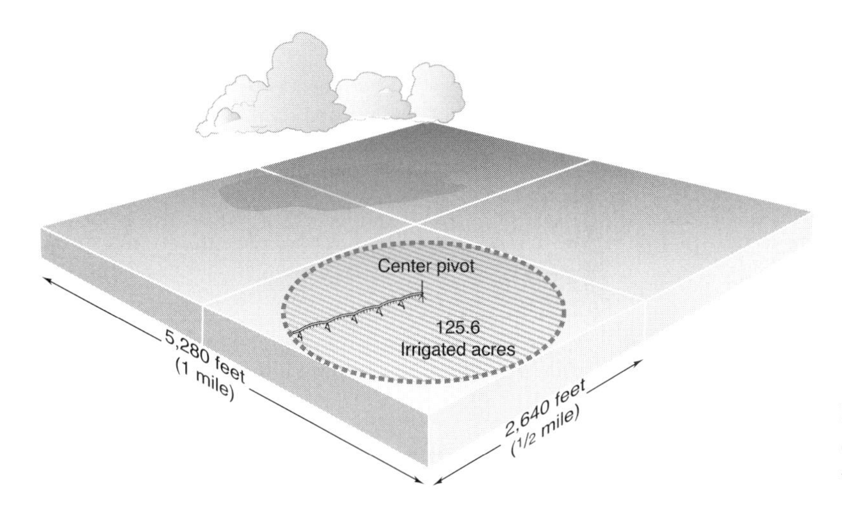

FIG. 6.10. Land irrigated by a typical center pivot system on a quarter section of land.

POLICY ISSUE

The preceding discussion of irrigation techniques presents a wide range of efficiencies. Gravity irrigation has been used for centuries because it is simple, relatively inexpensive, and does not require mechanization. In contrast, drip irrigation is a recent development that uses much less water but requires high initial investments.

As a policymaker, what factors would you need to take into account if a statewide policy was proposed to require all irrigators to use the most efficient technology available? Would it be appropriate for the government to require irrigators to adopt new practices without providing financial incentives? Should inefficient water delivery systems be prohibited even though it could mean reduced food production, loss of jobs, and loss of farms? Should irrigation efficiency be a concern of government regulation?

CHAPTER SUMMARY

Human requirements for water have led to the development of elaborate water delivery systems. The three case studies presented in this chapter show a wide range of water supply systems developed in the United States. The result of these water diversion and conveyance projects is an improved, more reliable, and far more extensive supply of water. The downside of these projects has been the permanent removal of water supplies from native watersheds, urban sprawl, and severe alterations to the natural environment, particularly in the Los Angeles example. The City of Boulder, Colorado, attempted to control urban sprawl by restricting the delivery of water and sewer services to outlying areas of the community. This led to the protection of unique natural features adjacent to the city, but it contributed to urban sprawl, traffic congestion, and air pollution in surrounding communities.

Irrigation development has followed a similar pattern of development. Irrigation began with simple gravity irrigation methods developed many centuries ago but in some locations evolved into modern center pivot and drip irrigation systems. The negative impacts of irrigation development have been the diversion of surface and groundwater supplies, water quality issues (discussed in Chapter 5), and the alteration of natural habitats (see Chapter 12). The use of water, for any purpose, is not without detrimental effects to the natural or developed environment.

QUESTIONS FOR DISCUSSION

1. Who was Ebenezer Howard, and what was the Garden City Movement?
2. Would you consider Howard to be a proponent of the free market system, or did he believe in government control? Which philosophy would you support?
3. What led New York City officials to develop water projects north of the city?
4. Why did Los Angeles acquire water from the Owens Valley?
5. Why did the City of Lincoln develop a water supply system from wells beneath the Platte River?
6. Explain the "Blue Line" in Boulder, Colorado.
7. Why is irrigation important in many regions of the world?
8. Describe the difference between gravity and sprinkler irrigation. What types of benefits or detriments does each have?
9. What types of environmental impacts could large-scale irrigation projects have on a region?
10. What role could water conservation play in urban or irrigation water development?

KEY WORDS TO REMEMBER

100th Meridian p. 167
center pivot sprinkler p. 175
check dam p. 172
crop yield p. 169
delivery canal p. 171
drip irrigation p. 177
dryland farming p. 169
farm ditch (field ditch) p. 171
flexible plastic pipe p. 173
furrow irrigation p. 171
gated-pipe p. 173
gravity irrigation p. 171
headgate p. 171
irrigation p. 167
irrigation efficiency p. 173
lateral sprinkler systems p. 175
salinity p. 174
set p. 173
sprinkler irrigation p. 174
syphon tubes p. 172
wastewater p. 173
wild flood irrigation p. 171

SUGGESTED RESOURCES FOR FURTHER STUDY

READINGS

Grumprecht, Blake. *The Los Angeles River—Its Life, Death, and Possible Rebirth*. Baltimore, MD: Johns Hopkins University Press, 1999.

Hoffman, Abraham. *Vision or Villainy—Origins of the Owens Valley—Los Angeles Water Controversy*. College Station: Texas A&M University Press, 1981.

Howard, Ebenezer. *Garden Cities of To-Morrow*. Cambridge, MA: MIT Press, 1965.

Israelsen, Orson W., and Vaughn E. Hansen. *Irrigation Principles and Practices*. 3rd ed. New York: John Wiley & Sons, 1962.

Koeppel, Gerard T. *Water for Gotham—A History*. Princeton, NJ: Princeton University Press, 2000.

Lavine, David. *Under the City*. Garden City, NY: Doubleday & Co., 1967.

Postel, Sandra. *Pillar of Sand: Can the Irrigation Miracle Last?*, Washington, DC: Worldwatch Books, Worldwatch Institute, 1999.

Reisner, Marc. *Cadillac Desert: The American West and Its Disappearing Water*. New York: Penguin Books, 1986.

WEBSITES

Los Angeles Department of Water and Power, "Home Page," July 2003. http://www6.ladwp.com/aboutwp/history/allabout/allabout.htm

Metropolitan Water District of Southern California, June 2003. http://www.mwd.dst.ca.us

Public Works/Utilities Department, City of Lincoln, Lincoln, Nebraska, "History of Lincoln Water System," June 2003. http://www.ci.lincoln.ne.us/city/pworks/water/history.htm

The City of New York, Department of Environmental Protection, "New York City's Water Supply System," October 2001. http://www.nyc.gov/html/dep/html/watersup.html

American Groundwater Trust, "Home Page," June 2003. http://www.agwt.org

VIDEOS

Cadillac Desert, PBS Home Video & Turner Home Entertainment, 4-Part series.

Chinatown. Directed by Roman Polanski, Paramount Studios, Hollywood, CA, 1974. 2 hrs., 11 min.

"Plumbing: The Arteries of Civilization," Modern Marvels Series, The History Channel. 50 min.

"Water Well Basics: Ground Water, Water Wells, and Home Water Systems." The American Ground Water Trust, Concord, NH, 1999. 15 min.

"The West—One Sky above Us," Episode 9, PBS Home Video & Turner Home Entertainment, The West Film Project, Inc. and Greater Washington Educational Telecommunications, Inc., 1996. 62 min.

REFERENCES

1. Howard, Ebenezer. *Garden Cities of To-Morrow.* Cambridge, MA: MIT Press, 1965.

2. Los Angeles Department of Water and Power, http://www.ladwp.com/aboutwp/history/allabout/allabout.htm, October 2000; and personal communication with Eric Tharp, Department of Government, Legislative & Public Affairs, Los Angeles Department of Water and Power, Los Angeles, California, December 27, 2000.

3. Metropolitan Water District of Southern California, http://www.mwd.dst. ca.us, August 2001.

4. Los Angeles Department of Water and Power, http://www.ladwp.com; and personal communication with Eric Tharp, December 27, 2000.

5. Grumprecht, Blake. *The Los Angeles River—Its Life, Death, and Possible Rebirth.* Baltimore, MD: Johns Hopkins University Press, 1999, 1–7.

6. City of Los Angeles Water Services, "Water Rates," http://www.ladwp.com/water/rates/sch_a.htm, October 2000; and personal communication with Eric Tharp, December 27, 2000.

7. "History of Lincoln Water System," Public Works/Utilities Department, City of Lincoln, Lincoln, Nebraska, http://www.ci.lincoln.ne.us/city/pworks/water/history.htm, August 2001; and personal communication with John Miriovsky, Superintendent of Production, Lincoln Water System, Lincoln, Nebraska, November 8, 2000.

8. Ibid.

9. Ibid.

10. Weidner, Charles H. *Water for a City.* New Brunswick, NJ: Rutgers University Press, 1974, 18.

11. Lavine, David. *Under the City.* Garden City, NY: Doubleday & Company, 1967, 58–60.

12. Weidner, 21–22.

13. The City of New York, Department of Environmental Protection, "New York City's Water Supply System," http://www.ci.nyc.us/html/dep/html/croton, October 2000; and personal communication with Geoffrey Ryan, Acting Director, Bureau of Public Affairs, New York City Department of Environmental Protection, New York City, New York, May 2, 2001.

14. *New Jersey* v. *City of New York,* 284 U.S. 585 (1931).

15. "Mayor Giuliani Inaugurates City Water Tunnel No. 3," http://www.cic.nyc.ny.us/html/om/html/98b/pr389–98.html, October 2001.

16. Personal communication with Geoffrey Ryan, May 2, 2001.

17. The City of New York, http://www.ci.nyc.us/html/dep/html/croton, October 2000; and personal communication with Geoffrey Ryan, May 2, 2001.

18. Gleick, Peter H. *The World's Water: 2000–2001.* Washington, DC: Island Press, 2000, 63–65.

19. "Cranberry Bog Site Selection," Nova Scotia Department of Agriculture and Fisheries, http://www.gov.ns.ca/nsaf/elibrary/archive/engine/facts/cransite.htm, August 2001.

20. "Cranberry Harvest & Virtual Bog Tour," Ocean Spray Cranberries, Inc., Lakeville-Middleboro, Massachusetts, http://www.oceanspray.com, August 2001.

21. "Salinization of Soil," Agriculture and Agri-Food Canada, http://res2.agr.gc.ca/publications/hs/chap08_e.htm, September 2003.

22. "Monitoring the White Death—Soil Salinity," Australian Academy of Science, Canberra, Australian Capital Territory, September 2003.

CHAPTER 7

DAMS

Water resources have been manipulated for urban and agricultural uses since ancient times. The results of such water diversions have been both extremely positive (in terms of economic growth, food production, and surface water recreation enhancement) and extremely negative (urban sprawl, loss of wildlife habitat, and destruction of river corridors). However, the importance of dams in this manipulation of the hydrologic cycle (by storing surface water during wet cycles for later use during dry periods) cannot be refuted.

Dams are the sentinels of the water development paradigm. Without dams, floodwaters would damage developed regions, irrigated lands would revert to dryland crops or native desert, and sprawling urban growth would be severely limited in arid regions such as Phoenix, Arizona; Los Angeles, California; Denver, Colorado; and other arid regions around the world.

In this chapter, we examine the underlying role that politics played in the development of dams in the United States, particularly in the first half of the twentieth century. In Chapter 6, we briefly discussed the need for reliable water supplies to promote the economic and population growth of an urban area. How does a dam create a more reliable supply of water for a city? Why did businesspeople, politicians, and others lend their support to dam construction projects for economic expansion? Why did many U.S. presidents, particularly Franklin D. Roosevelt, actively promote the construction of dams in the United States in the 1930s? Why did the Chinese government pursue construction of the Three Gorges Dam, the largest dam in the world, at the end of the twentieth century? Keep these questions in mind as you read this chapter.

DAM BASICS

PURPOSES OF DAMS

Dams have been used to regulate rivers for centuries. The ancient civilizations constructed dams made of earth, rocks, logs, and other simple materials to redirect rivers for flood control and irrigation. Dam projects of the twentieth century were generally built for multiple purposes, including flood control, irrigation, recreation, navigation, and hydroelectric generation.

Dams and reservoirs have enhanced the health and economic prosperity of citizens around the world. However, dam construction comes with a price: altered natural and human environments. Reduced streamflows, degraded water quality, and impacts on migrating fish are among the serious problems caused by dams throughout the United States and the world for centuries.

Why do we need to understand the basic principles of dams? Very few major rivers in the world do not have a dam crossing its course. Dams are a basic, fundamental management tool used to control, regulate, and deliver water for a variety of purposes (see Table 7.1 and Figure 7.1). Many people are fascinated by dams, whereas others abhor these monolithic structures. Regardless of one's personal opinions, it is important to understand the basic features of dams when studying the topic of water.

TABLE 7.1 Primary Purposes of Dams in the United States, 2001

Primary Purpose	% of Total	Number of Dams
Recreation	33.8	26,152
Flood control	15.6	12,088
Fire and farm ponds	13.7	10,589
Irrigation	9.5	7,392
Water supply	9.4	7,297
Other	8.1	6,279
Undetermined	3.5	2,647
Hydroelectric	2.9	2,280
Fish and wildlife	1.4	1,046
Mining (tailings)	1.3	991
Debris control	0.5	396
Navigation	0.3	250
Total	100%	77,407

Source: U.S. National Inventory of Dams, U.S. Army Corps of Engineers, January 2001.

How are dam sites selected? In some locations, a **flood-control dam** is built upstream of a city or developed area in order to protect downstream property and human life from flood devastation. Dam designers (usually experts in civil engineering) determine the size of such a flood-control structure on the basis of a Probable Maximum Flood (PMF) event, perhaps a Q500 flood discussed in Chapter 3. Dams for irrigation projects are generally constructed at an elevation high enough to deliver irrigation water to cropland entirely by gravity. Dams for **hydroelectric power generation** are located at a site where the difference in elevations between the surface of the new reservoir and the outlet to the downstream river is adequate to power electrical-generating turbines.

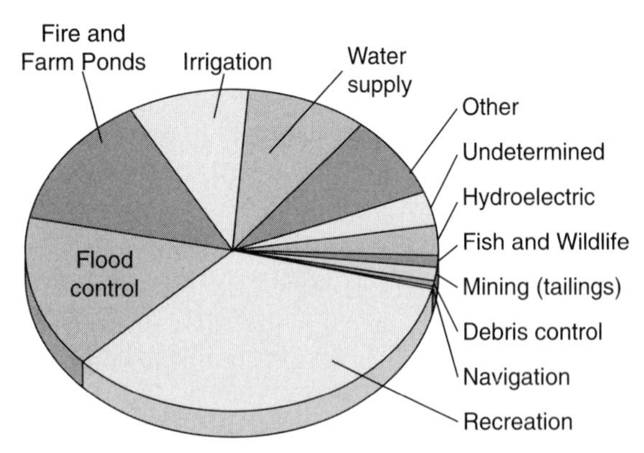

FIG. 7.1. Primary purposes of dams in the United States.

Almost every major river in the United States is now regulated by dams, locks, or diversion structures. Currently, the United States has over 75,000 dams that are more than 6 feet (1.8 m) in height. Reservoirs created by these structures cover more than 3 percent of the total land area of the country. The ages of these dams are presented in Table 7.2, with the information presented in a histogram in Figure 7.2. In any given year, 60 percent of the flow in all rivers and streams in the United States can be stored behind a dam. On the Colorado River alone, four years of flow can be stored behind dams on that extensive river system. (1)

COMPONENTS OF DAMS

All dams have certain characteristics in common (Figure 7.3). The **face** of a dam is the exposed surface of the structure that contains materials such as rockfill, concrete, or earth. The face can

TABLE 7.2 Age of Dams in the United States, 2001

Time Period	Completed Number
1990–2001	2,557
1980–1989	5,017
1970–1979	13,076
1960–1969	19,310
1950–1959	11,388
1940–1949	4,053
1930–1939	3,716
1920–1929	2,252
1910–1919	1,907
1900–1909	2,127
1800–1899	2,509
1700–1799	22
1600–1699	1
Undetermined	9,472
Total	77,407

Source: U.S. National Inventory of Dams, U.S. Army Corps of Engineers, January 2001.

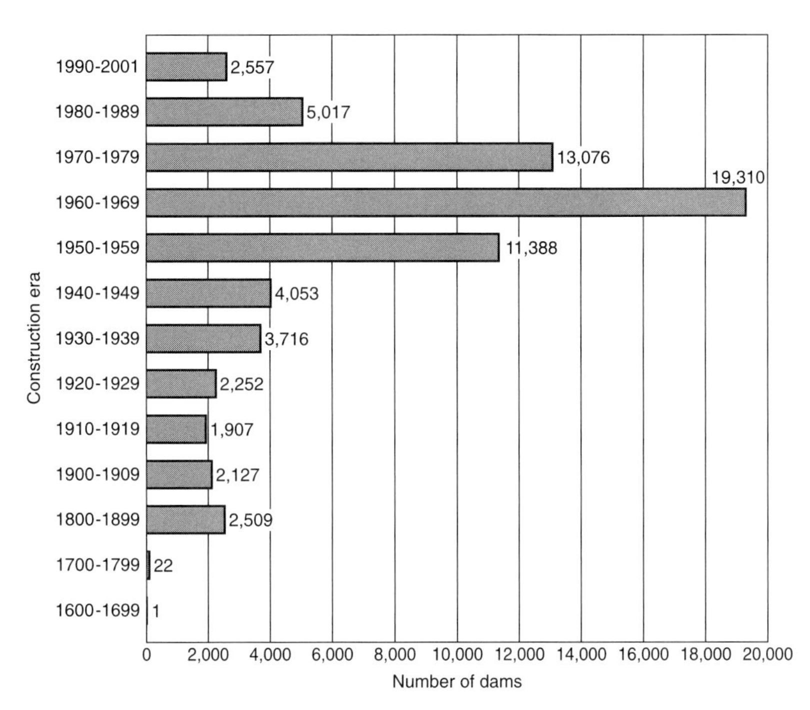

FIG. 7.2. Age of dams in the United States, 2001. Note that the peak construction period was between 1950 and 1979. Since that time, dam construction has declined significantly due to environmental concerns and changing societal values. Would you predict that dam construction will continue to decline in the future, or will it increase due to population growth? Why?

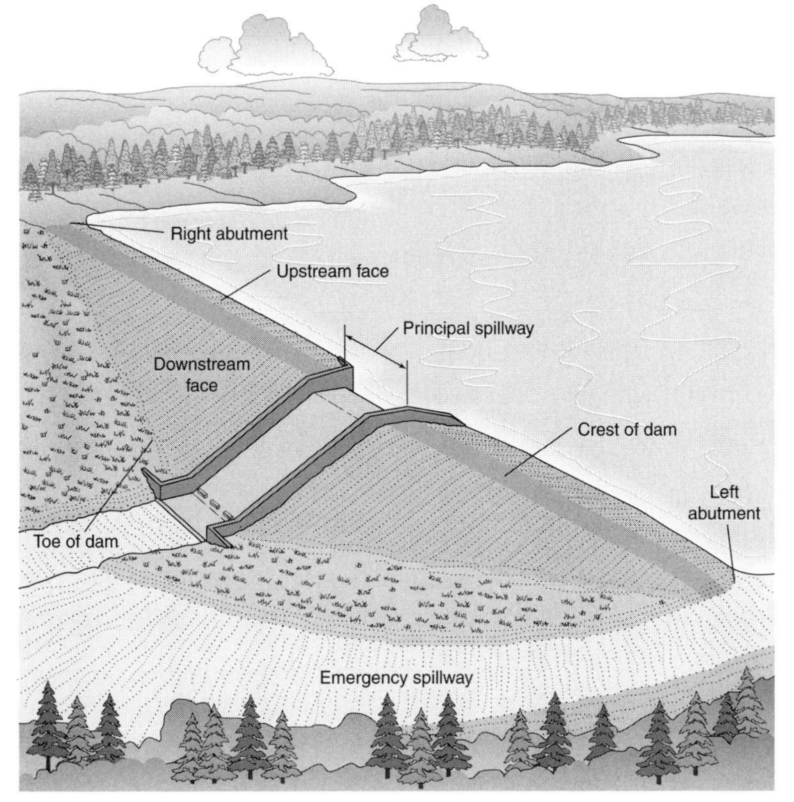

FIG. 7.3. Principal parts of a dam. The emergency spillway is located away from the toe of the dam so that erosion from flowing water will not harm the structure during a flood event. Notice, also, that no trees are allowed to grow on this earthen structure illustration. Tree roots could cause pathways for reservoir water to seep through the dam.

be on either side of a dam and is referred to as the upstream or downstream face of a dam. **Abutments** are the sides of the dam structure that tie (or extend into) canyon walls or contact the far sides of a river valley. Abutments are called left or right by facing downstream from the reservoir and viewing the dam (i.e., left abutment, right abutment). The top of a dam is called the **crest,** while the point of intersection between the downstream face and the natural ground surface is called the **toe.** The crest usually has a road or walkway along the top that follows the entire length of the dam. The length of a dam (also called the **dam axis**) is measured from abutment to abutment. A **parapet wall** is often built along the dam crest for ornamental, safety, or wave-control purposes. The wall may be a few feet (or meters) high and is usually made of concrete. The **dam foundation** is the excavated surface on which a dam is constructed.

An **outlet** is an opening used to discharge water from a reservoir to a river for a particular purpose, such as irrigation or hydroelectric generation. It is generally made of steel or concrete and usually extends out into the reservoir. An **outlet gate** is a metal door used to regulate the water discharge through a reservoir outlet. Most dams have a **spillway,** which is a wide chute placed at a calculated elevation, to allow excess water to flow past the dam in a safe manner. Spillways can be as simple as a low, flat spot carved beyond an abutment, or they can be an elaborate concrete apron (channel) to divert flood flows over an abutment or through a dam. Probable Maximum Precipitation (PMP) and Probable Maximum flood (PMF) calculations (discussed in Chapter 3) are determined for the upstream watershed to assess maximum flood flows and the necessary spillway capacity.

TYPES OF DAMS

There are three primary types of dams: gravity concrete, concrete arch, and earthen embankment (see Figure 7.4). The geology, topography, and streamflow at a site dictate the type of dam appropriate for a particular location.

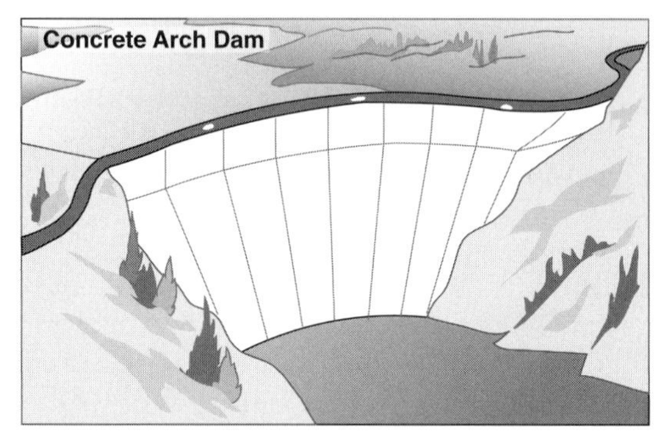

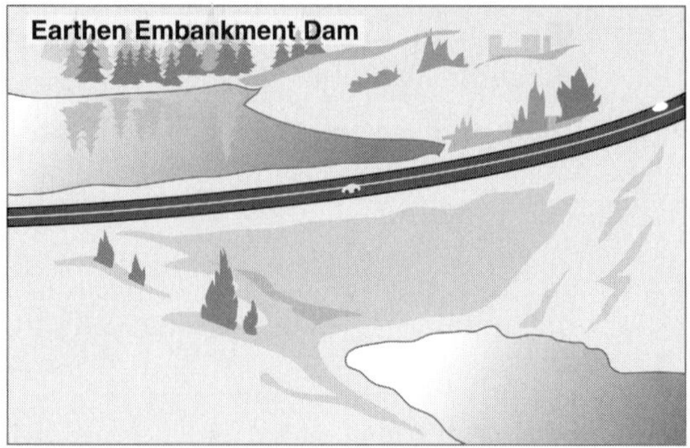

FIG. 7.4. Basic dam designs. Most dams in the world are the earthen embankment type and can last for centuries if properly constructed.

Dams made of concrete can be categorized as gravity or arch dams depending on the design. A **gravity concrete dam**—such as Grand Coulee Dam and discussed later in this chapter—is a solid, concrete structure that uses its mass (weight) to hold back water. It requires massive amounts of concrete, especially at the base of the structure, to provide the weight necessary to withstand the tremendous head pressure generated by impounded water behind the dam. A gravity concrete dam may have triangular supports, called **buttresses,** on the downstream side of the dam to strengthen it and to distribute water pressure to the foundation of the dam. Dam construction in areas of high seismic activity may require the use of buttress dams. The upstream face of a dam may have a broad slope to add mass to the base of the dam.

A **concrete arch dam**—such as Hoover Dam—has a curvature design that arches across a canyon and has abutments embedded into solid rock walls. Water pressure pushes against the curve of a concrete arch dam and compresses the material inside the structure. This pressure actually makes a dam more solid and dissipates head pressure into the canyon walls. Arch dams require less concrete but must have solid rock walls as anchors for the abutments. Concrete dams are very expensive to build when compared to earthen embankment structures.

Earthen embankment (or **earthfill**) **dams**—like Kingsley Dam in Nebraska—are structures in which more than 50 percent of the total volume of the dam consists of compacted earth material. Large earthen dams have an impervious core of clay, or other material of low permeability, that prevents reservoir water from rapidly seeping through or beneath the foundation of the structure. Most large earthen dams also have drains installed along the downstream toe of the dam. These small, parallel conduits channel any seepage water inside the earthen dam to locations downstream and away from the toe of the dam. This prevents water from building up inside the structure and eventually eroding material within the dam.

The base of an earthen dam is very broad when compared to the crest of the dam. A slope of 1:2 on the downstream slope (1 unit vertical for every two units horizontal) and 1:3 on the upstream slope are fairly common. The upstream face of the dam is often covered with **rip-rap** (randomly placed layers of concrete, stones, or large rocks) to prevent erosion caused by wave action. Earthen embankment dams are usually the most economical to build in small watersheds or across broad valleys. Almost 80 percent of all dams in the world are earthen dams.

DAM OPERATIONS

Dams are operated by individuals called **dam tenders** (also called *operators* or *reservoir managers*). Their job is to make releases of water, perform maintenance on the dam, and monitor water storage levels. Reservoir releases are based on predetermined purposes (such as flood control, recreation, or irrigation) at various reservoir capacities. Dam tenders of flood-control dams prefer to keep the water level as low as possible in order to retain the space needed to capture surface runoff during storm events or from melting snow in the spring.

Operators of irrigation dams try to keep a reservoir as full of water as possible until irrigation supplies are needed for crops. Managers of recreation reservoirs prefer a constant water level so that boat docks and concession stands are not flooded or shorelines lowered and turned to mud. Hydropower plant operators often release large volumes of water for short durations to generate power during "peak" electrical demand periods. What downstream effects would these various dam operations cause?

All reservoirs have storage space located at the bottom that cannot be released by gravity. This is called **dead storage** (or **dead capacity**) and will eventually fill with silt and other sediments. **Inactive storage** (or **inactive capacity**) is the amount of water stored just above the dead capacity. Under normal conditions, water in inactive storage is used only for purposes such as fisheries, water quality, or other nonconsumptive uses.

Water can be evacuated from this space in an emergency or when a reservoir must be drained to make repairs on a dam or **appurtenances** (outlet pipes, gates, etc.).

If a reservoir has multiple uses, the next layer of use above the inactive storage space is called the **joint use capacity.** Within this storage zone, water may be held for several purposes such as irrigation, recreation, power generation, or industrial uses. It is also called the **active conservation pool** (or **active storage capacity**) of a reservoir.

In flood-control projects, all space above the dead capacity and inactive capacity is called the **flood pool,** or **flood-control capacity.** If a flood-control project has multiple purposes, then only space above the joint use capacity will be used for flood control. The area above the joint use capacity or the flood pool (if there is a designated flood pool in the reservoir) is called the **surcharge capacity.** This is the portion of reservoir capacity used to pass predicted floods through the reservoir. Surcharge capacity can also be considered **temporary storage capacity,** or **temporary flood pool.**

The top of the surcharge capacity is also the **maximum water surface** (or **maximum water pool**) elevation of a reservoir. This is the highest acceptable water surface elevation for a reservoir and is based entirely on dam safety considerations. Elevations are expressed as "feet (or meters) above mean sea level," which is the elevation of the ocean halfway between high and low tides. The **total capacity** of a reservoir includes all storage space in a reservoir from the top of the maximum water surface within a reservoir down to the bottom of the dead storage space. **Live capacity** is the total amount of reservoir storage that can be released by gravity. It equals the total capacity of a reservoir minus the dead capacity.

Freeboard is the difference in elevation between the dam crest and the maximum water surface of the reservoir. Adequate freeboard is critical to protect the dam crest from erosion due to wave action or sudden water inflows into the reservoir that could overtop (flow over) a dam. If water spilled over a dam crest and across the downstream face, severe erosion could cause a dam to breach and collapse in a matter of hours (Figure 7.5).

The **firm yield** of a reservoir is the amount of water that can be stored with relative certainty during projected hydrologic events. Water providers rely on firm yield estimates to determine available water supplies during extended dry periods. Total reservoir storage capacity and firm yield can be significantly different numbers.

EXPERT ANALYSIS

Using a topographic map of your area, locate the major dams in your watershed. Determine the types of dams constructed and analyze why these construction sites were selected. What are the primary purposes of the dams in your watershed? Was construction funded by local, state, or the federal government? Was there

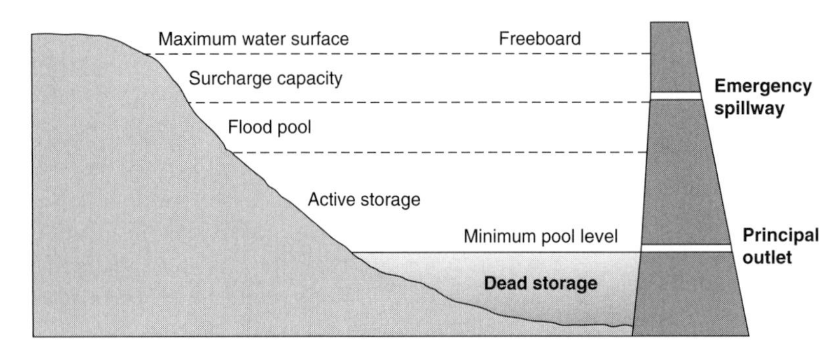

FIG. 7.5. Classification of principal storage zones in a cross section of a multipurpose reservoir.

local opposition to any of the projects? Why or why not? Has the primary purpose of any dams in your watershed changed since they were constructed?

CASE STUDY

Hoover Dam and Lake Mead—Nevada/Arizona

http://www.usbr.gov/lc/hooverdam

Hoover Dam was completed in 1935 at a repayment cost of $165 million to U.S. taxpayers. It is a concrete gravity arch dam that rises 726 feet (221 m) above bedrock in the Black Canyon of the Colorado River, about 35 miles (56 km) southeast of Las Vegas, Nevada. Hoover Dam's first water management priority is flood control.

Water stored in Lake Mead, the reservoir created behind the dam, is also used for irrigation, industrial and municipal use, hydropower, recreation, and fish and wildlife habitat maintenance. Hoover Dam is one of 11 major dams across the Colorado River and its tributaries between western Wyoming and southern California. It has also been called Boulder Canyon Dam (because it is part of the Boulder Canyon Project, since Boulder Canyon, several miles upstream of Black Canyon, was the original site proposed for the dam). Congress passed legislation permanently naming it in honor of President Herbert Hoover in 1947. Lake Mead is named after Dr. Elwood Mead, the Commissioner of the Bureau of Reclamation during the dam's planning and construction. (2)

The weight of concrete in the dam alone provides enough mass to offset the 45,000 pounds per square inch (3164 kg/cm^2) of water pressure at the base of the dam. However, the arch design was probably included to provide additional safety or to reduce the dam's construction cost. The base of Hoover Dam is 660 feet (201 m) thick. The dam's crest stretches 1244 feet (379 m) across Black Canyon. The dam is 45 feet (14 m) thick at the crest and provides the roadway for U.S. Highway 93. (3)

Statistics on Lake Mead are staggering. Its original maximum storage capacity was 32,350,000 acre-feet (10.5 trillion gal, or 39.9 billion m^3) at the maximum water surface, elevation 1229.0 feet (374.6 m) above mean sea level. However, a 1963–1964 sedimentation survey showed that the maximum storage capacity had decreased to 29,755,000 acre-feet (9.7 trillion gal, or 36.7 billion m^3). As a practical matter, storage does not exceed elevation 1221.0. (4) The lake is approximately 581 feet (177 m) deep at its maximum depth and about 110 miles (177 km) in length at its maximum capacity; as such it is the largest artificial lake in the United States today. Storage space in Lake Mead is allocated as follows:

Lake Mead's lower 255 feet (78 m) of storage capacity (2,378,000 acre-ft, which is 774.9 billion gal, or 2.9 billion m^3) is allocated as dead storage (a sediment catchment area from elevation 640 to elevation 895).

The next 188 feet (57 m) is allocated as inactive storage (10,024,000 acre-ft, which is 3.3 trillion gal, or 12.4 billion m^3) from elevation 895 to elevation 1083.

The next 136.6 ft (41.6 m) is the active conservation or joint use pool (15,853,000 acre-ft, which is 5.2 trillion gal, or 19.6 billion m^3), and is used for flood control, municipal and industrial water supply, irrigation, and power generation from elevation 1083 to elevation 1219.6.

The upper 9.4 feet (2.9 m) is used exclusively for flood control (1.5 million acre-ft, which is 488.8 billion gal, or 1.9 billion m^3) from elevation 1219.6 to elevation 1229.0. (5)

Opponents of the project, concerned about such a large volume of water stored behind a single structure, argued that the weight of water in the reservoir would cause earthquakes. (Others simply predicted that a large flood would wash out the dam and destroy everything downstream.) Skeptics had reason to be concerned. Almost 10 percent of the entire land area in the United States drains into Lake Mead. The reservoir was designed to store two years of the entire average flow of the Colorado River and would cover approximately 160,000 acres, or 250 square miles (64,752 ha). (6)

Work on Hoover Dam required that the entire flow of the Colorado River be diverted around the construction site. Four bypass tunnels were drilled and blasted through the canyon walls to channel all river water around the dam. Two tunnels were located on the Arizona side of the canyon and the other two on the Nevada side, all parallel to the Colorado River. Each tunnel was 56 feet (17 m) in diameter, approximately 4000 feet (1219 m) long, and lined with a 3-foot-thick (1 m) layer of concrete to prevent erosion. This may seem like an excessive amount of reinforcement, but huge chunks of a similar liner in the spillway tunnels at the upstream Glen Canyon Dam were lost due to *cavitation* (the process of erosion caused by moving water) during high water releases in 1983. Holes were drilled through the cement lining at Hoover Dam to allow workers to pump grout into the void spaces between the blasted rock wall and the lined tunnel. (7)

After the tunnels at Hoover Dam were completed, a 98-foot-high (30 m) cofferdam (a temporary dam made of earth and rock) was erected to divert the entire flow of the Colorado River into the four tunnels. Then the canyon walls had to be cleared of unstable rock to ensure there would be no rockfalls during construction of the facilities below, or in the future, after the dam was completed. Workers, called *high-scalers,* suspended by safety harnesses or boatswain's (bosuns) chairs at the ends of ropes strung down from the canyon rim, chiseled, blasted, and pried loose rocks and boulders. For over two years, 400 men dangled hundreds of feet (100 m or more) in the air with no safety nets to complete the most dangerous part of the construction project. Many fell to their deaths. (8)

FIG. 7.6. Notice the concrete blocks that are being poured during construction of Hoover Dam and the tremendous width of the structure at its base. Individual blocks varied in size between 60 feet square (18.3 m^2, or the size of a large house) at the upstream face of the dam to about 25 feet square (7.6 m^2, or the size of a two-car garage) at the downstream face. Pouring of concrete into any one block was limited to only 5 feet (1.5 m) in height in 72 hours to ensure stability. After cooling, grout (a cement and water mixture) was pumped into voids created between the blocks (caused by the contraction of the cooling concrete) to form a monolithic (one-piece) structure.

The dam was poured nonstop over a period of two years, night and day, in a series of trapezoidal columns (irregular-shaped boxes) formed by 25- to 65-foot-square (8- to 20-m^2) blocks (see Figure 7.6). Concrete was placed in 5-foot (1.5-m) horizontal "lifts," or depths, in each block, which was then cooled before another block could be placed on top of or adjacent to it. This construction technique was used to allow heat generated by the concrete to dissipate.

Curing concrete creates heat as it dries, and will crack if it is not allowed to dry quickly and properly. Engineers estimated that one continuous, solid pour of concrete at Hoover Dam would have required 125 years to cool. Concrete also contracts as it cools and would have caused stress fractures in such a large mass of material. The structure would have been rendered useless if the concrete had been allowed to cool naturally.

To solve the cooling problem, engineers essentially built a very large refrigerator to speed up the concrete-cooling process. A web of 1-inch (2.5-cm) hollow steel pipes were placed within each block before the concrete was placed in that block. After the pour was completed, river water was pumped through the pipes to cool the concrete. Chilled water from a big refrigeration unit just downstream of the dam was then circulated inside the network of pipes, lowering the temperature of the concrete to the required temperatures. Later, the hollow pipes were cut off and grouted (filled with a liquid mixture of cement and water) to seal all openings within the concrete column. (Although the concrete in Hoover Dam continues to cure today, it reached a safe temperature to prevent cracking during construction.)

Two 150-foot-wide (46 m), 650-foot-long (198 m) spillways were built to protect the dam, one on each side of the canyon walls. Each spillway has the capacity to carry 200,000 cfs (5663 cms) of water (almost twice the normal flow of the Columbia River). These spillways were designed to discharge excess floodwaters into the outer 56-foot-diameter (17 m) bypass tunnels used during construction. Four 32-foot-diameter (10 m) intake towers (outlets) in Lake Mead allow water to be released from the reservoir for power generation, irrigation, or municipal/industrial uses.

Most large concrete dams have walkways (called galleries) inside the structures that lead to various operational or inspection sites within the dam. These corridors may be located at several levels, are well-lighted and ventilated, and provide unique opportunities for observations deep within the dam. At Hoover Dam, and many other sites, fine hairline cracks are closely monitored by structural engineers to determine whether any are lengthening or widening. Cracks in concrete are common, so it is not unusual to see very fine cracks in concrete dams. A crack may only increase fractions of an inch (or millimeters) in decades and is a sign of normal pressures within a dam.

Hydropower at Hoover Dam (see Figure 7.7) is generated by 17 main turbines (modern versions of ancient waterwheels). The hydroelectric plant at Hoover Dam can generate over 2000

FIG. 7.7. The dramatic concrete arch design of Hoover Dam securely holds the impounded waters of Lake Mead. Power generators are located out of view at the base of the dam. Notice the two water intake towers just above the crest of the dam near the abutments. U.S. Highway 93 can be seen at the left side of the photo along the rim of Black Canyon. The highway across Hoover Dam was closed to all large truck and RV traffic immediately after the events of September 11, 2001, due to the threat of terrorism.

megawatts of electricity (enough to provide the daily electrical needs of a city of over one million residents). The costs incurred for construction of the dam as of October 1937 have been repaid to the U.S. Treasury, with interest, from power generation revenues. Some of the costs incurred after 1937 have also been repaid, while more recent costs will not be repaid for as long as 40 years. The Hoover Dam power plant was the world's largest hydroelectric installation until 1949 when it was surpassed by Grand Coulee Dam. It remains one of the largest in the United States. Hydropower revenues continue to pay for the operation and maintenance of the dam, as well as for the costs of replacing equipment, machinery, and other items needed to ensure the dam's safe and effective operation. (9)

A CLOSER LOOK

A highlight for many visitors to Hoover Dam is the artwork located throughout the project. Norwegian-born sculptor Oskar J.W. Hansen did much of the work, including a terrazzo plaza inlaid with a celestial map, sculptures, wall plaques, and bas-relief (three-dimensional) work in concrete on elevator towers that represent water uses provided by the dam. Other artists and architects contributed the terrazzo floors throughout the dam and power plant, which are embedded with illustrations that honor Native American tribes of the region, as well as decorative doors and railings.

The most prominent sculpture is a bronze piece called "The Winged Figures of the Republic," located near the Visitors Center. It bears a plaque commemorating the 96 men who died during construction of Hoover Dam. It reads:

They died to make the desert bloom. The United States of America will continue to remember that many who toiled here found their final rest while engaged in the building of this dam. The United States of America will continue to remember the services of all who labored to clothe with substance the plans of those who first visioned the building of this dam.

The Hoover Dam was built during a brutal period of history. The stock market crash of 1929, coupled with the devastating drought across the Midwest in the 1930s, crushed the economic viability of millions of families. For many, dam construction projects provided relief from these severe conditions. The promise of a paycheck led men to willingly dangle their lives hundreds of feet in the air, along the walls of a desert canyon, to build a government dam project. Would modern workers do the same? Would the general public support funding for such a project today?

Some readers find dam construction projects such as Hoover Dam remarkable, enlightening, and fascinating. Others, however, are appalled by the economic, human, and environmental costs of damming a river. What is your reaction, at this point, regarding the construction of dams to store water for human uses?

CASE STUDY

Kingsley Dam and Lake McConaughy—Nebraska

http://www.cnppid.com/history2.html

Earthen dams require very different construction techniques from concrete structures. In 1936, the Central Nebraska Public Power and Irrigation District began construction of Kingsley Dam near Ogallala, Nebraska. It took five years to complete the dam that created Lake McConaughy, an irrigation, hydropower, and flood-control project for residents of southwest Nebraska. The dam was built across the North Platte River Valley at a cost of $43,540,510. The project was named to honor two local businessmen who were strong supporters of the effort to build a dam on the North Platte.

At the center of Kingsley Dam is an impervious clay and steel core. A solid steel wall of interlocked sheet steel was pile-driven across the river valley to depths of 30 to 160 feet (9 to 49 m) until it was hammered into impervious Brule clay. This created a solid steel sheet barrier along the dam site. Then the core was expanded (widened) by compacting Brule clay and other impervious material around the steel wall. The remainder of the earthen dam consists of sand, gravel, and soil. Kingsley Dam extends 3.5 miles (5.6 km) across the broad North Platte River Valley and has a height of 162 feet

(49 m). The base of the dam has a width of 1100 feet (335 m) and tapers to only 28 feet (8.5 m) wide at the crest where Nebraska Highway 61 crosses over it. (10)

In addition to hydropower and irrigation water supply, Lake McConaughy is a very popular recreation area. Owing to persistent wind in the region, wind surfing and sailing are common. The reservoir has a capacity of 1,948,000 acre-feet (634.8 billion gal, or 2.4 billion m^3), is 22 miles (35 km) long, and has a maximum width of 4 miles (6.4 km). Maximum depth of water is 167 feet (51 m). Kingsley Dam is one of seven major reservoirs constructed across the North Platte River in Nebraska and Wyoming. (11)

A major difference between a concrete and earthen dam is the width of the structure at the base. Unlike Hoover Dam which was built between the sheer canyon walls of the Colorado River, Kingsley Dam stretches across a broad, relatively flat prairie valley. The dam had to be very long to fill the valley with material to hold back all flows, including flood events, of the North Platte River. In addition, the dam abutments had to attach to geologic material that was stable and would provide a secure anchor.

In 1972, wind-generated waves caused considerable damage to the crest of Kingsley Dam. The elevation of water in Lake McConaughy was at 3266.4 feet (995.6 m) above mean sea level when a severe May windstorm hit the area from the northwest. Ten-foot (3-m) waves were generated by the storm as 30 to 40 mph (48 to 64 km/hr) winds whipped across the length of the generally west-to-east reservoir. Severe erosion occurred to the top 25 feet (8 m) on the upstream slope of the dam and to a 150-foot-long (46 m) section of Highway 61 that crossed the axis of the dam. Many people feared the dam would fail, and some downstream residents were evacuated. (12)

Immediately, sand, gravel, and 48-inch (122-cm) hexapods (jack-shaped cement riprap) were placed on the upstream dam face to dissipate the energy of crashing waves and to temporarily repair erosion damage (see Figure 7.8). Additional rock was shipped from

FIG. 7.8. This photo was taken shortly after repairs were started along the freeboard of Kingsley Dam after the severe damage caused by wind and wave erosion in May 1972. Note the concrete tetrahedrons (or hexapods) at the water's edge of Lake McConoughy and the damaged parapet wall in the foreground.

Wyoming quarries and placed on the dam, and fortunately, the emergency was soon declared over. Later, an engineering consulting firm recommended a reduction in the maximum surcharge pool to elevation 3250.0 feet (990.6 m) above mean sea level during the storm-prone periods of March 1–May 15 and October 1–December 31 in order to create more freeboard on the upstream face of the dam. This dropped the maximum water level by 16.4 feet (5.0 m) from historic levels. Additional recommendations were made regarding implementation of coastal engineering principles for wave protection rather than historic slope protection techniques. This was an interesting concept for a water body in the middle of Nebraska but quite appropriate given the seriousness of the wind and waves incident in May 1972. (13)

Many reservoirs around the world, particularly in the Central Plains of the United States and Canada, followed the above recommendations and reduced the elevation of surcharge capacities in reservoirs during windy months, especially if winds generally blow parallel down the length of a reservoir toward a dam. The riprap on the upstream face of Kingsley Dam is now marked with paint at regular intervals, and aerial photos are regularly taken of the dam face. With this information, the location of each painted rock is entered into a computer database to determine whether any movement occurs in later photos. (14)

A CLOSER LOOK

Four years after the severe damage to Kingsley Dam in Nebraska, the failure of Teton Dam in Idaho sent shock waves throughout the world's engineering community. At a hearing held later that year in 1976 to determine the cause of the Teton Dam's collapse, Congressman Leo J. Ryan of California described the Teton Dam failure as "one of the most colossal and dramatic failures in our national history." (15) Eleven people died and 25,000 were left homeless by the wall of water that roared down the Teton Valley.

The 307-foot-high (94 m) earthen Teton Dam was completed across the Teton River in November 1975. The dam had a crest length of 3100 feet (945 m) and a base width of 1700 feet (518 m). It included a hydropower plant and a complex network of deep trenches filled with concrete and a grout curtain of cement in the foundation beneath the core of the dam. The concrete curtain was intended to increase dam stability and provide an impervious barrier against piping. (**Piping** is the internal erosion of an earthen dam that occurs when water seeps through and carries away soil particles from the embankment or abutment of a dam. This erosion can create a cavity or "pipe" within the dam that can erode backward toward the reservoir.) The dam was constructed by the U.S. Bureau of Reclamation and was considered state-of-the-art at the time.

Teton Dam was designed to provide flood control, recreation, and irrigation water for residents of the Teton River Valley in eastern Idaho. Excitement was high when dam construction was completed and the reservoir started to fill with water on October 3, 1975. On March 23, 1976, the Teton Project Office received permission to increase the filling rate of the new structure from 1 foot (0.3 m) per day to 2 feet (0.6 m) per day. The filling process was closely monitored.

About two months later, on Thursday, June 3, 1976, two small springs were found 600 and 900 feet (183 and 274 m) beyond the toe of the dam with a flow rate of 40 and 60 gallons per minute (gpm), or 0.15 and 0.23 m^3 per minute, respectively. The next day another spring emerged about 150 feet (46 m) downstream from the toe of the dam near the right abutment. It flowed at about 20 gpm (76 l per minute, or 0.08 m^3 per minute). The situation was monitored carefully by engineers and was thought to be normal seepage through the earthen dam.

On Saturday, June 5, 1976, at around 8:00 A.M., muddy water was seen flowing from within rip-rap on the right abutment. The flow rate was estimated to be approximately 20 to 30 cfs (8977 to 13,465 gpm, or 0.57 to 0.85 cms) and was located about 15 feet (4.6 m) above the streambed. By 9:00 A.M., the flow had increased to approximately 40 to 50 cfs (17,953 to 22,442 gpm, or 1.1 to 1.4 cms). Another small pipe of water emerged at the right abutment, near the contact point with the valley wall, about 130 feet (40 m) below the dam crest. At approximately 9:30 A.M. a wet spot appeared on the downstream face about 20 feet (6 m) from the right abutment. The material around the spot quickly liquified and started to flow. Embankment material eroded. Two bulldozer operators frantically tried to fill the enlarging weak spot with rip-rap but became stuck in the collapsing material. Both managed to escape.

Emergency officials were alerted at 10:43 A.M. At about the same time, a whirlpool developed in the reservoir near the right abutment, just a few yards from the upstream face of the dam. The swirling vortex indicated that water was rapidly exiting the lake through a natural outlet. The gaping hole had grown to more than 25 feet (8 m) in diameter, and the flow of seeping water was increasing.

At 11:57 A.M. the dam broke. Eyewitnesses said the western third of the dam disintegrated, and the wall of water seemed to hang in the air for a second, then came crashing down the canyon. An estimated 246,000 acre-feet (80.2 billion gal, or 303.4 million m^3) of water swept through the Upper Snake River Valley. The town of Wilford, Idaho, was obliterated, and 11 people lost their lives. Damages were later placed as high as $2 billion. (16)

Subsequent government hearings identified poor site selection, due to high permeability and porosity of geologic material beneath the dam, as the probable cause of the Teton Dam failure. Although the government was not legally responsible, President Gerald Ford released hundreds of millions of federal dollars to compensate for some of the losses incurred by local residents. Subsequently, the role of the Bureau of Reclamation shifted from construction of dams to management of existing water resources projects. The events of June 5, 1976, played a major role in the revised mission of the Bureau of Reclamation (and will be discussed further in Chapter 9).

FIG. 7.9. Grand Coulee Dam is a gravity concrete dam. Banks Lake, at the top of the photo, is approximately 250 feet (76 m) higher in elevation than Franklin D. Roosevelt Lake located immediately behind Grand Coulee Dam. Water must be pumped into Banks Lake, but the gain in elevation allows gravity delivery of irrigation water along the Columbia Plateau.

CASE STUDY

Grand Coulee Dam and Franklin D. Roosevelt Lake—Washington State

http://www.usbr.gov/pn/grandcoulee

Grand Coulee Dam (Figure 7.9) is a multipurpose, concrete gravity dam constructed across the Columbia River in central Washington. The Bureau of Reclamation started construction in 1933, and the main, original dam was completed in 1941. The dam is the largest concrete structure in the United States and one of the largest in the world. Grand Coulee Dam is 5223 feet (1592 m) long, 500 feet (152 m) wide at the base, 30 feet (9 m) wide at the crest, and cost $300 million to build. The geologic material beneath the base of the dam is solid granite. (17)

Franklin D. Roosevelt Lake, the reservoir created behind the dam, holds 9,562,000 acre-feet (3.1 trillion gal, or 11.8 billion m^3) of water at elevation 1,290.0 feet (393.2 m) above mean sea level. The reservoir has an active storage capacity of 5,232,000 acre-feet (1.7 trillion gal, or 6.5 billion m^3). The crest of the dam's central spillway is at elevation 1290.0 feet (393.2 m) above mean sea level. (18)

The reservoir stretches over 150 miles (241 km) upstream along the Columbia River to the Canadian border with British Columbia. The dam's height was limited to 550 feet (168 m) to prevent reservoir water from stretching across the international boundary. Grand Coulee Dam is one of 11 major dams constructed across the Columbia River in the United States and obtained its name from the empty riverbed created by catastrophic Ice Age floods. (19)

Cofferdams were used to divert the flow of the Columbia River around the construction site. Howcvcr, part of the cofferdam system began to leak during construction, then failed, and allowed 90 cfs (40,395 gal per minute or 2.5 cms) of the Columbia River to flow into the construction site. Workers immediately began hauling tons of rock and other material to fill the gap. Simultaneously, pumps worked at full throttle to remove the flood of water from the worksite inside the temporary dams. The river continued to gain on the workers until a substance called bentonite, which expands to over 10 times its normal size when wet, was poured by the ton into the break. The bentonite quickly expanded and stopped the leak. (20)

Engineers cooled the concrete at Grand Coulee Dam with the same technology used at Hoover Dam a few years earlier. Based on the estimate that 150 years would be needed for concrete at Grand Coulee to cool naturally, 2000 miles (3219 km) of refrigeration pipes were installed throughout the dam. The frigid water lowered the concrete temperatures to acceptable levels within a month. (21)

Twenty-one hydroelectric generators were originally installed in the dam, and later, additional generators brought Grand Coulee Dam to its current rated generating capacity of 6,809,000 kilowatts of electrical power (Figure 7.10). Initially, the primary beneficiaries of power were the Hanford Nuclear Reservation located in southeastern Washington near Richland, and the factories that manufactured aluminum for the military aircraft industry in the Northwests (particularly Boeing in Seattle) during World War II. (22)

FIG. 7.10. Hydroelectric turbines at Grand Coulee Dam. Water from the Columbia River flows through a series of louvers, or wicket gates, located around the turbine inlet. Water inflows can be regulated by opening or closing the louvers. This allows an operator to maintain a constant turbine speed even under varying electrical loads.

These turbines are connected to an electric generator by a long shaft. Each generator has a large spinning *rotor* and a stationary *stator.* The spinning rotor creates a magnetic field and induces electrical current (electricity) in the stator's copper windings.

The largest hydroelectric complex in the world is at Itaipu Dam on the Parana River between Paraguay and Brazil. Its 18 turbines produce 12,600 megawatts of electricity, nearly double that of the Grand Coulee facility.

TABLE 7.3 Comparison of Hoover, Grand Coulee, Kingsley, and Teton Dams (all dimensions in feet)

Dam	Type	Height	Length	Base Width	Crest Width	Cost
Hoover	Concrete arch	726	1,244	660	45	$165M
Grand Coulee	Concrete gravity	550	5,223	500	30	$300M
Kingsley	Earthen	162	18,480	1,100	28	$ 43M
Teton	Earthen	307	3,100	1,700	30	$ 85M

The irrigation component of the Grand Coulee Dam project was completed after World War II. Pumping stations were constructed to lift water out of the Franklin D. Roosevelt Lake and into a reservoir called Banks Lake for gravity delivery into canals. Irrigation water is presently delivered to about 670,000 acres (271,140 ha) in the Columbia Basin as far as Pasco, Washington, a region some 125 miles (201 km) to the south that receives approximately 10 inches (25 cm) of precipitation annually. (23)

POLICY ISSUE

Every dam construction project has detractors, and Grand Coulee Dam was no exception. President Franklin D. Roosevelt was a strong supporter of public works projects during the 1930s and was a champion of hydropower development. He was also a Democrat. Francis Dugan Culkin, a Republican congressman from Roosevelt's home state of New York, tried to stop funding for Grand Coulee, arguing that the only customers in that region of the United States were jack rabbits and rattlesnakes, and they were not inclined to purchase electricity. (24)

Today, the negative impacts of dams on endangered species, such as salmon, would be insurmountable obstacles for any new proposed project across the Columbia River. The protection of wildlife, particularly species with severely declining populations, is an important national policy and would have quickly stopped Grand Coulee Dam construction plans.

What if the economic downturn of the 1930s had not occurred and President Roosevelt's efforts to create jobs were not necessary to bolster the U.S. economy? Would the population distribution of the United States be different today without the construction of projects such as Hoover and Grand Coulee dams? Consider the role that climate plays in the spatial distribution of precipitation in the United States. If Hoover Dam had not been built, would large population centers only exist east of the 100th Meridian? What other methods of water acquisition and conservation could have been used to support population growth in the West without the construction of large dams for water supply? A comparison of the four dams discussed in this chapter is shown in Table 7.3.

BENEFIT-COST ANALYSIS

Construction of large dams involves elaborate economic analysis to determine whether costs justify the benefits to be gained. About 100 years ago, President Theodore Roosevelt argued that all federal dam projects should be repaid from benefits, such as income from hydroelectric production, charges for irrigation water, or fees from municipal and industrial water users.

Whenever the Government constructs a dam and lock for the purpose of navigation there is a waterfall of great value. It does not seem right or just that this element of local value should be given away to private individuals of the vicinity, and at the same time the people of the whole country should be taxed for the local improvement.

It seems clear that justice to taxpayers of the country demands that when the government is called upon to improve a stream, the improvement should be made to pay for itself, so far as practicable. (25)

This concept was later enacted into law, and methods were developed to determine the economic feasibility of proposed federal projects. The science of benefit-cost analysis evolved over the next 60 years.

Benefit-cost analysis is the ratio of the present value of project benefits to the present value of project costs. Benefit-cost analysis has been used to determine the economic feasibility of projects for the past 60 years. Values are assigned to project purposes such as flood control, irrigation, industrial and municipal water supply, hydropower, recreation, and any other benefits associated with construction of the water storage project. Each project benefit is assigned a dollar value, based on the year accrued, and then totaled.

Generally, direct benefits can readily be quantified, such as the projected value of crops that will be grown with the irrigation water or the amount of revenues that will be generated from concession operators of marinas at a reservoir created by the dam. It is much more difficult to assign dollar values to indirect benefits since these include items such as the value of saving a life downstream from future floods or the dollar value of new businesses created in nearby communities owing to the availability of less expensive electricity. Economists spend years developing acceptable benefit computations for a water project.

On the cost side, all direct expenses are listed, including but not limited to, construction costs, future operation and maintenance expenses, and repayment interest. Again, indirect cost values are difficult to determine because these include items such as water quality degradation, loss of wildlife habitat, and reduction in river-based recreation that will be inundated (flooded) by the new reservoir over the life of the project.

After all direct and indirect costs and benefits are determined, a ratio is developed—such as 2:1. A 2:1 ratio means that for every two dollars of benefits generated by the project, one dollar will be expended on construction and operation of the dam and reservoir facility. A ratio greater than 1:1 means the project has a **positive benefit-cost ratio.** If the number is less than 1:1, such as 1:2, the benefits will equal only one dollar for every two dollars of expenses, and the project has a **negative benefit-cost ratio.**

This explanation has been greatly simplified since it does not include the concept of the present value of money. When benefits and costs are estimated, many figures will be for activities that will occur years in the future. The value of a dollar will be greater in 10 years because the owner of the dollar could place it in a savings account today and gain interest. Similarly with benefit-cost analysis, the value of a cost or benefit 10 years in the future must be discounted (reduced) to equal the dollar value of the cost or benefit today. This is called determining the **present value** of money. All benefits and costs must be expressed in present value to provide fairness when comparing the economic benefits, such as recreation revenues over 50 years, to the cost of construction that occurs over three years.

EXPERT ANALYSIS

Determine the present value of the following costs of dam construction. Consider Year 1 as the present year. All values are in thousands, and the interest (also called the **time value of money** or the **discount rate**) is 8 percent simple interest compounded annually.

Year 1	Year 2	Year 3	Total
$100,000	$150,000	$200,000	$450,000

The present value cost of Year 1 = $100,000

The present value cost of Year 2 = $150,000 divided by 1.08% = $138,889 ($138,889 invested at 8% simple interest for one year = $150,000)

The present value cost of Year 3 = $200,000 divided twice by 1.08% = $171,468. This means that $171,468 invested at 8% simple interest will be worth $200,000 after two years.

Therefore, the present value cost of $450,000 in this example is $410,357.

$100,000 + $138,889 + $171,468 = $410,357

This is the concept of the present value of money. If this were a benefit-cost analysis, the same technique would be used to obtain the present value of benefits.

The selection of the value for the discount rate is extremely important. A higher rate may cause the benefit-cost ratio to be positive, while a slightly lower discount rate may provide a negative benefit-cost ratio.

IMPACTS OF DAMS

Dams change the character of rivers. Water released from behind a dam has a different temperature than native river water and varies based on the depth of the outlet gates. Water released from the bottom of a reservoir will generally be much colder than water released from near the reservoir surface. These thermal variations can be extremely stressful to aquatic species downstream.

Dams act as rigid barriers to migratory fish such as salmon and shad, and can wipe out entire aquatic populations. Dams on the Columbia River, in particular, are under attack by groups that have been working to protect upstream spawning grounds for Pacific salmon. Although fish ladders have been installed and fingerlings (infant fish) are being grown in hatcheries, the adverse effects of dams on fish migration are very serious. We will discuss this issue in detail in Chapter 12.

Water releases from a reservoir may be uniform or erratic, from the top or the bottom of a reservoir, and can cause significant habitat changes downstream. Uniform flow releases can reduce historic sediment loads, allowing "cleaner" water to scour the downstream riverbed and banks for miles. Erratic, large-volume releases usually transport high sediment loads and can harm aquatic plants and wildlife.

Water released through outlets near the reservoir surface may be low in dissolved oxygen and nutrients in the summer but high in salts and nutrients in the fall. Outlets near the bottom of a reservoir generally release water with low oxygen levels in the summer but relatively high levels in the fall. Water flow and water quality variations can be fatal to downstream aquatic wildlife.

Most rivers have a natural, annual cycle of high and low flows. By contrast, a dammed river has artificial cycles that can be both beneficial and detrimental. Some aquatic and wildlife species may thrive in areas where new wetlands are formed and river flows are stabilized (fluctuations reduced). The change in river water temperature below a dam may actually be beneficial for some purposes. Anglers are often seen fishing downstream of dams due to improved conditions for certain species of fish such as trout. Paradoxically, reduced flows may allow willows, cottonwood trees, or tamarisk *(Tamarix ramosissima)*, a nonnative salt-tolerant brushy plant in the desert regions of the western United States and Australia, to encroach along riverbanks below dams. These plants can reduce flow velocities and significantly alter wildlife habitat.

Sediment buildup in reservoirs is another serious problem for water managers because it results in a loss of reservoir storage capacity (see Figure 7.11). Desert thunderstorms, upstream construction activities, poor agricultural practices, and floods can all generate large quantities of sediments. Prior to the construction of dams, sediment was transported downstream until river velocities decreased and sediment loads were deposited. Dams completely change this natural system by capturing most of this silt in the reservoir behind a dam. Many reservoirs have already lost significant amounts of storage capacity due to siltation. Occasionally, a solution

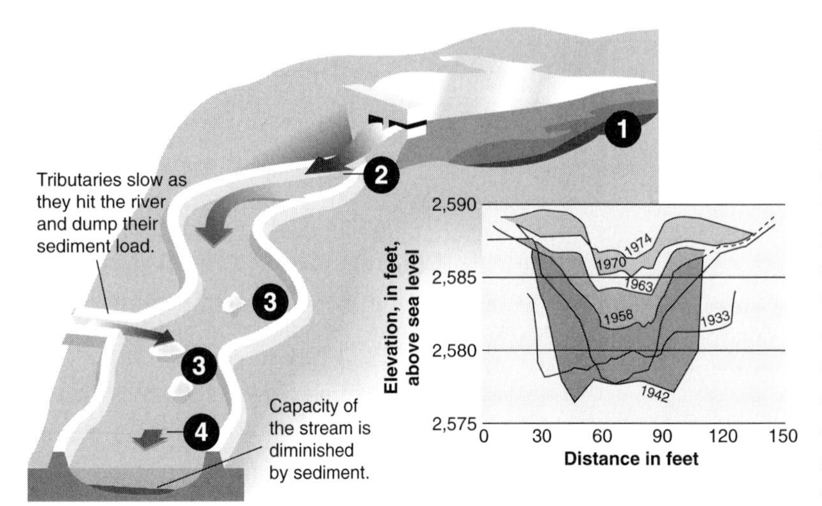

FIG. 7.11. Dams change the behavior of rivers, causing relatively still water behind a dam ❶ to drop much of its sediment load. Downstream of a dam ❷, rapidly moving water released through outlet pipes causes the river channel to erode. Farther downstream ❸, the opposite occurs and silt from surrounding land can pile up ❹ and cause islands and sandbars. This cross-section is at Presidio, Texas on the Rio Grande River.

is to simply increase the height of a dam in order to increase reservoir capacity.

How can the problem of sediment buildup be solved? There is generally no easy or inexpensive answer. Removing sediment deposits mechanically is not cost effective behind large dams since dredging with cranes, or pumping silt from a reservoir, could cost tens of millions of dollars. In addition, new sediments would be deposited immediately from upstream sources, and the process of sediment removal would have to continue indefinitely.

It is generally agreed that sediment loading in reservoirs is best solved by reducing the amount generated in the first place. Watershed scientists can estimate sedimentation rates by dividing a watershed into sub-basins and then identifying and monitoring areas of high sediment loading. Land-use management practices can be implemented at construction sites, on agricultural lands, along streambanks, and other identified sources to reduce erosion.

The removal of smaller dams is becoming a more common method of mitigating the negative impacts of dams. One of the first to be removed in the U.S. was the Edwards Dam on the Kennebec River near Augusta, Maine, in 1999. At issue were the environmental costs of lost fish migration upstream past the dam versus the benefits of hydropower generation from the 24-foot high (7.3 m) structure. The controversy over removing the Edwards Dam reached a peak in 1997 when the Federal Energy Regulatory Commission (FERC) denied a new permit for the dam. The owners of the dam (Edwards Manufacturing Company and the City of Augusta) would have been required to build a $9 million fish ladder to allow migrating fish to pass the dam on their way upstream. See Chapter 9 to learn more about FERC and the Edwards Dam issue.

GUEST ESSAY

Forced Urbanization: The Three Gorges Dam Relocation Process

by **Colin Flahive,** graduate, Anthropology and Asian Studies Departments at the University of Colorado in Boulder

A Denver native, Colin Flahive attended the University of Colorado in Boulder. There he received Bachelor's degrees in both Asian Studies and Anthropology. His interest in China and the Three Gorges Dam began in 1998 when he first traversed the Asian continent over a six-month journey. After several return visits, Colin decided to settle in Dali, China, where he now runs his own restaurant, Salvador. When he is not studying, working, or traveling, his time is spent rock climbing, skiing, and mountain hiking.

The Three Gorges Dam, a concrete gravity dam, spans a length of 1.2 miles (1.9 km) and has a height of 608 feet (185 m). It now blocks the mighty Yangtze River, a river with a watershed that houses one-third of China's population and supplies one-half of the country's food. Inundation will take until 2009 to reach the reservoir target level of 574 feet (175 m). This will fill a reservoir equal in length to Lake Superior, flooding 373 miles (600 km) upstream behind the dam site (see Figure 7.12).

Among the benefits of the dam is a hydroelectric capacity large enough to supply 10 percent of China's future energy needs. It will produce up to 85 billion kilowatt-hours of electricity per year, providing a clean alternative to heavily polluting coal-fired power plants. Another benefit will be increased river navigability; the dam will allow easy access for commercial shipping into a potentially industry-rich central China. In addition, the dam will provide flood protection to an area devastated by past flooding; nearly 300,000 died from floods in the area during the last century alone.

Although the dam may be essential to China's future prosperity, many experts believe that the repercussions of the project may outweigh its benefits. Environmentally, the consequences will be enormous. Stagnation of water within the reservoir will significantly alter the natural habitat by increasing water temperatures and saline levels due to sun exposure and evaporation. The dam will also block the rapid water that naturally flushes the river of sewage and human waste. This polluted water will significantly harm area wildlife, threatening the demise of multiple endangered species of fish and birds. In addition, the Yangtze River is one of the siltiest rivers in the world. The stagnant reservoir will cause most of the nutrient-rich minerals to settle to the river bottom. Farming downstream of the dam will suffer significant decreases in soil quality, making the use of chemical fertilizers essential in floodplains once abundant with quality soil.

FIG. 7.12. This scene is in Wanxian, the largest of the relocation cities affected by the Three Gorges Dam, called *Sanxia Ba* in China (*San* meaning "Three," *Xia* meaning "Gorge," and *Ba* meaning "Dam"). The electronic sign counted down the days until completion of the project. In the middle of the sign you can read the date June 1, 2003, which was the scheduled completion date for the Three Gorges Dam. Officials began filling the massive reservoir on that date, but it will take until 2009 for the entire project to be completed and the reservoir filled.

Perhaps the most tragic, yet least acknowledged, consequence of the Three Gorges Dam will be the impact on China's cultural and social structures. Between 1.2 and 2 million residents now living within the reservoir's path are being forcibly relocated upland to newly constructed urban environments. Many will have to leave lands that have been in their families for generations. Ancestral burial grounds and farmlands flooded by the reservoir will have to be abandoned. Traditionally in China, this would be considered the most perverted of ethical crimes. But even though it means ignoring filial responsibilities, many are willing to move to achieve economic and social mobility.

The Chinese government has promised to give the displaced people financial aid, housing, and career opportunities in a modern urban society. Many residents of the area see this as an opportunity to "free" themselves from their land in order to play an active role in global culture. However, the dam project will be at least 100 percent over budget, and corruption at the higher levels means that those who are relocated will likely be poorly compensated. Poor compensation combined with competitive city life and the breakup of a traditionally rural social system will prove to be detrimental to many people both culturally and psychologically.

China has a large cultural gap between urban and rural populations. Although both populations will have to be relocated as a result of the dam project, it will be the rural people who will suffer the most. For urban residents their entire cities will be moved uphill out of the way of inundation; therefore, they will suffer only a change of location. Farmers and other rural residents will suffer changes in both location and lifestyle. They will be forced to relocate into an urbanized culture that is alien to them (see Figure 7.13).

One rural family that I met in the area typifies the relocation situation. The family lives within the inundation area, which means their home and farmland will be flooded over the next decade. When the father was asked where he, his wife, and two daughters would be moving to, he pointed to a skyline of tall buildings under construction in the distance. For generations, home for this family has been their ancestral lands, and they have subsisted almost entirely on hard work and trade within their small extended village. Their closest neighbor is barely out of shouting distance, and family life is their social basis of support. But within the next decade, they will have to leave their home and move into one of the multistory apartment buildings atop the hill surrounded by other

FIG. 7.13. This tributary of the Yangtze River flows through the narrow canyon called Xiao Sanxia (Lesser Three Gorges) and will be flooded after completion of the Three Gorges Dam. Tourism in the area, a major part of the local economy, will be seriously affected.

crowded offices and apartment buildings. Adapting to an urban lifestyle will be difficult for them and for thousands of others.

Many to be relocated have known only farming in the past and will be pressured into industrial labor. Over 74,130 acres (30,000 ha) of prime agricultural land will be lost to inundation, and farming in upland areas will be difficult inasmuch as most fertile and flat land presently exists as riverbanks. Land terracing on the steep slopes uphill will develop the only agricultural land in the area. This will be both costly and risky, for the area is prone to landslides. Many farmers will have no choice but to join the industrial sectors of the newly built cities.

Declines in agricultural output will necessitate the importing of grains and produce from outside city limits. Families and towns once somewhat self-subsisting will become increasingly dependent on the government for support. Traditionally, the family or clan was the basis of social and economic exchange. Over the years, these social networks, based on a system of equal exchange for goods and favors, allowed villages to remain relatively independent of government support. Relocation and urbanization will break up these reciprocal relations, which means that many will be cast out on their own to achieve social and economic security. Even within the borders of the socialist People's Republic of China (PRC), capitalism exists in the cities as economic competition.

Since the government controls the media, the extent of resistance to the Three Gorges relocation will probably never be known. But despite the danger, some people have courageously spoken out against the dam project. In one of the rural relocation villages, for example, three elderly farmers produced evidence of corruption in the allocation of relocation funds. For their effort, the police confiscated all of their family's belongings, and the three are currently facing three- to seven-year prison terms. Dai Qing, the most vocal critic of the dam, has also suffered the consequences of activism. After her first book was published revealing corruption and incompetence within the dam project, she was sentenced to 10 months of imprisonment and served six of these months in solitary confinement. Nonetheless, upon her release, Dai Qing continued her opposition, and today she remains the leader of resistance to the Three Gorges Dam.

Forced relocation, and the accompanying forced urbanization in the Three Gorges region, will metaphorically sever the culture's social and ancestral roots. Much as a tree with severed roots will have wilted leaves, so too will the relocated populations suffer great cultural stress.

OUR ENVIRONMENT

Unlike China and other countries, the United States has experienced a major policy shift against dams. Environmental impacts, relocation issues, and subsidies for irrigation water use have led many to support efforts to remove existing dams.

Former Department of Interior Secretary Bruce Babbitt was a strong proponent of removing dams that had outlived their usefulness. In 2000, Secretary Babbitt formed a "dam-busting tour" to oversee the removal of several dams around the country. In California, the Seltzer Dam (a nearly 100-year old structure) was removed from across Deer Creek 150 miles (241 km) north of Sacramento. The dam had impounded such large amounts of sediment that it had become useless, according to California State Resources Secretary Mary Nicholls. The cost of its removal was approximately $6 million, and it is expected to restore about 13,000 salmon annually. (26)

Nearly two dozen dams were removed across the United States in 1999, and approximately 20 additional dams were removed in 2000. Opponents of dam removal argue that it could negatively affect navigation and that it could reduce power generation in the Pacific Northwest alone by 4 percent. (27)

What role does poverty play in the construction or removal of a dam? For example, would dams be removed in the United States if poverty was prevalent in areas served by water storage projects? Would the Three Gorges Dam have the same level of support from the Chinese government if income levels and living standards were higher? Should income and standard of living be the dominant factors determining a government's policy on water resources management?

DAMS AND LOCKS FOR NAVIGATION

Navigation and the transport of goods on rivers remains a major industry in the United States today. America's river highways have been dammed, diked, widened, cleaned, dredged, lined, and redirected for decades to improve navigation and reduce flooding. Barges loaded with grain wind their way around sandbars, snags, and sawyers, and carry materials to ocean terminals such as New Orleans, or inland ports as far north as St. Paul, Minnesota, a distance of some 2000 river miles (3219 km).

When did the federal government become involved in federal navigation issues? Although the Gallatin Report of 1808 proposed a comprehensive plan for improving the nation's transportation network of highways, canals, and rivers, it was not until 1824, when the U.S. Supreme Court heard the case of *Gibbons v. Ogden,* that federal involvement for navigation improvements was declared appropriate under the Interstate Commerce Clause of the U.S. Constitution. That same year Congress authorized work to remove sandbars and snags on the Ohio and Mississippi rivers.

The U.S. Congress funded lock construction on the Mississippi River in the early 1900s in response to the rapid growth of population and agriculture in the region. At the time, the Mississippi River was confined by sandbars and shallow sections, and was unable to handle large-volume barge traffic. Barges with drafts of 8.5 feet (2.6 m) could transport goods only partway up the Mississippi. Barge traffic could not operate in the shallower upper reaches unless goods were transferred to smaller vessels (called breaking the bulk) that didn't require such deep water. When breaking the bulk was required, shipping costs increased significantly and discouraged shipping on the upper river.

By 1929, a 9-foot (3 m) channel had already been completed along the Ohio River, but the Mississippi proved to be more difficult. In some locations, the floodplain varied from 0.5 to 2 miles (0.8 to 3.2 km) in width across proposed dam and lock sites. Ice in the winter was another major concern for structures proposed for the river channel. The river's surface often froze during the winter, and water simply flowed beneath the ice. With the spring thaw, river ice broke apart and floated downstream until stopped by an obstacle. Ice jams could be massive, and structures had to be built to withstand enormous pressures. Dam spans of 80 feet (24 m) or more were needed between supporting piers in order to allow ice jams to flow through. With the Great Depression, the improvement project on the Mississippi River stalled until 1933, when President Roosevelt signed the National Industrial Recovery Act, which provided $51 million for the work.

Before construction began, scale models of proposed construction elements were built in a hydraulic laboratory at the University of Iowa to determine the coefficients of discharge of weirs, gates, and tubes. Infiltration rates were tested to determine the effects on dam foundations. Dam stability was assessed for various river discharges, and sediment studies were conducted. The use of physical models to test engineering calculations was a new procedure in hydraulic engineering. Work started with removal of snags, damming of side channels, and construction of small, protruding dams (called **jetties**) that channeled water into narrower sections of the river. Damming of the side channels increased the discharge in the main channel of the river, although it also dried up historic wetlands along the shores. The jetties concentrated flows in a deeper, more narrow channel, or thalweg, of a river. This process also helped remove sediment deposits from the thalweg of the riverbed while creating sediment mounds immediately downstream of jetties. This provided a deeper channel for barge traffic by reducing sediment loading in the shipping channel of the river.

Low concrete gravity dam structures, with locks for navigation, were used to reduce the amount of water that would be impounded, as well as to reduce lowland flooding of homes and farms located in the river floodplain. The river locks—110 feet (34 m) wide and 600 feet (183 m) long—allowed long barges to pass through the structures.

In 1938, the final two dams and sets of channel locks were completed to create a continuous 9-foot (3 m) depth all the way to St. Paul, Minnesota. Thus was completed a massive construction project of 24 locks and dams between Red Wing, Minnesota, and St. Louis, Missouri. Barge traffic on the Mississippi doubled in just a few years and doubled yet again by 1960. Today the system of locks and dams enables barge traffic to carry nearly 80 million tons (72.6 million metric tons) of goods annually on the Upper Mississippi. (28)

In many locations along the Mississippi, the U.S. Army Corps of Engineers has installed a flexible concrete revetment system along the riverbanks to reduce flood erosion. Concrete blocks are approximately 1×2 foot squares ($0.3 \times 0.6\ m^2$) and 5 inches (13 cm) thick, and are held together with wire and cables along miles (km) of river shoreline. Flexibility is required to allow the revetment "blanket" to settle as the slope is underscoured (eroded from below the revetment).

Another major policy issue that enhanced navigation was the creation of the Tennessee Valley Authority (TVA) in the southeast region of the United States. In 1933, Congress enacted the TVA Act establishing a government agency to build dams for flood control, navigation, recreation, hydropower, and other uses on the Tennessee River and its tributaries. Twenty-four gravity concrete dams were built, channels dredged, and navigational locks constructed to improve the economy of the region, which includes Alabama, Kentucky, Virginia, North Carolina, Georgia, Mississippi, and Tennessee. River shipping volume has increased over a hundredfold since construction of the projects.

The concept of improving navigation along the Tennessee River was not new. Long before, in 1824, Secretary of War John C. Calhoun recommended a series of proposals to President James Monroe that would improve the nation's transportation system of roads, canals, and river channels. Many early improvements were made, but uncontrolled flooding constantly washed them out. The development of TVA flood-control dams 100 years later finally eliminated the problem of destructive floods. Today, barge traffic from the Ohio River can navigate hundreds of miles (or hundreds of kilometers) up the Tennessee River system. (29)

SIDEBAR

[The Tennessee Valley Authority should be] . . . a corporation clothed with the power of government but possessed of the flexibility and initiative of a private enterprise. It should be charged with the broadest duty of planning for the proper use, conservation, and development of the natural resources of the Tennessee River drainage basin and its adjoining territory for the general social and economic welfare of the nation. (30)

President Franklin D. Roosevelt, 1933

GUEST ESSAY

Navigation on the Tennessee River

Dr. M. Ted Nelson

by **Ted Nelson, Manager, Navigation Program, Tennessee Valley Authority, Knoxville, Tennessee**

Ted Nelson is the Program Manager for Navigation at the Tennessee Valley Authority. He has degrees in geography and regional planning from the University of Alabama and earned the Ph.D. from the University of Tennessee. He served three years with the U.S. Army in Europe as a geographic intelligence officer. He has been employed with the Tennessee Valley Authority since 1975 where he has served in technical and management positions. Dr. Nelson also serves on several university advisory boards and is a member of the National Transportation Research Board.

The Tennessee Valley Authority (TVA) is an anomaly in the American political economy. It was created in 1933 during the Great Depression as an experiment in government and regional planning based on a river basin. TVA was given the broad and ambiguous mission of improving the social and economic welfare of the Tennessee River Valley, an area encompassing all of the State of Tennessee and parts of six other southeastern states. The mission was vague, but TVA was given three specific tools to accomplish it—navigation, flood control, and electric power.

Improvement of the navigability of the Tennessee River is the first purpose mentioned in the TVA Act. It was an important mission for two primary reasons. First, including navigation in the TVA Act provided a constitutional basis for TVA. There was no precedent for creating a federal electric power or flood-control agency. However, the Interstate Commerce Clause of the United States Constitution had provided the rationale for a strong federal role in improving the nation's waterways since 1824 when the U.S. Army Corps of Engineers was authorized to improve the Mississippi and Ohio rivers for commercial navigation.

Second, the founders of TVA recognized the importance of good transportation in improving the economy of the region. The Tennessee River had served as an important communication and transportation artery since the earliest Native American settlements in the Valley; but the river was shallow and hazardous, in many places dropping 515 feet (157 m) over its 650-mile (1046 km) length. It sufficed for canoes and flatboats, but as vessel technology changed, the river could not support modern navigation.

Before TVA, private companies and local communities had made several small, underfunded, ineffective attempts to improve the Tennessee River. But the job was too big, and the resources were inadequate. After a century of effort, controlling depths on the Tennessee River averaged just a couple of feet (less than 1 m), which was far too shallow for modern commercial navigation. TVA's plan was to build a series of multiple-purpose dams on the Tennessee River and its tributaries which would accomplish all the objectives envisioned in its enabling legislation—navigation, flood control, hydroelectric power, water supply, recreation, and irrigation. The comprehensive plan, published in 1936, called for the construction of seven high dams to create a continuous navigation pool on the mainstem of the Tennessee River stretching from its headwaters at Knoxville to its confluence with the Ohio River near Paducah, Kentucky.

Today, TVA manages one of the most technically advanced and complex multiple-purpose river systems in the world. It coordinates a system of 49 dams stretching from Paducah, Kentucky, to the western slopes of the Appalachian Mountains, occupying a basin of 41,000 square miles (106,190 km^2). It includes an 800-mile (1287-km) navigation system with 14 locks located at 9 dams (see Figure 7.14). This includes a 650-mile (1046-km) main channel which connects Knoxville to the Ohio River and 150 miles (241 km) of commercially navigable tributaries. The navigation system has a minimum project depth of 11 feet (3.4 m) to provide safe and efficient transportation for 9-foot (2.7-m) draft navigation. The minimum width is 300 feet (91 m).

Ten main and four auxiliary locks make it possible for both commercial and recreational vessels to pass easily from one reservoir to another. Much like an elevator, the locks raise or lower barges and other boats from one water level to the next. Once a vessel is inside a lock, the gates close and water is pumped in or drained out of the lock. On average, this process takes about 45 minutes. When the water level inside the lock is equal to the level of the next reservoir, gates at the other end of the lock are opened and the vessel continues its voyage.

TVA built and manages the dams and locks that comprise the Tennessee River's multiple-purpose, integrated system. However, the daily operation and maintenance of the navigation

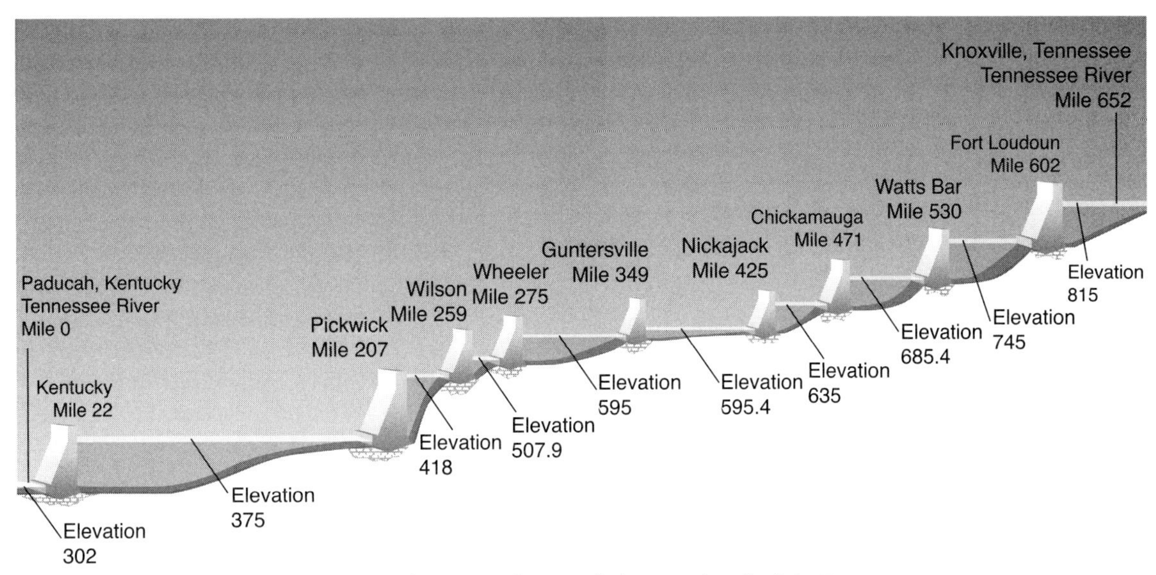

FIG. 7.14. Main-river dams form a staircase of reservoirs that stretch the entire length of the Tennessee River. From its beginning near Knoxville, Tennessee, the Tennessee River drops 513 feet (156 m) before it reaches the confluence with the Ohio River at Paducah, Kentucky.

system is a partnership among TVA, the U.S. Army Corps of Engineers, and the U.S. Coast Guard. In order to ensure consistency of operational processes across the entire inland waterway system, the Corps of Engineers operates all TVA locks and cost-shares maintenance responsibilities. The Coast Guard, as on all U.S. waterways, has primary responsibility for the installation and upkeep of all navigation aids on the Tennessee River, including buoys, lights, and other markers.

The Tennessee River system generally is in good condition. However, like the rest of the nation's inland waterway system, parts of it are aging and wearing out. The average age of the locks is over 50 years, exceeding the planned economic life of the locks. Two problem areas require attention on the Tennessee River. The first is at Kentucky Lock, the busiest and farthest downstream lock on the system, located 22 miles (35 km) above its confluence with the Ohio River. Kentucky Lock is in good physical condition, but its efficiency is declining. It is a 110 × 600-foot (34 × 183-m) lock that can theoretically handle a maximum of about 38 million tons (34.5 million metric tons) annually. It currently handles about 33 million tons (29.9 million metric tons) of goods each year, or about 35,000 barges. In addition, about 800 recreational boats go through it annually. That heavy use puts a great deal of pressure on the lock and causes congestion, or traffic jams, especially at peak periods. The result is that lock maintenance and transportation costs are increasing rapidly. Commercial vessels using Kentucky Lock can encounter delays of up to 60 hours during certain periods. A towboat pushing an average tow of 15 barges costs about $500 an hour to operate. These costs are, of course, passed on to the consumer, reducing the transportation advantage the waterway was intended to provide.

Chickamauga Lock, on the other end of the system near Chattanooga, Tennessee,¡ has a twofold problem. The main problem is a structural deficiency called concrete growth—a

FIG. 7.15. Chickamauga Lock and Dam, located on the Tennessee River near Chattanooga, Tennessee, is a major lock in the TVA navigation system. On February 3, 2004, the U.S. Congress authorized planning and construction of a new lock immediately adjacent to the existing facility. Go to http://www.lrn.usace.army.mil/pao/chickamaugalock for detailed information on the Chickamauga Lock Replacement Project.

phenomenon caused by a reaction between the alkali in the cement and the carbonates in the limestone aggregate used to make the concrete. The reaction causes the concrete to swell, resulting in cracking and structural deformation. The problem is so severe that the lock may have to be closed in about 5 years. Ten years may seem like a long time, but the process of planning, funding, and building a new lock takes at least that long. The cost of a new lock is estimated to be about $200 to $300 million, depending on what size lock is selected. In addition, the lock is very small by modern navigation standards. It was designed and built in the 1930s when smaller barges were in use, and it can accommodate only one modern size barge at a time (see Figure 7.15). This substantially reduces the efficiency of the lock, erasing some of the advantage of water transportation cost savings on the upper Tennessee River.

Despite the isolated problems, the Tennessee River navigation system has proved to be a valuable asset to the Tennessee Valley region and to the nation as well. It has been a major factor in reducing transportation rates charged by other modes, increasing commodity movements by Valley shippers, and attracting high-wage industry to the region. More than 50 million tons (45.4 million metric tons) of diverse commodities move on the Tennessee River each year, saving shippers and consumers over $400 million in transportation costs. Coal accounts for the majority of traffic on the Tennessee River—about 40 percent. A large portion of that traffic is delivered to TVA coal-fired steam plants on the Tennessee and Cumberland rivers. The rest goes to other utilities and manufacturing plants in the southeastern United States. Construction materials such as stone, sand, and gravel comprise the second largest group of commodities and are exported from the limestone-rich quarries of the Tennessee Valley. Other major groups of commodities shipped on the Tennessee River include grains, chemicals, iron and steel, and forest products.

Grain shipments are a good example of the value water transportation brings to the region. Normally, industrial location theory would suggest that grain processing is a classic case of location near source of input rather than market because the manufacturing process reduces the bulk of the product. However, the availability of low-cost water transportation provides the advantage of location near markets. The A. E. Staley plant on the upper Tennessee River near Knoxville is an example. This plant imports large quantities of corn by barge from the Midwest to convert to high-fructose corn syrup, a product used extensively in the soft drink and baking industries. Location near large southeastern U.S. markets gives the company a distinct advantage over its competitors.

Recreation is a rapidly growing use of the waterway. More than 28,000 recreational craft use the locks in an average year. The increasing number of inter-reservoir trips by long-distance recreationists, waterborne snowbirds, and tour vessels stimulates local tourism and helps create opportunities for riverside communities to redevelop their waterfronts.

But the waterway does not only benefit the Tennessee Valley region; it is a national resource that benefits the nation's economy. About 80 percent of the goods shipped on the Tennessee River either originate or terminate outside the region, encompassing origins and destinations in just about every state in the country.

SIDEBAR

Transportation continues to remain free of charge on the Tennessee River system in part thanks to the Northwest Ordinance of 1787. This congressional act was passed for the territories of the United States located northwest of the Ohio River, and stated:

The navigable waters leading into the Mississippi and St. Lawrence, and the carrying places between the same, shall be common highways and forever free, as well to the inhabitants of the said territory as to the citizens of the United States, and those of any other States that may be admitted into the confederacy, without any tax, impost, or duty therefor.

Northwest Ordinance, July 12, 1787

CHAPTER SUMMARY

Dams have been constructed across rivers since ancient times to divert water for irrigation and to control raging floodwaters. Early dam construction materials were basic items: rocks, soil, and brush. Although today's methods are much more sophisticated, they still use basic geologic materials such as rock or soil for construction material.

Dams have greatly altered settlement patterns, economic development, and natural habitat, not only in the proximity of the dam project, but also at the end point where water may be delivered. These impacts have been both positive and negative. Current national public policy is protective of natural environmental features and has effectively halted major dam construction in the United States. This has also led to the removal of some existing dams to return rivers to more natural conditions.

QUESTIONS FOR DISCUSSION

1. Why are dams important features of current water management programs?

2. How is a dam site selected?

3. What are the major components of a dam?

4. Discuss the differences in construction techniques of concrete and earthen dams.

5. What natural and human factors led to the construction of Hoover, Kingsley, and Grand Coulee dams in the 1930s?

6. Describe the magnitude of the Three Gorges Dam project with respect to the residents in the area.

7. Consider the relocation methods used in China for the Three Gorges Dam project. Would the citizens'

response be different in other regions of the world? How and why?

8. Discuss the impact of dams and locks on economic viability for the area served by the Tennessee Valley Authority.

9. What if political support for creation of the Tennessee Valley Authority had not existed in the U.S. Congress in the 1930s? What other alternatives could have been considered to promote human health and economic growth in the area without expensive construction projects (dams, locks, and river dredging)?

10. Consider your personal views on dams. How did you develop these feelings and opinions? Do you believe the federal government was correct in constructing massive dams in the 1930s? Should some dams be removed in the future?

11. Are you a proponent or an opponent of future dam construction? Why or why not?

KEY WORDS TO REMEMBER

SUGGESTED RESOURCES FOR FURTHER STUDY

READINGS

Corthell, Elmer L. *History of the Jetties*. New York: John Wiley & Sons, 1881.

Cullen, Allan H. *River in Harness—The Story of Dams*. Philadelphia: Chilton Co., 1962.

Heat-Moon, William Least. *River-Horse: Across America by Boat*. New York: Penguin Books, 1999.

Larson, John Lauritz. *Internal Improvement: National Public Works and the Promise of Popular Government in the Early United States*. Chapel Hill: University of North Carolina Press, 2001.

Qing, Dai. *The River Dragon Has Come! The Three Gorges Dam and the Fate of China's Yangtze River and Its People*. Armonk, NY: M.E. Sharpe, 1998.

Reisner, Marc. *Cadillac Desert—The American West and Its Disappearing Water*. New York: Penguin Group, 1993.

WEBSITES

U.S. Army Corps of Engineers, St. Paul District. "Old Man River." http://www.mvp.usace.army.mil/history/old_man_river, June 2003.

Central Nebraska Public Power and Irrigation District, June 2003. http://www.cnppid.com

U.S. Department of the Interior, Bureau of Reclamation, "Grand Coulee Dam," June 2003. http://www.usbr.gov/dataweb/dams/wa00262.htm

U.S. Department of the Interior, Bureau of Reclamation, "Hoover Dam," June 2003. http://www.usbr.gov/lc/hooverdam

Tennessee Valley Authority, Home Page June 2003. http://www.tva.gov

U.S. Army Corps of Engineers, Nashville District. "Chickamauga Lock Replacement Project," March 2004, http://www.lrn/usace.army.mil/pao/chickamaugalock

VIDEOS

"The Building of the Hoover Dam." *Modern Marvels Series,* The History Channel, 1999. 58 min.

"Grand Coulee Dam." *Modern Marvels Series,* The History Channel, 1994, 48 min.

"Tennessee Valley Authority." *Modern Marvels Series,* The History Channel, 1995. 50 min.

"Three Gorges: The Biggest Dam in the World." The Discovery Channel, 1998. 52 min.

REFERENCES

1. Collier, Michael, et al. *Dams and Rivers.* U.S. Geological Survey Circular 1126, Denver, CO, June 1996.

2. U.S. Bureau of Reclamation, "Facts & Figures on the Colorado River and Hoover Dam," http://www.lc.usbr.gov/pao/faq.html, November 2000; and personal communication with Robert V. Walsh, External Affairs Officer, Bureau of Reclamation, Lower Colorado Region, Boulder City, NV, June 1, 2001.

3. Ibid.

4. Ibid.

5. Ibid.

6. Ibid.

7. Ibid.

8. Reisner, Marc. *Cadillac Desert—The American West and Its Disappearing Water.* New York: Penguin Group, 1993, 128.

9. Ibid.

10. "Lake McConaughy State Recreation Area," Nebraska Game and Parks Commission, http://www.ngpc.state.ne.us/parks/bigmac.html, August 2001.

11. Central Nebraska Public Power and Irrigation District, http://www.cnppid.com; and personal communication with Mike Drain, Central Nebraska Public Power and Irrigation District, Holdrege, NE, May 30, 2001.

12. Keith County News, Ogallala, NE, May 4, 1972.

13. Quinn, M. L. "Managing Waves on Nebraska's Lake McConaughy." *Canadian Water Resources Journal* 7, No. 2 (1982): 63–89.

14. Central Nebraska Public Power District, http://www.cnppid.com; and personal correspondence with Jeff Buettner, Communications Officer, Central Nebraska Public Power and Irrigation District, Holdrege, NE, November 1, 2000.

15. U.S. Congress, House of Representatives, Subcommittee of the Committee on Government Operations. Teton Dam Disaster: Hearings Before a Subcommittee of the Committee on Government Operations, House of Representatives. 94th Cong., 2nd Sess., 5, 6, and August 31, 1976. Washington, DC: U.S. Government Printing Office, 1976.

16. "The Teton Basin Project," U.S. Department of Interior, Bureau of Reclamation, http://dataweb.usbr.gov/html/teton1.html, November 10, 2000.

17. U.S. Department of the Interior, Bureau of Reclamation, "Grand Coulee Dam," http://www.usbr.gov.cdams/dams/grandcoulee.html, August 2001.

18. Ibid.

19. Ibid.

20. Cullen, Allan H. *River in Harness—The Story of Dams.* Philadelphia: Chilton Co., 1962, 84.

21. Ibid., 34–35.

22. U.S. Department of the Interior, Bureau of Reclamation, "Grand Coulee Dam." http://www.usbr.gov/coulee.html; and personal communication with Craig Sprankle and Dave Lyngholm, Grand Coulee Power Office, Bureau of Reclamation, Grand Coulee, WA, May 25, 2001.

23. Ibid.

24. Cullen, 82.

25. *Congressional Record* 36, Washington, DC: U.S. Government Printing Office, March 4, 1903, 3,071.

26. Howard, John, Associated Press. "Dam on Salmon Stream Coming Down." *The Denver Post,* October 7, 2000, 4A.

27. Ibid.

28. U.S. Army Corps of Engineers, St. Paul District, "Old Man River," http://www.mvp.usace.army.mil/history/pamphlets/old_man_river, October 2000.

29. Kyle, John H. *The Building of TVA*. Baton Rouge: Louisiana State University Press, 1958, 5.

30. H.R. Document No. 15, 73rd Congress, 1st Session, 1933.

CHAPTER 8

WATER ALLOCATION LAW

- ANCIENT WATER ALLOCATION LAW
- WATER ALLOCATION LAW (1200–1799)
- WATER ALLOCATION LAW (1800–1847)
- WATER ALLOCATION LAW (1848–1899)
- WATER ALLOCATION LAW (1900–PRESENT)
- GROUNDWATER DOCTRINES
- INTERSTATE COMPACTS
- FEDERAL RESERVED WATER RIGHTS

Through the centuries, civilizations have tried to develop appropriate water allocation laws to promote economic development, public health, and, in recent history, environmental protection. In humid regions of the world, water allocation rules are generally based on the concept of sharing, whereas most arid locations use a rigid, priority-based system of water allocation. This chapter will show the evolution of water allocation from a landownership perspective, called the Riparian Doctrine, to the concept of water ownership as a private property right, called the Doctrine of Prior Appropriation.

Conflict over scarce water resources continues today and will increase as the world's population grows and environmental concerns deepen. In addition, conflicts of the past (see Figure 8.1), are being repeated in many regions of the world today. The issues often remain the same: inadequate water supplies, increasing demands, and diversity of perspectives and values. Control of water, particularly in arid regions, is a basis of wealth, power, economic growth, and political influence. Consider your own value sets and viewpoints on water resources and the natural environment as you read this chapter.

FIG. 8.1. Frederic Remington, "Fight for the Waterhole," oil on canvas, 1903. The painting is on display at the Museum of Fine Arts, Houston, Texas. In some ways, we continue to guard precious water resources today as we have in the past.

ANCIENT WATER ALLOCATION LAW

CODE OF HAMMURABI

The earliest cradle of human civilization also nurtured the development of water allocation law. The reign of King Hammurabi (1795 B.C.–1750 B.C.) of Babylonia was characterized by his title, the "King of Justice." His kingdom, also known as Mesopotamia, was the area between the Tigris and Euphrates rivers in present-day Iraq, Syria, and Turkey. Many kings before Hammurabi (pronounced ham-moo-**rahb**-bee) unsuccessfully attempted to organize the laws of the region.

. . . for it was an axiom of royal prudence that, if a king paid no heed to justice, his subjects would rebel and his kingdom be laid waste, his destiny be reversed and misfortune follow close upon him. (1)

The Babylonian Laws: Volume II
Reprinted by permission of Oxford University Press

King Hammurabi brought together all aspects of Babylonian law and ordered scribes to etch his laws on tablets of diorite, a very hard stone, to preserve them for future generations. The work was called the **Code of Hammurabi,** and portions of the original tablets are on display at the Louvre Museum in Paris (Figure 8.2). The entire Code (a body of legal rules expressed in written form) contains some 300 sections of law that range from water allocation to marriage to theft. (2)

Rain was sparse in Mesopotamia and fell primarily during the winter months. Average annual rainfall was only 8 inches (20 cm), although multiple years of drought were common. During the time of King Hammurabi, spring snowmelt from the Armenian Mountains was captured and diverted into low-lying lands near irrigated fields. An elaborate system of dikes was maintained to channel spring floodwaters into storage reservoirs and then into irrigation canals for delivery to fields. If managed properly, irrigation water would be available throughout the dry summer months. If irrigators were careless and didn't maintain their dikes, however, excess water could flood neighboring fields and destroy crops. If lack of maintenance caused damage to a neighbor's field, the Code required payment of just compensation for lost crops:

FIG. 8.2. Stele of Hammurabi, bas-relief sculpture, on display at the Louvre Museum, Paris. King Hammurabi was famous for developing a code of law for Mesopotamia and Sumeria. Here, Hammurabi is either receiving or offering the law to Shamash, the Sun God. This artifact was discovered in Susa, an ancient city located in the province of Khuzistan in southwestern Iran, in 1901. The Code of Hammurabi created a strong centralized control of water and was primarily aimed at preventing carelessness and waste.

Sec. 56. If a man has released waters and so has let the water carry away the works on his neighbour's field, he shall pay 10 gur [a unit of measurement] *of corn for every bur* [unit of land] *flooded. (3)*

Code of Hammurabi
The Babylonian Laws: Volume II
Reprinted by permission of Oxford University Press

JUSTINIAN CODE

Although the Code of Hammurabi was one of the earliest sets of laws, the Romans were the first civilization to view law as a science. They also organized and established extensive bodies of law for Roman citizens and their property. (4) In A.D. 528 Roman Emperor Justinian I (483–565) ordered the compilation of all existing Roman laws, including those for water allocation, which had evolved over the previous 13 centuries. This set of Roman law was called the **Justinian Code** (also known as the *Corpus Juris Civilis*, or Body of Civil Law) and was the most elaborate system of law in the world up to that time. (5) It was derived by organizing laws passed by legislative bodies, edicts from the emperor, and interpretation of these laws by judges for specific cases of over one thousand years. It is the basic premise for all modern civil law.

Ancient Riparian Doctrine The **Riparian** (from the Latin *ripa* for "bank" or "shore") **Doctrine,** also called the **Common Law of Water,** was developed within the Justinian Code of the sixth century and provided the framework for water allocation throughout the Roman Empire. The Riparian Doctrine states that water in a stream belongs to the public for use by fishermen and for navigation, and cannot be controlled by private individuals. However, the owner of land along a stream owns the property to the water's edge and in some cases may own the underlying property to the center of the stream. This riparian landowner was allowed to make *de minimus* (reasonable) use of water in the stream for milling, domestic, and agricultural purposes as long as navigation was not injured. A riparian landowner was required to return any diverted water back to the stream unchanged in quantity or quality.

The public use of the banks of a river is part of the law of nations, just as is that of the river itself. All persons therefore are as much at liberty to bring their vessels to the bank, to fasten ropes to the trees growing there, and to place any part of their cargo there, as to navigate the river itself. But the banks of a river are the property of those whose land they adjoin; and consequently the trees growing on them are also the property of the same persons. (6)

Justinian Code (A.D. 533)

The Visigoths, Germanic invaders in Spain during the sixth century A.D., took the Justinian Code and the Code of Hammurabi one step further. They established rules by Royal Order that prohibited the construction of any dams or weirs across streams that could hinder fish migration or navigation. (7) Recceswinth (pronounced **reks**-swinth), Visigothic king of Spain in the seventh century, developed a 12-volume legal code called the *Forum Judicum*, which was a compilation of Spanish, Roman, and German law. (8)

The Arabic concepts of an *acequia* (irrigation ditch), *zanja* (a smaller ditch), *charca* (pond), and *acena* (water mill) were introduced in Spain and Portugal during the eighth and ninth centuries A.D. by the invading Moors and were later transferred to the New World. Spanish rulers encouraged the development of community irrigation projects because it meant increased tax revenues for the Crown.

By the end of the ninth century A.D., water allocation rules and practices were slowly evolving into an extensive body of law. Rulers were empowered, and judged, by their ability to create an orderly society based on fair and sound laws. King Hammurabi was one of the earliest to tackle this job and was followed by leaders in Rome and Spain. The concept of riparianism was being expanded, and navigation, fisheries, and irrigation were primary concerns. The "hydraulic societies" of ancient Babylonia and Rome were also developed in Egypt, China, and other locations around the world as described in Chapter 1. Rulers acquired wealth (and kept subjects well-fed) as irrigation provided both food and wealth in ancient times.

WATER ALLOCATION LAW (1200–1799)

SPANISH WATER LAW

Following the lead of the Roman emperor Justinian, Alfonso the Wise (1212–1284), the Spanish king of Castile, developed a set of laws for his kingdom. In 1263 he completed the compilation of existing Spanish law into *Las siete partidas.* This extensive body of Spanish law held that all water, land, and minerals belonged to the Royal Crown and private ownership could only exist through a special grant (i.e., land grant) from the sovereign (the king or queen). However, natural rainfall or diffuse water flow not caused by human intervention could be used without permission from the sovereign.

In 1516, the year that Ferdinand II (1452–1516) died and his grandson Charles I (1500–1558) assumed the Spanish throne, the king of Spain ordered the recompilation of all Spanish law in the ***Recopilación de leyes de los Reynos de las Indias*** (Law of the Indies), but it was not completed until 1680. It organized the laws of all Spanish possessions around the world, including water allocation law. Irrigation was the primary water use during this period, especially in the New World. The Royal Crown encouraged the sharing of irrigation water because productive land produced more taxes for the Treasury.

King Carlos II (1661–1700) of Spain addressed the use of streams for navigation and irrigation, and digging of wells, in 1680 as follows:

Partida 3, title 23, and law 8, and title 32, law 18: The river not being navigable, any inhabitant of the town through which it passes may extract part of its water and construct a ditch to irrigate his land or run his mill or water mill, or for any other object he may wish. He must, however, do it without prejudice of the communal use or the function the town council may have given it. If the ditch must cross another's property, or crown or council land, he must have the permission of the owner, king, or council.

Partida 3, title 28, law 8: On a navigable river, no one may open an acequia or canal that impedes navigation. One that is already built, be it old or new, must be blocked up or destroyed at its owner's cost, because public utility must be preferred to private.

Partida 3, title 32, law 19: Any landowner may open a spring or well in his house or property, even if his neighbor's spring or well is depleted or completely exhausted because of it. He shall, nevertheless, have the right to prevent work or demand that it be blocked or destroyed when the former did it without necessity or in malice. (9)

Las siete partidas del rey don Alfonso El Sabio (The Wise), 1263
From *Law of the Land Grant: The Land Laws of Spain and Mexico*, © Los Sanchez Limited Partnership

No one could divert water for irrigation without a grant from the Crown. However, the use of water for domestic purposes was unlimited. Spanish water systems at settlements in Mexico, and subsequent development to the north in America, utilized the Arabic *acequia* to distribute water between users. A *mayordomo* (ditch rider or irrigation system superintendent) was selected by local water users to maintain and supervise the operation of the *acequia*. By 1700, there were an estimated 60 *acequias* operating in present-day New Mexico alone, and 400 more were developed by the 1800s. This system would later be copied by European settlers in Colorado, Wyoming, California, and other western states.

Acequias were an interesting addition to Spanish law. Historically, the king or queen handled all issues of law, landownership, and allocation of water in Spain. However, since the New World was such a great distance from Madrid, decisions regarding irrigation were granted to the locals. These water management decisions belonged to the *ayuntamiento* (town council). The governor of the Spanish state was then advised regarding landownership as well as surface and groundwater distribution and allocation decisions. The *ayuntamiento* allocated water between towns on the same river and between competing uses such as irrigation and domestic uses. The group could even limit the population of an area (somewhat similar to the limitation on growth established by the City of Boulder, Colorado, discussed in Chapter 6) and the period of time that water would be delivered during the irrigation season. Water priorities were given for irrigation, although everyone was entitled to drinking water during a drought.

> **SIDEBAR**
>
> Spanish water law was somewhat based on priority of use. Spanish settlement in the New World occurred in areas with intermittent streams and low rainfall. Water scarcity was a problem, and the general rules of the Riparian Doctrine did not work in the dry climate.

Many of the water allocation concepts used by the Spanish *ayuntamiento* were developed centuries earlier by the native Indians who were already irrigating in Mexico and present-day Arizona and New Mexico. These sound water practices did not go unnoticed by Spanish rulers:

> *We ordain that they* [the Spanish settlers in the New World] *keep the same order the Indians followed in distribution, practice, and apportionment of water among the Spaniards to whom the land may be allotted and assigned. About this, the natives themselves, who previously had it as their responsibility, shall mediate regarding what lands in their judgment shall be irrigated. They shall give each person the water he ought to have, successively from one to the next, under penalty that they take the water away from anyone who takes and employs it by his own authority till all those below him irrigate the land designated. (10)*
>
> *Recopilación de Indias*, Law XI, Title XVII, Book IV. Emperor Carlos and the Empress Regent in Valladolid, November 20, 1536. From *Law of the Land Grant: The Land Laws of Spain and Mexico*, © Los Sanchez Limited Partnership

Pueblo water rights are another unique water allocation system rooted in Spanish law. The king of Spain often granted the use of water that flowed through a town to its citizens. Later, treaties between the United States and Spain awarded some of these pueblo water rights to the communities that held this right. These Spanish water rights are still used today in a few locations. The City of Los Angeles has a pueblo water right for the use of all water from the Los Angeles River, including the alluvial groundwater of the river valley, for any municipal purpose. In addition, the city can expand that right as urban growth occurs, even though it may infringe on other water rights awarded by the State of California. Pueblo water rights also exist in San Diego, California, and in the town of Las Vegas, New Mexico. (11)

ENGLISH COMMON LAW (1200–1799)

Simultaneous with the Spanish settlement in the New World, the English and French were creating settlements along the humid eastern seaboard of America. Annual precipitation generally averaged over 30 inches (76 cm) and created a similar water availability situation as in Western Europe, the previous home of most settlers. Most water allocation disputes were settled by using the rules of **English Common Law.** This body of law had evolved from the Justinian Code of Roman law, Teutonic law (early English settlers from Scandinavia), and **French Common Law,** which was a combination of Roman, Teutonic, and Visigothic law. Common law was developed over the years by court rulings that were handed down to settle specific disputes between individuals. These rulings were then applied to future cases of a similar or "common" nature within that jurisdiction. Common Law is also referred to as *case law*.

Mill Acts of the Eastern United States Some of the earliest water allocation laws in the United States came from the construction of mills powered by flowing water in the original colonies. Mills received favored treatment from most local governments in America, beginning as early as 1600, owing to the increased commerce provided to a community. This preferential treatment included financial assistance, free local labor for construction of the mill and appurtenant water supply canals, and the grant of water rights on local streams to power the waterwheels. These incentives often caused conflicts with local navigation since reduced stream discharge, and the construction of mill dams to impound and divert water, created barriers to navigation and fish migration.

A mill dam was usually constructed of logs, rough-cut wood planks, stone, or other native materials. Water was then diverted into a head race—a canal that carried river water directly to the "buckets" of the waterwheel. The mill dam (also called a *weir*) was anchored into the riverbank with abutments that were designed to allow excess water to spill over securely placed materials. This spillway prevented river water from undermining and piping through the abutments, and it also allowed excess water to flow past the mill dam.

The mill dam created a reservoir called a *millpond*. The millpond could back up a short distance, or it could extend upstream for miles. Since the power generated by falling water multiplied as the head pressure increased, mill owners wanted large mill dams but riparian landowners upstream dreaded these dams. Millponds often flooded upstream riparian lands but allowed storage of water during periods of high river flows for use during lower flow events. This extended the operating season of a mill during a local drought.

POLICY ISSUE

Mill operations began in the early 1600s and peaked during the 1800s along the eastern seaboard of the United States. Mill owners of that era were completely dependent on the hydrologic cycle for their supply of water power. Although millponds helped extend valuable water supplies through storage, water sources from other watersheds, groundwater, or large upstream reservoirs were not developed or utilized as a supplemental supply.

What technological, hydrologic, and economic factors may have contributed to a lack of incentives to supplement water supplies for mills in the eastern United States? Consider the extensive system of aqueducts developed by the Romans (see Chapter 1), and the expansion of the New York City water delivery system from the Croton River Basin in 1842 (see Chapter 6). Why didn't mill owners use these same water resource development models? What if a state or the federal government had subsidized construction costs of water projects for mills? How could this have changed water law in the eastern United States?

A mill developer was often required to obtain a "water right" from a river if water needs exceeded the normal diversion right for a riparian property owner. Local officials generally ignored flooding of upstream riparians caused by millponds, as well as reduced river discharge for downstream riparians, in order to encourage the building of mills and the expansion of local commerce. (12)

In 1669, the Maryland General Assembly passed the **Maryland Mill Act** to encourage continued construction of watermills in that state. One of the first mill acts in America, it entitled a developer to acquire an 80-year lease on 10 acres (4 ha) of private, riparian property on both sides of a stream. The Maryland Mill Act legalized and encouraged the construction of mill dams on rivers in the state as long as "damages" were paid to injured riparian landowners. The value of damages was determined by a jury of 12 people appointed by the local sheriff. This law greatly enhanced the legal and economic status of mills in the State of Maryland. (13)

The Massachusetts Legislature passed the **Massachusetts Mill Act of 1714**, which awarded mill owners the right to construct mill dams with little regard to the effect of flooding on upstream riparian landowners. Compensation was required for damages, and again, a local jury was appointed to determine injury. The citizens of Massachusetts supported the concept that mills were more important to the public good than damages to a few upstream landowners. (14)

The height of mill dams was sometimes raised several times in order to increase head (the elevation of water in the stream) and to store water at nighttime for power production the following day. These larger millponds caused reservoir water to flood more extensive areas of upstream riparian lands. Farmers whose lands were flooded were occasionally compensated with small payments for financial injuries. Mill owners were powerful and highly regarded members of a community because they provided flour, sawed lumber, created jobs, and greatly enhanced the local economy. Farmers argued that they, too, were important to local communities because they provided food for residents; nonetheless, they were usually forced to take a back seat to mill owners and existing water law. Farmers often drained swamps (wetlands) to develop more farmground for cultivation, but this action met with stiff opposition from mill owners. Drainage of wetlands created uneven water flows for mills because the stored water in wetlands tended to even out the flow regime of a river. (The dispute over the proper use of wetlands created severe tensions between industry and agriculture. However, the environmental benefits of wetlands were largely ignored in this water conflict):

> *. . . where any person hath already erected, or shall erect any water mill on his own land, or on the land of any other person by his consent legally obtained, and to the working of such mill, it shall be found necessary to raise a suitable head of water, and in so doing any lands shall be flowed* [flooded] *not belonging to the owner of such mill, it shall be lawful for the owner or the occupant of such mill to continue the same head of water to his best advantage. (15)*
>
> Massachusetts General Laws (1796)

POLICY ISSUE

Prime agricultural bottom lands (floodplains) were routinely (and legally) flooded by mill owners in Maryland, Massachusetts, and other eastern states during the 1700s. What recourse did flooded landowners have? They could appeal to the local sheriff and obtain annual "injury" payments from a jury of their peers. Today, we routinely debate the definition of the public interest whenever a large water diversion or storage project is proposed. Fish migration, flooded property, and injury to downstream water users are major concerns that must be addressed before a water project can be started.

If you were a member of a local jury in Maryland in 1705, how would you determine injury to a flooded landowner caused by a mill dam?

1. Would you consider the period of time floodwaters stood on the land?
2. What importance would you give to trees and other vegetation destroyed by the standing water?
3. Would you require damages for dead fish on the flooded property?
4. Would you require payment for damages to eroded soils?
5. Would you require payment for damages if the farmer's drinking well was contaminated by floodwaters?
6. How would you place a value on the above damages?

An important concept that was added later allowed the sale of the right to build a mill dam to others. This became a valuable, transferable property right in Massachusetts. The Massachusetts Mill Act also extended the right of eminent domain (condemnation) to mill owners. In effect, this law allowed a mill owner to condemn upstream riparian lands as long as fair compensation was paid. Under this law, many productive farms were flooded, but many other states east of the Mississippi also adopted these same principles for mill owners.

The condemnation process, allowed in the Maryland Mill Act of 1669, was repealed by the Maryland General Assembly in 1766, partially owing to concern for fisheries. Mill dams created barriers to migrating fish such as shad, herring, and salmon. During the migration, local residents would often line the riverbanks with nets to catch hordes of fish; they became greatly alarmed by declining fish populations in the 1700s caused by mill dams. When voters intensely lobbied legislators to do something about the issue, the Maryland Legislature passed an Act for the Preservation of the Breed of Fish in 1768. (16)

In 1771, Pennsylvania declared the Lehigh and Delaware rivers to be common highways for the purposes of navigation and fixed a £20 penalty for the erection of mill dams that impeded navigation. Later, the construction of wing dams (structures that only partially extended across a watercourse) required legislative authorization. Navigation and fishing interests had sufficient political power to alter local water law against mill owners who had historically wielded most of the political influence in these states. (17)

During the early years of the United States, federalism promoted the concept of state's rights in the area of water allocation. Individual states adopted their own water laws that met the needs of their local climate and economy, as well as the will of the people. Federal law regarding free passage for navigation was followed, but each state had ultimate control over surface and groundwater use. Generally, this concept continues today.

WATER ALLOCATION LAW (1800–1847)

CODE NAPOLÉON

After the French Revolution, France began to organize and codify its laws since many were contradictory or outdated. In 1804, the **Code Napoléon** (also called the *Code Civil des Français* or the French Civil Code) was created. Napoléon I (1769–1821), as emperor of France, ordered the compilation of French law, which was based on the Justinian Code, Visigothic, and English Common Law. Accordingly, basic legal principles were defined for property ownership, individual rights, and commerce. The Code defined riparian water rights, the ownership of streambeds, navigation rights, and the forfeiture of property rights. (18)

RIPARIAN DOCTRINE (1800–1847)

The Code Napoléon provided some guidelines for the continued development of riparian law in the United States, and, in particular, for emerging water law in the future State of Louisiana, which had a strong French influence. During the early 1800s, individual states in the eastern United States developed an elaborate system of water law through state legislation and judicial interpretation. Disputes of that era were primarily between mill owners, fish migration proponents, and navigational interests. Riparian water law was the basis of all legal rulings in the eastern United States inasmuch as water supplies were generally adequate for most uses. The principles of water law were based on English Common Law and included the following basic concepts:

1. Riparian water rights extended to the center of nonnavigable streams.
2. Navigable streams—rivers that could be traveled by floating vessels such as boats—were owned by the general public (also called the public domain) and could not be obstructed.
3. The right to develop mills and mill dams belonged to the riparian landowner on either side of the stream and could be lawfully transferred when the property was sold.
4. Excess water could not be diverted from a stream and had to be returned unimpaired in both quantity and quality.
5. Injured riparian landowners had to be compensated for injuries.

Protection of fishing rights and fish migration rights continued to be a problem in the early 1800s in the New England states. Fishermen were known to destroy mill dams to protect fishing areas, provoking much controversy and tension. "The role of fishing rights as an obstacle to milling enterprise is not to be underrated, especially as regards New England." (19)

A CLOSER LOOK

The interests of dams, navigation, and migrating fish have been in conflict for centuries. The Magna Carta, adopted in A.D. 1215 in England, required that all permanent kydells (fish weirs) be removed from the Thames River in England to allow navigation. (20) A mill in Braintree, Massachusetts, lay idle for years after impacted fishermen destroyed the dam. (21)

In the 1820s, three efforts were made to build a watermill at Weymouth, Massachusetts, but local voters stopped the town government's sale of riverfront property to a mill developer. In another incident, dam construction on the Housatonic River in Connecticut was delayed 25 years because of its feared impact on the shad fishery. As a compromise, mill developers often proposed building fishways to allow fish to migrate past a dam. In severe cases, a mill owner might also agree to end water diversions entirely during the fish run. (This is quite similar to current issues of salmon migration on the Columbia River, and will be discussed in Chapter 12.)

Economically, these concessions by mill owners could be devastating. Rhode Island industrialized faster than Connecticut partly because of relaxed laws that originally required dams to release water during fish spawns. Laws to protect fish were an economic hardship on mill owners since the loss of water meant a loss of water power for manufacturing processes.

Tyler* v. *Wilkinson The first court decision regarding the Riparian Doctrine in the United States was rendered in Rhode Island in 1827. In ***Tyler v. Wilkinson,*** U.S. Supreme Court Justice Joseph Story ruled that no one could reduce or obstruct the flow of water in a stream to the detriment of a downstream riparian landowner. He defined reasonable use as the use of water from a stream without injury to other water users but did not define injury in any detail to allow interpretation on a case-by-case basis. (22)

Tyler v. Wilkinson was a landmark ruling, with all eastern states in the United States later adopting a similar method of water allocation. The owner of property that borders a stream does not have an ownership right to the waters of the stream. Instead, the landowner has the right to make reasonable use of water in the stream by virtue of the location of property ownership. Riparian landowners cannot be harmed by the unreasonable water use of others and cannot injure others by their own unreasonable use of water. This right to use water remains with the riparian parcel of property and cannot be separated from the land. Although landowners do not own a water right, they are given the right to use water from the stream. The right to use something without ownership is called a **usufructory property right.**

As human populations increased, demand grew for power generation from mills for industrial purposes. More and more mills crowded together at key river locations that had significant drops in elevation. Often, water supplies were inadequate to power the combined diversions of multiple mills, and apportionment (sharing) of water and priority of use were established. Some mill owners settled disagreements over water use by developing voluntary joint use agreements along the course of a river. Flow measurements were a very important part of these agreements. British-born James B. Francis, superintendent for the Proprietors of the Locks and Canals on the Merrimack River in Massachusetts from 1837 to 1885, conducted elaborate flow measurements to ensure fair water delivery to the big milling corporations of Lowell, Massachusetts. He was called the Chief of Police of Water, and held the position for nearly 50 years. (23) The Francis School of Engineering at the University of Massachusetts at Lowell is named in his honor.

Diversion rates were established for the water mills in Lowell, and were based on the horsepower potential of the river (see Figure 8.3). Since water power potential is determined by the gradient of a river, reaches with significant elevation drops became attractive construction sites for milling operations (see Figure 8.4).

Power potential of a river =
total river flow (discharge) × gradient between
the source and mouth of a river

River gradients of 5 to 10 feet per mile (0.9 to 1.9 m/km) are common along many rivers in New England. Local rivers such as the Saco, Androscoggin, Kennebec, and Penobscot have downhill gradients of over 1100 feet (335 m) in only 140 miles (225 km), or almost 8 feet of drop per mile (1.5 m/km). (24) By contrast, rivers west of the Appalachian Mountains, such as the Ohio or the Mississippi, have gradients of less than 6 inches per mile (9.5 cm/km) over a distance of 1000 miles (1600 km).

WATER-POWER OF EASTERN NEW ENGLAND.

Table giving details regarding the mills and power at Lowell, Massachusetts.

Name of corporation.	Designation on map.	Goods manufactured.	Number of spindles. (*a*)	Number of looms. (*a*)	Water taken from—	Water discharged into—	Head and fall in feet. (*b*)	POWER OWNED.		Steam-power in horse-power. (*a*)	Quantity of water in cubic feet per second.	No. shares in Proprietors of Locks and Canals.
								Mill-powers.	Gross horse-power.			
Merrimack Manufacturing Company.	A.	Cotton......	153, 552	4, 267	Merrimack canal	Merrimack river....	30 (33. 5).	$24\frac{20}{30}$	2, 097	6, 000	616. 667	740
Hamilton Manufacturing Company.	B.	....do	59, 816	1, 597	Hamilton canal.......	Lower level of Pawtucket canal.	13 (14)	16	1, 360	1, 200	968. 000	480
Appleton Company.......	C.	....do	45, 000	1, 228	....do	....do	13 (14)	$8\frac{16}{30}$	725	750	516. 267	256
Lowell Manufacturing Company	D.	Cotton and woolen....	22, 750 (worsted and wool). 2, 000 (cotton).	317 (power carpet). 75 (lasting).	Merrimack canal.....	....do	13 (14)	$8\frac{13}{30}$	714	1, 040	508. 200	252
Lowell Machine Shop (*c*)...	E.	....	....	....	Merrimack canal and Machine-shop basin.	....do	13 (14)	$3\frac{9}{30}$	280	375	199. 650	99
Middlesex Company.......	F.	Woolen.....	18, 640	250 (broad).	Lower level of Pawtucket canal.	Concord river.......	17 (17)	$5\frac{23}{30}$	490	125	262. 383	173
Boott Cotton Mills.........	G.	Cotton......	127, 000	3, 600	Eastern canal.........	Merrimack river....	17 (19)	$17\frac{23}{30}$	1, 519	1, 000	812. 933	536
Massachusetts Cotton Mills	H.	....do	119, 528	3, 658	....do	Merrimack and Concord rivers.	17 (19)	$24\frac{18}{30}$	2, 085	950	1116. 267	736
Tremont and Suffolk Mills.	I..	... do	94, 000	2, 700	Northern canal.......	Lawrence basin.....	13 (14)	13	1, 105	1, 500	786. 500	390
Lawrence Manufacturing Company.	J.	....do	100, 000	2, 360	Lawrence basin.......	Merrimack river....	17 (21)	$17\frac{9}{30}$	1, 470	1, 000	787, 150	519
Total	....	....	742, 286	20, 052	....	....	....	$139\frac{11}{30}$	11, 845	13, 940	....	4, 181

a These data are from *Annual Statistics of Manufactures in Lowell and Neighboring Towns, January*, 1882. Published by the Lowell *Vox Populi*.
b Figures in parentheses are actual falls (see page 32).
c This establishment consumes per year 1,100 tons wrought-iron, 8,500 tons cast-iron, and 200 tons steel—in all 9,800 tons of metal.

FIG. 8.3. This table of water power, contained in the 1880 U.S. Census, shows a high level of sophistication for allocating river flows among mills at Lowell, Massachusetts. In particular, notice the figures under Power Owned and Gross horsepower.

FIG. 8.4. Line shafts and belting in the weaving room of the Amoskeag Mill located along the Merrimac River in Manchester, New Hampshire. Mills like this were found throughout New England in the 1800s. The Amoskeag Company was the largest textile producer in the world by 1915 when it had 17,000 workers in 30 mills like the one pictured here. After World War I, Southern competition and labor unrest crippled the company. On Christmas Eve, 1935, the Amoskeag Company permanently closed its doors.

WATER ALLOCATION LAW IN THE WESTERN UNITED STATES (1800–1847)

Although water laws of the eastern United States dealt with abundant quantities of water, allocation of water in the arid regions of Mexico and the future states of New Mexico, Arizona, and California followed the indigenous practice of sharing scarce water resources for irrigation. Spanish control of the region ended in 1821 and was subsequently governed by a number of different Mexican administrations for the next 30 years. However, the *acequia* continued to be a focal point of water allocation decisions.

American exploration of the West increased rapidly during this time period. Zebulon Pike (1779–1813), born in Trenton, New Jersey, crossed the arid region in 1806–1807, and Stephen H. Long (1784–1864), a native of Hopkintown, New Hampshire, followed later in 1820. The myth of the Great American Desert was promoted by maps generated by these and other explorers who were accustomed to the humid climate of the eastern United States. Predictions that the land would never be settled to any great extent were derived from observations made of poor soils and a lack of adequate water supplies. Henry M. Brackenridge wrote in his book, *Views of Louisiana*, in 1817:

The prevailing idea, with which we have so much flattered ourselves, of these western regions being like the rest of the United States, susceptible to cultivation, and affording endless outlets to settlements, is certainly erroneous. The [Indian] *nations will continue to wander over those plains, and the wild animals, the elk, the buffaloe, will long be found there; for until our country becomes supercharged with population, there is scarcely any probability of settlers venturing far into these regions. A different mode of life, habits altogether new, would have to be developed. (25)*

Explorers did notice the small irrigation projects developed by Spanish missionaries in the Southwest and borrowed the concept to irrigate small plots of vegetables near army outposts in the 1830s. Competition for water was almost nonexistent in the West because irrigation projects were controlled by missionaries, the Mexican government, or small military outposts. Population numbers were also extremely low.

The scale of irrigation in the West changed dramatically when the Mormons, led by Brigham Young (1801–1877), arrived in the Salt Lake Valley of Utah on July 24, 1847. Realizing that the Valley was barren, the immigrants immediately set out to construct irrigation ditches from City Creek near Temple Square to raise food. Their ultimate goal was to make the region entirely self-sufficient.

Brigham Young's rule that "No man has the right to waste one drop of water that another man can turn into bread" set forth the principle of beneficial use for the group. No one could divert more water to their land than was needed for their beneficial use. Mormons somewhat followed the viewpoint that had been developed by Mohammed, who saw water as a religious charity. Access to water was a given right in all Muslim communities. (26)

In the Koran, Mohammed states:

Excess water is not withheld in order to prevent herbage from growing. Book 36, Number 36.25.29

Do not withhold the surplus water of a well from people. Book 36, Number 36.25.30

Similarly, Brigham Young stated on September 30, 1848:

There shall be no private ownership of the streams that come out of the canyons, nor the timber that grows in the hills. These belong to the people: all the people. (27)

Construction of each water delivery system and the allocation of water were handled locally and systematically. As in the Spanish *acequias*, each landowner shared in the work of ditch construction and maintenance based on his total landholdings. Like the Spanish *mayordomo*, the local Mormon religious leader, called the "Bishop," and later county-appointed watermasters, allocated water. This practice also followed the teachings of the Koran:

Dam them [rivers] *systematically, so that the water is diverted into each property in turn up to ankle level, starting upstream.* Book 36, Number 36.25.28

The concept of shared water use continued in Utah for almost 30 years before state law adopted a priority system of water ownership and use.

WATER ALLOCATION LAW (1848–1899)

RIPARIAN DOCTRINE

Riparian water law did not change greatly during the second half of the nineteenth century. By that time, mill owners in the humid regions of the eastern United States had already developed an elaborate system for sharing water in congested milling areas. In 1851, however, the U.S. Supreme Court defined the term *navigable* as any stream that could be used for interstate or foreign commerce. In 1871, the legal definition was modified to state that any stream that could be used for navigation in its ordinary condition was considered navigable. If it could be used for interstate or foreign commerce, it was under federal control; if it could only be used within a state for navigation, it was under the control of that state. This change primarily affected rivers located east of the 100th Meridian since the vast majority of rivers west of the meridian were intermittent streams and could not be navigated. (28)

THE DOCTRINE OF PRIOR APPROPRIATION (1848–1899)

The California Gold Rush American water law underwent incredible change during the second half of the nineteenth century in the western United States, as a result of the California gold rush, which began in January 1848 along the American River east of Sacramento. James W. Marshall found traces of gold in the millrace at Sutter's sawmill and created an unprecedented international legal and political phenomenon. The Mexican-American War was over, but the California Territory was still governed by a military governor of the U.S. Army. Technically, the Mexican government retained title to all land in the region. Since no American property rights existed, private individuals could not legally settle or purchase land, and the eastern concept of riparian water rights could not be used. Gold mining continued anyway as tens of thousands of prospectors moved into the region in a matter of months. (29)

With no legal system in place for allocating mining claims and water use in California, miners developed their own set of rules. Each mining camp created mining districts and mining codes. This was not a new practice since prospectors came from around the world and simply used allocation rules for gold, silver, and mercury exploration in Peru, Chile, Bolivia, Colombia, Brazil, and Mexico. Mining law had been established for hundreds of years in these locations. (30) Gold miners from North Carolina, coal and iron miners from Pennsylvania, as well as coal experts from Great Britain, also provided knowledge regarding mining custom and law. (31)

The procedures for obtaining and working gold mining claims were quickly defined to allow prospectors the right to dig on a piece of land, generally 10 or 20 feet (3 to 6 m) wide and perhaps 50 feet (15 m) long. Each mining camp's set of regulations was unique but generally followed these seven principles:

1. Mining claims were limited in size.
2. A miner had to "stake" a claim with wooden or steel poles to show intent to mine.
3. A miner obtained the exclusive right to work a claim.
4. A claim had to be filed with the appropriate local official, either at the mining camp or at the county courthouse, if one existed.
5. A miner had to work a claim diligently, or it would be forfeited to another miner.
6. Water for each claim was based on a priority system of "first come, first served."
7. Claims could be bought and sold.

The development of these water "rules" presented a challenge. Most streams of the mining districts had inadequate flow to meet the needs of the thousands of mining claims created. Many prospectors were aware of the *acequia* system used to the south and also knew that mining claims situated away from a stream would not receive water if the Riparian Doctrine of water appropriation was used. The unique political situation of Mexico and the Territory of California forced miners to develop a priority system of water distribution not used before in the United States. This new method closely followed the seven principles of mining claims. The local mining and water use codes were enforced by vigilante committees.

In 1850, California became a state and was now entitled to establish its own water laws. The departure from a riparian system used in humid regions made sense in this dry climate but created problems for downstream riparian landowners who had already settled the area. In April 1850, the California Legislature declared that both the Riparian Doctrine and English Common Law would be the law of the state as long as it did not conflict with state or federal law. Three years later, in 1853, Justice Heydenfeldt of the California Supreme Court ruled in ***Irwin* v. *Phillips*** that gold miners could divert water from a stream under the priority system, even though it was detrimental to downstream riparian landowners:

The most important are the rights of miners to be protected in the possession of their selected localities, and the rights of those who, by prior appropriation, have taken the waters from their natural beds, and by costly artificial works have conducted them for miles over mountains and ravines, to supply the necessities of gold diggers, and without which the most important interests of the mineral region would remain without development. So fully recognized have become these rights, that without any specific legislation conferring or confirming them, they are alluded to and spoken of in various acts of the Legislature in the same manner as if they were rights which had been vested by the most distinct expression of the will of the law makers. (32)

Irwin v. Phillips

This ruling of the California Supreme Court established the concept of the **Doctrine of Prior Appropriation,** which allows a water user to divert water from a stream for delivery and use on nonriparian lands. The use of water is limited to the amount beneficially used and includes a priority date in a basinwide system of appropriation. Diversion quantities are based on a flow rate of cubic feet per second (cfs). This right to use water, called a **water right,** can be sold, leased, or moved as long as downstream senior water rights (water rights with older priority dates) are not injured. Nonuse of a water right can result in the reversion of that right back to the priority system (the concept of "use it or lose it").

SIDEBAR

Why was water needed to mine gold? Early gold seekers in California and in other western states used metal pans and screened wooden boxes (called sluice boxes) to remove flecks of gold from riverbed sediments of mountain streams. Water was used to wash off dirt, sand, and gravel sediments that contained gold. However, if a claim did not adjoin a stream, a small ditch was usually dug to divert and transport water. Some claims were located miles (kilometers) from a stream and required the construction of wooden flumes (aqueducts) to transport water. These were common in the West since large quantities of timber were readily available. Construction usually involved a group of prospectors working cooperatively and was a precursor to the construction of irrigation systems at lower elevations. (33)

A water right is acquired by diverting water from a stream for a beneficial use. A priority date is based on the concept of **"first in time, first in right"** and is acquired by filing papers with the local courts or appropriate state agencies. The first person to appropriate and place water to beneficial use has a superior water right and is a **senior appropriator** to all that come later, called **junior appropriators.** Junior appropriators are barred from diverting water if insufficient water exists for a senior appropriator downstream.

The California Legislature created water law that included a mix of both the Riparian Doctrine and the Doctrine of Prior Appropriation. This unique combination of water law, the so-called **California Doctrine,** allows water users in the humid regions of the state to follow riparian law, while those in water-scarce areas use prior appropriation law. This dual system of water allocation has been adopted by eight other states west of the 100th Meridian (Hawaii, Kansas, Nebraska, North Dakota, Oklahoma, Oregon, South Dakota, and Washington) that have both humid and dry regions within state boundaries. (34)

A CLOSER LOOK

Without laws or rules, people generally follow the natural law of first come, first served. Patrons at a restaurant are normally seated in the order of arrival, with people at the front of the line seated first, while latecomers must wait for an available table. If the restaurant runs out of seating capacity, people at the end of the line will either have to wait or come back another time when tables are available.

Seating at an open-seating concert follows a similar practice. Concert goers at the front of the line, who arrived first, will get to select prime seats, perhaps near the stage, while latecomers, at the rear of the line, may be left with few options. First come, first served is the allocation method used in these situations and is a widely accepted societal practice for the allocation of limited resources such as seats.

Water rights in the California mining camps evolved in the same way. Each mining camp reached an agreement for the allocation of scarce water supplies and claims for mining (Figure 8.5). A priority system was used, and the first miner to establish a claim received the right to divert enough water from a stream needed to work that claim. The next prospector who filed for a claim received the second water right on the stream, and so on.

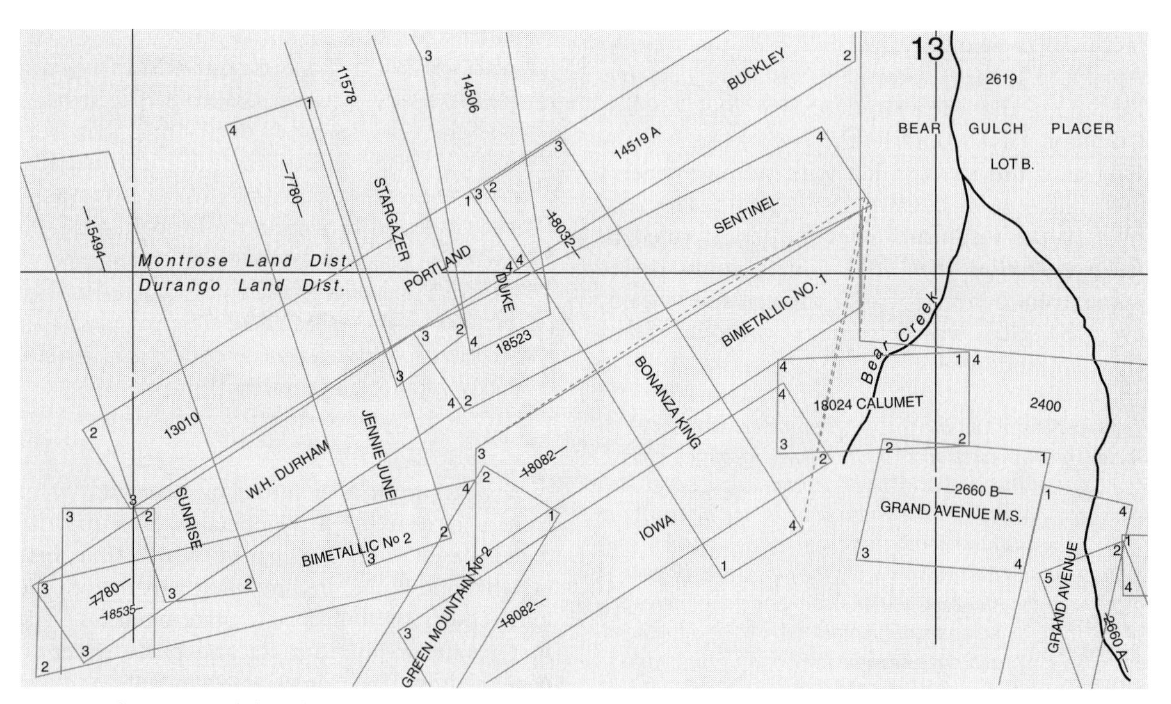

FIG. 8.5. This Connected Sheet from the U.S. Department of Interior, Bureau of Land Management records, identifies the geographic position of individual mining claims under the General Mining Law of 1872. The jackstraw arrangement of claims appears to be a problem until one applies the "first in time, first in right" doctrine. Claim A, located yesterday, for example, is senior and has rights to the entire area within the claim's boundaries. Claim B, located sometime later and overlapping parts of Claim A, has rights only to the portion of Claim B that is not within the boundaries of Claim A. This shows claims 4 miles (6.4 km) south of Telluride, Colorado, within Township 42 North, Range 9 West, New Mexico Principal Meridian.

The Colorado Gold Rush Colorado experienced a similar California-type gold rush in the Rocky Mountains in 1859 (Figure 8.6). Once again, no formal water law existed in the Colorado Territory other than the eastern Riparian Doctrine. Mountain streams were small and flowed intermittently, yet thousands of miners had filed claims on nonriparian lands west of Denver. Aware of the relative success of the California priority system of water law established just a decade earlier, Colorado adopted a similar scheme. Claim sizes were patterned after those in California and ranged from 20 to 50 feet (6 to 15 m) wide to 100 feet (30 m) long for individual prospectors, and up to 250 feet (76 m) wide and 250 feet (76 m) long for mill sites. (35)

Water allocation practices used in the Colorado mining camps were quickly adopted by irrigators on the plains to the east. In 1859, the same year gold was first discovered in Colorado, David Wall dug an irrigation ditch from Clear Creek to irrigate 2 acres (0.8 ha) of land that he had purchased in the dry foothills just west of Denver. Wall was a frustrated gold miner who had struck out in both California and Colorado. However, aware that miners craved fresh vegetables, he hoped to earn a little money by growing them for the nearby mining camps. His idea turned golden, and he earned over $2000 (equal to about $40,000 today) the first year from the sale of produce from the 2-acre (0.8 ha) plot. Wall neglected to file for a water right on Clear Creek at the local government office or he would have obtained the Number One water priority right on the South Platte River Basin of Colorado. Regardless, his idea of irrigation quickly spread north along the Front Range.

In 1861, the Territorial Government of Colorado legalized the rapidly expanding irrigation practice of diverting water from streams for delivery to nonriparian lands. The Colorado Supreme Court formalized the acquisition of rights-of-way for irrigation ditch construction a year later. This interpretation of existing law prevented intervening landowners from stopping construction of an irrigation ditch across property, owned by others, as long as fair compensation was paid to the landowner.

FIG. 8.6. Placer mining probably somewhere in Colorado. Prospectors are shoveling gravel and other sediments from a streambed into a sluice box. In later years, huge crushing machines were transported up steep mountain roads to mining camps to replace this labor-intensive process.

A CLOSER LOOK

Early water allocation methods in the West were often contentious. The first large-scale community irrigation system in America was developed at the Union Colony settlement (today the City of Greeley, Colorado) in 1870. The irrigation system was called the Greeley Number Three Ditch and was constructed along the floodplains of the Cache la Poudre River. The first ditch rider was David Boyd, a graduate of the University of Michigan. His job was to plug gopher holes along the ditch and to allocate irrigation water to 200 parcels of land. That first year was "ash dry" and caused many delivery problems because the area only averaged 12 to 14 inches (30 to 36 cm) of precipitation annually.

A few summers later, in 1874, the weather was again very hot and dry. Irrigators under the Greeley Number Three Ditch were running out of water and couldn't understand why the flow in the Cache la Poudre River was so low. Several irrigators rode their horses upstream to investigate. To their horror, they saw several newly constructed irrigation ditches diverting water from the Cache la Poudre River near the new town of Fort Collins. The Greeley irrigators knew their downstream community was doomed if something wasn't done quickly, so they threatened to destroy the upstream "junior" ditches to protect their downstream "senior" water right. The Fort Collins irrigators didn't like that proposal, so a meeting was called at a neutral schoolhouse midway between the two communities.

The discussion at the school was lively. B. H. Eaton (a Civil War veteran from Ohio who would later become the governor of Colorado) and General Robert A. Cameron (also a Civil War veteran and a native of Brooklyn, New York) proposed a compromise to appoint a disinterested person to divide water from the Cache la Poudre River according to greatest need. This idea was not widely accepted, and the delegates at the meeting "hurled defiance in hot and unseemly language." (36) The debate escalated, with the Greeley irrigators threatening to dig new ditches upstream of the diverters in Fort Collins to choke off their water supply.

Then the meeting got ugly. Someone stood up and yelled, "Every man to his tent! To your rifle and cartridges!" Fortunately, Mr. Eaton and General Cameron calmed the crowd, but the meeting ended without a solution. As luck would have it, rain fell a few days later and reduced the need for irrigation water. It was apparent, however, that the state needed a formal system of water allocation. Within two years, this group of irrigators was instrumental in promoting the adoption of the Doctrine of Prior Appropriation in Colorado. (37)

In 1876, the Colorado Constitution included these statements:

The water of every natural stream not heretofore appropriated within the state of Colorado is hereby declared to be the property of the public and the same is dedicated to the use of the people of the state subject to appropriation as hereinafter provided.

Colorado Constitution of 1876, Article XVI, Section 5

The right to divert unappropriated waters of any natural stream to beneficial uses shall never be denied.

Colorado Constitution of 1876, Article XVI, Section 6

This strict use of the Doctrine of Prior Appropriation, which is called the **Colorado Doctrine,** completely eliminated the use of the Riparian Doctrine in Colorado. The right to divert unappropriated water is given to any resident of the state and provides a strong entrepreneurial system of water development. The Colorado Doctrine has been adopted only by other very dry states in the West such as Arizona, Idaho, Montana, Utah, Nevada, New Mexico, and Wyoming.

The U.S. Supreme Court, in a decision written by Justice George Sutherland, interpreted the Desert Lands Act of 1877 to allow states to also develop their own water laws on federal lands:

What we hold is that following the [Desert Lands] *act of 1877 if not before, all non-navigable water then a part of the public domain became "publici juris," subject to the plenary control of the designated states, including those since created out of the territories named, with the right in each to determine for itself to what extent the rule of appropriation or the common-law rule in respect to riparian rights should obtain. (38)*

The 1882 landmark court case of ***Coffin* v. *Left Hand Ditch Company*** put the Doctrine of Prior Appropriations to the test in Colorado. As in the California case of *Irwin* v. *Phillips,* a downstream riparian landowner claimed injury and sued an upstream irrigation company that was diverting river water to a different basin. The Colorado Supreme Court ruled that the irrigation company was entitled to divert and use water on nonriparian lands under the Doctrine of Prior Appropriation. This ruling eliminated the use of the Riparian Doctrine from Colorado water law. (39)

EXPERT ANALYSIS

Water distribution under the Doctrine of Prior Appropriation follows a rigid set of priority dates and discharge rates for any water diversion from a stream. Authority to divert water depends on the water rights held by the potential diverter. In the following example, six diverters would like to exercise their right to divert water from the South Platte River in Colorado. It's August 6, 2005 (the peak of the demand season for irrigation water), river discharge is declining, and irrigation water demand is increasing. Who will be required to shut his or her river headgate and cease diversions from the river?

It's 10:00 A.M., and the total flow in the river is 331 cfs (213.9 million gal per day, or 9.4 cms) at point A, the upper location of this river reach (Figure 8.7).

Priority Ranking	Priority Date	Water Right (cfs)	Location
1	May 1, 1863	55	C
2	June 9, 1863	62	A
3	August 5, 1864	58	B
4	May 3, 1866	85	F
5	April 29, 1870	158	D
6	April 30, 1870	28	E
Total		**446 cfs**	

The example shows that only the first four diverters, with priority dates between May 1, 1863, and May 3, 1866, will be allowed to divert their full water right from the South Platte River on August 6, 2005. The priority of April 29, 1870 will only receive 71 cfs since the four more senior water right holders are entitled to a total of 260 cfs from the river (55 + 62 + 58 + 85 = 260 cfs). Since the river has a total of only 331 cfs, all that remains for priority ranking number 5 is 71 cfs (331 cfs − 260 cfs = 71 cfs). What happens to priority number 6? Correct, they get nothing.

This allocation method has been used for over 100 years in western states that follow the Doctrine of Prior Appropriation. Today, unless state or federal law requires that maintenance flows remain in a river for environmental purposes, all water can still be diverted in the above situation. The Doctrine of Prior Appropriation historically gave no right of use for nonconsumptive uses such as piscatorial (fish) needs, aquatic habitat, or other beneficial environmental purposes. However, the protection of the environment and other modifications to the Doctrine of Prior Appropriation have increased dramatically in the past several decades and now allow for the use of water for nonconsumptive uses. (This subject will be discussed at length in Chapter 12).

An appropriator can enlarge an existing diversion structure but must apply for a new water right or permit. This new water right will be junior to all prior rights, even though the original ditch may have existed for many years. Therefore, a ditch may have a water right of May 1, 1863, and a second right (or priority) of August 14, 1871. All water rights senior to the August 14, 1871, date will need to be satisfied before the ditch can legally divert water under its second, but more junior, priority. A water right (or in some states a water permit) is issued by a court or state water agency and generally includes a flow rate in cubic feet per second or miner's inches. The right to store water in a reservoir is generally based on total acre-feet or on a gage height for a predetermined water elevation level. In general, water rights for direct use in ditches and canals were obtained decades before reservoir rights were developed. Dam construction required more money and resources than were available to early settlers, and associated water rights for reservoirs generally have priority dates junior to irrigation ditches.

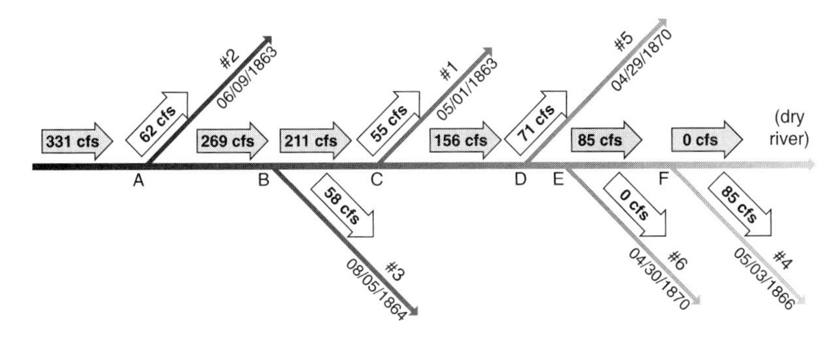

FIG. 8.7. Illustration of priorities for above example.

A CLOSER LOOK

The miner's inch is an old unit of water measurement used in mining camps of the western United States and Canada during the 1800s. A **miner's inch** is defined as the rate of water discharge through an orifice (opening) in a channel 1-inch square (6.45 cm^2) under a specified head of water. Each state or province defined the quantity of a miner's inch differently:

1 miner's inch = 0.020 cfs (9.0 gpm or 0.0006 cms) in Idaho, Kansas, Nebraska, New Mexico, North Dakota, South Dakota, and Utah
= 0.025 cfs (11.2 gpm or 0.0007 cms) in Arizona, California, Montana, and Oregon
= 0.026 cfs (11.7 gpm or 0.0007 cms) in Colorado
= 0.028 cfs (12.6 gpm or 0.0008 cms) in British Columbia

Today, only a few irrigation companies continue to use the miner's inch to measure water to their shareholders.

A water right can be leased, sold, mortgaged, or bequeathed to others in the same manner as other property such as a house or vehicle. A water right is a usufructory property right since an individual has the right to the use of water while not actually owning water in a river. The owner of the right to use water holds a valuable asset. River water is owned by the people of a state and often creates conflicts over the definition of its best use (also called the public good).

POLICY ISSUE

Water development during the 1800s in the western United States was extensive, expansive, and extraordinary. In addition, water developers rarely considered the negative effects of surface water diversions on the local environment. Why were developers of that era so unconcerned about the potentially destructive effects on wildlife and habitat?

Consider some of the historical events of the 1800s. The Louisiana Purchase and the acquisition of land from Mexico in the first half of the nineteenth century created tremendous opportunities for Americans to move from crowded, impoverished urban conditions along the East Coast of the United States to the unsettled West. The discovery of gold was another major catalyst for this migration, although the promise of free land for homesteaders was probably an even greater lure.

Unfortunately, the land available for settlement and cultivation was quite harsh in most locations. Water supplies were scarce, and successful dryland farming was nearly impossible. The immediate concern for most settlers was survival, as witnessed by the Mormon immigrants to the arid Salt Lake Valley in 1847. Their first duty was to construct irrigation ditches to grow food.

What if the original western settlers had paid more heed to the water needs of the natural environment? What if early irrigators had reduced or limited the amount of water diverted from rivers to protect adequate flows for downstream environmental purposes, such as fish habitat or wetlands preservation? What changes in state or federal water policy would have been necessary in the 1800s to protect the environment? Would such government policies have been feasible during that era? What role did expanding population play in the development of state water allocation law during the 1800s?

WATER ALLOCATION LAW (1900–PRESENT)

RIPARIAN DOCTRINE

Today the Riparian Doctrine is the basis of existing water law in 31 states in the eastern United States as well as numerous foreign countries. It evolved over the centuries through Royal edicts, legislative acts, and judicial interpretation, and it forms a vast body of law and rights for riparian landowners. It's important to remember that the Riparian Doctrine was developed in humid regions where water was abundant, irrigation was basically nonexistent, and water allocation did not cause major problems for individual water users. Riparian water rights are passed on to subsequent landowners through the sale of land, and not by the sale of privately owned water rights.

The Riparian Doctrine has evolved into two basic principles: (1) Reasonable Use; and (2) Correlative Rights.

The **Reasonable Use Principle** means that a riparian landowner can divert and use any quantity of water he or she chooses, for use on riparian lands, as long as these diversions and uses do not interfere with the reasonable use of other riparian landowners as characterized by the principle of *damnum absoque injuria,* "harm without injury." To stop upstream water diversions or to obtain payments for damages, an injured riparian must prove injury to his property, through the actions of another riparian. Proof of injury is presented either in a court of law or at an administrative hearing before a state water agency official.

The **Correlative Rights Principle** requires that riparian landowners must share the total flow of water in a stream. The proportion of use allocated to each riparian is based on the amount of waterfront property owned along a stream and creates equal rights for riparians (see Figure 8.8). There is no priority of water use in this system. The Correlative Rights Principle provides a minimum, reasonable amount of water to all water users along a stream. During a drought, allocations are prorated so that everyone receives a proportionate share of the discharge in the stream.

EXPERT ANALYSIS

How much water can each entity in the following example divert? It's August in the early 1900s, and the flow of water in an eastern river is declining. A power plant, three mills, and a municipality need water. This state uses the Correlative Rights Principle of the Riparian Doctrine, and no other riparian water users divert water along this river. The total flow in the river is 310 cfs (200 mgd, or 8.8 cms) at point A. The power plant was just placed into operation, while Mill #1 has been operating for over 100 years.

FIG. 8.8. Fertile farm ground, like this land adjacent to the St. Lawrence River near Les Eboulements, Quebec, was surveyed into long, narrow riparian lots so that all farmers would have frontage along the river. Such a system of land division provided access for transportation of goods on the river and is also common in France.

	River Frontage Property		
Entity	Miles	Km	Location
Mill #1	1.0	1.6	A
Mill #2	0.5	0.8	B
City	6.0	9.7	C
Mill #3	1.0	1.6	D
Power Plant	1.5	2.4	E
	10	16.1	

To solve this problem, first develop a proportion of water use for each entity based on the amount of river frontage property owned. (The proportions will be the same for use of kilometers.)

Mill #1 owns 1 mile out of a total of 10 miles of river frontage property (1 mile divided by 10 miles = 10% of the total river frontage).

Mill #2 owns 0.5 mile out of a total of 10 miles of river frontage property (0.5 mile divided by 10 miles = 5% of the total river frontage).

The city owns (or has within its control) 6 miles of river frontage property (6 miles divided by 10 miles = 60% of the total river frontage).

Mill #3 owns 1 mile of river frontage or 10% of the total.

The power plant owns 1.5 miles of river frontage (1.5 miles divided by 10 miles = 15% of the total river frontage).

Next, multiply each percentage times the total water discharge available in the river, which is 200 mgd (million gallons per day):

Mill #1 receives 10% of the total river flow	(10% times 200 mgd = 20 mgd)
Mill #2 receives 5% of the total river flow	(5% times 200 mgd = 10 mgd)
The city receives 60% of the total river flow	(60% times 200 mgd = 120 mgd)
Mill #3 receives 10% of the total river flow	(10% times 200 mgd = 20 mgd)
The power plant receives 15% of the total river flow	(15% times 200 mgd = 30 mgd)
	100% = 200 mgd

The date of construction for any of these uses has no significance regarding water distribution under the Riparian Doctrine.

In contrast, this sharing method of allocation is never used under the Doctrine of Prior Appropriation because a strict priority system is based on first in time, first in right. A strict priority may seem unfair and severe to other water users under a priority system, but imagine that you're back at the open-seating concert discussed earlier. You've waited in line for hours to get a great seat after spending the night in a sleeping bag outside the concert venue. The admission gates finally open, and you find excellent front row seats for you and your friends. However, just as the main act is ready to come on stage, an usher tells you that half of your group needs to leave because the place is full and a few latecomers need seats. What would be your reaction?

Under the Riparian Doctrine, courts have ruled that it is illegal to detain water during the daytime, and then release it at night, if there is injury to downstream riparian users that also require water during daytime hours. (40) It is also unreasonable to release water in a flash, so that downstream users are faced with extremes of small pulses of water followed by short-duration floods. However, a riparian user can make use of water in any manner as long as no other riparian water user objects. For example, if a dam floods upstream property, the owner of the dam can continue this practice if no other riparian objects within a prescribed time—somewhere between 7 and 18 years. Under this scenario, the dam owner would obtain a new right called a *prescriptive easement* and could legally continue historic water use practices if no one objected or proved injury.

Under the Riparian Doctrine, a stream can be treated like an enemy, and appropriate action can be taken during floods to protect property even if temporary dikes, berms, or channels cause temporary injury to a neighbor's land. Also, a riparian landowner has no liability for water that naturally flows across his or her land and is not required by law to control such natural flows of water.

As a result of increased demands, drought, and expanding water use conflicts, some eastern states are abandoning the riparian system and are adopting a permit system that distributes water resources under the Correlative Rights Principle. A permit system can allocate water to riparian landowners based on use, need, percolation rates, and climatic factors, and can be changed or revoked by the state.

DOCTRINE OF PRIOR APPROPRIATION (1900–PRESENT)

Climate has led to the adoption of various forms of the Prior Appropriation Doctrine by all states west of the 100th Meridian. The strict form of this appropriation doctrine, called the Colorado Doctrine, is now used in Alaska, Arizona, Colorado, Idaho, Montana, Nevada, New Mexico, Utah, and Wyoming. A combination of the Riparian and Prior Appropriation Doctrine law, called the California Doctrine, is used in California, Kansas, Nebraska, North Dakota, Oklahoma, Oregon, South Dakota, Texas, and Washington. Hawaii uses a unique set of water laws based on ancient Hawaiian allocation principles, which use a variation of the priority system based on type of water use.

The amount of water provided with a water right under the Doctrine of Prior Appropriation is based on the amount of water historically diverted and placed to beneficial use. In the case of David Wall, the gold miner who grew vegetables in Colorado in 1859, this doctrine allowed him to legally divert as much water as needed to flood irrigate his 2-acre plot of land. However, he could not divert any additional water for other uses, or for the expansion of his irrigated ground, without obtaining a new, more junior water right.

Under the Doctrine of Prior Appropriation, an irrigator who files for the use of water to irrigate 100 acres (40 ha), but only irrigates 20 acres (8 ha) the first year, must diligently continue to develop irrigation of all 100 acres if he or she wishes to preserve the original appropriation date (called a filing date). An irrigator (or other water developer) is required to appear before a judge or an appropriate administrator to prove **diligence** on the water project by demonstrating continued efforts toward completion of the project. This allows water development and construction to continue, particularly on large reservoir planning and construction projects, which can take decades for completion.

A CLOSER LOOK

Stock Ownership in Private Irrigation Companies Most early irrigation canals in the western United States in the 1800s were constructed by consortiums of irrigators and/or investors. After an irrigation ditch (also called a canal) was completed, an irrigation company was usually formed to operate and maintain the system. An irrigation company is a legal entity that issues shares in the company, represented by stock certificates, to allocate ownership of water to individual irrigators (and to reimburse investors). This ownership of stock is similar to stock sold by companies on the New York Stock Exchange.

Stockholders are required to pay annual assessments to provide capital for repairs, employee salaries, and other operational and maintenance expenses. A ditch superintendent or ditch rider typically manages the day-to-day operations of an irrigation system, similar to the mayordomo system used in Spain and Mexico. The shareholders elect a board of directors each year at an annual meeting. This board makes major financial and other management decisions regarding the operation and maintenance of the ditch system, similar to the ayuntamiento system.

A water right, or several rights, are owned by the irrigation company to allow diversion of water from a river (if the state utilizes the Doctrine of Prior Appropriation). The ownership of shares gives an individual irrigator the right to divert a proportional share of water from the ditch. For example, ownership of one share of water, in a ditch that has a total of 100 shares, allows the shareholder to divert 1 percent of the water from the ditch on any day during the irrigation season. No priority of water use exists between the shareholders within the irrigation canal company, and everyone shares the shortages. The ditch superintendent allocates water, on a daily basis, based on share ownership, to irrigators who are members of the ditch company.

EXPERT ANALYSIS

Suppose an irrigation company issued 100 shares of stock when it was incorporated, and today you own one share. The company's water right on the river allows the diversion of 200 cfs (129 mgd or 5.7 cms) of water when in priority. However, it's the beginning of August, and the flow in the river is low. A downstream irrigation company has a priority date that is senior to your ditch and has just placed a "call" on the river, demanding that all upstream junior diverters reduce the amount of water they're diverting from the river (or stop diverting water completely). This action will allow additional water to flow downstream to the senior appropriator to allow the calling ditch to divert its full decree.

Your ditch company superintendent has just been ordered by the state water official, in charge of diversions from the river, to reduce your company's diversions by half until the water situation improves. (You are not required to completely shut off since your water right is fairly senior.)

After the reduction, the flow in your ditch is only 100 cfs (65 mgd or 2.8 cms). The average ditch loss (seepage loss through the bottom and sides of the ditch, often called *shrink*) is 25 percent. All shareholders must share in the ditch shrink to ensure that everyone (particularly those at the end of the ditch) receive adequate irrigation water. How many cubic feet per second of water will you receive with your one share of water stock?

Facts

1. The total flow in the ditch today is 100 cfs.
2. All shareholders must share in the 25% shrink—100 cfs minus the 25% shrink = 75 cfs.
3. You own one share out of a total of 100 shares = 1% ownership of all available water in the ditch.

The amount of irrigation water that you will receive on your farm today = Total flow of water available in the ditch times your percent ownership of shares in the ditch = 75 cfs × 1% = 0.75 cfs (0.48 mgd or 0.02 cms).

EXPERT ANALYSIS

The City of Greeley wishes to purchase your water rights from a 160-acre (65 ha) farm along the Cache la Poudre River in Colorado. Historically, 150 acres (60 ha) were irrigated, with the remainder of the farm used for buildings, yard, driveways, and fences. Corn, pinto beans, and alfalfa were grown in a three-year crop rotation, each on 50-acre (20 ha) fields. The 10-year average annual consumptive use was 1.6, 1.2, and 2.0 acre-feet, respectively, for each crop. (Consumptive use, as noted earlier, is the amount of water consumed by a plant and is "lost" from the local hydrologic system.)

Your farm was flood irrigated with an average irrigation efficiency of 50 percent. It is located in an alluvial valley, and senior water rights are located downstream. How much water can the City of Greeley divert after your water right is purchased and transferred (moved to a new point of diversion) for municipal use?

First, determine the historic consumptive use of each crop:

50 acres corn @ 1.6 acre-feet per acre =
80 acre-feet average annual consumptive use

50 acres pinto beans @ 1.2 acre-feet per acre =
60 acre-feet average annual consumptive use

50 acres alfalfa @ 2.0 acre-feet per acre =
100 acre-feet average annual consumptive use

Therefore, total historic consumptive use available from your farm equals 240 acre-feet (78 million gal, or 296,036 m^3) of average annual consumptive use.

Since the irrigation efficiency on your farm was 50 percent, a total of 480 acre-feet of irrigation water was applied (240 acre-ft divided by 0.5 = 480 acre-ft). This means that 50 percent of the irrigation water applied was consumed by the crop or evaporated. The other 50 percent of applied irrigation water either percolated underground or flowed off the end of the field as surface water runoff and returned to the Cache la Poudre River as return flow. Therefore, 50 percent of the irrigation water applied to the crops (or 240 acre-ft) would be required by state law to remain in the river to maintain historic return flows for downstream senior diverters. The city could only transfer the other 240 acre-feet to its drinking water treatment plant for municipal use.

In this analysis, the city's future diversion rate of the transferred irrigation water must be the same as the historic consumptive use pattern of the three crops. For example, the total daily consumptive use rate for your crops may have been 1.0 cfs during June, 2.8 cfs during July, 3.1 cfs during August, and 1.0 cfs during September. The diversion rates for the drinking water treatment plant must mimic the historic consumptive use on the farm. In addition, no diversions of this new municipal water right could be made during other times of the year if irrigation water was not historically diverted and consumed by crops during those periods (such as during the winter months).

In a real-world water transfer case, the above figures would be scrutinized by water resource engineers, water attorneys, and state water officials to protect downstream senior water users. Disputes are common and often require a judge or state water administrator to serve as an arbitrator.

What if the city, in the above example, was located upstream of the irrigator's 160-acre farm? How would the 240 acre-feet of consumptive-use water be transferred to the city's drinking water treatment plant? How would intervening senior water rights be protected? In this situation, the city would be required to cooperate with a state water administrator to develop a new method of water diversion called an **exchange.** An exchange allows a water user to divert water from a river, at a new upstream location, as long as intervening water rights are not injured. In the preceding example, the City of Greeley could divert 2.8 cfs (1.8 mgd or 0.08 cms) of water daily from the Cache la Poudre River during July, as long as 2.8 cfs were flowing in the river at all diversion points between the city's diversion point and the original **point of diversion** from the river for the farm's irrigation ditch. An exchange allows a water user to transfer a water right upstream to a new point of diversion as long as no downstream water rights are injured (see Figure 8.9).

Efficiency of water use has long been a criticism of the Doctrine of Prior Appropriation. The concept of use it or lose it forces the owner of a water right to divert as much water as possible for beneficial use. If the diversion of water is not maximized, the user could be forced to forfeit ownership of some water under state abandonment laws. Efficiency is not rewarded in this priority system based on seniority because any unused water belongs to the next senior water right downstream. Some states, such as Oregon, have considered the concept of incentives for water conservation. Conserved water, called **salvaged water,** can be sold, along with its priority date, to other water users in the system. Supporters of this legal concept argue that, for example, reducing evaporation from a flood-irrigated farm by reducing the total water applied

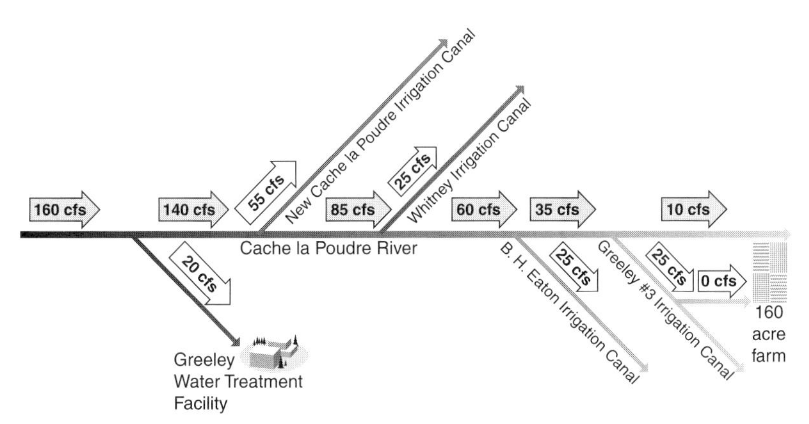

FIG. 8.9. In this hypothetical example, the City of Greeley would like to exchange 20 cfs from water rights it purchased along the Greeley #3 Irrigation Canal, upstream to its water treatment facility. It may do so as long as there is at least 20 cfs at all intervening points along the Cache la Poudre River. Notice that the 160-acre farm receives 0 cfs. That would be the case if the owner sold all of his or her water rights to the city. Only dryland crops could be grown on that farm in the future.

to a field would help stretch water supplies in water-short regions. Opponents of the concept argue that salvaged water would reduce historic return flow patterns if historic water use practices were altered by financial incentives created by the sale of salvaged water.

POLICY ISSUE

You live southwest of Pocatello, Idaho, and irrigate 360 acres (146 ha) of land along the Snake River. You have a good priority date that provides water during most of the growing season. Two neighbors irrigate land upstream of your farm, under a different irrigation canal system, and have a priority date senior to your water rights. The Idaho Legislature is considering adoption of a salvaged water law that would allow your neighbors to sell conserved consumptive-use water to municipalities upstream. You believe this law will reduce the amount of water available to you from upstream historic return flows. You plan to testify at a committee hearing of the legislature in the morning. How would you develop your testimony to convince legislators to defeat the proposal for salvaged water?

What if your upstream neighbors supported the proposed law? What arguments could they use in their testimony before the legislative committee? How could you and your neighbors develop a compromise?

GROUNDWATER DOCTRINES

To whomever the soil belongs, he owns also to the sky and to the depths.

English Law, 152 Eng. Rep. 1235 (1843).

Lawmakers in the 1800s did not completely understand the properties of groundwater movement, and not surprisingly, few U.S. laws regarding groundwater allocation or use existed during that time. Under Spanish law, a landowner could dig a well and remove groundwater with total disregard for the effects on neighboring wells. This concept, known as the **Rule of Capture,** places no restrictions on groundwater use. This reasoning followed the strict **Rule of Absolute Ownership,** which states that a landowner owns everything on his or her property from the land surface, up to the heavens, and down to the center of the Earth.

Landownership was the original basis of groundwater law, similar to riparian law. Later, as increased groundwater pumping caused more well-to-well interference, many areas adopted the Reasonable Use Principle of surface water (discussed earlier in this chapter). In addition, the Correlative Rights Principle of surface water (also discussed earlier) was introduced to promote sharing. Recharge rates and groundwater infiltration were also later factored into groundwater law.

Some states treated groundwater as an underground stream and issued permits to allow for a reasonable quantity and time of use. However, it was extremely difficult to determine the availability of exact quantities of groundwater, particularly in high-use areas or during periods of drought. Disputes were usually resolved on a case-by-case basis under Common Law. (41)

In the United States, all groundwater law is developed by individual states (see Table 8.1). Many states have adopted the Correlative Rights Principle to provide allocations of groundwater to all users, thus reducing or restricting pumping. Most states limit the number of wells drilled in a given location by placing distance requirements between wells. This attempts to reduce well-to-well interference and conflict. The use of Correlative Rights requires accurate information regarding the location and quantification of groundwater, transmissivity, recharge rates, and direction of groundwater movement. Some states regulate the amount of groundwater that can be pumped by issuing diversion permits.

A few western states use the Doctrine of Prior Appropriation to allocate groundwater, awarding priorities based on the date a well was drilled

TABLE 8.1 Groundwater Doctrines by State

Rule of Capture	Riparian Rights			Prior Appropriation	Both Riparian and Prior Appropriation
Texas	Alabama	Maine	North Carolina	Alaska	California
	Arkansas	Maryland	Ohio	Arizona	Hawaii
	Connecticut	Massachusetts	Pennsylvania	Colorado	Kansas
	Delaware	Michigan	Rhode Island	Idaho	Nebraska
	Florida	Minnesota	South Carolina	Montana	North Dakota
	Georgia	Mississippi	Tennessee	Nevada	Oklahoma
	Illinois	Missouri	Vermont	New Mexico	Oregon
	Indiana	New Hampshire	Virginia	Utah	South Dakota
	Iowa	New Jersey	West Virginia	Wyoming	Washington
	Kentucky	New York	Wisconsin		
	Louisiana				

and groundwater was put to beneficial use. Pumping rates and groundwater withdrawals are restricted to a given flow rate from the pump or to a limited number of irrigated acres. Expansion of this water right is illegal because it could injure other senior surface rights (in an alluvial system) or other groundwater rights.

Arizona adopted an innovative and highly regarded program in 1980 following passage of the Groundwater Management Act. The act reallocated groundwater resources in the state based on use and has a goal of "safe yield" by the year 2025. (Safe yield is a condition whereby the amount of groundwater withdrawn equals total aquifer recharge.) This goal will be reached by greatly reducing, or in some locations completely eliminating, groundwater withdrawals for agricultural purposes. It is hoped that these restrictions on groundwater use will help meet the needs of population growth in the state (see http://www.adwr.state.az.us/AZWaterInfo/groundwater/code.htm). This innovative legislation received the Ford Foundation's award for state and local government in 1986.

POLICY ISSUE

Artesian groundwater allocation is a particular problem in many areas because groundwater pumping can dramatically decrease artesian pressure. Should an artesian well owner be entitled to a particular elevation (head) of groundwater inside his well? Should priority be given to a senior well owner if a junior well user caused the reduction in artesian pressure? Should a junior well user be required to pay for redrilling a well if a senior well owner is forced to redrill her well to a greater depth? Most states have developed laws that require all owners to share in reduced groundwater levels and to pay their own costs for deepening wells. Do you agree that this is fair and appropriate?

INTERSTATE COMPACTS

Most large river basins do not lie wholly within the borders of a single state. Interstate water agreements, called **interstate compacts,** are binding agreements between states, ratified by the U.S. Congress, to share the flow of water in a river. If states cannot reach agreement, federal courts are asked to determine a fair apportionment scheme of interstate rivers through a court decree.

Interstate compacts were started under the Articles of Confederation, but the first interstate compact for water allocation came much later with the 1922 Colorado River Compact. Over 30 interstate water compacts exist today in the United States and are used for water allocation, flood control, planning, and pollution control (Table 8.2).

SELECTED RIVER COMPACTS

The Colorado River Compact of 1922 In the early 1920s, California's growth was seen as a threat to other states that also relied on the Colorado River. In June 1922, the U.S. Supreme Court ruled that the Doctrine of Prior Appropriation applied to all water users in the West, regardless of state lines. This meant that a fast-growing state could apportion (claim) all available water for beneficial use and leave little for the slower-growing upstream states that didn't require large water supplies at the time. California is the final water user in the United States along the 1500-mile-long (2414 km) Colorado River and contributes very little flow to the river. Negotiators for upstream states, such as Colorado and Utah, believed that the surface water contributions provided by their rivers and streams should allow them a greater share of Colorado River water for future needs (see Figure 8.10).

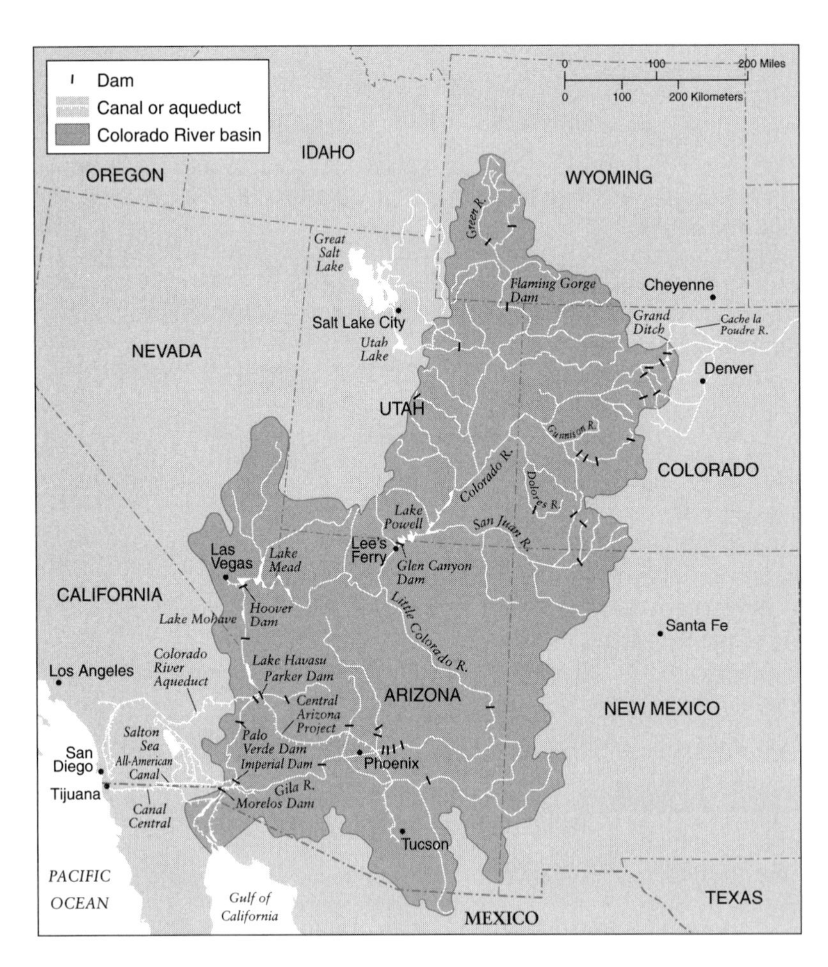

FIG. 8.10. The Colorado River Basin. Note the large number of dams on the Colorado River and its tributaries. Much of the basin is sparsely populated, but several canals and aqueducts transport water great distances to growing population centers such as Phoenix, Los Angeles, Denver, and Las Vegas. The Colorado River provides water to over 25 million people and helps irrigate some 3.5 million acres (1.4 million ha) of cropland. See http://www.water.ed.org/coloradoriver.asp for more information on the Colorado River and its compact.

TABLE 8.2 Selected Interstate Compacts in the United States, 1783–2004

River	States Involved	Year Adopted	Primary Purpose(s)
Delaware	New Jersey, Pennsylvania	1783	Navigation
Potomac	Maryland, Virginia	1785	Navigation and fishing
Savannah	Georgia, South Carolina	1788	Navigation
Colorado	Wyoming, Colorado, Utah, New Mexico, Arizona, Nevada, California	1922	Quantity
La Plata	Colorado, New Mexico	1923	Quantity
South Platte	Nebraska, Colorado	1923	Quantity
Rio Grande	New Mexico, Oklahoma, Texas	1927	Quantity
Rio Grande	Colorado, New Mexico, Texas	1939	Quantity
Ohio	Illinois, Indiana, Kentucky, Ohio, New York, Pennsylvania, Virginia, West Virginia	1939	Quality
Potomac	Maryland, Pennsylvania, Virginia, West Virginia, District of Columbia	1940	Quality
Belle Fourche	South Dakota, Wyoming	1943	Quantity
Cheyenne	South Dakota, Wyoming	1943	Quantity
Republican	Colorado, Nebraska, Kansas	1943	Quantity
Costilla	Colorado, New Mexico	1945	Quantity
North Platte	Colorado, Nebraska, Wyoming	1945	Quantity
Snake	Idaho, Wyoming	1948	Quantity
Arkansas	Colorado, Kansas	1949	Quantity
Pecos	New Mexico, Texas	1949	Quantity
Connecticut	Connecticut, Massachusetts, New Hampshire, Vermont	1949	Flood control
Yellowstone	Montana, North Dakota, Wyoming	1951	Quantity
Canadian	New Mexico, Oklahoma, Texas	1951	Quantity
Sabine	Louisiana, Texas	1953	Quantity
Tennessee	Tennessee, Alabama, Georgia, Mississippi, North Carolina, Virginia, Kentucky	1955	Quality
Columbia	Idaho, Montana, Nevada, Oregon, Utah, Washington, Wyoming	1955	Water development
Klamath	California, Oregon	1957	Quantity
Bear	Idaho, Utah, Wyoming	1958	Quantity
Delaware	Delaware, New Jersey, New York, Pennsylvania	1961	Water development
Upper Niobrara	Nebraska, Wyoming	1962	Quantity
Arkansas	Kansas, Oklahoma	1965	Quantity
Wheeling Creek	Pennsylvania, West Virginia	1967	Quantity
Susquehanna	Maryland, New York, Pennsylvania	1970	Quantity, quality, and flood control
Big Blue	Kansas, Nebraska	1971	Quantity
Tennessee	Alabama, Georgia, Kentucky, Mississippi, North Carolina, Tennessee, Virginia	1972	Quantity
Red River	Arkansas, Louisiana, Oklahoma, Texas	1980	Quantity
Alabama-Coosa-Tallapoosa (ACT)	Alabama, Florida, Georgia	1999	Quantity
Apalachicola-Chattahoochee-Flint (ACF)	Alabama, Florida, Georgia	Ongoing	Quantity

Delph Carpenter (1877–1951), a water lawyer from Greeley, Colorado, proposed that individual states within the Colorado River Basin should negotiate an interstate compact to share in the total flow of the Colorado River. (Compacts had already been established in the 1700s on the Delaware, Potomac, and Savannah rivers for navigation purposes as shown in Table 8.2.) Carpenter believed that each state should receive an entitlement (allocation) to a given volume of water, thus preventing California from taking more than its fair share of water from the Colorado River. He was also very concerned about federal intervention into state water rights issues and feared that, if the individual states didn't reach a solution, the federal government would establish federal law for the Colorado River.

At first, the delegates tried to apportion a set amount of water to each state based on the amount of irrigated land contained in each. Attempts to quantify these figures were disastrous, however, and almost brought negotiations to a breaking point. Fortunately, discussions continued. Eventually, various states were drawn into common interests, and these viewpoints evolved into the Upper versus Lower Colorado River Basin States.

After months of negotiations, an agreement was reached. The total annual flow in the Colorado River would be divided at Lee's Ferry, in northern Arizona. Colorado, New Mexico, Utah, and Wyoming were placed in the Upper Basin allocation, while Arizona, California, and Nevada became the Lower Basin States. Each basin received 7.5 million acre-feet (2.4 trillion gal, or 9.3 billion m^3) per year. The Lower Basin states were allowed to increase their allotment by 1 million acre-feet (325.9 billion gal, or 1.2 billion m^3) at a later date as population growth occurred. This incentive was given by Upper Basin States to the Lower Basin negotiators to sweeten the deal. Upper Basin States were required to deliver 7.5 million acre-feet past Lee's Ferry over a 10-year period to average years of high and low flows. If a particular 10-year flow period is significantly below average, the Upper Basin States are still required to deliver 7.5 million acre-feet—a figure based on hydrologic data provided by the federal government.

The Colorado River Compact was signed on November 24, 1922, in the Palace of the Governors in Sante Fe, New Mexico, and was ratified by the U.S. Congress in 1923.

POLICY ISSUE

The creation of the Colorado River Compact was based on hydrologic data provided by the federal government. Climatic variations, discussed in Chapter 2, are common in watersheds and can range greatly in some locations between relatively wet and dry cycles. If the long-term average flow of the Colorado River is considered, it appears that the Colorado River Compact may have been based on hydrologic data from a relatively wet weather cycle.

What incentives would water officials from the Colorado River Basin States have to renegotiate the Colorado River Compact if it could be definitively proven that the figures used in 1922 were from a wetter than normal period? What problems could renegotiation of the Compact create? The Colorado River Compact uses a 10-year river flow average (sometimes called a floating average) to adjust base water allocation quantities. What problems could be caused for urban planners, irrigators, or wildlife managers if a floating average was used on all interstate water allocation compacts? What benefits could be derived from the use of such averages?

The Niagara River Water Diversion Treaty of 1950 Extensive water power development at Niagara Falls created a huge demand for water diversions from the Niagara River. As discussed in Chapter 1, the area around Buffalo, New York, became a magnet for industrial development due to the plentiful supply of hydropower in the region. Upstream river diversions became substantial and led to reduced flows over the Falls. The effect of these diminished flows on tourism and the natural beauty of the area was of great concern.

On February 27, 1950, a Treaty was signed between Canada and the United States to regulate the amount of water that could be diverted from the Niagara River. The Treaty also designated that no diversions would occur from the Niagara River (for power generation purposes) if less than 100,000 cubic feet per second (64.6 billion gal per day, or 2,832 cms) was flowing over Niagara Falls between the hours of 8 A.M. Eastern Standard Time (EST), and 10 P.M. (EST) between April 1 and September 15th. The Treaty also stated that at least 100,000 cfs was required, between 8 A.M. and 8 P.M. the remainder of the year. During the nighttime hours, diversions from the Niagara River can increase substantially. This can cause a nighttime reduction in river discharge over the Falls by as much as 50 percent. (42)

The Delaware River Compact of 1961 In 1925, New York City developed plans to divert water from the Delaware River in New Jersey for use in the city. Negotiations were attempted with downstream states over allocation of riparian rights on the river but were unsuccessful. The city proceeded with the project but was stopped by objections filed by the State of New Jersey and other downstream states, and eventually the dispute landed in the U.S. Supreme Court. The Court ruled in 1931 that water in the Delaware River should be apportioned (shared) based on "needs" and not on the length of the riparian shoreline of the affected states (New York, Delaware, New Jersey, and Pennsylvania). The Court awarded the City of New York 1230 acre-feet per day (400 mgd, or 1.5 million m^3 per day), but it was not perpetual. Later, an interstate compact, signed in 1961, created a commission to develop long-range water plans and an equitable allocation of Delaware River water between the states. In addition, the compact gave the Commission the power to declare an emergency and change the proportion of water allotment (a concept many would consider impossible in the arid western United States).

FEDERAL RESERVED WATER RIGHTS

The federal government owns vast amounts of land in the United States, particularly in national parks, monuments, forests, and wildlife refuges in the West. Since the 1950s, the federal government and members of Indian tribes have been obtaining new water rights that were not historically available or used. These new senior water rights have forced existing water right holders to modify long-standing water use practices. **Federal reserved water rights** are rights obtained by the federal government when a federal reservation, such as a national park or national monument, was formed. By law, the establishment of these federal reservations implicitly provided water rights to meet the needs of such holdings. The primary conflicts have arisen over priority dates, quantification of water rights, and points of diversion.

The **Winters Doctrine,** also called **Indian Reserved Rights,** originated in the court case *Winters v. United States,* which centered on the issue of diversions from the Milk River located on the Fort Belknap Indian Reservation in northern Montana. (43) The reservation was established in 1888 for the Gros Ventre and Assiniboine Bands (or Tribes). Around 1905, however, non-Native Americans began diverting water from the Milk River, leaving little water for the future needs of the Fort Belknap Reservation. Irrigation had already been developed on the Reservation in the 1880s to divert approximately 10,000 miner's inches (250 cfs, 161.6 mgd, or 7.1 cms) of Milk River water for irrigation of 30,000 acres (12,141 ha) of grain, grass, and vegetables. Irrigation water from the Milk River was vital for food production on the Reservation, and eventually the Reservation superintendent obtained help from the U.S. Attorney General's Office. This took place during the presidency of Theodore Roosevelt, a strong advocate of preserving natural resources on federal lands. (44)

The U.S. Attorney General's Office argued that Native Americans were entitled to irrigation water from the Milk River due to the riparian nature of the Reservation. It maintained that upstream diverters (including Henry Winters, a local irrigator involved in the litigation) were making an unreasonable use of water from the river and were injuring downstream riparian water users. Attorneys for the State of Montana disagreed, arguing that Montana was a prior appropriation state. As such, riparian landownership had nothing to do with water rights in Montana. The Court ruled that the residents of the Fort Belknap Reservation were entitled to water from the Milk River because of the federal government's intent to reserve necessary water rights for use on the Reservation. The date the water right was awarded was the date the Treaty was signed. This gave the Reservation a water right of May 1, 1888, far senior to the 1905 priority date of Mr. Winters and other upstream irrigators. (45)

> ... [the Gros Ventre and Assiniboine People] *hereby cede and relinquish to the United States all their right, title, and interest in and to all the lands embraced within the aforesaid* [lands] ... reserving to themselves *[emphasis added] only the reservation herein set apart for their separate use and occupation. (46)*
>
> Treaty between the United States and the Gros Ventre, Piegan, Blood, Blackfeet, and River Crow Tribes—May 1, 1888.

Other treaties between the federal government and Native American tribes also emphasized the intent of reserved rights:

> *... commencing at the mouth of the Red Creek or Red Fork of the Arkansas River; thence up said creek or fork to its source; thence westwardly to a point on the Cimarone River, opposite the mouth of Buffalo Creek; thence due north to the Arkansas River; thence down the same to beginning, shall be, and is hereby, set apart for the* absolute and undisturbed use *[emphasis added] and occupation of the tribes who are party to this treaty.*
>
> Treaty of the Little Arkansas between the United States and the Cheyenne and Arapaho Indian Tribes—October 14, 1865.

Since most Reservations were established to force tribal members to adopt farming practices, water rights awarded under the Winters Doctrine are usually substantial. In addition, most Reservations were established before irrigation projects were developed in the West, and most Winters Rights have very senior priority dates. (The term *Winters Rights* can be used to describe federal or Indian reserved rights.) *Winters* v. *United States* was the first time the federal government stated that Reservations had implicit water rights. Many tribes are continuing to litigate to affirm water rights on Reservations. Treaties are given the status of supreme law in the U.S. Constitution and preempt state law if there is a conflict. (47)

The Winters Doctrine later formed the basis for reserved water rights on federal lands such as national parks, forests, monuments, wildlife refuges, and military installations. The quantity and beneficial use of these water rights have been litigated extensively. In 1977, the U.S. Supreme Court ruled that these claims do not permit unlimited use of water resources on federal land. Rather, the Court held that the "implied reservation of water rights doctrine reserves only that amount of water necessary to fulfill the purpose of the reservation and no more." (48)

OUR ENVIRONMENT

In the next chapter, we will consider the powerful organizations that develop and distribute water resources throughout the United States. Many federal water agencies were created in the early 1900s for the primary mission of developing projects to deliver water for urban and irrigation uses. More recently, agencies such as the U.S. Environmental Protection Agency have been formed to protect

the environment from the negative impacts of such projects. How important is the protection of existing state water laws when considering implementation of goals and objectives for environmental protection? Should new, innovative water allocation schemes be explored to enhance the use of water while adequately protecting the environment? How would water user groups (such as irrigators) respond to a call for changes in existing water law if the new law negatively affected their operations? How might an irrigator respond to a demand by a state or the federal government to relinquish water rights to protect endangered wildlife downstream? Should existing water use practices take precedence over the needs of the environment?

GUEST ESSAY

Stream Subflow and Water Rights

by **John Regan, GIS Manager**
Pima County Department of Transportation
Tucson, Arizona

John Regan has 21 years of professional GIS experience and is currently a GIS manager for the Pima County Department of Transportation Technical Services Division Special Projects group. He works on various high-profile GIS projects, including the Sonoran Desert Conservation Plan, for the Board of Supervisors and County Administrator's Office. Prior to holding this position, John taught GIS at the University of Arizona's School of Renewable Natural Resources on both the graduate and undergraduate levels while managing a variety of GIS projects for the School's Advanced Resource Technology Lab. John holds a Master of Science degree in Regional Planning and a Bachelor of Arts in Anthropology, both from the University of Arizona.

Scientists have known for years that there is often a hydrologic connection between surface and groundwater, but water law has failed to recognize that relationship until recently. In the western United States, surface waters are subject to the doctrine of prior appropriation and beneficial use, while groundwater is not appropriable in most states and may be pumped by the owner of the land above the groundwater under the doctrine of reasonable use. An affirmation of the Gila River II case by the Arizona Supreme Court provided rules on what constitutes "subflow" of surface streams. Well owners meeting specific criteria while pumping groundwater from the saturated portion of the Holocene alluvium are using surface water and are subject to the Gila River II adjudication. Gila River II is summarized below. Language about chemical composition, water-level elevations, and gradations is left out because of lack of data to evaluate these conditions.

1999 Arizona Supreme Court, In re: The General Adjudication of All Rights to Use Water in the Gila River System and Source, 175 Ariz.382,384–86, 857 P,2d 1236, 1238–40 (1993), (Gila River II); affirmed.

1. "Subflow" shall be defined as the saturated floodplain Holocene alluvium and is adjacent to and beneath a perennial or intermittent stream.
2. There must be a hydraulic connection to the stream from the saturated "subflow" zone.
3. The alluvial plain must be of a perennial or intermittent stream and not an ephemeral stream or part of the alluvial plain of a tributary aquifer even if there is an alluvial connection.
4. No well located outside the lateral limits of the "subflow" zone will be included unless the "cone of depression" caused by its pumping has now extended to a point where it reaches adjacent "subflow" zones.

The Court left it up to the Arizona Department of Water Resources (ADWR) to determine specific parameters of the subflow zone, cones of depression diameters, and which wells may have a *de minimus* effect on the stream. If ADWR determines that a well is using subflow, the burden of proof is shifted to the well owner to show that it is either outside the subflow zone or is not pumping subflow. At this point, GIS may prove useful by contributing to the resolution of conflicts between a well owner and the findings of a state regulatory agency.

GIS can also visualize the intent of this legal ruling by producing maps and reports displaying the results of a spatial analysis based on Court-provided criteria. Some assumptions for the extent of the saturated zone and cones of depression will be made. It can be adjusted later when and if better data becomes available.

To begin the GIS analysis, perform the following steps: (1) frame the question; (2) locate and prepare the available data; (3) choose an approach for the analysis; (4) perform the analysis; and (5) interpret the results.

1. *Frame the question.* Which wells in this watershed are subject to the subflow ruling?
2. *Locate and prepare the data.* Four data layers are necessary: watershed, geology, streams, and wells. Determine what attributes are present in the data. Reselect specific features in the layers, essentially removing information not required for the analysis. This will simplify processing. The watershed layer will specify the drainage basin of concern, reducing the larger data sets to only the study area. The geologic feature of interest is the Holocene alluvium. Use only the perennial and intermittent types from the stream's layer. Well types are domestic and irrigation. Each well type will have a different cone of depression diameter. Reduce the area even further by clipping the geology, streams, and wells to just a rectangle around these features. See Figure 8.11 for the results of these operations.
3. *Choose an approach for the analysis.* Select a method of analysis based on the original question. A combination of buffering and overlay operations will be performed, and Boolean logic will be used to reselect the final information that answers the original question. In this case, a map of the wells affected by the subflow ruling will be the final product, and from this a list of the well registration numbers will be generated.
4. *Perform the analysis.* Execute the appropriate commands to answer the question being posed. Assume that the saturated zone of the younger alluvium extends 300 feet (91 m) on each side of the stream centerline. Buffer the stream layer 300 feet on each side. Next, reselect only domestic and irrigation wells and buffer those points. Assume that domestic wells have smaller pumps so their cone of depression will be 200 feet (61 m) in diameter, while the irrigation wells will have a cone of 400 feet (122 m). Figure 8.12 displays the

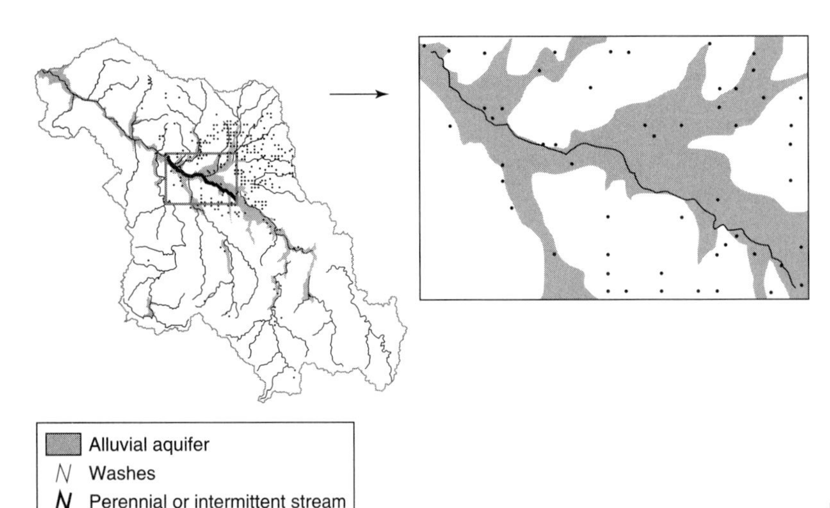

FIG. 8.11. Watershed clipped to stream boundary of the Gila River.

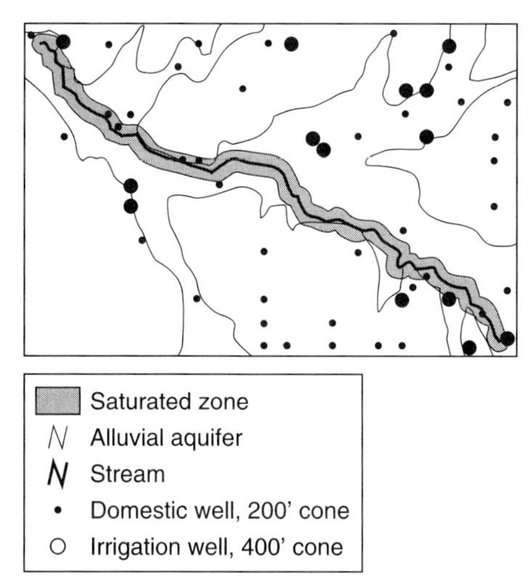

FIG. 8.12. Buffered stream and wells.

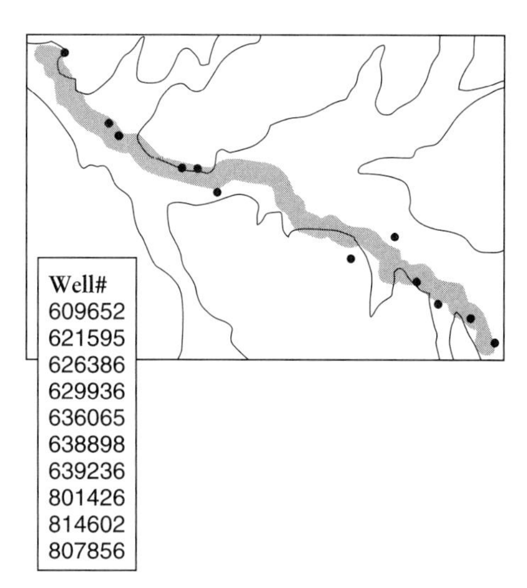

FIG. 8.13. Wells subject to subflow ruling of Gila River II.

result of these operations. The resulting layers are topologically combined, and the final subset of interest is reselected using the Boolean logic expression RESELECT WELL_INSIDE5100 and STREAM_INSIDE 5100 and ROCKS5'Qal'. This results in well buffers inside the stream buffer and touching the Holocene alluvium being included in the final selection. These wells will be considered impacted by the subflow ruling. See Figure 8.13 for the final results.

5. *Interpret the results.* Look closely at what came out of the analysis and make sure the results make sense. Never assume the software automatically gave you the correct answer. Notice that several wells seem to be outside the Holocene alluvium. Their cones touch the alluvium and the saturated zone, so that the wells are included. How the question was answered depends on how well the analysis was put together and, of course, on the accuracy and completeness of the data. Finally, generate a map from the new data layers and then extract the tabular data required for the report. Rerun the analysis as better data becomes available.

CHAPTER SUMMARY

The allocation of water resources has been a critical economic and societal issue since ancient times. Early in its development, water law focused on equity between water users. For example, an irrigator in ancient Babylonia was expected to reimburse a neighbor with a payment of corn if excess water flooded a neighbor's field. In England and France, property owners along rivers were allowed to utilize surface water as long as unused water was returned to the stream for use by others located downstream. In New England, mill owners were required by law to share in the power potential of water flowing past industrial areas. These legal systems were developed to provide a fair system of water use for irrigation, navigation, milling, and urban uses.

Today, court proceedings, legislative declarations, interstate compacts, and court decrees fill voluminous libraries with water allocation decisions. However, a completely new perspective has emerged in recent decades. Wildlife and related habitat, water quality, and recreation needs are becoming major issues in water law. Chapter 12 will discuss the issues of the environment and endangered species.

QUESTIONS FOR DISCUSSION

1. How did the Riparian Doctrine evolve from the Code of Hammurabi?
2. What role did climate play in the evolution of the Doctrine of Prior Appropriation?
3. What method of water allocation did the Mormon settlers of the Salt Lake Valley use in the mid-1800s?
4. Why were gold miners in California and Colorado concerned about water rights?
5. Contrast the early water allocation methods of the Mormon settlers in Utah to those of settlers of the Union Colony and Fort Collins in Colorado in the 1870s.
6. Explain the purpose of a water stock certificate in an irrigation company.
7. How do correlative rights contrast to the reasonable use principle?
8. Why are interstate compacts an important legal concept in the development and management of water resources between states?
9. What negative effects could the hydrologic cycle have on the operation of the Colorado River Compact?
10. Do you agree that states, and not the federal government, should establish groundwater laws used in the United States? How might groundwater law be different if the federal government was in control of its use?

KEY WORDS TO REMEMBER

California Doctrine p. 224
Code of Hammurabi p. 212
Code Napoléon p. 218
Coffin v. *Left Hand Ditch Company* p. 226
Colorado Doctrine p. 226
Correlative Rights Principle p. 229
diligence p. 231
Doctrine of Prior Appropriation p. 223
English Common Law p. 215
exchange p. 233
federal reserved water rights p. 239
"first in time, first in right" p. 223
French Common Law p. 215
interstate compact p. 236
Irwin v. *Phillips* p. 223
junior appropriator p. 224
Justinian Code p. 213
Maryland Mill Act (1669) p. 216
Massachusetts Mill Act (1714) p. 216
miner's inch p. 228
point of diversion p. 233
Pueblo water rights p. 215
Reasonable Use Principle p. 229
Recopilación de leyes de los Reynos de las Indias p. 214
Riparian Doctrine (Common Law of Water) p. 213
Rule of Absolute Ownership p. 234
Rule of Capture p. 234
salvaged water p. 233
senior appropriator p. 224
Tyler v. *Wilkinson* p. 219
usufructory property right p. 219
water right p. 223
Winters Doctrine (Indian Reserved Rights) p. 239

SUGGESTED RESOURCES FOR FURTHER STUDY

READINGS

Baxter, John O. *Dividing New Mexico's Waters, 1700–1912* Albuquerque: University of New Mexico Press, 1997.

Burton, Lloyd. *American Indian Water Rights and the Limits of Law.* Lawrence: University Press of Kansas, 1991.

Council of State Governments. *Interstate Compacts—1783–1956.* Chicago, Illinois, July 1956.

Dzurik, Andrew A. *Water Resources Planning.* 3rd ed. Lanham, MD: Rowman & Littlefield, 2003.

Grigg, Neil S. *Water Resources Management.* New York: McGraw-Hill, 1996.

Hunter, Louis C. *A History of Industrial Power in the United States, 1780–1930.* Volume One: *Waterpower in the Century of the Steam Engine.* Charlottesville: University Press of Virginia, 1979.

Radosevich, G. E., et al. *Evolution and Administration of Colorado Water: 1876–1976.* Littleton, CO: Water Resources Publications, 1976.

Rohrbough, Malcolm J. *Days of Gold: The California Gold Rush and the American Nation.* Berkeley: University of California Press, 1997.

U.S. Department of the Interior, Bureau of Indian Affairs. *Laws and Treaties.* Volume II. Washington, DC: U.S. Government Printing Office, 1904.

Wiley, Peter, and Robert Gottlieb. *Empires in the Sun.* New York: G. P. Putnam's Sons, 1982.

Worcester, Donald E. *Forked Tongues and Broken Treaties.* Caldwell, ID: Caxton Printers, 1975.

WEBSITES

"Water Rights," State Water Resources Control Board, California Environmental Protection Agency, June 2003. http://www.waterrights.ca.gov/WRINFO

"Colorado Water Knowledge," Colorado State University, February 2002. http://www.waterknowledge.colostate.edu

"The Water Page," June 2003. http://www.africanwater.org/rivers_regions.htm

"Colorado River Project," Water Education Foundation, March 2004. http://www.water-ed.org/coloradoriver.asp

REFERENCES

1. Driver, G. R., and John C. Miles (trans.). *The Babylonian Laws.* Volume II. London: Clarendon Press, 1955, 5.

2. Ibid., 27–29.

3. Ibid., 16.

4. Kolbert, C. F. *Justinian—The Digest of Roman Law.* Harmondsworth, Middlesex, England: Penguin Books, Harmondsworth, 1985, 7.

5. Ibid., 9.

6. Sandars, Thomas Collett. *The Institutes of Justinian with English Introduction, Translation, and Notes.* Westport, CT: Greenwood Press, 1970, 91–92.

7. Matthews, Olen Paul. *Water Resources—Geography & Law.* Washington, DC: Association of American Geographers, 1984, 24.

8. *Columbia Electronic Encyclopedia.* New York: Columbia University Press, wysiwyg://53/http://infoplease.lycos.com/ce6/people/A0841292.html, December 2000.

9. Guim, Juan B. *Diccionario razonado de legislación y jurisprudencia, nueva edición . . . con nuevos articulos . . . sobre el derecho americano . . .* Paris and Mexico: Libreria de Ch. Bouret, 1888. Jane C. Sanchez, *Law of the Land Grant: The Land Laws of Spain and Mexico,* http://home.sprintmail.com/sanchezj/1-title.htm, August 2001.

10. Ibid.

11. Matthews, 39–40.

12. Hunter, Louis C. *The History of Industrial Power in the United States, 1780–1930,* Volume 1: *Waterpower in the Century of the Steam Engine.* Charlottesville: University Press of Virginia, 1979, 33–35.

13. Ely, James W., Jr. *Property Rights in American History—From the Colonial Period to the Present.* New York: Garland Publishing, 1997, 109–111.

14. Ibid., 109.

15. Massachusetts General Laws, Chapter 75, paragraph 1 (1796), http://www.courts.state.me.us/98me202A.htm, December 2000.

16. Ely, 126.

17. Hunter, 142.

18. Matthews, 25.

19. Hunter, 145. A portion of the text (145, lines 14–15) was reprinted with permission from Hagley Museum and Library, Wilmington, Delaware.

20. Matthews, 25.

21. Hunter, 145.

22. See *Tyler* v. *Wilkinson,* 24 F. Cas. 472 (D.R.I. 1827) No. 14, 312.

23. Hunter, 157–158.

24. Ibid., 120–121.

25. Brackenridge, Henry M. *Views of Louisiana.* Baltimore, MD: Schaeffer & Maund, 1817.

26. Carr, Donald E. *Death of Sweet Waters.* New York: W. W. Norton, 1966, 83–84.

27. Dunbar, Robert G. *Forging New Rights in Western Waters.* Lincoln: University of Nebraska Press, 1983, 12.

28. Mather, John R. *Water Resources—Distribution, Use, and Management.* New York: John Wiley & Sons, 1984, 279–280.

29. Rohrbough, Malcolm J. *Days of Gold—The California Gold Rush and the American Nation.* Berkeley: University of California Press, 1998, 7–16.

30. Monaghan, Jay. *Chile, Peru, and the California Gold Rush of 1849.* Berkeley: University of California Press, 1973.

31. Paul, Rodman W. *California Gold—The Beginning of Mining in the Far West.* Lincoln: University of Nebraska Press, 1947, 45–48.

32. *Irwin* v. *Phillips,* Supreme Court of California, 5 Cal. 140, 1853.

33. King, Joseph E. *A Mine to Make a Mine—Financing the Colorado Mining Industry, 1859–1902.* College Station: Texas A&M University Press, 1977, 3–6.

34. Mather, 285.

35. Cox, Terry. *Inside the Mountains—A History of Mining Around Central City, Colorado.* Boulder, CO: Pruett Publishing Co., 1989, 18.

36. Boyd, David. *A History of Greeley and the Union Colony.* Greeley, CO: 1890.

37. Ibid.

38. *California Oregon Power Company* v. *Beaver Portland Cement Company,* 295 U.S. 142 (1935).

39. *Coffin* v. *Left Hand Ditch Company,* 6 Colorado 433 (1882).

40. Mather, 281.

41. Ibid., 289–290.

42. New York Power Authority, "Niagara River Water Diversion Treaty," http://niagara.nypa.gov/info/niagarariver_waterdiversion.htm, June 23, 2003.

43. *Winters v. United States,* U.S. Supreme Court, (45) 207 U.S. 564 (1908). Also see http://ceres.ca.gov/theme/env_law/water_law/cases/Winters_v_US.html, August 2001.

44. Burton, Lloyd. *American Indian Water Rights and the Limits of Law.* Lawrence: University Press of Kansas, 1991, 18–19.

45. Ibid., 20.

46. Agreement with Indians of the Gros Ventre, Piegan, Blood, Blackfeet, and River Crow Tribes, Montana, Acts of the Fiftieth Congress, First Session, Chapter 213, 25 Statute 113, Article 1, May 1, 1888.

47. Cohen, F. *Handbook of Federal Indian Law.* New York: AMS Press, 1972.

48. *United States* v. *New Mexico,* 38 U.S., 696, 700 (1977).

CHAPTER 9

FEDERAL WATER AGENCIES

U.S. ARMY CORPS OF ENGINEERS (USACE)
U.S. BUREAU OF RECLAMATION (USBR)
U.S. GEOLOGICAL SURVEY (USGS)
U.S. FISH & WILDLIFE SERVICE (USFWS)
NATIONAL PARK SERVICE (NPS)
BUREAU OF LAND MANAGEMENT (BLM)
U.S. ENVIRONMENTAL PROTECTION AGENCY (USEPA)
NATURAL RESOURCES CONSERVATION SERVICE (NRCS)
U.S. FOREST SERVICE (USFS)
FEDERAL ENERGY REGULATORY COMMISSION (FERC)
NATIONAL MARINE FISHERIES SERVICE (NMFS)
FEDERAL EMERGENCY MANAGEMENT AGENCY (FEMA)

In the early years of the country, before we had railroads, telegraphs and steamboats—in a word, rapid transit of any sort—the States were each almost a separate nationality. . . . But the country grew, rapid transit was established, and trade and commerce between the States got to be so much greater than before, that the power of the National government became more felt and recognized and, therefore, had to be enlisted in the cause of the institution. (1)

Ulysses S. Grant, 1885

The relationship of water resources development and management as viewed by the federal and individual state governments has been marked by transitions of power between local and federal control. In the early years of the nation, in the late 1700s and early 1800s, the federal government had only a limited role in water development. Some engineering was performed to improve navigation on major rivers, but federal involvement in flood control was minimal. Water allocation law was developed by individual states, and water development projects were generally funded by local landowners or outside investment groups.

As the need for irrigation projects increased in the western United States, and especially as devastating floods swept away entire communities in both humid and dry locations, settlers called for the federal government to assist with large-scale water projects. These projects varied from flood-control structures along the Mississippi and Ohio rivers to massive irrigation projects along the Colorado and Columbia rivers. The population of the United States was expanding, creating the need for a larger federal role in water resources development and management.

This chapter focuses on the primary federal agencies involved in water resources development, management, and protection in the United States. The discussion of each agency includes a brief overview, historical perspective, and current policy issues. As you read, notice how federal legislation evolved through time. Follow the changing roles of water resources administration—from the development era of the early 1900s to increased environmental protection after the 1960s. Also notice how, with time, agencies reorganized and some duties began to overlap (Figure 9.1).

SIDEBAR

For an excellent listing of federal water resources employment opportunities and salary ranges, go to http://www.awra.org/impact/jobs/html

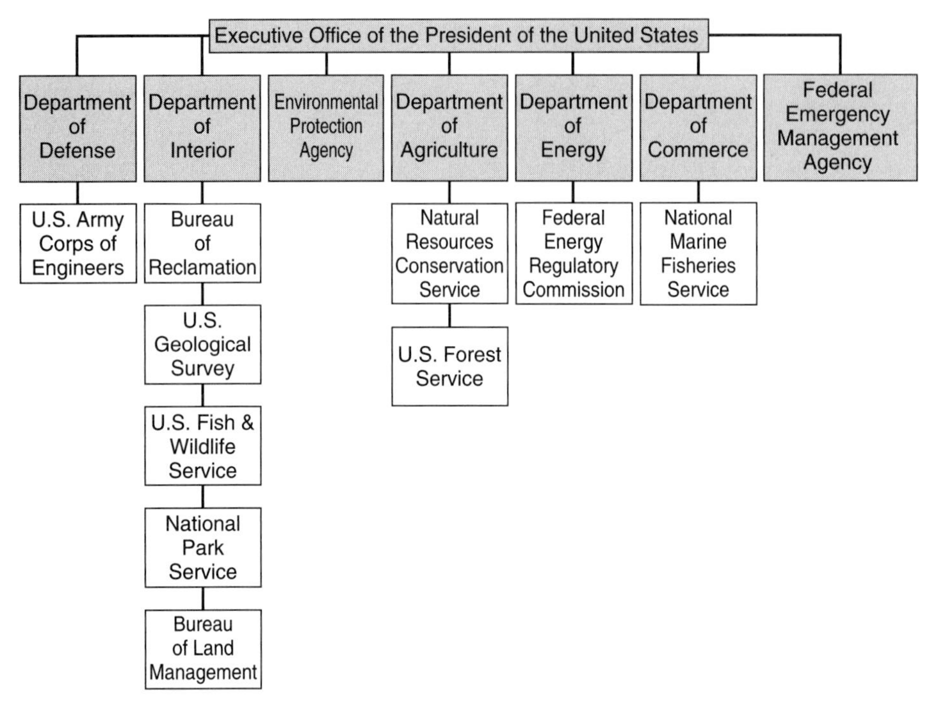

FIG. 9.1. Selected water resources agencies of the federal government.

U.S. ARMY CORPS OF ENGINEERS (USACE) http://www.usace.army.mil

OVERVIEW

The **U.S. Army Corps of Engineers** (also called the **Corps**, or **USACE**) is the nation's oldest water resource agency and is located within the U.S. Department of Defense. Its primary water resources activities are flood control and navigation improvement, although wetlands protection and environmental restoration are recent additions to its mission. Most employees of the Corps are civilians, but they are overseen by U.S. Army officers. In the past, the Corps has routinely constructed dams, boat locks, dikes, jetties, and other navigational and riverbank protection projects. These improvements in flood protection and commerce enhancement have saved billions of dollars. However, the Corps has sometimes been criticized for the cost, both financial and environmental, of many of these construction projects.

BRIEF HISTORY AND DUTIES

The U.S. Army Corps of Engineers can trace its history to the Continental Congress in 1775 when America's first chief army engineer, Colonel Richard Gridley (1710–1796) of Massachusetts, was authorized to build fortifications at Breed's Hill (later the site of the Battle of Bunker Hill near Boston). (2) After the Revolutionary War, President Jefferson sent emissaries from the United States to France. There, they studied engineering and management techniques at the Ecole Polytechnique in Paris (the same school attended by Henry Darcy of groundwater renown discussed in Chapter 4), and then returned to America with new engineering ideas for adoption. In 1802, Congress created the U.S. Military

Academy at West Point and placed the U.S. Army Corps of Engineers in charge of training the nation's new federal engineers.

In 1808, Secretary of the Treasury Albert Gallatin submitted the **Gallatin Report** to Congress to identify a plan for direct federal involvement in the construction of national roads (the early precursor to a national interstate highway system) and the development of federally funded waterways. In addition to building bridges, national roads, and Army outposts in the West, the Corps removed snags and built levees to improve river navigation on eastern rivers.

The federal government's role in water management was unclear in the early 1800s. What little flood control existed took the form of primitive, privately funded levees built along the lower Mississippi River in Louisiana. No coordination existed, and a dike constructed along one side of the river often flooded nondiked lowlands (or lower elevation dikes) on the opposite side. Since the Mississippi and other navigable rivers were used for interstate commerce, many argued that the federal government should change its policy and assist with the cost of maintaining navigation channels. Any benefit, even if only local, it was argued, would contribute to the national prosperity. (3) However, others debated that it was unfair to use federal funds for the parochial benefit of smaller river basins such as the Ohio River. Debate continued regarding the appropriate role of the federal government in navigation and flood control.

Navigation Duties As early as 1726, local farmers and communities attempted to effect navigation improvements along the Mississippi River. Methods were crude, such as dragging iron harrows across sandbars to promote scouring during high river flows. Flooding problems continued, but federal involvement was negligible for nearly a century. Then, in 1824, the U.S. Supreme Court ruled in *Gibbons* v. *Ogden* that the federal government had supremacy over all interstate navigable waters in the United States, and should encourage and protect interstate river commerce. That same year, Congress adopted the **General Survey Act** for the survey of roads and canals of national importance. This established the permanent involvement of the U.S. Army Corps of Engineers in inland waterway improvements. Also in 1824, Congress provided $75,000 for federal navigation improvements along the Ohio and Mississippi rivers. This was one of the first federal appropriations for a U.S. Army Corps of Engineers navigation project. (4) Unlike earlier appropriations, which dealt with harbor projects, this appropriation provided funds to improve and maintain inland waterways. This was the beginning of the Corps' permanent involvement in inland river improvements.

Of primary concern were the mouth of the Mississippi River and the reaches of the Mississippi, Missouri, and the Ohio rivers that ran over shoals, particularly around Louisville, Kentucky, and just upstream of St. Louis, Missouri. These areas continually filled with sediment and other obstructions because gradients decreased and the river's velocity slowed as it entered these locations. It has been estimated that more than 400 steamships sank on the Missouri River alone during the 1800s, with half of them hitting snags. The average life of a nineteenth-century paddle wheeler was less than two years. (5)

Finally, at the end of the 1800s, the **Rivers and Harbors Act of 1899** (which included Section 13, called the **Refuse Act**) gave the U.S. Army Corps of Engineers authority to regulate and control all construction along navigable rivers in the United States. The Act also provided the Corps with a new role by making it a crime to dump refuse into navigable streams without the Corps' permission. This ban on dumping was specifically intended to prevent obstructions, such as garbage, in waterways used for navigation. Although this law directed a major shift in responsibility, the Corps' ability to interpret the Act as a public health or wetlands protection measure was limited by early rulings of the U.S. Supreme Court. It was not until 1973 that the Court ruled that pollution protection was a proper function of the 1899 federal law. (6)

TABLE 9.1 Domestic Traffic for Selected U.S. Inland Waterways, 2001

Waterway	Location	Tons (millions of short tons)
Mississippi River and tributaries[a]	Central U.S.	1051.8
Ohio River	Central U.S.	242.5
Gulf Intracoastal Waterway	Gulf Coast	112.2
Tennessee River	Southeastern U.S.	47.9
Columbia River and tributaries	Idaho, Oregon, and Washington	40.0
Monongahela River	Pennsylvania and West Virginia	38.1
Morgan City—Port Allen route	Louisiana	23.3
Black Warrior and Tombigbee rivers	Alabama	18.9
Missouri River	Iowa, Kansas, Nebraska, and Missouri	9.7
Tennessee—Tombigbee Waterway	Alabama and Mississippi	6.8
Atlantic Intracoastal Waterway	Atlantic Coast	2.5

[a]Excludes the Missouri, Monongahela, Ohio, and Tennessee rivers.

Source: U.S. Army Corps of Engineers, Navigation Data Center, Alexandria, Virginia, http://www.iwr.usace.army.mil/ndc/wcsc/wtwytraffic.htm, February 2004.

The Rivers and Harbors Act continued federal funding of navigation improvements that had been ongoing since the end of the Civil War. Since most navigable streams existed in the humid eastern states, budget requests for the U.S. Army Corps of Engineers were strongly supported by eastern lobbyists and legislators. A group called the **National Rivers and Harbors Congress** was formed in 1901 to promote navigation projects east of the 100th Meridian. Their goal was to increase federal spending to improve navigation for barge traffic.

Today, navigation occurs in 41 states in the United States and in all states east of the Mississippi River. The Corps owns or operates navigation locks at 230 sites, of which 46 include hydropower generation. Table 9.1 shows the tonnage for the busiest waterways in the United States, and Table 9.2 presents the major commodities transported. Figure 9.2 presents a

TABLE 9.2 U.S. Waterborne Traffic by Major Commodities, 2001

Commodity	Tons (millions of short tons)	Percent
Petroleum and petroleum products	369.7	35
Coal	227.7	22
Crude materials	213.7	20
Food and farm products	96.5	9
Chemical and related products	71.1	7
Primary manufactured goods	40.3	4
All manufactured equipment	19.9	2
All other	3.5	1
TOTAL	1042.5	100

Source: U.S. Army Corps of Engineers, Navigation Data Center, Alexandria, Virginia, http://www.iwr.usace.army.mil/ndc/wcsc/pdf/wcusnat/01.pdf, February 2004.

SIDEBAR

The busiest lock in the United States is the Ohio River Lock Number 52 in Illinois, which moved 96 million tons (87.1 metric tons) in 2001. The locks that serviced the greatest numbers of pleasure craft in 2001 were the Hiram M. Chittenden Locks at Seattle, Washington (48,646 vessels) and the Chicago Lock at Chicago, Illinois (35,961 vessels). Oregon's John Day Lock has the highest lift at 110 feet (34 m). This compares to the collective 404 feet (123 m) of lift provided by all 29 locks on the upper Mississippi River. (7)

typical U.S. Army Corps of Engineers navigation chart.

A CLOSER LOOK

The aging of the nation's navigation system has created controversy for federal water and environmental policies. Over half of the locks operated by the U.S. Army Corps of Engineers have exceeded their 50-year design lives and will require minor repairs, rehabilitation, or, in some cases, total replacement. The oldest operating locks in the United States are Kentucky River Locks 1 and 2, built in 1839.

Numerous groups have argued that old dam and lock systems should be permanently removed or significantly altered to restore historic wetlands and wildlife habitat. (8) Others have countered that navigation projects have saved lives and played a crucial role in establishing the infrastructure needed for the nation to grow and prosper. (9) This is the same, classic debate that has arisen over the removal of dams (see Chapter 7).

What is your opinion? If you were a member of Congress, what additional information would you need to decide if additional funding should be given to the U.S. Army Corps of Engineers to rehabilitate an aging lock and dam system? What alternatives should be considered?

Flood-Control Duties The Mississippi River flooded extensively in 1849 and 1850, leading ultimately to the creation of the Mississippi River Commission in 1879 to coordinate flood-control and navigation activities. Major floods occurred again in 1912 and 1913, and clearly proved the need for increased federal involvement in the region. Congress responded by approving the first federal flood-control act in 1917 for protection along the lower Mississippi

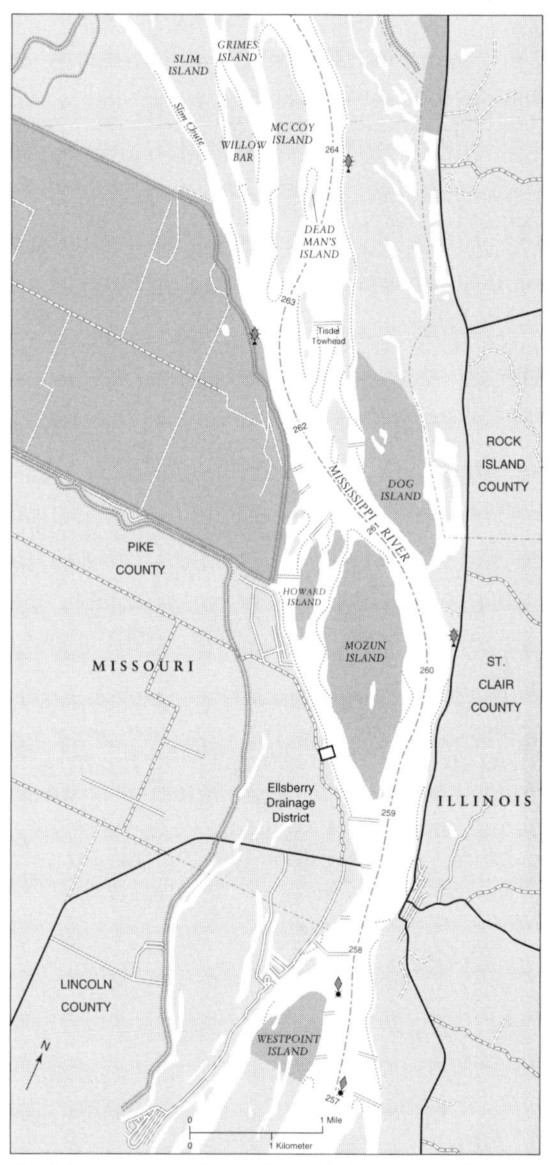

FIG. 9.2. Navigation charts of the U.S. Army Corps of Engineers like this one, which shows an area north of St. Louis, Missouri, provide barge captains and others with a clear map for safe passage through the nation's waterways (see http://www.mvr.usace.army.mil/navdatas).

The diamond-shaped symbols along the river represent locations of light- and day-marks as well as lighted buoys that serve as navigational aids. Numerous jetties can also be seen extending from the left and right riverbanks that direct flows toward the thalweg (deepest channel) of the river. Note the numerous braided channels of the Mississippi in this area.

River and the Sacramento River in California. The Sacramento River was a special case in the West because of continual problems with flooding. Congressional approval for the Sacramento River was based on a flood-control plan already developed by the Corps and the State of California in 1910–1911. (10)

Federal funding for flood control was increasing, but it took the Great Mississippi River Flood of 1927 (Figure 9.3) to elevate the federal government's flood protection activities to a higher level. The peak of the great flood exceeded 2.5 million cubic feet per second (1.6 trillion gal per day or 70,792 cms), which was over twice as great as the peak of the Mississippi River flood in 1993, which carried about 1 million cfs (646.3 billion gal per day or 28,317 cms). Floodwaters killed over 200 people in 1927 and left approximately 600,000 homeless. Responding to the enormous public outcry, Congress passed the Flood Control Act of 1928. (11)

The **Flood Control Act of 1928** included channel improvements and levee construction along the Mississippi River and its tributaries. The Act also directed the Corps to investigate the construction of dams across tributaries of the Mississippi to control upstream flood flows. In addition, the new law authorized the federal government to pay substantial costs for maintaining all levees along the Mississippi River. Prior to passage of the Flood Control Act of 1928, Congress disguised flood-control benefits as navigational improvements in order to appease conservative members who continued to question the constitutionality of federal involvement in flood control. (12)

SIDEBAR

In 1933, the Tennessee Valley Authority (TVA) became the agency responsible for flood control and navigation along the Tennessee River in the economically depressed southeastern United States. In addition, the TVA was authorized to consider social programs to reduce poverty, improve sanitation, create education programs, and promote economic development. The Tennessee Valley was a region of low-income, limited jobs and experienced frequent flooding. Since the U.S. Army Corps of Engineers was handling flood control on the main stem of the Mississippi River, President Franklin D. Roosevelt and Congress believed it was a logical step to try a new approach on a tributary. The TVA was an American experiment in social reorder that reflected changing federal programs during the drought, Depression, and New Deal Era of the 1930s.

FIG. 9.3. Refugees huddle in tents pitched on the levees at Greenville, Mississippi. A street can be seen fronting the levee. Almost the entire town was submerged by Mississippi River waters during the tragic flood events of 1927. See "Fatal Flood," *American Experience,* Public Broadcasting Service (PBS), at http://www.pbs.org/wgbh/amex/flood/filmmore/index.html for an excellent resource on the local and federal response to one of America's greatest natural disasters.

The **Flood Control Act of 1936** expanded the Flood Control Act of 1928 by recognizing that flood control was an appropriate federal responsibility nationwide, and it directed the Corps to do most of the work. In addition, it required the U.S. Army Corps of Engineers to use benefit-cost analysis (see Chapter 7). A flood-control project was required to have a positive benefit-cost ratio or Congress would not provide funding.

A CLOSER LOOK

Poor cost estimates plagued the U.S. Army Corps of Engineers for decades and continue to do so even today. Inaccurate estimates are often caused by inflation that occurs during the course of construction and is exacerbated if federal appropriations from Congress are late. A terse exchange in 1961 between U.S. House of Representatives Appropriation Committee Chairman Clarence Cannon of Missouri and Colonel Clarence Renshaw of the Corps provides insight into the political animosity that can exist between politicians and federal agencies. The following discussion regards an inland waterway from the Delaware River to the Chesapeake Bay of Maryland:

CHAIRMAN CANNON: "Colonel Renshaw, why are you asking for $1 million more for this project this year than you asked last year, and why are you asking $2 million more this year than you asked for 2 years ago, and may we expect that by next year you will be asking for $3 million more?

COLONEL RENSHAW: First, let me say, sir, we believe that the estimate we now have will complete the project.

CHAIRMAN CANNON: You made jumps of $1 million a year, and there is certainly a wide difference when you increase 2 years in succession at the rate of $1 million a year. There must be something wrong here. (Discussion off the record.)

CHAIRMAN CANNON: There is poor engineering to begin with, or there is some factor in it that is not at first apparent to the Committee.

COLONEL RENSHAW: Of the $1 million increase this year over last year, $375,000 of that was due to change in conditions in the foundations.

CHAIRMAN CANNON: About one-third is due to changes in the foundation?

COLONEL RENSHAW: Yes, sir . . .

CHAIRMAN CANNON: . . . There seems to be a lot of loose engineering here." (13)

As is evident in this exchange, underground foundation conditions for dams are extremely hard to predict and often result in unforeseen problems and expenses. Detractors of dam projects can use cost overruns in these situations as "political ammunition" in an attempt to make a federal agency "look bad."

The 1936 Act also authorized the U.S. Department of Agriculture to investigate small, upstream nonstructural flood-control practices that would reduce erosion and flooding. Such practices included planting cover crops like alfalfa, clover, or hay on hillsides to reduce erosion and encourage groundwater recharge. The involvement of another federal agency in flood control created political tensions that continued into the 1950s. Each year, representatives from the Corps and the Department of Agriculture fought for federal appropriations for agency projects. The Department of Agriculture believed that upstream land conservation practices were just as beneficial as the construction of large, on-channel dams. The Corps naturally disagreed. Congressional disputes were common as the two federal agencies tried to align political support for their conflicting flood-control agendas. (14)

Major floods occurred again in 1943 and 1944, and in response Congress passed the **Flood Control Act of 1944.** Interagency river basin commissions were later created for the Missouri, Columbia, Arkansas, and others, although these agencies did not have the same authority or scope as the Mississippi River Basin Commission created in 1879. (15)

The Flood Control Act of 1944 included authorization for the construction of the controversial Garrison Dam across the Missouri River in North Dakota. The dam site had been investigated earlier by the Bureau of Reclamation for an irrigation project, but it was determined to be uneconomical for that purpose. However, the U.S. Army Corps of Engineers, believing that flood control would provide a positive benefit-cost ratio, successfully lobbied for inclusion of the project in the 1944 Act. (16) History had been very cruel to Native Americans in the

FIG. 9.4. Secretary of Interior J. A. Krug signs the contract on May 20, 1948, whereby the Fort Berthold Indian Tribal Business Council sold 155,000 acres (62,727 ha) of reservation land in North Dakota for construction of the Garrison Dam and Reservoir Project. George Gillette (left foreground), chairman of the tribal council, covers his face and weeps. Gillette said in a signed statement, "The members of the tribal council sign this contract with heavy hearts. Right now the future doesn't look good to us."

vicinity of the proposed Garrison Dam. In 1851, the U.S. government signed a treaty with the Three Tribes (Arikaras, Mandan, and Hidatsa) that gave a 19,500 square mile (50,505 km^2) reservation of North Dakota land in perpetuity to the tribes. It was called the Fort Berthold Reservation, but it was later reduced to only 1005 square miles (2603 km^2) through changes largely unknown by and without the consent of the Three Tribes. (17)

The Garrison Dam Project was completed in 1956. The dam is 11,300 feet (3444 m) long and 210 feet (64 m) high, and was the largest earthen dam in the world at the time. The dam created Lake Sakakawea, the third largest reservoir in the United States in terms of storage volume at 23.8 million acre-feet (7.8 trillion gal, or 29.4 billion m^3). The lake flooded most of the productive land within the Fort Berthold Reservation. Native American homes were relocated, and the U.S. government initially paid the tribes $5,105,625 for the 242 square miles (627 km^2) of flooded property. Later this figure was increased to $12.5 million, a little over $80 per acre ($200 per ha). Tribal members were distraught over the sale of their ancestral hunting grounds (see Figure 9.4) but could do little to stop the federal government. (18)

Wetlands Protection Duties Congress approved the **Louisiana Swamp Land Act** in 1849 to encourage draining and cultivation of lowland wetlands. The federal government granted the State of Louisiana all swamp and overflowed lands within the state to aid in the reclamation of those lands for cultivation and settlement. The following year, the **Swamp Lands Act of 1850** was approved to extend the law to other public-land states in the Union. Ten years later, Congress enacted the **Swamp Lands Act of 1860** to extend the Swamp Lands Act even farther to Minnesota and Oregon. These federal laws were intended to reduce waterborne diseases, increase cultivation of fertile farm ground, and improve construction opportunities for roads by providing federally owned swamp land to states. These properties were then sold to people willing to drain and reclaim land for cultivation or settlement.

FIG. 9.5. *L'eau Mare Mensonge* (Water Lily Pool) by Claude Monet, 1899, shows his wetland complex at Giverny, located near the Seine River about 40 miles (64 km) northwest of Paris. The contrast between Monet's artistic exploration of wetlands as an Impressionist versus the U.S. Congress's destruction of wetlands through the various Swamp Land Acts (both occurring in the mid- to late 1800s) is striking. Monet utilized tributaries of the Seine River to maintain a constant flow of water into his lily pool garden. (Natural stagnant conditions in the pond inhibited aquatic plant growth.) Monet was required to first obtain authorization for diverting water from the Ru from the Giverny Town Council. See *Monet's Giverny*, by William H. Gerdts (New York: Abbeville Press, 1993), for more on Monet and his wetland garden.

This mid-nineteenth century American policy of destroying wetlands along rivers and floodplains (contrast with Monet's accomplishment in France, Figure 9.5) provided a significant economic benefit to riparian landowners with wetlands. Other policy goals were to stop the rampant spread of disease such as malaria and typhoid, to encourage economic growth, and to promote settlement west of the Mississippi River.

A century later, federal public policy toward wetlands was completely reversed. In 1972, the **Clean Water Act Amendments** (these were amendments to the Clean Water Act of 1948) were approved by Congress and addressed water quality and wetlands protection. The U.S. Army Corps of Engineers and the U.S. Environmental Protection Agency were given the responsibility, under **Section 404** of the Act, to protect the nation's wetlands. The Act requires the Corps to review all development plans that could alter or destroy wetlands on public or private property. In addition, anyone who wished to discharge dredged or fill material into the navigable waters of the United States was required to obtain an appropriate permit from the U.S. Army Corps of Engineers. Nationwide permits, issued by the Corps under Section 404, are a common umbrella permit issued for construction and farming activities in wetlands where only minimal adverse effects to the aquatic environment will occur. In other situations, replacement wetlands can be required to mitigate damaged or destroyed wetlands. (For a listing of major events in the history of the USACE protection of wetlands and other activities, see Table 9.3.)

POLICY ISSUE

Wetlands responsibilities are of critical importance to the U.S. Army Corps of Engineers. However, a 1998 court ruling created uncertainty regarding the Corps' jurisdictional rule in wetlands protection. Section

TABLE 9.3 Major Events Affecting the Role of the U.S. Army Corps of Engineers in Water Resources Management

Year	Event
1775	Continental Congress authorizes the chief army engineer to construct fortifications at Breed's Hill.
1808	Gallatin Report encourages federal involvement in waterway development.
1824	U.S. Supreme Court rules federal government has supremacy over interstate navigable waters in *Gibbons* v. *Ogden*.
1824	**General Survey Act** appropriates $75,000 for navigation improvements along the Ohio River.
1849, 1850, and 1860	**Swamp Land Acts** passed to encourage draining of wetlands in floodplains.
1879	**Mississippi River Commission** formed.
1899	**Rivers and Harbors Act** (Section 13 included the Refuse Act) passed by Congress to prevent obstructions to navigation caused by garbage dumping.
1901	Rivers and Harbors Congress formed.
1917	**Flood Control Act of 1917** passed by Congress to protect the lower Mississippi River and the Sacramento River in California.
1928	**Flood Control Act of 1928** passed by Congress and directed the Corps to construct flood-control structures on the Mississippi River and its tributaries. This flood-control plan is still substantially used today.
1936	**Flood Control Act of 1936** passed by Congress. Required the use of benefit-cost analysis. Also established the first designation of flood control as a national duty for the U.S. Army Corps of Engineers. It also included involvement of the U.S. Department of Agriculture's Soil Conservation Service (today called the Natural Resources Conservation Service).
1944	**Flood Control Act of 1944** passed by Congress. Created interagency river basin commissions.
1972	**Clean Water Act Amendments** give the Corps responsibility for protecting the nation's wetlands with 404 permits.
1993	The USEPA adopts the Tulloch Rule for wetlands protection.
1997	The U.S. District Court for the District of Columbia rules the USEPA exceeded its authority under the Clean Water Act to regulate "incidental fallback" of dredged materials from wetlands.
1998	The U.S. Court of Appeals for the District of Columbia Circuit affirmed the district court's decision regarding "incidental fallback."
2001	The Clinton Administration finalized a rule to reestablish USEPA/USACE jurisdiction over all dredge and fill activities in wetlands.
2001	The Bush Administration further refines wetlands regulations.

404 of the Clean Water Act authorizes the U.S. Army Corps of Engineers to issue permits for the discharge of dredged or fill material into waters of the United States. On August 25, 1993, the U.S. Environmental Protection Agency (USEPA) issued a regulation called the Tulloch Rule (58 FR 45008), which defined the term *discharge of dredged materials*. This broad regulation was challenged by various development groups, including the American Mining Congress and other trade associations.

On January 23, 1997, the U.S. District Court for the District of Columbia ruled that the USEPA had exceeded its authority under the Clean Water Act to regulate "incidental fallback" of dredged materials (soil or other material that inadvertently fell into waters of the United States during construction activities in or adjacent to a wetland). The federal government appealed the ruling, and on June 19, 1998, the U.S. Court of Appeals for the District of Columbia Circuit affirmed the district court's decision. On January 10, 2001, the Clinton Administration finalized a rule to reestablish USEPA/USACE jurisdiction over all dredge and fill activities in wetlands. The new rule redefined "discharges of dredged or fill material" and provided guidelines for mechanized activities in wetlands. (19) The Bush Administration further defined these guidelines in 2001.

U.S. BUREAU OF RECLAMATION (USBR)

http://www.usbr.gov

OVERVIEW

The **Bureau of Reclamation** (also called **USBR, Reclamation,** or **BuRec**) is located within the **U.S. Department of Interior** (also called **DOI** or **Interior**). DOI was created in 1849 to administer federal land sales, Indian affairs, and military pensions but soon became involved in emigration routes and geographic surveys of the West. The agencies under DOI—especially the Bureau of Reclamation—evolved to become the primary developers of water resources west of the 100th Meridian in the twentieth century.

The Bureau of Reclamation has been the nation's premier water development agency since the early 1900s. Its primary responsibility has been to develop irrigation projects to promote settlement of the arid western states. The USBR constructed over 200 irrigation projects during the twentieth century, including major construction efforts such as Hoover and Grand Coulee dams (Figure 9.6). Such water storage projects produce irrigation water for agriculture, electrical power for economic development, and water supplies for urban areas.

Investment for completed USBR projects totals approximately $11.0 billion and provides irrigation, household, and industrial water to over 30 percent of the population in the West. About 5 percent of the land area west of the 100th Meridian is irrigated, with USBR projects providing irrigation to about 20 percent of those lands (approximately 15,625 square miles or 40,469 km^2). The agency has 58 power plants on-line and generates over 42 billion kilowatt-hours of electricity annually. (20) The USBR is the nation's second largest producer of hydroelectric power. Water stored in USBR reservoirs equals 245 million acre-feet (79.8 trillion gal, or 302.2 billion m^3), which produces 60 percent of the nation's vegetables and 25 percent of the country's fruits and nuts. (21)

BRIEF HISTORY

John Wesley Powell (1834–1902), a nineteenth-century explorer, educator, scientist, and philosopher, is widely considered to be the father of the Bureau of Reclamation (Figure 9.7). Powell was a professor of geology at Illinois Wesleyan University and later led expeditions down the Colorado. On May 24, 1869, he and a nine-member crew left Green River, Wyoming, to explore the lower reaches of the Colorado River. Public excitement was high over the prospect of exploring the unknown canyons of the Colorado River. (The Union Pacific Railroad even transported Powell's wooden boats from Chicago, where they were made, to Green River at no charge.) After 99 days on the Green and Colorado rivers, and a harrowing journey through the Grand Canyon, Powell and his crew reached the mouth of the Virgin River. That location today is at the bottom of Lake Mead behind Hoover Dam.

Powell's second float trip down the Colorado River in 1871 resulted in the first-ever topographic map of the region and an extensively

FIG. 9.6. U.S. Bureau of Reclamation Projects. Irrigation water supply is the primary purpose of these dam, reservoir, and canal projects, although flood control and hydropower are added benefits. Note the relative absence of Bureau of Reclamation projects east of the 100th Meridian. Go to http://www.usbr.gov/dataweb/projects/ for an excellent overview of all Bureau dams, projects, and powerplants.

FIG. 9.7. John Wesley Powell, Civil War Union officer and Illinois geology professor.

documented account of his journey. Powell became a national hero and wrote the famous *Report on the Lands of the Arid Region of the United States for Congress* in 1878. In his report, he discussed the potential for water resources development and argued that irrigation was the key to settlement in this arid region. He encouraged Congress to construct large-scale irrigation storage and delivery projects since farmers would not have adequate funds to build such massive irrigation systems. Powell further argued that the federal government must take a leadership role in irrigation water delivery to farms and should create federal irrigation districts in the West. (22)

In his *Report on the Lands of the Arid Region of the United States*, Powell wrote:

To a great extent, the redemption of all these lands will require extensive and comprehensive plans, for the execution of which aggregated capital or cooperative labor will be necessary. Here, individual farmers, being poor men, cannot undertake the task. For its accomplishment a wise prevision, embodied in carefully considered legislation, is necessary. It was my purpose not only to consider the character of the lands themselves, but also the engineering problems involved in their redemption, and further to make suggestions for the legislative action necessary to inaugurate the enterprises by which these lands may eventually be rescued from their present worthless state. (23)

Just one year earlier, in 1877, Congress approved the **Desert Land Act** and authorized the sale of 640 acres (260 ha) of federally owned land for $1.25/acre ($3.09/ha) (without water rights) to any U.S. citizen who would homestead and develop irrigation.

A decade after Powell's report to Congress, Congress authorized a study to investigate reservoir storage sites for irrigation and to measure the streamflow of rivers in the West. Stream gaging was a new science, and Powell's staff at the U.S. Geological Survey (he was the second director of the agency and will be discussed later in this chapter) held one of the first training camps during the winter of 1888–1889 in New Mexico. Official streamflow measurements began the next spring and became a major activity of that agency, a function that continues today.

National Irrigation Congresses The call for additional irrigation spread throughout the West and reached an almost feverish pitch in the late 1800s and early 1900s. The Great Plains of the central United States experienced a devastating drought in 1890, as a result of which thousands of settlers lost crops and, ultimately, their farms to local bankers. As the national economy floundered, many feared that the Manifest Destiny of the American people (a phrase used to describe the need for westward expansion) was at risk. William Smythe, an editor with the *Omaha Bee,* wrote about the devastation caused by drought in Nebraska and began to investigate the potential for irrigation in his state:

Irrigation seemed the biggest thing in the world. It was not merely a matter of ditches and acres, but a philosophy, a religion, and a program of practical statesmanship rolled into one. There was apparently no such thing as ever getting to the bottom of the subject, for it expanded in all directions and grew to importance with each underfoldment. (24)

Smythe's enthusiasm and promotion of irrigation led to a statewide irrigation convention, held in Lincoln, Nebraska, in 1891. A **National Irrigation Congress** was later organized in Salt Lake City, Utah, that same year. The irrigation movement grew, and a second national congress (attended by the now-famous John Wesley Powell) was held in Los Angeles in 1893. Annual Irrigation Congresses were later organized around the country, with the ninth occurring in Chicago in 1900. Enthusiasm for irrigation was high, and the national congresses were generally promoted by railroads, developers, and western business associations to encourage federal funding of irrigation projects in the West. (25)

A CLOSER LOOK

John Wesley Powell was widely considered to be an eccentric. Although he became a national hero after his voyage down the Colorado River, many of his ideas on western water management seemed bizarre in his day. For example, when Powell attended the second National Irrigation Congress in Los Angeles in 1893 and spoke to the crowd about his perceptions of irrigation in the West, he argued that western water supplies could at best provide irrigation water to only 3 to 5 percent of the lands west of the 100th Meridian. He argued against transporting water from one watershed to another because doing so could leave the basin of origin with severe water shortages (recall the City of Los Angeles raid on the Owens Valley discussed in Chapter 6). Powell promoted the idea that state boundaries should follow watershed boundaries (the ridges separating watersheds) to avoid interstate water disagreements. If this notion had been followed, each state would have had its own river or drainage area, which would not have been shared with any other state.

Powell's comments were met with disdain and ridicule by delegates of the irrigation congress in Los Angeles. The *caveat emptor* attitude of the era (let the buyer beware) promoted by land speculators, investors, and other businessmen was completely contradictory to many of Powell's ideas. Immigrants were needed for western settlement, and prospects of irrigated land throughout the West fueled expectations of both settlers and land speculators. Powell's recommendations were therefore largely ignored. In 1894, Powell retired from government service and moved to Maine.

The Reclamation Act of 1902 In 1901, Congressman Francis G. Newlands of Nevada introduced legislation that authorized the federal government to construct irrigation projects in the western states and territories. Political support was tenuous at first, until the bill was amended to include a provision to protect existing vested (legally owned) water rights in each state. The bill, called the **Reclamation Act of 1902,** was approved by Congress and signed into law by President Theodore Roosevelt. The Act created the Reclamation Service, which was placed within the U.S. Geological Survey in the Department of Interior. The agency was later renamed the U.S. Bureau of Reclamation and separated from the U.S. Geological Survey in 1923.

The primary purpose of the Bureau of Reclamation was to encourage settlement of lands west of the 100th Meridian through construction of irrigation projects. Projects included dams for water storage reservoirs, diversion dams, and delivery canal systems to farms. The original intent was to provide water only to farms less than 160 acres (65 ha) in size that were owned by a single landowner. This provision in the law was designed to prevent landownership monopolies by large landholders or corporations. Project construction funds had to be repaid by irrigators, with interest, to a revolving fund over a period of 10 years. That time period was extended to 20 years in 1914 and to 40 years after the Dust Bowl years of the 1930s made repayment more difficult.

The Bureau of Reclamation had early success with the Newlands Project (named after Congressman Newlands who helped create the agency) in western Nevada. Construction began in 1903 to irrigate approximately 400,000 acres (161,880 ha) but was later reduced to approximately 73,000 acres (29,543 ha). (26) In 1910, Shoshone Dam was completed in northwest Wyoming. At 328 feet (100 m), it was the world's tallest dam at the time. Following John Wesley Powell's advice from 30 years earlier, the USBR created local government agencies, called irrigation districts, to administer the day-to-day operations of the Reclamation projects. Irrigation districts served a role somewhat similar to that of the *acequias* located in the Southwest United States and Mexico. Each Bureau of Reclamation irrigation district hired a superintendent, similar to the mayordomo, to administer the day-to-day water deliveries and maintenance of the irrigation system.

The early activities of the USBR were compared in 1917 to

> *. . . the building of the dikes that protect Holland from the sea; the great road systems of the Romans; the building of the Panama Canal; and the leveeing of the Mississippi River. The work of the Reclamation Service is of this same gigantic character.* (27)

U.S. Bureau of Reclamation activity peaked with the construction of Hoover Dam, a project proposed decades earlier by John Wesley Powell. Construction was authorized in 1928, the same year the Flood Control Act of 1928 authorized the U.S. Army Corps of Engineers to construct flood-control structures east of the 100th Meridian. The Hoover Dam Project was the culmination of efforts to implement the Colorado River Compact of 1922 (discussed in Chapter 8).

SIDEBAR

Approval of federal funds for the construction of Hoover Dam in December 1928 (a project for western water users) was equalized by the Flood Control Act of 1928 passed by Congress in May of that same year (for flood control in eastern states). It's not unusual for Congress to appropriate federal funds for one region of the country in exchange for monies to another region in the same year. Legislation must be supported by a majority of members in Congress, and political support often involves tradeoffs, compromise, and broad allocation of federal funds.

The pattern of massive Bureau of Reclamation irrigation project development continued. In 1970, the Commissioner of the Bureau of Reclamation reported that the USBR had constructed 276 reservoirs with a total storage capacity of 134 million acre-feet (43.7 trillion gal, or 165.3 billion m^3). Of these, 49 had hydropower plants that generated 7 million kilowatts of electricity and irrigation water for approximately 6 million acres (2.4 million ha) of dry, western lands. A total of 18 million acres (7.3 million ha) were irrigated in the West at that time. (28)

The Anti-Dam Construction Era During the 1970s, the U.S. Department of Agriculture administered programs to subsidize (pay) farmers to not grow surplus crops such as corn. This federal farm program helped maintain corn prices at higher than market levels, thereby providing assistance to financially struggling farmers. To many, it was a totally unreasonable government policy for Congress to subsidize federal water development projects to expand irrigated acres while, at the same time, subsidizing farmers through the U.S. Department of Agriculture. (29)

Public perception and support for the Bureau of Reclamation declined significantly during the 1970s. Ralph Nader wrote in *Damming the West* (1973) that the agency had outlived its usefulness. The National Water Commission wrote a pamphlet called *Disasters in Water Development* in 1973 describing negative aspects of massive water projects. The USBR's Teton Dam failure in 1976, described in Chapter 7, created major concerns regarding dam safety. After the Teton Dam disaster, the Commissioner of the Bureau of Reclamation, Gilbert G. Stamm, tried to bolster the historical role of the agency. He testified before Congress that Reclamation had spent $9.4 billion for the cultivation of 9 million acres (3.6 million ha) that yielded crops worth $5 billion and domestic water for 17 million people. (30)

Commissioner Stamm's testimony was not enough. The growing negative sentiment toward big government water projects encouraged President Jimmy Carter to create a "Hit List" in 1977. The list eliminated funding for 18 proposed federal dam construction projects at a savings of $2.5 billion. The reasons cited for the cuts were dam safety, economics, and environmental concerns. (31) The "Hit List" changed the future direction of the Bureau of Reclamation. For a listing of the USBR's major historical activities, see Table 9.4.

POLICY ISSUE

In 1993, the Bureau of Reclamation declared that, since it had achieved its goals of increasing settlement in the West, it now had a new mission: to manage existing water projects to promote conservation and to develop partnerships with customers, states, and tribes. As part of this new relationship, the USBR has transferred the ownership and operation of many dams and delivery systems to local water agencies. For many, the big dam construction era in the United States has come to an end.

Recently, existing Bureau of Reclamation projects have also come under close environmental scrutiny. In 2001, Reclamation announced that, because of the requirements of the Endangered Species Act (see Chapter 12) and a "critically dry year," it would not be able to make water deliveries from Upper Klamath Lake to irrigators under the Klamath Project. Congress authorized construction of the Klamath Project in 1905 to deliver irrigation water to the dry lands of southern Oregon. The area is located in the rain shadow of the Cascade Mountains and receives only 12 to 14 inches (30 to 36 cm) of average annual precipitation. The Klamath Project serves 1400 farms and provides irrigation water to 210,000 acres (84,984 ha) of irrigated alfalfa, barley, oats, wheat, potatoes, and sugar beets. (32)

During the extremely dry summer of 2001, Klamath irrigators watched their crops wither and die as water from Upper Klamath Lake was held for the benefit of the Lost River sucker *(Deltistes luxatus)*, the shortnose

TABLE 9.4 Major Water Resources Events Affecting the Bureau of Reclamation

Year	Event
1869	John Wesley Powell floats down the Colorado River. Becomes a strong proponent of irrigation in the West.
1877	**Desert Land Act** passed by Congress. Authorizes the sale of federal lands for $1.25/acre to settlers.
1878	John Wesley Powell writes *Report on the Lands of the Arid Region of the United States for Congress.*
1890	Drought begins in the Great Plains.
1891	First National Irrigation Congress held in Salt Lake City, Utah.
1902	**Reclamation Act** creates the present-day Bureau of Reclamation.
1910	Shoshone Dam in Wyoming completed by the Bureau of Reclamation.
1928	Congress approves funds for construction of Hoover Dam.
1933	Drought once again hits the Great Plains.
1976	Teton Dam fails.
1977	President Jimmy Carter's Hit List stops construction of 18 western dam projects.
1993	The Bureau of Reclamation declares it has achieved its goal of increasing settlement in the West.
2001	Surveillance cameras placed on headgates of Klamath Irrigation Project in Oregon to prevent illegal irrigation water diversions.

sucker *(Chamistes brevirostris)*, and the Southern Oregon/Northern California coasts' coho salmon *(Oncorhynchus kisutch)*. Federal officials stated that environmental laws took precedence over long-standing federal irrigation policy in the area. Most irrigators were stunned by that declaration, and a visit by Oregon governor John Kitzhaber in April 2001 drew over 5000 extremely concerned people to a town hall meeting. The conflict over federal water policy continues to escalate and now includes surveillance cameras on irrigation headgates along the Klamath River to prevent an illegal taking of water by local irrigators.

U.S. GEOLOGICAL SURVEY (USGS)

http://www.usgs.gov

OVERVIEW

The U.S. Geological Survey (USGS) is a scientific agency of the federal government and like the Bureau of Reclamation is located within the Department of Interior. The USGS has offices in every state to measure and monitor surface and groundwater characteristics in cooperation with local and state governments, universities, and other federal agencies. The USGS is responsible for presenting nonbiased technical data regarding streamflow, groundwater levels, and other aspects of earth and life sciences. It also prepares technical reports on the status of the nation's water quantities and quality. The agency has no regulatory or water development duties.

BRIEF HISTORY

The **U.S. Geological Survey** was created by Congress in 1879 in response to a recommendation from the National Academy of Sciences that the western federal scientific surveys be consolidated into a single bureau. (33) The early mission of the USGS was to classify public lands held by the federal government and to examine the geological and mineral resources of such lands. This charge originated with the Land Ordinance of 1785 when the colonies decided to give ownership of all lands west of the Allegheny Mountains to the federal government. This was intended to eliminate fights between states over control of unsettled lands to the west. The landholdings of the federal government increased greatly with the Louisiana Purchase in 1803. The U.S. Congress agreed that these public lands should be sold to generate revenue for the national treasury and to encourage settlement in the West.

An early duty of the USGS, and one that continues today, was stream gaging (discussed in Chapter 3). As mentioned earlier, the first U.S. stream gaging station was on the Rio Grande River near Embudo, New Mexico, in 1889. Within two years, a gage was placed on the Potomac River at Chain Bridge near Washington, D.C. By 1895, discharge measurements were being conducted in 27 states.

Until about 1907, reclamation and irrigation activities and policies of the federal government were centered in the USGS, but were subsequently turned over to the Bureau of Reclamation. Since that time, the USGS has focused on monitoring, assessment, and research related to water resources in all 50 states.

Today, the USGS operates and maintains over 7000 stream gaging stations, which represent over 85 percent of the total gaging stations in the United States and its territories. (34) The remaining sites are operated primarily by state or local government agencies.

Modern USGS streamflow gages serve multiple purposes, including:

- Flood prediction and monitoring
- Water supply forecasting and management
- Drought monitoring and management
- Water quality monitoring and management (e.g., calculation of contaminant loads)
- Compliance with international treaties
- Water rights adjudication
- Recreation information
- Management of aquatic habitat
- Basic hydrologic research (35)

These data are gathered to provide citizens, local, state, and federal agencies, universities, and others with real-time hydrologic data. (Real-time data from USGS streamflow gages are available at http://water.usgs.gov/realtime.html.)

OUR ENVIRONMENT

The U.S. Geological Survey has been instrumental in developing the Adaptive Management Program to protect the downstream river ecosystem below Glen Canyon Dam in Arizona. Since its completion in 1963, Glen Canyon Dam has changed the flow of water and sediment down the Colorado River and through Grand Canyon National Park. USGS scientists have asessed the changes in sedimentation and sandbars below the dam caused by reduced streamflows and dam operation procedures. At risk is the preservation of habitat used by native fish, as well as associated riparian vegetation.

In 1996, a controlled flood was created by releasing approximately 45,000 cfs (29.1 billion gal/day, or 1274 cms) of water for seven days. The Bureau of Reclamation, National Park Service, and the USGS cooperated in this experiment to determine if higher volume releases would facilitate the reconstruction of sandbars and other desired features downstream. Four streamflow gaging stations were equipped with satellite telemetry, and approximately 2200 pounds (998 kg) of nontoxic red dye were injected into the Colorado River to measure the velocity of the floodwaters through the Grand Canyon. The dye was used to determine how quickly it moved downstream and mixed with river water. This information helped calibrate USGS computer simulation models of the Colorado River.

It's important to note the level of cooperation between numerous federal agencies on this flood flow study. What incentives did the USGS have to promote this type of research along the Colorado River? What incentives and disincentives did the Bureau of Reclamation have to participate? Do you expect that the National Park Service and the U.S. Environmental Protection Agency supported this experiment? Why or why not? See the USGS and USBR websites for additional information on this ongoing project.

The USGS also studies the nation's groundwater resources, including:

- Mapping of the extent and characteristics of major aquifers

- Monitoring of water levels in wells and aquifers
- Development and use of computerized groundwater flow models to answer "what-if" questions related to the effect of management actions on groundwater levels and movement
- Remediation (cleanup) of contaminated groundwater
- Study of salt-water encroachment rates on coastal aquifers
- Study of the movement of contaminants from the land surface, through the unsaturated zone, through aquifers and into wells

In addition, the USGS conducts research on the effects of human activities on water quality in lakes, rivers, estuaries, and aquifers. A wide variety of reports, maps, fact sheets, and webpages are produced to guide resource management decisions by local, state, and federal agencies. The USGS also keeps track of statistics on water use by various categories and on chemical contaminants in rain, snow, and dry fallout.

Another important duty of the USGS is land surveys. The entire United States has been measured, surveyed, and mapped by cartographers (map makers) to clearly present geographic information on maps. The USGS produces topographic maps ("topo" maps) that show land and water features, elevations, and structures at various scales of detail. The Appendix provides a more in-depth look at USGS topographic maps.

POLICY ISSUE

The U.S. Geological Survey has played key roles in water resources management and policy across the country, including:

1. Development of municipal water supplies in Myrtle Beach and Hilton Head in South Carolina; Savannah, Georgia; much of Florida; Albuquerque, New Mexico; Tacoma, Washington; Lincoln, Nebraska; Kansas City and Independence, Missouri; Rochester, Minnesota; Vancouver, Washington; and Washington, D.C.
2. Management and restoration of water resources in sensitive aquatic habitats in the Everglades of Florida and the Atchafalaya Basin in Louisiana.
3. Elimination of lead from household plumbing in New Jersey.
4. Determination of sources of fecal contamination along recreational beaches in Michigan and urban streams in northern Virginia.
5. Determination of sources of nitrate contamination to the Mississippi River and the Gulf of Mexico Hypoxic Zone.
6. Determination of susceptible aquifers in Virginia, based on groundwater age dating and water quality sampling.
7. Determination of pesticide contamination probability for aquifers in New Jersey and Washington.
8. Phasing out of the gasoline additive MTBE based largely on occurrence studies by the USGS. (36)

Do you believe it is important to have a federal water resources agency devoted strictly to scientific research? How should such an agency interact with other federal agencies involved in water development?

U.S. FISH & WILDLIFE SERVICE (USFWS) http://www.fws.gov

OVERVIEW

The **U.S. Fish & Wildlife Service (USFWS)**, yet another agency within the Department of Interior, is the only federal agency whose primary responsibility is to conserve, protect, and enhance fish, wildlife, plants, and their habitats. As such, it has become an extremely important, and at times controversial, government agency through its efforts to protect endangered species and associated habitat. The USFWS is at present involved in numerous debates over species protection, river discharges, and minimum flow requirements, as well as habitat maintenance and protection.

BRIEF HISTORY

The U.S. Fish & Wildlife Service was created in the Department of Interior in 1940 but had its origins as the U.S. Commission on Fish and Fisheries in the Department of Commerce in 1871. An early mission of the Commission was to investigate declining numbers of the nation's fish resources. A Division of Economic Ornithology and Mammalogy also existed which studied the food habits of migratory birds as agents of pest control in agriculture.

Federal Bird Reservations (later called National Wildlife Refuges) were created in the early 1900s for the protection of migratory birds; today they are administered by the USFWS. The first Federal Bird Reservation was created on Pelican Island, Florida, by President Theodore Roosevelt in 1903. Later, the Migratory Bird Treaty Act was passed in 1918 for the protection of birds passing between Canada and the United States. The Act also established hunting seasons for migratory waterfowl. Setting these frameworks is still a major part of what the USFWS does today. (37)

The **Fish and Wildlife Coordination Act**, passed by Congress in 1958, required that wildlife conservation be considered during any federal water resource development project. The goal of the legislation was to create a consultation process with the USFWS to determine the potential impacts of federal water projects on wildlife. However, the Act was not very successful because a dollar value was placed on the displacement of fish and wildlife, and these "costs" had to compete with the dollar values placed on economic and human "benefits" of water projects. Wildlife values, within this benefit-cost analysis, were usually too low to warrant protection under the 1958 Act.

Landmark legislation was finally passed in 1973 with the approval of the **Endangered Species Act (ESA)**. President Richard Nixon signed the legislation into law giving the USFWS responsibility for administering the Act. The USFWS is charged with protecting endangered species and their habitat from destruction. The ESA has sometimes been very controversial (see Chapter 12).

This is the legacy I would like to leave behind: I would like to stop the ridicule about the conservation of snails, lichens, and fungi, and instead, move the debate to which ecosystems are the most recoverable, and how we can save them, making room for them and ourselves. (38)

Mollie H. Beattie, Director,
U.S. Fish & Wildlife Service, 1993–1996

POLICY ISSUE

The U.S. Fish & Wildlife Service is currently at the center of many wildlife protection controversies across the country. For example, the agency has required the delivery of additional flows in the Platte River of central Nebraska in order to protect the endangered whooping crane, least tern, and piping plover. Water for these endangered species is protected either by prohibiting new water depletions hundreds of miles (hundreds of km) upstream in Colorado and Wyoming or by requiring a one-for-one replacement of new water depletions. (A water developer may be required to release 1 acre-foot of water into the Platte River for every acre-foot developed upstream.) This will be the focus of a case study in Chapter 12.

A new policy issue is emerging regarding fish hatcheries operated by the U.S. Fish & Wildlife Service. Many hatcheries were built during the dam-building era to mitigate the loss of fisheries associated with water project construction. Because the ongoing maintenance of these facilities was rarely considered at the time, the USFWS is now faced with a crumbling infrastructure of hatcheries badly in need of maintenance, a federal obligation to provide mitigation fisheries, and no budget to do either. The General Accounting Office recently recommended congressional action to require federal water development agencies or beneficiaries to reimburse the USFWS for all hatchery operation and maintenance expenses associated with these projects. (39) This budget request is expected to be controversial since water development agencies generally prefer to retain their funds for other purposes.

OUR ENVIRONMENT

The U.S. Fish & Wildlife Service has the responsibility to conserve, restore, enhance, and manage the nation's fishery resources and aquatic ecosystems. This has been an integral part of the USFWS core mission for over 130 years. In 1871, Congress established the National Fish Hatchery System to provide additional domestic food fish for an increasing human population. At that time, overharvesting, pollution, and habitat loss were seriously reducing fish populations in many regions of the country. (40)

Today, the USFWS has 70 fish hatcheries in 34 states, 7 Fish Technology Centers, and 9 Fish Health Centers. These facilities continue to produce fish for restocking in lakes, reservoirs, rivers, and streams to provide food and recreational opportunities for anglers (see Figure 9.8). In addition, some fish hatcheries are used to help in the recovery of species protected under the Endangered Species Act (discussed in Chapter 12). The USFWS works closely with state-operated fish hatcheries, tribes, and the private sector on habitat restoration programs, such as those for Great Lakes lake trout, Atlantic Coast striped bass, Atlantic salmon, and Pacific salmon. (41)

Visit a state or federally operated fish hatchery in your region. What issues exist for the successful operation of the facility—lack of funds, diseases such as whirling disease, degraded ecosystems which increase fish mortality rates, and so on? Has the operation of the hatchery changed recently due to budget cuts or disease? Go to http://fisheries.fws.gov/ for additional information.

NATIONAL PARK SERVICE (NPS)

http://www.nps.gov

OVERVIEW

The **National Park Service** (also called the **NPS** or **Park Service**) is also located in the U.S. Department of Interior and was created by an Act of

FIG. 9.8. A U.S. Fish & Wildlife Service biologist distributes food to fish in a hatchery raceway. Wild fish feeding habits are studied to mimic natural food sources for fish reared in captivity. Go to the U.S. Fish & Wildlife Service's "National Fish Hatchery System" website at http://fisheries.fws.gov/FWSFH/NFHSintro.htm for a wide range of additional information.

Congress signed by President Woodrow Wilson in 1916. The NPS has the primary responsibility of managing the nation's national park system. Currently, 384 areas with 130,937 square miles (339,128 km^2) of land are contained within national parks, with most located east of the 100th Meridian. However, the largest national parks are located west of the 100th Meridian. It was inevitable that water issues would eventually surface within the boundaries of national parks since many contain large, environmentally sensitive watersheds.

BRIEF HISTORY

George Catlin originated the idea of national parks in the United States, but Frederick Law Olmsted, John Muir, Steven T. Mather, and others made the concept a reality. As early as 1832 during his travels in the Dakotas, Catlin called for the protection of unspoiled lands for future generations. In 1864, the U.S. government donated Yosemite Valley to California as a state park. In 1872, Yellowstone National Park was created by an Act signed by President Ulysses S. Grant and was the first national park in America. Administration of the park property was given to the Department of Interior, although management was turned over to the Department of War (now the Department of Defense) between 1886 and 1916.

Creation of other national parks continued in earnest in the late 1800s and early 1900s with the designation of Sequoia, Mount Rainier, Crater Lake, and Glacier National parks. Military parks, especially Civil War battlefields, were established during the 1890s, and national monuments became part of the national collection beginning in 1906. Military parks were managed by the War Department until 1933. (42)

Tourism was an important component of these early parks, and railways were sometimes constructed to provide easy access to national parks in the West. Luxury hotels were also built within the boundaries of some parks to encourage tourist visitation, although most lodging accommodations could not be described as luxurious.

One of the earliest controversies for the Bureau of Reclamation occurred inside a national park, just a few years before the National Park Service was created. In the early 1900s, the City of San Francisco was searching for water storage sites and developed plans to dam the Hetch Hetchy Valley inside Yosemite National Park. Intense public opposition developed, led by John Muir, the noted artist and conservationist. Congress authorized the Bureau of Reclamation to construct the dam in 1913, and O'Shaunessey Dam and Hetch Hetchy Reservoir were completed in 1922. The controversy around this project led to charges that Reclamation created the worst disaster ever to come to any national park. (43) Many argue today that the dam should be removed to allow the Hetch Hetchy Valley to return to its natural condition.

Years after the Hetch Hetchy Project was completed, federal water managers developed plans to build dams inside other national parks and national monuments. These proposals included locations in Dinosaur National Monument in Colorado and the Grand Canyon in Arizona, but they were defeated in part because of the lingering ill-will toward Hetch Hetchy in California and the growing environmental movement across the country. In some situations, several national parks were created because proposals to dam rivers failed, and conservation advocates turned to the National Park Service to ensure that the dam would never be built. Buffalo National River in Arkansas was created by Congress in 1972, in part for this reason. (44)

In 1976, the U.S. Supreme Court ruled in *Cappaert* v. *United States* that national parks received a federally reserved water right when a park was created (discussed in Chapter 8). This reserved water right could be used for the original purposes of the park, such as fire control, maintenance of streamflows, and protection of natural resources within park boundaries. The ruling has also been used to curtail groundwater pumping outside national park boundaries if it can be shown that pumping negatively affects wetlands or other water systems within a park. This curtailment can occur even if the groundwater pumping is by a private individual. (45)

SIDEBAR

The National Park Service is currently conducting studies at Cape Cod in Massachusetts and Cape Hatteras National Seashore in North Carolina to determine the effects of groundwater withdrawals on wetlands and maritime forests caused by groundwater pumping outside park boundaries. In Montana, the National Park Service and the State of Montana developed a water rights compact to protect the natural water resources within Yellowstone National Park. The NPS needed to protect drinking supplies within the park, flows in the Yellowstone River, and the hydrothermal system (geysers, boiling pots, and thermal springs) from depletion.

POLICY ISSUE

National parks are in danger of being loved to death by a mobile and recreation-minded society. Entrance limits at parks, such as at Yosemite National Park, are now used to reduce visitor impacts on trails, wildlife, and facilities. In part because of these increased pressures, the National Park Service has been diligently pursuing protection of existing federal reserved water rights and continues to seek establishment of such rights in additional national parks. This creates tremendous tensions among other senior water rights holders who have not been subjected to these federal water needs in the past. Water users along the Black Canyon of the Gunnison National Park near Montrose, Colorado, are currently struggling with this issue.

BUREAU OF LAND MANAGEMENT (BLM) http://www.blm.gov

OVERVIEW

The **Bureau of Land Management (BLM)** was created in 1946 and is part of the Department of Interior. It administers federal public lands, located primarily west of the 100th Meridian, and is responsible for the management of 412,500 square miles (1.1 million km^2) of land, about one-eighth of the total land area of the United States. This represents approximately 40 percent of all federal lands (1.0 million square miles or 2.6 million km^2) that are owned by the American people. Most BLM lands are located in the western United States and Alaska. (46)

SIDEBAR

It is the mission of the Bureau of Land Management to sustain the health, diversity, and productivity of the public lands for the use and enjoyment of present and future generations. (47)

BRIEF HISTORY

The history of the Bureau of Land Management goes back to the Land Ordinance of 1785 and the Northwest Ordinance of 1787. These laws provided for land surveys to encourage settlement west of the Appalachian Mountains. In 1812, the General Land Office was established in the U.S. Department of Treasury to administer the sale of these federal properties to settlers. The **Homestead Act of 1862** and later the **Mining Law of 1872** allowed citizens of the United States to settle and acquire public lands, as well as to explore and mine for minerals on other federal properties. The 1872 Mining Law, still in effect today, also set standards for acquiring mining rights, recording claims, and establishing mill sites. In 1934, Congress passed the **Taylor Grazing Act** which created the U.S. Grazing Service to manage cattle grazing on federal rangelands (generally semiarid, short-grass prairie). The primary goal of these Acts was the disposal (sale or grant) of lands to encourage settlement and economic development. Limited regard was given to management of public lands to protect water resources. (48)

In 1946, the General Land Office and the U.S. Grazing Service were combined into the Bureau of Land Management. At the time of the BLM's creation, over 2000 unrelated and often conflicting laws existed regarding the management of public lands. However, it was not until the **Federal Land Policy and Management Act (FLPMA)** was passed by Congress in 1976 that the Bureau of Land Management received a clear congressional mandate.

POLICY ISSUE

The Bureau of Land Management, the largest land administrator in the United States, manages grasslands, forests, high mountains, arctic tundra, and deserts. These federally owned properties include timber, livestock forage, fish and wildlife habitat, scenic and recreation resources, wilderness areas, mineral resources, and archaeological and historical resources. The BLM must manage these lands for multiple uses such as recreation, wildlife protection, and economic production including grazing, mining, and forestry. (49)

Land and water conservation have become major management issues of the BLM. Reduced grazing on sensitive lands, shoreline protection from erosion along rivers and streams, and the elimination of mining impacts on water quality are all significant policy issues facing the BLM. In addition, forest management is being developed to protect watersheds and water quality. The BLM is struggling with the responsibilities of sustained yields from federal lands while protecting environmental resources. This often creates conflict with other federal agencies, such as the U.S. Environmental Protection Agency and the U.S. Fish & Wildlife Service.

U.S. ENVIRONMENTAL PROTECTION AGENCY (USEPA) http://www.epa.gov

OVERVIEW

The **U.S. Environmental Protection Agency (USEPA)**, an independent agency of the federal government, was created in 1970. Prior to this time, federal environmental protection efforts were uncoordinated and often nonexistent. Formation of the USEPA combined a patchwork of existing programs from the Department of Interior, Department of Health, Education, and Welfare, the Food and Drug Administration, and the Department of Agriculture.

The USEPA is a regulatory agency of the federal government with legislative authority to impose substantial monetary fines, or even jail sentences, for noncompliance with federal environmental laws. This enforcement authority gives the USEPA substantial power for protection of the environment. Shortly after it was formed, the USEPA filed suit against the cities of Detroit, Cleveland, and Atlanta for polluting rivers with municipal sewage. The mayors of these three cities had six months to bring their cities into compliance with federal environmental regulations or face court action. Water quality of the Detroit, Cuyahoga, and Chattahoochee rivers has improved greatly since 1970.

BRIEF HISTORY

Environmental protection in the United States has historically been sporadic, with the federal government taking only a small role, or no comprehensive approach, until the mid-1900s. Congress passed the **Water Pollution Control Act of 1948** and directed that technical assistance and funds be provided to states to protect water quality. The **Federal Water Pollution Control Act of 1956** went a step further and provided funding for pollution studies and development of local wastewater treatment plants.

The **Water Quality Act of 1965** created the first federal water quality standards program by Congress to assess pollution levels in lakes and streams across the country. The Act required states to carry out this water quality assessment program and to stop illegal discharges into lakes and streams. This law intended to set a water quality standard for each water body and prohibited discharges that exceeded those standards. Since standards were extremely difficult to identify and implement, however, the federally mandated program had only limited success. It has been estimated that over $20 billion was spent trying to implement this law. (50)

In 1969, the **National Environmental Policy Act (NEPA)** was passed to authorize the federal government to become the "protector" of earth, air, land, and water in the United States. The law stated that the intent of Congress was to "create and maintain conditions under which man and nature can exist in productive harmony . . . and to assure for all Americans safe, healthful, productive, aesthetically and culturally pleasing surroundings." (51) This landmark piece of environmental legislation was signed into law by President Richard Nixon.

The basis of NEPA is the requirement for an **environmental impact statement (EIS)** to consider all environmental consequences, or "impacts," of a proposed project. This involves public notification and participation in the planning process for any federally funded project or one that requires a federal permit (such as a hydropower license from FERC discussed later in this chapter). President Nixon stated that the decade of the 1970s was the time for America to repay its debt to the past by reclaiming the purity of its air, water, and our living environment. "It is," he said, "literally now or never." (52)

The Act also created a **Council on Environmental Quality,** which has the responsibility of reporting to the president on the state of the environment, overseeing federal agency implementation of the environmental impact assessment process, and acting as a referee when federal agencies disagree over the adequacy of such assessments. The Council is located within the Executive Office of the President.

In 1972, the **Federal Water Pollution Control Act Amendments** (also known as the **Clean Water Act Amendments**) were passed. This was a landmark, comprehensive piece of federal legislation designed to protect both interstate and intrastate waters, including lakes, rivers, estuaries, and wetlands. The Clean Water Act Amendments strengthened the nation's water quality standards system, made it illegal to discharge pollution without a permit, and encouraged use of the most achievable pollution control technology available. (53)

Discharge permits, required by the Clean Water Act, list the maximum concentrations of specific chemicals that can be placed in a water body. This approach continues today and had an original goal of making all waters of the United States "fishable and swimmable" by 1983. (The term *fishable* means water suitable for fish and all supporting aquatic organisms, while *swimmable* refers to water safe for recreation.) Massive construction programs were also implemented to build municipal wastewater treatment plants, and over $20 billion was spent between 1972 and 1992. (54) The Clean Water Act is discussed in more detail in Chapters 5 and 11.

The **Safe Drinking Water Act of 1974** was passed to protect the nation's public drinking water supplies from pollution and communicable waterborne diseases. Federal drinking water standards were established to protect the public health, and the USEPA was designated as the agency responsible for carrying out the program. (This Act will be discussed in Chapter 11.)

The **Resource Conservation and Recovery Act (RCRA) of 1976** directed the USEPA to regulate land disposal of hazardous wastes. It requires monitoring of such wastes from the time it is created until it is disposed of to prevent hazardous waste from entering the environment. Municipal wastewater plants fall under this law since sludge generated by such plants is considered potentially hazardous waste. In 1984, the Act was amended to allow USEPA to regulate land disposal operations and leaking of underground storage tanks (such as those found at gasoline stations that could pollute groundwater). A state may take over this program if it meets criteria acceptable to the USEPA.

In 1977, the USEPA was given additional duties with the passage of amendments to the Clean Water Act of 1972. The **Clean Water Act Amendments of 1977** strengthened controls on toxic pollutants and allowed states to assume responsibility for federal programs. Its primary principles were as follows:

1. Improvement of water quality to protect aquatic organisms and recreation by 1983.
2. Elimination of the discharge of pollutants into waters by 1985.
3. Prohibition on the discharge of toxic pollutants.
4. Construction of publicly owned wastewater treatment plants.
5. Development of regional water treatment planning processes.
6. Development of new technology to eliminate the discharge of all pollutants into navigable waters, wetlands, or other contiguous areas, and the oceans. (55)

The Clean Water Act also implemented discharge permitting processes, called **National Pollutant Discharge Elimination System (NPDES) Permits,** to enforce water quality rules. (These processes will be discussed in Chapter 11.)

Section 404 of the Clean Water Act of 1977, as discussed earlier, gives the U.S. Army Corps of Engineers the responsibility to prevent wetlands destruction. A Section 404 Permit is required before any dredging or filling can occur in a wetland. Originally, normal farming, ranching, and silviculture (forestry) activities were exempted from the permit requirement, but these activities have been placed under increased scrutiny in recent years.

During the late 1980s, the first President Bush implemented a national goal of "no net loss of wetlands" whereby the Corps was directed to protect all wetlands from destruction, even if located on private property. Even though the U.S. Army Corps of Engineers is responsible for issuing 404 Permits, the USEPA is the federal agency ultimately responsible for the program under the Clean Water Act. Therefore, the USEPA can veto the issuance of a permit, can delegate the 404 permitting process to states if a state wishes to implement more stringent requirements, and is responsible for the evaluation of federal projects that may be exempt from Section 404 permitting processes. The U.S. Fish & Wildlife Service and the U.S. Forest Service also have important advisory roles to the USEPA during review of a 404 Permit request. (We will discuss wetlands and 404 Permits in more detail in Chapter 12.)

The **Comprehensive Environmental Response, Compensation, and Liability Act (CERCLA) of 1980** (also called **Superfund**) authorized the USEPA to address environmental issues involved in accidental spills and releases of hazardous waste from landfills or old dump sites. USEPA is allowed to inspect a site without a search warrant. This authority has led to the arrest and prosecution of individuals and companies that were illegally dumping pollutants.

The **Safe Drinking Water Act Amendments of 1996** were passed by Congress to give existing drinking water regulations more flexibility. Additional funding was provided to communities, programs were developed to protect source water (untreated water from groundwater, rivers, and lakes), and risk assessment was added to drinking water standards. Each state was directed to create a Source Water Assessment and Protection (SWAP) Program to evaluate the safety of all public drinking water supplies.

POLICY ISSUE

The U.S. Environmental Protection Agency has replaced the Bureau of Reclamation as one of the most powerful (some would argue it is the most powerful) federal water agency at the present. The USEPA mission of environmental protection, along with its ability to levy substantial fines, places it in a key position to affect change.

The USEPA is developing more stringent water quality parameters for drinking water, pollution levels in rivers, and wastewater discharge. Water diversions can greatly affect water quality in each of these areas by reducing the amount of water available for dilution of pollution ("the solution to pollution

TABLE 9.5 Major Federal Legislation Affecting Water Resources Activities of the U.S. Environmental Protection Agency

Year	Event
1899	**Refuse Act** (also known as the Rivers and Harbors Act) protects navigable waterways from pollution.
1948	**Water Pollution Control Act** provides technical assistance and funds to the states to promote efforts to protect water quality.
1956	**Federal Water Pollution Control Act** provides for pollution studies and development of local sewage plants.
1965	**Water Quality Act** creates first federal water quality standards program.
1969	**National Environmental Policy Act (NEPA)** authorizes federal agencies to become the "protector" of earth, air, land, and water resources in the United States.
1970	Congress creates the U.S. Environmental Protection Agency.
1972	**Federal Water Pollution Control Act** amended to require permits for the direct discharge of pollutants into waters of the United States, and provides comprehensive protection for both interstate and intrastate waters, including lakes, rivers, streams, estuaries, and wetlands. These amendments are also known as the Clean Water Act Amendments of 1972.
1974	**Safe Drinking Water Act** protects nation's drinking water supplies from pollution and communicable waterborne diseases.
1976	**Resource Conservation and Recovery Act** directs the USEPA to regulate land disposal of hazardous wastes.
1977	**Clean Water Act** amended by Congress; strengthens controls on toxic pollutants and allows states to assume responsibility for federal programs; Section 404 authorizes the U.S. Army Corps of Engineers to prevent wetlands destruction.
1980	**Comprehensive Environmental Response, Compensation, and Liability Act (CERCLA** or **Superfund)** directs the USEPA to address environmental problems caused by accidental spills and releases of hazardous waste from landfills or old dump sites.
1996	**Safe Drinking Water Act Amendments** passed by Congress. Source Water Assessment and Protection (SWAP) Program created and all states required to assess public drinking water supplies.

is dilution"). These topics are discussed in detail in Chapters 5 and 11. For a listing of federal laws relating to the USEPA's water resources activities, see Table 9.5.

NATURAL RESOURCES CONSERVATION SERVICE (NRCS)

http://www.nrcs.usda.gov

OVERVIEW

The **Natural Resources Conservation Service** (also called the **NRCS**) is located within the **U.S. Department of Agriculture (USDA).** The USDA was created by President Abraham Lincoln in 1862 to assist the nation's farmers. Water resources management within the USDA was limited until after the Bureau of Reclamation was formed. These two agencies were located in different federal departments, and they immediately came into competition for federal funds to improve agricultural production in the West. Although the USBR (and the U.S. Army Corps of Engineers) generally won extensive funding during the first half of the twentieth century, the USDA, through the Watershed Protection and Prevention Act of 1954, was authorized to investigate and construct small on-farm reservoirs to reduce downstream flooding.

The NRCS was originally called the Soil Erosion Service and then renamed the Soil Conservation Service (SCS) in 1935. The purpose of the

NRCS is to protect soil resources for agricultural production, although water quality issues have become more important in recent years.

BRIEF HISTORY

The passage of early soil erosion legislation had an interesting beginning during the 1930s. Dust storms in the Midwest were severe, and it was not unusual for eroded soils to be carried by strong winds thousands of miles through the atmosphere to the Atlantic Ocean. The first agency head of the Soil Erosion Service, Hugh Bennett, was testifying before the Senate Public Lands Committee in May 1934 when a sudden dust storm darkened the Washington, D.C., sky. The meeting was temporarily halted so that everyone could view the terrible spectacle outside the windows of the Senate Office Building. After a short recess, Committee members returned to the hearing room, and the legislation was approved shortly thereafter. (56)

SIDEBAR

Hugh Bennett was an astute public servant and knew the politics of Washington well. He also paid close attention to the weather during his term as head of the Soil Erosion Service. A few days before he testified to Congress regarding his agency's budget in May 1934, he learned of a major Southern Plains duststorm heading up the Ohio Valley toward the East Coast. Mr. Bennett asked that the hearing be delayed a day in hope that the dust storm would arrive during his testimony. It did, and he received his funding from Congress.

The Soil Conservation Service was renamed the Natural Resources Conservation Service in 1994 to better represent its expanding mission in the protection of all natural resources. The NRCS partners with local conservation districts, state land agencies, and individual farmers to develop conservation programs, including the protection of water quality from inefficient farming practices. The agency also provides water supply forecasts for western states by conducting snow depth and water content surveys, as described in Chapter 2.

POLICY ISSUE

The NRCS continues to promote small-scale, on-farm improvements to reduce erosion and water quality degradation. Financial incentives are available through the agency, such as EQIP (Environmental Quality Incentives Program) established in 1996. EQIP provides matching funds for on-farm conservation activities such as wetlands restoration, soil erosion prevention, and wildlife habitat protection. The NRCS provides conservation education programs for teachers and the agriculture community, and continues the agrarian values promoted by President Lincoln. The role of the NRCS in water resources management continues to evolve.

U.S. FOREST SERVICE (USFS)

http://www.fs.fed.us

OVERVIEW

The **U.S. Forest Service (USFS)**, also located within the U.S. Department of Agriculture, has the primary goal of managing the country's national forests. Currently, the USFS controls 298,437 square miles of land (772,953 km^2) or 23 percent of all federal lands in national forests and grasslands in 44 states. This equals 8.5 percent of the total land area in the United States.

BRIEF HISTORY

The U.S. Forest Service was established by Congress in 1905 to provide quality water and timber for the nation's benefit. However, the USFS's origins go back much earlier to 1881 when Congress created the Division of Forestry within the Department of Agriculture. The original goal of the Division of Forestry was to provide information to Congress regarding the condition of the nation's forests. In 1891, Congress granted the

Office of the President the authority to create forest reserves on existing federally owned lands. This policy was enlarged by passage of the Organic Administration Act of 1897. One of the stated purposes of the Act was

to improve and protect the forest within the reservation, or for the purpose of securing favorable water flows, and to furnish a continuous supply of timber for the use and necessities of the citizens of the United States. (57)

Gifford Pinchot was the first head of the Division of Forestry in 1881 and promoted the concept of "wise use." He believed timber resources were meant to be harvested and used, but that land must be protected, trees replanted, and resources conserved for future generations. The Division of Forestry's activities overlapped those of the Department of Interior and the Department of Agriculture until Congress created the U.S. Forest Service in 1905.

President Theodore Roosevelt more than doubled the acreage within U.S. forest reserves in the early 1900s, which increased to 235,937 square miles (611,078 km^2) by 1907. Concerned about this rapid growth, Congress renamed the reserves "national forests." It also took over responsibility for future additions from the Executive Office of the President.

A CLOSER LOOK

For decades U.S. presidents have attempted to reduce duplication of federal programs, but too often with limited success. During the 1920s and 1930s, debates arose over which federal agency should house the U.S. Forest Service. President Herbert Hoover's efforts to move it from the Department of the Interior to the Department of Agriculture were defeated in the early 1930s. President Franklin Roosevelt continued the controversy by transferring lands from the USFS to the National Park Service. In 1947, President Harry Truman established a Commission to review the organization of all federal land agencies, but no recommendations were enacted.

In the 1950s, President Dwight Eisenhower, after establishing a committee to review the same question considered by the Truman Commission, agreed with Hoover's idea to move the USFS to the U.S. Department of Agriculture. In the early 1960s, President John F. Kennedy encouraged federal agencies to develop a cooperative attitude between federal land agencies, and a Commission was established to that end. In 1970, under President Richard Nixon, a plan was floated to create a federal department of natural resources, which would merge the Bureau of Land Management and the U.S. Forest Service with certain functions of the U.S. Army Corps of Engineers and the U.S. Department of Interior, but no action was taken. President Jimmy Carter established a comprehensive review of all federal agencies when he came into office in 1977, but his development of the "Hit List" for western water projects, along with other perceived "negative" proposals, created little change in Congress. In 1985, President Ronald Reagan proposed swapping lands between the BLM and the U.S. Forest Service to improve administration and efficiency, but this idea, too, eventually failed.

POLICY ISSUE

The definition of forest management has become a matter of debate within the U.S. Forest Service recently as timber harvesting, protection of old growth forests, and the need to protect water quality and quantity increase. The USEPA and the U.S. Fish & Wildlife Service consult with the U.S. Forest Service to review forest management plans for issues related to environmental protection as well as water quality issues. As forest management evolves, this debate is expected to continue among these federal agencies, Congress, and U.S. voters.

FEDERAL ENERGY REGULATORY COMMISSION (FERC) http://www.ferc.gov

OVERVIEW AND BRIEF HISTORY

The Federal Power Commission was formed in 1920 and was renamed the **Federal Energy Regulatory Commission** (**FERC,** or **Commission**) in 1977. It is an independent agency located within the U.S. Department of Energy and has a five-member board appointed by the president and confirmed by the U.S. Senate. The Commission regulates nonfederal hydroelectric projects that meet at least one of the following criteria:

- Projects are located on navigable rivers.
- They are located on nonnavigable waters over which Congress has Commerce Clause jurisdiction, were constructed after 1935, and affect the interests of interstate or foreign commerce.
- They use surplus water or water power from a federal dam (usually a U.S. Army Corps of Engineers or Bureau of Reclamation dam).

The Commission regulates most of the nonfederal hydroelectric power production, which represents about 56 percent of the nation's hydroelectric capacity. Most of the remaining 44 percent is federally developed, primarily by the U.S. Army Corps of Engineers, the Bureau of Reclamation, and the Tennessee Valley Authority. (58)

The Commission issues two forms of authorization to construct and operate a hydroelectric project: licenses and exemptions from licenses. A project includes all lands, water, and facilities needed to carry out project purposes. A typical hydroelectric project consists of:

- a dam, a reservoir, and a penstock (pipe) that diverts water from a reservoir to the turbine
- a powerhouse containing the turbine and generator
- a channel or pipe returning the diverted water downstream
- a transmission line connecting the project to the power grid
- the lands encompassing the above facilities
- the necessary water rights to operate the project

Projects can consist of multiple facilities (e.g., two or more dams and reservoirs, and more than one powerhouse).

A licensee can be a state, municipality, U.S. corporation, U.S. citizen, or an association of U.S. citizens. Licenses are issued for terms up to 50 years. The Commission issues licenses for projects that, in the Commission's judgment, are best adapted to improve or develop a waterway for beneficial public purposes. Before reaching a final decision on a license application, the Commission must explore all issues relevant to the public interest and give equal consideration to developmental (power) and nondevelopmental (e.g., recreation, wildlife, fishery, etc.) values. Comments are requested from federal, state, and local agencies, nongovernmental organizations, and local citizens. Also consulted are the U.S. Fish & Wildlife Service, U.S. Environmental Protection Agency, state fish and wildlife agencies, and other relevant agencies.

An exemption from licensing is generally intended for small projects with minimal environmental impacts. The Commission may exempt from some or all licensing requirements two types of projects where new capacity is being added:

- constructed conduits, generally irrigation works, that are issued for nonmunicipal projects under 15 megawatts (MW) and municipal projects under 40 MW; and
- 5 MW projects, which are projects proposing additional capacity (5 MW or less) and using an existing dam or natural water feature. (59)

Exemptions are issued in perpetuity, subject to mandatory conditions set by federal and state fish and wildlife agencies. The exemptee must already own the necessary land or, for a project on federal lands, must obtain a use permit from the land management agency.

POLICY ISSUE

The relicensing of existing hydroelectric projects comprises the majority of current license applications pending at FERC. (60) The processing of these applications frequently involves contentious environmental issues, such as minimum streamflows, fish mortality, and changes in project operation. In rare cases, such as the Edwards Dam in Maine, FERC has ordered the removal of a water storage facility for environmental reasons.

In 1991, owners of the Edwards Dam applied for a 50-year FERC license to expand electrical generation capacity from 3.5 to 11.5 megawatts. However, Maine Governor John McKernan called for removal of the dam, which was built in 1837, in order to promote fishery restoration along the Kennebec River. The Maine Legislature agreed with Governor McKernan's position and passed a resolution calling for removal of the dam. In 1996, the U.S. Fish & Wildlife Service specified a $9 million fish passage design as a condition of the FERC permit. The dam owners objected to this requirement, and in 1997 FERC ordered removal of the dam at the owner's expense. In 1999, ownership of the dam was transferred to the State of Maine; it was later breached and demolished. (61)

NATIONAL MARINE FISHERIES SERVICE (NMFS) http://www.noaa.gov/nmfs

OVERVIEW AND BRIEF HISTORY

The **National Marine Fisheries Service** (**NMFS** or **NOAA Fisheries**) is part of the National Oceanic and Atmospheric Administration (NOAA), which is located within the U.S. Department of Commerce. The NMFS is responsible for the protection of most marine mammals in the United States. It was founded in 1871 as the U.S. Commission of Fish and Fisheries and was originally created to investigate fish declines in New England waters. It was one of two bureaus in the U.S. Fish & Wildlife Service that was established by the Wildlife Act of 1956 (the other being the Bureau of Sport Fisheries and Wildlife). The NMFS was split off and placed in the Department of Commerce in 1970 during a major governmental reorganization. The NMFS currently manages 3.4 million square miles (8.8 million km^2) of ocean and coastal area. (62)

POLICY ISSUE

The National Marine Fisheries Service is extensively involved in the protection and reintroduction of endangered salmon in the Pacific Northwest. This program has created great controversy with dam operators along the Columbia River and its tributaries, as well as with the general public. In addition, the NMFS is working to protect the coho salmon on the Klamath River in southern Oregon. As discussed earlier, this effort is creating tremendous conflict with irrigators in the area. (These issues will be discussed further in Chapter 12.)

FEDERAL EMERGENCY MANAGEMENT AGENCY (FEMA)

http://www.fema.gov

OVERVIEW AND BRIEF HISTORY

The **Federal Emergency Management Agency (FEMA)** is an independent agency that reports directly to the president. Its mission is to plan for, respond to, and assist in recovery from disasters. The Agency can trace its roots to the Congressional Act of 1803, which was the first piece of federal disaster response/recovery legislation in the United States. (63) The 1803 Act was passed in response to a major fire that destroyed much of the commercial district in Portsmouth, New Hampshire. Over the next 175 years various federal agencies were created to assist with natural and human-caused disasters; efforts soon overlapped and were uncoordinated. (64)

In 1979, President Jimmy Carter signed an executive order creating FEMA and merged the activities of many other federal assistance programs. Since its formation, FEMA has assisted with earthquakes, hurricanes, floods, and other disasters, both natural and man-made. One of the first disasters which FEMA responded to was the contamination disaster at Love Canal near Buffalo, New York, in the late 1970s. An excellent source of information on Love Canal can be found at the State University of New York at Buffalo website: http://ublib.buffalo.edu/libraries/projects/lovecanal

Additional responsibilities of FEMA are the development of floodplain maps and administration of the **National Flood Insurance Program (NFIP).** The NFIP provides federally backed flood insurance to communities that agree to adopt and enforce floodplain management ordinances (primarily zoning laws). In 1968, the U.S. Congress enacted the **Flood Insurance Act** to

encourage state and local governments to make appropriate land use adjustments, to constrict the development of land which is exposed to flood damage and minimize damage caused by flood losses, and guide the development of proposed future construction, where practicable, away from locations which are threatened by flood hazards. (65)

NFIP floodplain hazard maps delineate 100- and 500-year floodplains on U.S. Geological Survey planimetric maps (maps representing only horizontal positions of features.) When possible, three floodplain conditions are shown: 100-year floodplain, existing conditions of the 500-year floodplain, and future development conditions of the 100-year floodplain (Figure 9.9). FEMA requires the regulation of all development within the 100-year floodplain. Mortgage lenders for homes and businesses require flood insurance if a structure is located within a FEMA 100-year flood zone designation. Local floodplain managers (typically a municipal or county planning agency) regulate development in the 500-year floodplain and may or may not require flood insurance.

Regulation of development in flood-prone areas will continue to evolve as growth encroaches onto floodplains. FEMA floodplain designations require flood insurance for all structures within the 100-year floodplain but do not require coverage beyond that flood zone.

The Mississippi River Flood of 1993 was considered a 500-year flood, and owners of structures located beyond the 100-year flood zone (and many within the zone) were entitled to disaster relief. During the Clinton Administration, FEMA spent approximately $1 billion to buy out 27,000 flood-prone structures, about half of those in the Midwest. The federal government even purchased and relocated the entire town of Valmeyer, Illinois, to higher ground to eliminate repeated flood insurance claims. Prior to the 1997 spring flood, FEMA spent $300,000 in advertising, asking people beyond the 100-year floodplain to purchase flood insurance, but only about 10 percent of flood victims had flood insurance coverage. A National Wildlife Federation study in the 1990s showed that 40 percent of all flood insurance payments went to repeat victims but represented only 2 percent of all policyholders. The owner of one house in Houston, Texas, valued at $114,480, received payments worth $806,591 from 16 floods in 18 years. (66)

Local, state, and federal officials, and taxpayers, are becoming aware of current problems with multiple flood damage payments to flood victims. Policy changes will continue to occur in the coming decades to reduce the need for disaster payments by relocating flood-prone homes, businesses, and communities, or by reducing or eliminating repetitive claims.

POLICY ISSUE

The Merriam-Webster Dictionary defines politics as "the art or science concerned with guiding or influencing governmental policy." Throughout this chapter you have read examples of politically motivated water resource policy decisions. For example, in 1808, the Gallatin Report was submitted to Congress for the purpose of identifying a plan for federal involvement in the construction of

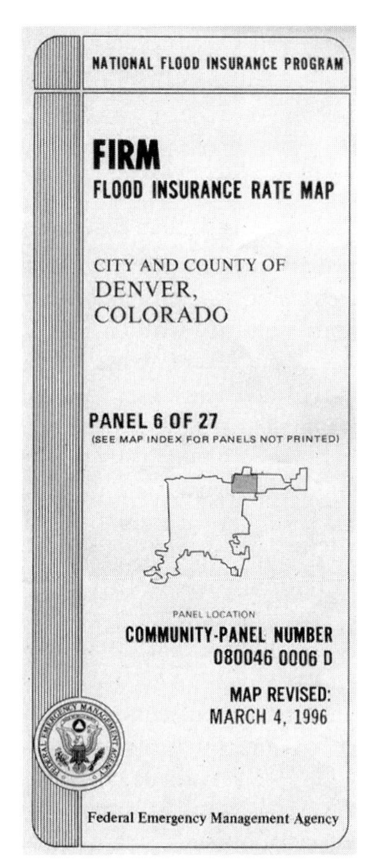

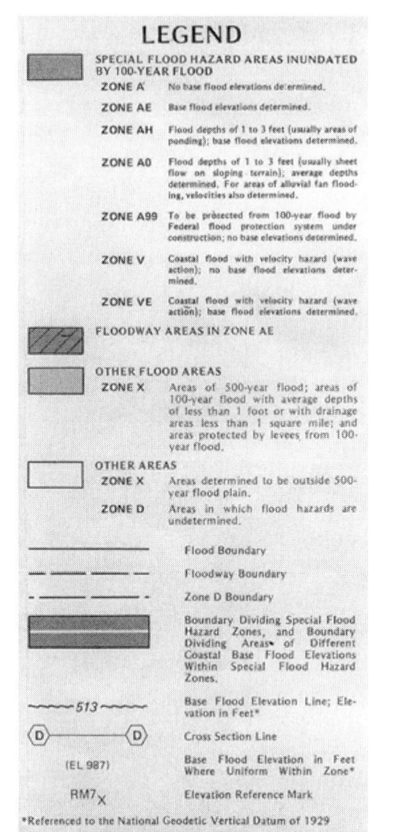

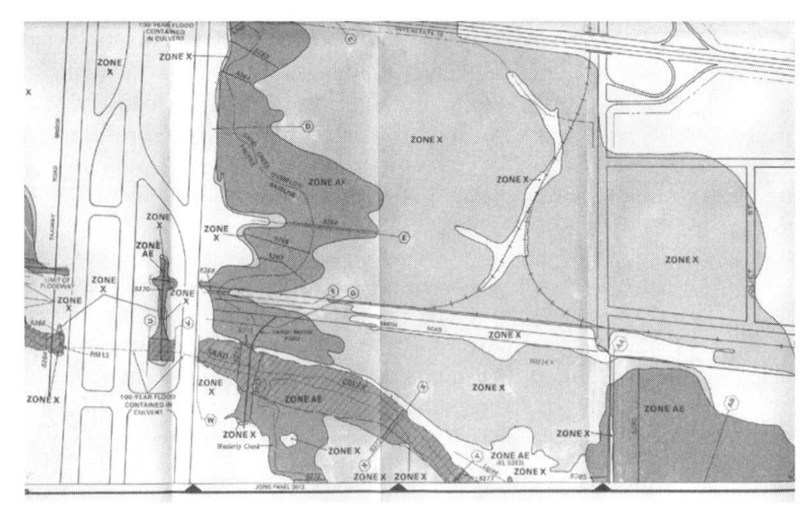

FIG. 9.9. FEMA floodplain map showing 100- and 500-year floodplains near old Stapleton International Airport in Denver, Colorado. Go to http://www.hazardmaps.gov/atlas.php for access to floodplain maps and other online hazard advisory maps including earthquakes and hurricanes.

Another interesting FEMA website can be found at http://www.gismaps.fema.gov. It provides Geographic Information Systems (GIS) maps of current and past declared disasters and tropical storms/hurricanes, and updated FEMA news.

The events of September 11, 2001, focused FEMA's activities on national preparedness and homeland security. The agency coordinated its activities with the newly created office of Homeland Security. FEMA created an Office of National Preparedness, which is responsible for training and equipping first responders for dealing with weapons of mass destruction.

roads and waterways. Interstate commerce was a driving force in the early construction of canals and navigable waterways, yet the U.S. Army Corps of Engineers was precluded from conducting much work on waterways in the early 1800s. This was partially true because of inadequate federal funds at the time, but their progress may have also been slowed owing to the concerns of elected officials in competing regions of the country. Representatives from the Ohio River Basin tried to obtain federal funds for channelization improvements on the Ohio River. Yet, others argued this was unfair and, since it only benefited the Ohio River Valley, it was too parochial for the use of funds from Washington.

The policy of "no net loss of wetlands" was promoted by the first Bush Administration, and was continued by the Clinton Administration in the 1990s. In 2001, the second Bush Administration attemped to refine, and in the opinion of some, loosen regulatory control over development near and in wetlands areas. What role did health, environmental, and economic development play in these political battles? Do you believe regionalism played a role in these debates, or was economics the root of these political fights?

The Klamath controversy, which reached its climax in 2002, is an excellent example of one federal agency promoting a historic policy, while another agency obtains political power to promote change. The U.S. Bureau of Reclamation built the Klamath Irrigation Project in 1905 in order to promote agriculture and economic growth in the dry lands of southern Oregon. However, through passage of the Endangered Species Act in 1973 and involvement of the U.S. Fish & Wildlife Service, the prior policy of water development was abrogated in the interest of protecting threatened and endangered species in the region. Millions of dollars of federal funds are now being spent to change a prior federal policy that promoted water development. Should local irrigators be compensated for the loss of their irrigation water, or should beneficiaries of federal subsidies have no recourse when public policy changes? What political power struggles had to occur to end the long-standing water development era in Congress? Do you believe this particular shift in public policy is good or bad? Why?

CHAPTER SUMMARY

Congress has given federal water agencies in the United States a wide range of legal duties. Historically, the primary purpose of these agencies was to promote the settlement of frontier territories. The concept of Manifest Destiny played a major role in the development of water policy.

After World War II, the federal government's focus shifted from water development to environmental protection and conservation. New agencies such as the U.S. Environmental Protection Agency were created, while others such as the U.S. Fish & Wildlife Service were given expanded missions. This contrasted greatly with the historic mission of development agencies such as the Bureau of Reclamation. The result has been conflict, sometimes between federal agencies due to opposing federal missions and at other times between existing federal policy and diverging constituencies.

The role of local and state water agencies is at times just as divergent as the role of federal agencies. Competition for scarce water supplies, ideological differences, and geographic variability often create conflict within states or regions. Chapter 10 will explore these issues further and will contrast the water resource duties of local, regional, state, and multistate water management agencies.

QUESTIONS FOR DISCUSSION

1. In the context of Manifest Destiny, contrast the historic activities of the Bureau of Reclamation with those of the U.S. Environmental Protection Agency.

2. The Council on Environmental Quality was created to coordinate federal environmental protection efforts, as well as to act as a "referee" when agencies disagree over the adequacy of environmental assessments. Is it appropriate for an agency within the executive branch of the federal government to serve in this role?

3. Federal agencies often compete for limited federal funds. Unfortunately, this sometimes occurs at the expense of coordinating efforts to protect the nation's water resources. What role does politics play in this competition between agencies? How could a federal agency utilize a local constituency to promote its own budget requests before Congress?

4. President Theodore Roosevelt added 235,937 square miles (611,078 km^2) of federal lands to U.S. forest reserves in the early 1900s. Congress responded by renaming the reserves "national forests" and taking over responsibility for future additions from the Executive Office of the President. Why would Congress become concerned over such an action of the executive? What constituency probably lobbied Congress extensively in the early 1900s for legislative reform over designation of lands into forest reserves?

5. In 1997, FERC ordered the owners of the Edwards Dam in Maine to remove the dam due to detrimental effects on migrating fish in the Kennebec River. The dam had been in place since 1837. Is it appropriate for the federal government to order the removal of private property if it negatively affects the environment? What environmental, economic, and health benefits could occur under this type of federal policy? What detrimental effects could this cause to the economy or to private property rights issues?

6. A National Wildlife Federation study in the late 1990s showed that 40 percent of all flood insurance payments went to repeat victims but represented only 2 percent of all policyholders. For example, the owner of one house in Houston, Texas, valued at $114,480, received payments worth $806,591 from 16 floods in 18 years. What federal legislation should be passed to stop this type of abuse of FEMA assistance?

7. The first National Irrigation Congress was held in Salt Lake City in 1891 and led to strong political support for federally funded irrigation projects in the West. Less than a decade later, the first of several National Rivers and Harbors Congresses was held. What was the relative political importance of these types of meetings? Why do you suppose the navigation and flood-control interests decided to replicate the political efforts of the irrigation supporters? Do similar groups exist today?

KEY WORDS TO REMEMBER

Bureau of Land Management (BLM) p. 268

Bureau of Reclamation (USBR, Reclamation, or BuRec) p. 257

Clean Water Act Amendments (1972) p. 255

Clean Water Act Amendments (1977) p. 270

Comprehensive Environmental Response, Compensation, and Liability Act (1980) p. 271

Council on Environmental Quality p. 270

Desert Land Act (1877) p. 259

Endangered Species Act (1973) p. 265

Environmental impact statement (EIS) p. 270

Federal Emergency Management Agency (FEMA) p. 276

Federal Energy Regulatory Commission (FERC) p. 274

Federal Land Policy and Management Act (1976) p. 269

Federal Water Pollution Control Act (1956) p. 269

Federal Water Pollution Control Act Amendments (Clean Water Act Amendments) of 1972 p. 270

Fish and Wildlife Coordination Act (1958) p. 265

Flood Control Act (1928) p. 252

Flood Control Act (1936) p. 253

Flood Control Act (1944) p. 253

Flood Insurance Act (1968) p. 277

Gallatin Report (1808) p. 249

General Survey Act (1824) p. 249

Homestead Act (1862) p. 268
Louisiana Swamp Land Act (1849) p. 254
Mining Law (1872) p. 268
National Environmental Policy Act (1969) p. 270
National Flood Insurance Program (NFIP) p. 277
National Irrigation Congress p. 259
National Marine Fisheries Service (NMFS or NOAA Fisheries) p. 276
National Park Service (Park Service or NPS) p. 266
National Pollutant Discharge Elimination System (NPDES) Permits p. 271
National Rivers and Harbors Congress (1901) p. 250
Natural Resources Conservation Service (NRCS) p. 272
Reclamation Act (1902) p. 260
Refuse Act (1899) p. 249
Resource Conservation and Recovery Act (1976) p. 270
Rivers and Harbors Act (1899) p. 249
Safe Drinking Water Act (1974) p. 270
Safe Drinking Water Act Amendments (1996) p. 271
Section 404 Permit p. 255
Superfund p. 271
Swamp Lands Act (1850) p. 254
Swamp Lands Act (1860) p. 254
Taylor Grazing Act (1934) p. 268
U.S. Army Corps of Engineers (Corps or USACE) p. 248
U.S. Department of Agriculture (USDA) p. 272
U.S. Department of Interior (DOI or Interior) p. 257
U.S. Environmental Protection Agency (USEPA) p. 269
U.S. Fish & Wildlife Service (USFWS) p. 264
U.S. Forest Service (USFS) p. 273
U.S. Geological Survey (USGS) p. 262
Water Pollution Control Act (1948) p. 269
Water Quality Act (1965) p. 269

SUGGESTED RESOURCES FOR FURTHER STUDY

READINGS

Barry, John M. *Rising Tide—The Great Mississippi Flood of 1927 and How It Changed America.* New York: Simon & Schuster, 1998.

Berkman, Richard L., and W. Kip Viscusi. *Damming the West—Ralph Nader's Study Group Report on the Bureau of Reclamation.* New York: Grossman Publishers, 1973.

Black, Peter E. *Conservation of Water and Related Land Resources.* 2nd ed. Totowa, NJ: Rowman & Littlefield, 1987.

Dunbar, Robert G. *Forging New Rights in Western Waters.* Lincoln: University of Nebraska Press, 1983.

Dzurik, Andrew A. *Water Resources Planning.* 2nd ed. Lanham, MD: Rowman & Littlefield, 1996.

Heat-Moon, William Least. *River-Horse: Across America by Boat.* New York: Penguin Books, 2001.

Ise, John. *Our National Park Policy—A Critical History.* Baltimore, MD: Johns Hopkins University Press, 1961.

James, George Wharton. *Reclaiming the Arid West—The Story of the United States Reclamation Service.* New York: Dodd, Mead & Co., 1917.

Morgan, Arthur E. *Dams and Other Disasters—A Century of the Army Corps of Engineers in Civil Works.* Boston: Porter Sargent Publisher, 1971.

Powell, John Wesley. *Report on the Lands of the Arid Region of the United States.* 2nd ed. Washington, DC: U.S. Government Printing Office, 1878.

Reuss, Martin. "Andrew A. Humphreys and the Development of Hydraulic Engineering: Politics and Technology in the Army Corps of Engineers, 1850–1950." *Technology and Culture* 26 (1985), 1–33.

Stegner, Wallace. *Beyond the Hundredth Meridian.* 3rd Printing. Boston: Houghton Mifflin Co., 1954.

Thompson, Stephen A. *Water Use, Planning, and Management in the United States.* San Diego, CA: Academic Press, 1999.

Wiley, Peter, and Robert Gottlieb. *Empires in the Sun—The Rise of the New American West.* New York: G.P. Putnam's Sons, 1982.

Worster, Donald. *A River Running West: The Life of John Wesley Powell.* New York: Oxford University Press, 2001.

WEBSITES

Delaware River Basin Commission, "A Brief History of Stream Gages," June 2003. http://www.state.nj.us/drbc/gage/history.htm

Federal Emergency Management Agency, FEMA. NFIP—National Flood Insurance Program, August 2001. http://www.fema.gov/fima/nfip.shtm

Federal Emergency Management Agency, ESRI/FEMA Project Impact Hazard Site, "Online Hazard Maps," June 2003. http://www.esri.com/hazards

Library of Congress, "Thomas: Legislative Information on the Internet," June 2003. http://thomas.loc.gov

The National Environmental Policy Act—A Study of Its Effectiveness after Twenty-five Years, Council on Environmental Quality, Executive Office of the President, Washington, DC, January 1997. See http://ceq.eh.doe.gov/nepa/nepa25fn.pdf, August 2001.

U.S. Army Corps of Engineers, Institute for Water Resources, June 2003. http://www.iwr.usace.army.mil/ndc/index.htm

VIDEOS

"Fatal Flood—Story of the Mississippi River Flood of 1927." Directed by Chana Gazit, A Steward/Gazit Productions, Inc., for AMERICAN EXPERIENCE, WGBH Educational Foundation, Boston, Massachusetts, 2001.

"Tennessee Valley Authority." *Modern Marvels Series,* The History Channel, 1995. 50 min.

REFERENCES

1. Grant, Ulysses S. *Grant: Personal Memoirs of U.S. Grant, Selected Letters 1839–1865.* New York: The Library of America, 1990, 774.

2. Personal communication with Dr. Marty Ruess, U.S. Army Corps of Engineers, Senior Historian in Water Resources, Alexandria, Virginia, May 21, 2001.

3. Ibid.

4. Thompson, Stephen A. *Water Use, Planning, and Management in the United States.* San Diego, CA: Academic Press, 1999, 33.

5. Heat-Moon, William Least. *River-Horse: Across America by Boat.* New York: Penguin Books, 2001, 215.

6. Dzurik, Andrew A. *Water Resources Planning.* 2nd ed. Lanham, MD: Rowman & Littlefield, 1996, 48; and personal communication with Dr. Marty Ruess, May 21, 2001.

7. U.S. Army Corps of Engineers, "Did You Know—Facts about Locks," Navigation Data Center, Alexandria, Virginia, http://www.wrsc.usace.army.mil/ndc/fcdidu2.htm, August 2001.

8. U.S. Army Corps of Engineers, http://www.wrsc.usace.army.mil/ndc/fedidu2.htm, January 2001.

9. Personal communication with Robert Bank, U.S. Army Corps of Engineers Headquarters, Washington, DC, May 9, 2001.

10. Personal communication with Dr. Marty Ruess, May 21, 2001.

11. Barry, John M. *Rising Tide—The Great Mississippi Flood of 1927 and How It Changed America.* New York: Simon & Schuster, 1998, 16; and U.S. Army Corps of Engineers. "The Mississippi River and Tributaries Project." New Orleans District, http://www.mvn.usace.army.mil/pao/bro/misstrib.htm, August 2001.

12. Personal communication with Dr. Marty Ruess, September 9, 2001.

13. U.S. House of Representatives, House Appropriations Committee, Part 4: Department of the Army, 1961, 69–71.

14. Thompson, 49.

15. Ibid., 51.

16. Morgan, Arthur E. *Dams and Other Disasters—A Century of the Army Corps of Engineers in Civil Works.* Boston: Porter Sargent Publisher, 1971, 47.

17. Ibid., 43.

18. Ibid., 47–50.

19. "Revisions to the Clean Water Act Regulatory Definition of Discharge of Dredged Material." *Federal Register* 64, No. 89. Washington, DC: U.S. Government Printing Office, May 10, 1999. Also found at http://www.epa.gov/OWOW/wetlands/tulloch.html, August 2001.

20. Personal communication with Peter Soeth, Public Affairs Specialist, U.S. Bureau of Reclamation, Denver, Colorado, March 22, 2001.

21. U.S. Bureau of Reclamation, Homepage, http://www.usbr.gov, January 2001.

22. Stegner, Wallace. *Beyond the Hundreth Meridian.* 3rd Printing. Boston: Houghton Mifflin Co., 1954.

23. Powell, John Wesley. *Report on the Lands of the Arid Region of the United States.* 2nd ed. Washington DC: U.S. Government Printing Office, 1878, viii.

24. James, George Wharton. *Reclaiming the Arid West—The Story of the United States Reclamation Service.* New York: Dodd, Mead & Co., 1917, xvi.

25. Wiley, Peter, and Robert Gottlieb. *Empires in the Sun—The Rise of the New American West.* New York: G. P. Putnam's Sons, 1982, 13.

26. Simonds, Joe. "The Newlands Project." U.S. Department of the Interior, Bureau of Reclamation, Bureau of Reclamation History Program, 1996, http://dataweb.usbr.gov/html/newlands1.html, August 2001.

27. James, 24.

28. Dunbar, Robert G. *Forging New Rights in Western Waters.* Lincoln: University of Nebraska Press, 1983, 58.

29. Berkman, Richard L., and W. Kip Viscusi. *Damming the West—Ralph Nader's Study Group Report on the Bureau of Reclamation.* New York: Grossman Publishers, 1973, 14–28.

30. Public Works for Water and Power and Energy Research Appropriation Bill, Hearings before a Subcommittee of the Committee on Appropriations, U.S. House of Representatives (95th Cong., 1st Sess., 1977), 225–226.

31. Personal communication with Dr. Marty Ruess, U.S. Army Corps of Engineers, May 21, 2001.

32. The Family Farm Alliance. Family Farm Water Review. "No Water: Klamath Project Gets Zero Supply," Issue No. 51, April 2001.

33. Personal communication with Glenn G. Patterson, U.S. Geological Survey, Reston, Virginia, May 26, 2001.

34. Delaware River Basin Commission. "A Brief History of Stream Gages," http://www.state.nj.us/drbc/gage/history.htm, October 2001.

35. Personal communication with Glenn G. Patterson, May 26, 2001.

36. Ibid.

37. Personal communication with Mitch Snow, Chief of Media Services, U.S. Fish & Wildlife Service, Washington, DC, May 22, 2001.

38. U.S. Fish & Wildlife Service. *The Road Back, Endangered Species Recovery—Success with Partners.* Washington, DC: U.S. Government Printing Office, no date, 49.

39. Personal communication with Mitch Snow, May 22, 2001.

40. U.S. Fish & Wildlife Service, "National Fish Hatchery System," http://fisheries.fws.gov/FWSFH/NFHSmain.htm, September 2003.

41. Ibid.

42. Personal communication with Dr. Dwight Pitcaithley, Chief Historian, National Park Service, Washington, DC, July 6, 2001.

43. Ise, John. *Our National Park Policy—A Critical History.* Baltimore, MD: Johns Hopkins University Press, 1961, 3.

44. Personal communication with Dr. Dwight Pitcaithley, July 6, 2001.

45. *Cappaert v. United States,* 426 U.S. 128 (1976).

46. U.S. Department of Interior, Bureau of Land Management Homepage, http://www.blm.gov, August 2001.

47. Ibid.

48. Ibid.

49. Ibid.

50. Dzurik, 51; and personal communication with Amy Zimmerling, U.S. Environmental Protection Agency, Washington, DC, June 8, 2001.

51. National Environmental Policy Act, Public Law 91–190, 42 U.S.C., 4321–4347, January 1, 1970.

52. U.S. Environmental Protection Agency, http://www.epa.gov/earthday/timeline/htm, February 2001.

53. Personal communication with Amy Zimmerling, June 8, 2001.

54. Thompson, 63.

55. 33 U.S.C. § 1251(a).

56. Natural Resources Conservation Service Homepage, http://www.nrcs.usda.gov, January 2001.

57. The Organic Act of 1897 (30 Stat. 34, as amended, 16 U.S.C. 475).

58. Personal communication with John Paquin, Program Analyst, Office of Energy Projects, Federal Energy Regulatory Commission, Washington, DC, October 1, 2001.

59. Ibid.

60. Ibid.

61. "A Brief History of the Edwards Dam," http://www.state.me.us/spo/edwards/timeline.htm, August 2001.

62. Personal communication with Mitch Snow, May 22, 2001.

63. Personal communication with Holly Harrington, Public Affairs Officer, Federal Emergency Management Agency, Washington, DC, May 24, 2001.

64. Federal Emergency Management Agency. "FEMA: History of the Federal Emergency Management Agency." http://www.fema.gov/about/history.htm, August 2001.

65. Flood Insurance Act, 42 U.S.C. 4001 (e)(1968).

66. National Wildlife Federation. "Higher Ground Report: A Report on Voluntary Property Buyouts in the Nation's Floodplains," http://www.nwf.org/floodplain/higherground/index.html, August 2001.

CHAPTER 10

LOCAL, REGIONAL, STATE, AND MULTISTATE WATER MANAGEMENT AGENCIES

Historically, the rights of the individual have been strongly protected in the United States and many other countries in the world. From the declarations made in the Bill of Rights to the federal settlement policies of the West, freedom of choice and opportunity for landownership have been highly regarded in America. However, as cities and regions became more congested, society demanded safe and reliable supplies of water, sanitary elimination of wastes, and protection from natural disasters such as floods.

In the early settlement days of the nation, settlers dug their own water supply wells and discarded wastes into ravines, rivers, or other unobjectionable locations. Too often, little or no attention was given to the effects of these activities on downstream landowners or the environment. Conflict increased as water shortages and water pollution became more common. Disease, the need for reliability, and improved technology ultimately led to the creation of local, regional, state, and multistate water agencies to serve communities.

LOCAL WATER AGENCIES

The term *grassroots* is often used to describe the desire of people to control issues of local importance. Individuals and groups generally prefer that local concerns be addressed at the community level, particularly in rural areas. Federal intervention in local water issues is often considered intrusive, but at the same time, federal assistance is generally welcomed for flood control, navigation, or water supply development. This dichotomy of views is an interesting subtext to the management of water resources at the local level.

MUNICIPAL WATER DEPARTMENTS

Municipal water departments are local agencies, operated within a town or city government, that provide drinking water to residents. Revenues are generally derived from fees (water bills, see Chapter 6) and local taxes. Many municipal water departments control or own raw (untreated) sources of surface or groundwater, a water treatment plant, and delivery systems (Figure 10.1). However, some municipal departments obtain raw or treated water from regional or state water providers. Staff generally includes operations, office, and maintenance personnel, and annual budgets that can range from under $250,000 to millions of dollars.

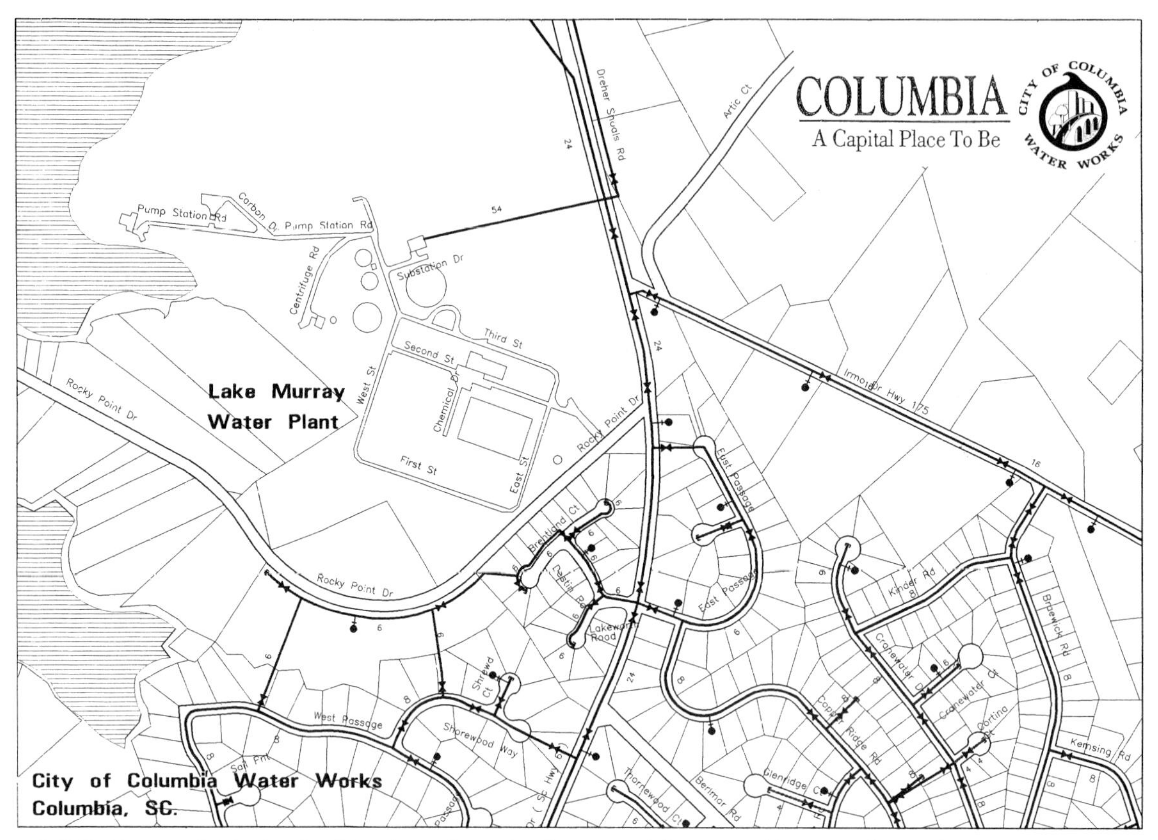

FIG. 10.1. The city of Columbia, South Carolina, is located in the center of the state and has a population of approximately 110,000. The Lake Murray Water Plant is one of two water treatment facilities in the city, and together they have a total capacity of 126 mgd (476,962 m^3/day). The city has approximately 1900 miles (3058 km) of waterlines that vary in diameter from 4 to 54 inches (10 to 137 cm) as shown on this map. Approximately 200 employees serve a total regional population of 350,000.

Historical Overview Early settlements (towns, cities, townships, and parishes) created the first municipal water supply departments in the United States. In the 1600s, eastern communities were located near plentiful water supplies, such as streams, lakes, or artesian springs. Water was generally carried to homes and small businesses in buckets. Water was plentiful in this humid region, the population was small, and pollution was a word not found in their vocabulary. However, as more settlers arrived, water delivery and waste removal became important. Many settlements were forced to dig additional wells or to find alternative locations for waste and other refuse. In larger communities, entrepreneurs sometimes formed private water supply companies to provide these services for a fee. In other communities, local residents directed the mayor, city council, town council, or other local governing body to use public funds (revenues from taxes) to construct municipal water systems.

In New York City, for example, early residents used groundwater from neighborhood wells. New York City's first public water well was constructed in 1666 under the direction of the English governor Richard Nicolls. The second public well, built in 1671, was lined with stone and located near the city tavern, jail, and court. The city's first executioner, Benjamin Johnson, was responsible for operation of the well since he lived so close by (he called the city tavern his home). The importance of sound, professional management of drinking water supplies was not well understood in the 1600s. In *Water for Gotham,* Koeppel states:

> *The killing business apparently was slow because six months into his January 1671 appointment Johnson was presenting bills* [to the local government] *for having the well built and in July was paid some 195 florins to cover his various disbursements . . .* [later] *he was convicted of running a theft ring and sentenced to languishing, severing an ear, and banishment.* (1)
>
> Gerard T. Koeppel, *Water for Gotham,*
>

Contamination of New York City's shallow wells soon became a problem. Human and horse waste, refuse, and waste from roaming pigs percolated into the shallow groundwater aquifer of the area and led to widespread outbreaks of cholera. Only the wealthy could afford to purchase artesian spring water carted by entrepreneurs to their homes. In 1832, a severe cholera epidemic in the city forced public officials to devise a better water system. Surface water supplies were developed from the Croton River in Westchester County to the north, and groundwater sources were eventually abandoned. (2)

Fire protection was also a major concern for cities during their early formation. Deadly fires in New York City in 1776 destroyed nearly one-quarter of all houses in the city, and there were other large fires in 1828 and 1835. These tragedies taught the communities that small, hand-dug wells were simply not capable of providing safe and adequate water supplies for a growing city. (3)

As the populations of towns and cities grew, municipal water departments (variously called the Department of Public Works, Department of Public Utilities, Water and Sewer Department, and Waterworks Department) were created to administer the day-to-day operations of water supply and wastewater removal and treatment. Modern municipal water departments are generally responsible for all aspects of water treatment, delivery, and wastewater treatment in a community. This responsibility includes construction and operation of dams, reservoirs, wells, transmission pipelines, drinking water treatment plants, and wastewater treatment plants. A municipal agency must comply with local, state, and federal environmental and water quality regulations, follow budgeting guidelines, and develop long-range water plans and forecasts to meet future demands. Contingency plans are developed for wet and dry cycles, and drought emergency plans are usually drawn up to meet critical water supply shortages.

Managers of municipal water departments are hired by the mayor, city council, or city manager, depending on the governmental organization of the community. If a city is large or has major water issues, a separate water board may be created. This board of directors, composed of residents knowledgeable about water resources, is appointed by the mayor or is elected by local residents. A water board may have ultimate control over all water issues in a community, or it may simply serve in an advisory capacity.

POLICY ISSUE

Municipal water departments are somewhat autonomous and independent from federal water agencies. Local water agencies are controlled by local water boards, town or city councils, or other grassroots organizational structures. However, local water departments must follow federal laws, particularly those related to drinking water standards and endangered species protection. Local water development activities cannot violate federal environmental laws enforced by the U.S. Fish & Wildlife Service, the U.S. Environmental Protection Agency, or the U.S. Army Corps of Engineers. Drinking water quality must meet federally mandated standards of the USEPA (see Chapter 11). Municipal water departments in the western United States often have storage reservoirs located on federal lands managed by the Bureau of Land Management, National Park Service, or U.S. Forest Service. Water diversion and delivery works in these locations may require consultation and cooperation with the USFS, NPS, BLM, USBR, USFWS, and the USEPA.

WATER AND SEWER DISTRICTS

Overview **Water and sewer districts** are very similar to municipal water departments except that they also handle sewage (wastewater) disposal. Water and sewer districts often provide services outside of town or city boundaries and may involve interagency agreements to service adjacent communities if excess capacity is available. Independent government agencies called water districts, water and sewer districts, metropolitan water and wastewater districts, sewer districts, and water and sanitation districts are sometimes formed to serve new growth regions that are beyond an existing municipal service area. These special districts are quasi-public entities created under state law and can be organized by land developers that do not have access to municipal services. Each state legislature determines the legal structure of how a water and/or sewer district is formed and operated.

EXAMPLE

Highline Water District, Kent, Washington

http://www.highlinewater.org

The Highline Water District is located in King County, just south of Seattle, Washington. It was formed in 1946 by a special election of local residents authorized by the laws of the State of Washington. The District currently serves approximately 68,000 people in seven cities. Its 17,400 connections include residential homes, apartments, businesses, schools, and other developments in a 26-square-mile (67-km^2) service area. The District also provides street lighting to a limited area within its boundaries. Prior to 1962, all water was supplied by groundwater wells operated by the District. Since then, most water has been provided by the City of Seattle Public Utility Department, and wells are used only for peaking (supplemental) needs during summer months. Surface water is obtained via a pipeline from a remote watershed in the Cascade Mountains to the west.

An elected board of directors represents consumers at the Highline Water District, and a staff of 40 administers an annual budget of approximately $7.3 million. Public meetings of the board of directors are held twice each month. The district continues to search for new sources of water (primarily groundwater obtained through wells) and implements a proactive maintenance program to ensure a reliable distribution system. (4)

EXPERT ANALYSIS

The boundaries of a water/sewer district are delineated by several factors, including climate, topography, water availability, the location of rivers, major highways, wetlands, and system capacity. In addition, some residents may have individual water and sewer systems (septic systems) that require no public infrastructure. The Metropolitan Water District (discussed in Chapter 6), for example, provides services to 17 million people in a five-county area of southern California. Contrast it to the Highline Water District in Kent, Washington, which provides drinking water to 68,000 people in seven cities. The factors listed above, as well as economies of scale (larger water/sewer systems are generally more efficient), local politics, and available funding sources determine the size and location of water/sewer district boundaries.

If you were a county planner, what factors would you consider in your review of a request by developers to create a new water and sewer district within your jurisdiction? Would you prefer to promote smaller, more grassroots agencies, or would you support a more regional approach to take advantage of economies of scale?

POLICY ISSUE

Water and sewer districts face many of the same issues with federal regulations that were identified for municipal water departments. Expansion plans may require permits from the U.S. Environmental Protection Agency, U.S. Army Corps of Engineers, or state health departments. Wastewater discharges are generally regulated by the USEPA and can require consultation with the U.S. Fish & Wildlife Service or the National Marine Fisheries Service if water quality issues are detrimental to aquatic wildlife.

LEVEE AND FLOOD-CONTROL DISTRICTS

Historical Overview **Levee districts** and **flood-control districts** are tax-funded, locally controlled agencies charged with preventing or minimizing flood damages. Flooding has been a problem in the United States since humans first arrived. Native Americans told ancestral stories of "great floods" that reached as far as the eye could see. Typically, they protected themselves by staying away from the dangerous flood flows. As European settlers arrived on the East Coast and later settled in the Mississippi and Ohio River valleys, floods caused death and destruction with increasing frequency. Too often, communities were located on the fertile floodplains of rivers and streams since cropland was very productive and water supplies were close by. Loss of life and property were often catastrophic during floods.

Initially, state and federal governments assigned flood-control responsibility to individual riparian landowners. This piecemeal approach to flood control was wholly inadequate and uncoordinated. After scores of devastating floods ravished watersheds, many states charged counties, townships, and parishes with responsibility for flood control. Although there was some modest success, massive floods along larger rivers such as the Mississippi and Sacramento led to calls for more federal involvement and financial assistance.

Levee Districts Following the Louisiana Purchase in 1803, settlers began to clear the rich bottomlands along the lower Mississippi River, particularly in the Delta Region of northwest Mississippi. Most new landowners constructed small levees 2 to 3 feet (0.6 to 0.9 m) high to control annual river overflows. Over the next 50 years, the Mississippi Legislature worked with counties in the Delta to construct levees 4 feet (1.2 m) in height along the entire length of the river through the area, but funding for construction was an ongoing problem.

Each levee district had its own board of directors called a *levee board*. These powerful groups of landowners were created by state legislation to construct larger earthen levees (dikes) along the banks of the Mississippi River. The landowners in each levee district paid taxes, and local landowners elected representatives to the levee boards to administer funds. The boundaries of a levee district were generally determined by the lands protected by levee district projects. A position on a levee board enabled individuals to influence the decision as to which lands would be protected—and which lands would be flooded—through the selection of levee locations.

The goal of levee boards was to protect the most property possible belonging to landowners paying taxes to the board. However, this was often detrimental to landowners who had property on the opposite side of the river but were not members of a levee district. Why? The primary goal of a levee board was to construct its levees as high as possible. At the peak of a severe flood, the tallest levee forced floodwaters to the other side of the river and across any dikes that were lower, even if by only a few inches or centimeters.

Levee board members would even dynamite "rival" levees that were causing problems for their own area (upstream or opposite bank levees that were higher or stronger) and during times of flood were forced to protect their structures around the clock with loaded guns. Levee boards were very powerful because of the protection they afforded to landowners.

SIDEBAR

The channelization of the Mississippi River and its tributaries was discussed for decades before significant action was taken:

The most general object I can think of would be the improvement of the Mississippi River and its tributaries.

Representative Abraham Lincoln

The statesmanship of America must grapple with the problem of this mighty stream; it is too vast for any state to handle; too much for any authority other than that of the nation itself to manage.

President James A. Garfield

We [the nation] must build the levees and build them stronger and more scientifically than ever before.

President Theodore Roosevelt

Floods during the Civil War extensively damaged the levee system along the Mississippi River, and few funds were available for repairs. In addition, many levees were deliberately destroyed to flood nearby communities. Other levees were destroyed to allow Union gunboats to move on interior streams to attack cities such as Vicksburg. Additional levees were simply blown up as a type of "scorched earth" policy by Northern troops. (5) These actions were remarkably similar to the acts of war conducted in China around 300 B.C. (see Chapter 1). After the Civil War, levee boards eventually began to function again and rebuild levees. Their primary goals were to build stronger, higher levees and to obtain funding from the federal government.

Levees were generally constructed to protect towns, or in some cases, the landholdings of influential district members. This led to a great controversy over whose property would be flooded for the greater good of a region. In 1881, President James A. Garfield argued that control of the Mississippi River was too great an effort for any one state to handle and would be a tremendous feat for the federal government to accomplish. However, the president did not believe that competing interests of local levee districts could overcome local politics and personal interests. (7) President Garfield died at the hands of an assassin after only eight months in office and did not live to witness the work of the Mississippi River Commission, formed two years earlier.

SIDEBAR

The Mississippi River historically flooded broad areas of the floodplain and inundated side channels that carried floodwaters throughout its valleys. The Mississippi created enormous problems for Northern troops trying to move through the region during the Civil War:

The rains had been so heavy for some time before that the lowlands had become impassable swamps, [General] *Sherman debarked his troops and started out to accomplish the object of the expedition; but the* [Mississippi] *river was rising so rapidly that the back-water up the small tributaries threatened to cut off the possibility of getting back to the boats, and the expedition had to return without reaching the railroad.* (6)

Ulysses S. Grant, 1862

The great Mississippi River flood of 1927 caused a breach at 226 different locations along the various levee district projects. Secretary of Commerce (and later President) Herbert Hoover described it as the greatest peacetime disaster in the history of the United States. Since there was little or no coordination between levee districts, the federal government was given the duty to handle flood protection in the future with passage of the Flood Control Act of 1928 (Figure 10.2). (8)

EXPERT ANALYSIS

The federal government's new role in flood control, as detailed in the Flood Control Act of 1928, was not uniformly accepted. Concerns were expressed over the federal government's increasing role in water issues to

FIG. 10.2. The earthen levee in this photo is located along the Mississippi River, downstream of New Orleans, and was constructed and maintained by the U.S. Army Corps of Engineers. The water elevation on the right side of the levee is several feet (or m) higher than the dry ground to the left side of the photo. The elevation difference can be almost 20 feet (6 m) during floods.

the detriment of the rights of individual states. Many were afraid that the U.S. Army Corps of Engineers would gain too much power and control of the Mississippi River. Others were concerned that the federal government would employ improper or inappropriate construction methods. Were their concerns valid? Is flood control an issue of greater national importance than drinking water or wastewater treatment (which is usually handled at the local level)? Would it be reasonable for Congress to authorize a federal agency to operate all local water and wastewater treatment plants to protect human health and the natural environment? Why or why not?

Flood-Control Districts Although flood control has become largely a federal responsibility in the United States, many regions of the country continue to support local flood-control districts. Most of these agencies, often found in urban areas, were created in the 1900s to construct and manage smaller-scale flood protection projects.

In Wyoming, for example, the state legislature authorized the formation of flood-control districts. A special election is usually required and then must be approved by the local board of county commissioners. Flood-control districts were authorized in the State of Washington by its legislature in 1935. These districts can borrow money, designate boundaries, and carry out the general duties of flood protection. In California, the state legislature created flood-control and water conservation districts in order to capture, contain, spread, and recharge floodwaters (in addition to the general protection of life and property). The county board of supervisors (elected officials who supervise county activities on behalf of the general public) is empowered to appoint a commission to oversee the activities of flood-control districts in California. All flood districts are required by federal law to meet National Pollution Discharge Elimination System (NPDES) permit requirements of the USEPA.

EXAMPLE

Pima County Flood Control District, Tucson, Arizona

http://www.dot.co.pima.az.us/flood

The Pima County Flood Control District was formed in 1978 in response to devastating flash floods that regularly occurred along the Santa Cruz River in Tucson. Summer

FIG. 10.3. Arroyos (Spanish for "creek" or "dry gulch") like this one in Arizona can quickly become raging torrents of runoff after summer thunderstorms.

thunderstorms in the dry, desert climate often trigger torrents of water along major and minor watercourses in Pima County (see Figure 10.3). The District has an annual budget of approximately $16 million and constructs flood-control structures such as levees, detention ponds, riverbank stabilization projects, and other drainage facilities. In addition, the District has purchased over 7000 acres (2833 ha) of flood-prone lands for open space and regulates construction of buildings and development in designated floodplains. The District has also developed a flood warning and emergency response program to alert residents of impending flood events in their area. (9)

The Pima County Flood Control District was formed as a result of legislation passed by the Arizona Legislature, which directed every county in the state to form a flood-control district. This was largely in response to flash floods on the Salt River in Phoenix. Flood-control districts in Arizona follow existing county boundaries, and the county Boards of Supervisors were given authority to select the board of directors (the governing body) of local flood-control districts. These districts are allowed to tax landowners within the district boundaries as a means of raising funds for operation, maintenance, and construction activities. The Arizona Legislature allowed cities to exempt themselves from flood-control districts if the existing municipal government performed flood-control functions.

EXPERT ANALYSIS

Why did the Arizona Legislature require every county in the state to create a flood-control district, and why do boundaries of the districts follow existing county lines? John Wesley Powell (discussed in Chapter 9) would have argued that flood-control district boundaries should follow watershed boundaries. If his theory of water management had been followed, what political problems would have been created? What flood-control benefits would have been achieved by using watershed boundaries instead of county boundaries?

POLICY ISSUE

As mentioned earlier, federal law requires flood-control districts to meet National Pollution Discharge Elimination System (NPDES) permit requirements of the USEPA as well as Federal Emergency Management Agency (FEMA) floodplain requirements. In addition, construction activities to control floodwaters must comply with federal environmental laws, including protection of endangered species and critical wildlife habitat.

State agencies such as transportation departments, departments of health and environment, and water management agencies must also be consulted.

MUTUAL DITCH AND IRRIGATION COMPANIES

Historical Overview **Mutual ditch and irrigation companies** are privately owned water stock companies organized to deliver irrigation water to shareholders and are found only west of the Mississippi River. The 1920 U.S. Irrigation Census described these water companies as

the most common form of organization for cooperative irrigation enterprises. . . . Water is apportioned on the basis of stock ownership, and the cost of annual operation and maintenance is raised by assessments on the stock . . . stock may be owned independent of land ownership, and it may be, and is at times, rented, the lessee receiving the water apportioned to the stock rented. (10)

The earliest irrigation efforts in the United States were conducted by groups of Native Americans in the Southwest. Construction and management of these ancient systems occurred through a management hierarchy within various tribes. Irrigation thrived for centuries but was later abandoned because of drought or was taken over by Spanish and Mexican settlers from the south.

In 1847, Mormon settlers came to the Salt Lake Valley of Utah and immediately developed irrigation ditches and reservoirs to grow food in the desert valley. Brigham Young served as leader in all aspects of Mormon life, including irrigation development. Since the Mormon religion places great value on community service for the common good, members joined together to construct community irrigation facilities. Irrigation development in early Utah, as noted earlier in Chapter 8, was patriarchal and involved the efforts of all settlers in one way or another.

The Utah method of irrigation development was closely followed by settlers of the Union Colony in Greeley, Colorado, in 1870. A community ditch (the Greeley No. 3 Ditch) was constructed by settlers, and irrigation water was delivered to plots of land bounded by wide streets (modeled on the street patterns of Salt Lake City). However, in contrast to the early Mormon method of water allocation in which crops from community plots were shared, settlers in Greeley had to pay a fee to obtain land and to acquire irrigation water. This fee entitled water users to own shares of stock in the Greeley No. 3 Ditch, and a corporation was later formed to create a mutual ditch company.

As more irrigation ditches were constructed west of the 100th Meridian, the organization and operation of mutual ditch companies became very important. Capital was needed for initial construction of diversion dams across streams and for delivery canals. Often, outside investors, corporations, or banks in foreign countries (such as the Bank of London or the Bank of France) provided startup funds for construction and then formed the mutual ditch companies as corporations that could sell stock. Potential irrigators were required to purchase shares of stock before water could be diverted to private lands. Profit was the ultimate motive of investors in creating mutual ditch companies.

Once an irrigation company was constructed and shares of water were sold, a management and water allocation system within the ditch needed to be developed. Most companies looked to the acequia culture of Mexico for their management scheme. As stated in Chapter 8, a ditch superintendent (mayordomo) was selected to carry out the day-to-day management activities on the ditch, while a board of directors (similar to the ayuntamiento) was selected by shareholders to decide when the ditch should be cleaned, to settle disputes, and to take care of other management issues.

The right to divert water from a mutual ditch was (and still is today) allocated based on shares of ownership. These shares are represented by stock certificates (similar to ownership in a company on the New York Stock Exchange) and entitle the owner to a proportionate share of water in the ditch. For example, if a shareholder owns one share of stock in a mutual ditch company (and there are 100 shares in the company) and 100 cubic feet per second (64.6 mgd, or 2.8 cms) of water is available in the ditch, the shareholder can divert up to 1 percent, or 1 cfs (0.6 mgd, or 0.028 cms), of water from the ditch. If the season becomes dry and only 50 cfs (32.3 mgd, or 1.4 cms) is flowing in the ditch, the shareholder can divert only 0.5 cfs (0.3 mgd, or 0.014 cms) of water for irrigation water for their crops. (Ownership of shares in a mutual ditch company allows shareholders to divert more than their proportionate share of water if other irrigators are not using their water shares at a given time. That is where the term *mutual* becomes important.)

POLICY ISSUE

Mutual ditch and irrigation companies face many of the same issues as municipal water districts. However, the presence of federally protected wildlife along or downstream of an irrigation ditch can severely restrict normal operational activities. For example, in Colorado the U.S. Fish & Wildlife Service issued a federal "guidance letter" in February 2000 to protect the rare Preble's meadow jumping mouse (*Zapus hudsonius preblei*). The guidance letter identified seasonal ditch maintenance activities that were prohibited, or restricted practices such as chemical spraying to eliminate weeds and underbrush, or burning of ditch banks for the same purpose. (Weeds and underbrush normally grow on the sides of irrigation ditches and canals, and can impede the flow of irrigation water.) The Preble's meadow jumping mouse hibernates among thickets during the normal spring ditch maintenance season and would have been destroyed if chemical spraying or fires were allowed.

REGIONAL WATER AGENCIES

OVERVIEW

Regional water agencies generally serve multicounty areas. These agencies usually administer large irrigation projects or watershed flood protection programs, or consolidate water management issues previously handled by multiple local agencies. A common goal is to minimize overlapping government authorities, to reduce government administrative costs, and to increase efficiency.

IRRIGATION DISTRICTS

Irrigation districts evolved during the late 1800s and early 1900s to provide a funding and administrative agency for operating and maintaining larger irrigation systems. Irrigation districts were typically authorized by western states to encourage economic expansion and to develop irrigated tracts of land too large for individuals or mutual ditch companies to develop and manage. Irrigation districts are entitled to collect property taxes (similar to school or fire districts) for purposes of funding construction and operation expenses.

The 1920 U.S. Irrigation Census described irrigation districts in this manner:

The fundamental purpose of the organization of irrigation districts is the obtaining of funds for the construction or purchase of irrigation works. The inclusion of land within a district renders the land subject to taxation for interest and principle of bond issues, and also for the expense of operation and maintenance. (11)

After passage of the Reclamation Act of 1902, the federal government utilized local irrigation districts to serve as local sponsors of federal irrigation projects. Initially, state and local officials were concerned that federal ownership and operation of water delivery systems would mean relinquishing control of water rights to the federal government. Since local irrigators wanted to monitor and control the day-to-day operations of water delivery, irrigation districts provided a safe and unique relationship between federal water development (the Bureau of Reclamation) and local irrigators. (12)

The Bureau offered guidance and financial support to local districts while carefully avoiding actions or policies that would contradict their support of local control. Federal projects continued to be constructed through the mid-1900s, and additional irrigation districts were formed to administer these federal systems after completion. Landowner taxes and federal funds continue to provide operating revenues for irrigation districts across the West.

EXAMPLE

Farwell Irrigation District, Farwell, Nebraska

http://www.usbr.gov/dataweb/html/farwell.html

The Farwell Unit of the Pick-Sloan Missouri Basin Plan is located in central Nebraska between the North and Middle Loup rivers. The irrigation project provides water to 50,000 acres (20,235 ha) of land, as well as flood protection, recreation, and fish and wildlife benefits. The project includes a diversion dam across the Middle Loup River, Sherman Dam and Reservoir (an earthen dam with storage capacity of 69,000 acre-ft, 22.5 billion gal or 85.1 million m^3), and almost 400 miles (644 km) of delivery canals to irrigated lands under the system (see Figure 10.4). Thirty-eight pumping plants are used to deliver water across topographic high points in the area to deliver irrigation water to district lands. The project was substantially completed in 1966. The Farwell District is governed by federal laws and U.S. Bureau of Reclamation policy. The District is managed by a staff of 20 and has a three-member board of directors elected annually by taxpayers of the District. Principal crops irrigated in the District are alfalfa, small grains, and corn. The land is highly productive. (13)

William Smythe, editor for the *Omaha Bee*, was instrumental in developing the relationship

FIG. 10.4. This aerial view is of the Arcadia Diversion Dam on the Middle Loup River in central Nebraska. It diverts river water into the Sherman Feeder Canal (seen at the right side of the photo) and then some 19 miles (30.6 km) to Sherman Reservoir for irrigation and recreation purposes.

between local irrigation districts and the Bureau of Reclamation. Through his efforts, the federal government agreed to relinquish control over operation of irrigation facilities to local districts. In exchange, irrigators were required to pay assessments for use of water from these federal projects, but they gained a voice in operation and maintenance issues through the local irrigation district. Subsidies kept costs down for irrigators.

Irrigation districts exist in most western states. The governing body is typically a board of directors elected by landowners of the district. Day-to-day operations are performed by staff members hired by the board. Generally, a manager is hired to supervise staff, while the manager is supervised by the board of directors. Board meetings are usually held monthly and are open to the public.

CONSERVANCY/CONSERVATION DISTRICTS

Conservancy districts (also called conservation districts, water conservancy districts, water conservation districts, and river basin conservancy districts) are located in many states including Ohio, New York, Maine, South Carolina, Georgia, Minnesota, California, Indiana, Illinois, Colorado, Utah, Nevada, Wyoming, New Mexico, Washington, and Missouri. These governmental agencies are political subdivisions of the state formed by local landowners to solve local water management problems. Conservancy districts are generally created to develop flood-control or water supply projects and often serve as a local sponsor for projects of the Bureau of Reclamation or U.S. Army Corps of Engineers.

Conservancy districts provide multiple services that cross numerous political jurisdictions such as city, county, school, and fire districts. These districts often conform to local watershed boundaries and are managed by a staff of employees and directed by a board of directors. Board members are either elected by local landowners or appointed by a district court judge.

In Ohio, the Ohio General Assembly approved the Conservancy Act in 1914 after a devastating flood in 1913. This law, the first conservancy law in the United States, has been followed by many others. The first conservancy district in the United States was the Upper Scioto Drainage and Conservancy District in Kenton, Ohio, in 1915. The Miami Conservancy District in Dayton, Ohio, was formed the same year. Both districts use local funds to construct flood-control programs. Fifty-seven conservancy districts have been created in Ohio since 1914. (14) In western states, numerous water conservancy districts were formed after the drought of the 1930s to provide watershed coordination of irrigation water development and conservation. Today, flood control and water quality management are becoming important duties of conservancy districts throughout the country.

EXAMPLE

Miami Conservancy District, Dayton, Ohio

http://www.miami-conservancy.org/

The Miami Conservancy District, located in Dayton, Ohio, provides flood protection, groundwater monitoring information, and water-based recreation opportunities to the people of the Great Miami River Basin. Staff of the District have been working for nearly a century to enhance the economic viability and quality of life of the region. Governed by a three-member board of directors, it is one of the oldest conservancy districts in the United States.

In many ways, the Miami Conservancy District is similar to the irrigation districts discussed earlier in this chapter—funding is derived from tax revenues, boundaries delineate service areas, and a board of directors provides oversight of activities. A key difference is that irrigation districts are primarily single-purpose agencies (the delivery of irrigation water), whereas conservancy districts often develop programs in flood control, water management, water quality protection, public education, erosion control, and other broad water-related issues.

The Miami Conservancy District, like many other water agencies across the country, publishes a newsletter. Most regional water agencies publish quarterly, semiannual, or annual newsletters to inform constituents (residents paying taxes to the district) of ongoing activities. The common goals of these newsletters are to alert readers to regional water issues, to provide a line of communication with board members and staff, and to solicit input from area residents regarding future activities of the district. (15)

NATURAL RESOURCES DISTRICTS—NEBRASKA

Prior to 1969, Nebraska had over 500 special-purpose districts that included irrigation districts formed in 1895, drainage districts created in 1905, soil conservation districts established in 1937, rural water districts originating in 1967, and other sanitary, watershed, and reclamation districts. These locally controlled agencies had duties of flood control, irrigation water delivery, drainage of wetlands, soil protection, drinking water supply, and watershed planning. Activities were funded through taxes paid by landowners located within the boundaries of these multiple special districts. It was not unusual for individuals to pay taxes to several districts that had overlapping activities. The result was a patchwork of government programs, potentially contradictory government assistance, and local control that was fragmented by numerous districts that followed county boundaries.

The Nebraska Legislature took a radical approach to water resources management in 1969 with passage of a law creating **natural resources districts** or **NRDs** (Figure 10.5). Twenty-four NRDs were created (today there are 23 due to a merger), with boundaries that follow Nebraska's watersheds. Large watersheds, such as the Platte River which flows across the entire state from west to east, are divided into several natural resources districts. The duties of hundreds of special districts have now been transferred to the NRDs. (16)

EXPERT ANALYSIS

In the 1960s, board and staff members of special districts in Nebraska were confronted with a unique dilemma—they were going to lose their jobs. Consolidation of the 500 special districts into 24 natural

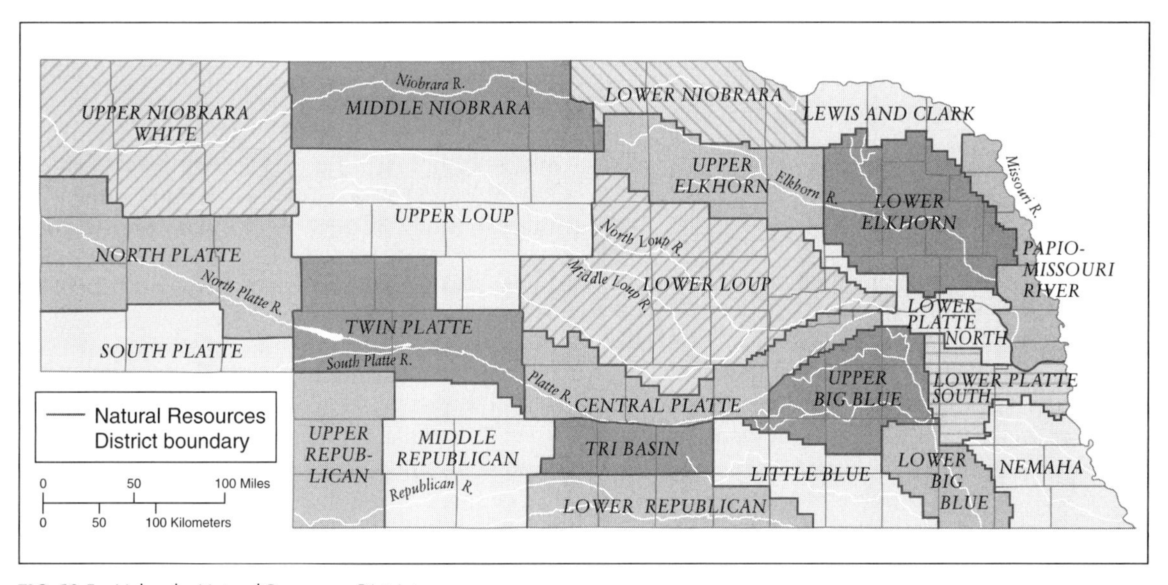

FIG. 10.5. Nebraska Natural Resources Districts.

resources districts meant that thousands of board members and staff would no longer be needed. Suppose you had just learned of the political movement to consolidate your small government agency into a large natural resources district. Also suppose that you had been a board member of your district for many years and had numerous projects in the planning stages. You are not a strong supporter of this new style of "big" government. What arguments could you make at an upcoming legislative hearing in the unicameral legislature in Lincoln to promote your position of maintaining the status quo?

EXAMPLE

Papio-Missouri River Natural Resources District, Omaha, Nebraska

http://www.papionrd.org

The Papio-Missouri River Natural Resources District (NRD) was formed in 1972 and serves six counties in eastern Nebraska. It has an elected board of 11 directors, a staff of 40, and an annual budget of approximately $24 million. It provides programs for flood control, outdoor recreation, rural domestic water supplies, reduced soil erosion, improved fish and wildlife habitat, and public education about natural resources.

The Papio-Missouri River NRD has built strong partnerships with other NRDs and other resources agencies at the state and federal levels. Ideally, regional water agencies develop strong working relationships with other natural resources agencies to improve communication, develop financial opportunities for development of large-scale construction and management projects, and promote government efficiency.

An additional activity of the Omaha-based NRD, as well as many other local water agencies, is natural resources education. The District has a full-time staff to serve the needs of the public through publication of newsletters and brochures, response to e-mail inquiries, and speaking engagements at water workshops, legislative functions, schools, community colleges, and universities. Education staff also develops materials for students and adults to better understand water-related issues of the area. Water events, such as water festivals and water fairs, are also held to encourage students and teachers to learn more about water resources in the region. (17)

GROUNDWATER MANAGEMENT DISTRICTS—KANSAS

Kansas has developed a system of local units of government, called **groundwater management districts (GMDs)**, to administer groundwater resources in the semiarid western half of the state. Enabling legislation was passed in 1972 to allow local residents to establish such districts for local management. Five groundwater management districts have been created west of the 100th Meridian to manage groundwater use. On average, the region receives approximately 17 to 20 inches (43 to 53 cm) of precipitation annually. The primary use of groundwater in this area is for irrigation, although municipal water supply is a growing issue. The districts are governed by local boards elected annually by landowners and water users within district boundaries.

EXAMPLE

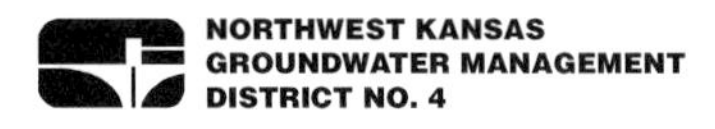

Northwest Kansas Groundwater Management District No. 4, Colby, Kansas

http://www.gmd4.org

The Northwest Kansas Groundwater Management District No. 4 (GMD No. 4) was formed in 1976 by local landowners and water users to address groundwater quantity and quality problems in all or parts of 10 counties in northwest Kansas. A groundwater management plan containing a number of policies and programs has been developed to reduce the use of limited supplies from the Ogallala Aquifer and to meet all other management objectives of the board. The District has 11 board members and an annual budget of approximately $400,000. (18)

Most public water management agencies are required to follow open records laws, and GMD No. 4 is no exception. The Kansas Open Records Act (K.S.A. 45–215 et seq., 2000 Session) requires that every public agency designate a "freedom of information officer" who must prepare and provide educational materials regarding the Open Records Act, assist the general public in any records dispute, respond to all requests for public information, and create a brochure that explains the rights of requesters, the public agency, and the agencies' procedures for providing public information.

EXPERT ANALYSIS

Water agencies are generally governed by open meeting and open records laws passed by state legislatures. **Open meeting laws** require that all board business be conducted in full view of the general public, and closed or secret meetings are usually illegal. The public must be informed of meeting times, place, and location. An agenda is usually prepared to provide information regarding topics of discussion, and public comment is both welcomed and encouraged.

Board meetings are usually operated under "Robert's Rules of Order" to provide a formal procedure to conduct meetings. "Robert's Rules" requires that board members be recognized by the chairperson before speaking and that formal motions—seconded by another board member—be made to take action on matters before the board. A vote of the board determines passage or failure of motions.

Open records laws require public agencies to allow public inspection of all agency records with the exception of personnel records, documents regarding ongoing negotiations or lawsuits, and some legal documents. States have varying standards regarding access to public records, but typically the secretary of a board of directors is designated to record minutes of all meetings and to make them available to the public. Many public agencies now post these minutes on agency websites along with other activities of the agency, upcoming meetings, and job announcements.

The State of Kentucky has an excellent explanation of its public records management law at http://www.kdla.state.ky.us/pubrec/MANWRIT.HTM.

WATER MANAGEMENT DISTRICTS—FLORIDA

Florida is a water-rich state with severe water supply problems. Even though it receives an average of 53 inches (135 cm) of precipitation annually, it can still be hit with severe drought. Historically, Florida's biggest concerns were with drainage and land reclamation. Wetlands, swamps, and other standing bodies of water bred mosquitos and other disease-carrying insects that caused great human suffering during settlement of the region. Large-scale efforts were made to drain these water features in the early and mid-1900s. Population growth placed further development pressures on Florida's unique water settings and ultimately led to reduced water levels in the Everglades, a huge lake/wetland complex in south-central Florida.

During the 1960s, a severe drought in south Florida caused water supply shortages. Fires in the Everglades were the result of lowered surface and groundwater levels. At the time, water resources were managed by single-purpose districts, such as water districts, sewer districts, and drainage districts. The Florida Legislature responded in 1972 by passing the Water Resources Act, which created six **water management districts (WMDs)** that cover the entire state (Figure 10.6). (Today these districts have been reorganized into five WMDs.) District boundaries follow hydrologic basin boundaries (similar to the natural resources districts in Nebraska) and provide regional control rather than state regulation of Florida's water resources issues.

Each water management district in Florida has a board of directors of nine members appointed by the governor and confirmed by the Florida Senate. Districts receive tax revenues to construct and operate water management projects. A district can regulate water use through

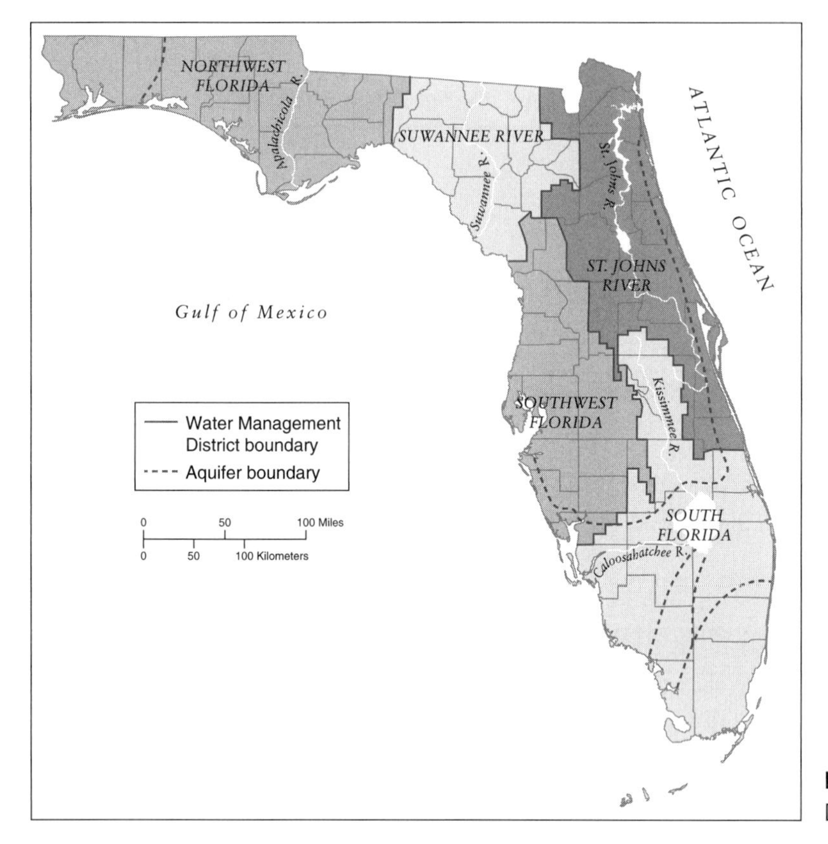

FIG. 10.6. Florida Water Management Districts.

permits, enter into contracts with federal agencies, develop drought water use plans, and formulate short- and long-range water management plans. (19)

EXPERT ANALYSIS

Lawsuits often follow the formation of regional water management agencies that tax landowners to raise revenues. The constitutionality of both the Florida Water Management Districts and the Nebraska Natural Resources Districts was challenged soon after their creation. Paying taxes, even if for a common cause, often meets severe resistance from landowners. Understandably, many people have a strong aversion to paying taxes. What other reasons would individuals have to oppose the concept of regional water management agencies?

POLICY ISSUE

This discussion of regional water agencies presented an overview of a wide variety of approaches used across the United States. A major goal of regional water agencies is to provide a broad-based approach to regional water issues. In some situations, such as the Klamath Irrigation District in Oregon discussed in Chapter 9, conflict with the federal government is intense. In other situations, a regional water agency can sometimes provide a larger forum for negotiation and problem-solving opportunities with officials at federal agencies. Do you believe that more regional water management agencies should be created, or is it better to utilize local government agencies to manage water resources?

STATE WATER AGENCIES

Every state in the United States has governmental departments that administer water resources programs. Even though every state's water agency structure is different, many similarities exist. **State water agencies** administer water quality programs, flood protection, drought planning, water allocation, and conservation efforts on a statewide basis. In some states, these agencies wield considerable political clout, while in other locations of the country a similar agency may serve in an advisory role only. History, climate, and topography all combine to shape the role of the government in water management.

The following discussion of water management agencies in Arizona and Rhode Island presents numerous contrasts and similarities. Other state-level water management agencies in other regions can be found on-line by using key words such as "Wisconsin" and "state government," "state water agencies," or "state water department."

THE STATE OF ARIZONA WATER AGENCIES

Arizona has numerous state agencies with authority over water resources. Water is of critical importance in the Cactus State which averages only 3 to 12 inches (8 to 30 cm) of precipitation each year. The Arizona Department of Water Resources (ADWR) (http://www.water.az.gov) is the primary agency responsible for administering all state water laws except those related to water quality. The ADWR investigates methods to increase water supplies to meet future demands, oversees the use of surface and groundwater under state jurisdiction (control), manages floodplain development, regulates nonfederal dams for safety purposes, and develops policies to promote conservation and equitable use of water. The agency also represents the State of Arizona in discussions with other members on the Colorado River Compact.

The Arizona Department of Environmental Quality (ADEQ) (http://www.adeq.state.az.us) is responsible for water quality and waste management programs. The Water Quality Division (WQD), a branch of the ADEQ, is responsible for ensuring the safety of drinking water supplied by public water systems and for issuing permits to limit the discharge of pollutants into surface water and groundwater in the state. The Waste Programs Division (WPD) of the ADEQ is responsible for regulating the disposal of hazardous wastes and use of underground storage tanks (such as fuel tanks beneath gas stations), and it encourages recycling, reuse, and other forms of pollution prevention.

The Arizona Geological Survey (http://www.azgs.state.az.us) provides impartial information to assist with the wise management of the state's land, water, mineral, and energy resources. It works closely with the U.S. Geological Survey and natural resources agencies of the state, as well as with universities, communities, and individuals throughout Arizona.

The Arizona Game and Fish Department (http://www.gf.state.az.us) is responsible for the conservation, enhancement, and restoration of the state's wildlife resources and habitats. It works closely with the U.S. Fish & Wildlife Service to administer programs designed to protect wildlife for present and future generations.

A CLOSER LOOK

What is the difference between a district, company, department, division, office, bureau, commission, and survey? These terms are used throughout local and state governments to identify an agency's management structure or role.

In general, local government agencies are called districts or companies. The term *district* means that an agency was formed under the authority provided by a state legislature. Districts have boundaries to delineate service areas and collect taxes from landowners in the district, and they have a board of directors selected by landowners or a district court judge. A district's staff members are considered government employees. In contrast, a company does not collect taxes, may or may not have boundaries, usually has a board of directors, and is less regulated by state law than a district is.

In state government, a department is generally a cabinet-level agency. This means the governor of a state appoints the heads of departments and has close contact with them regarding policy and the direction of day-to-day activities. A division, office, or bureau, such as the Office of Water or the Division of Wildlife, is usually a government agency within a state-level department. In Rhode Island, the Office of Drinking Water Quality is located within the Department of Health, whereas in Colorado, the Division of Wildlife is located within the Department of Natural Resources. The director (manager) of a division or office is usually selected and hired by the director of the appropriate cabinet-level department.

A commission or association generally implies an organization that covers a large geographic area. Board members may be appointed by governors or member federal agencies, boundaries may be rigid or nonexistent, and funds may be provided by a combination of state, federal funds, or membership fees from private groups or corporations. Duties may be broad or very specific, but probably include few enforcement mechanisms.

A survey is generally a state agency that conducts topographic, geologic, or other land-based investigations that are patterned after the U.S. Geological Survey. The name probably derives from the topographic surveys conducted by John Wesley Powell and others during the 1800s.

What water management agencies are located in your area, state, province, or region? Are any water commissions active, or is most management at the local level? How is your state/provincial/territorial government organized to manage water resources? Has the organization of water management changed over the past 20 years? Has local control increased or decreased?

THE STATE OF RHODE ISLAND WATER AGENCIES

Although Rhode Island has an average annual precipitation of 42 inches (107 cm), water resources management is still a critical issue in the Ocean State because environmental protection and water quality improvement have become important concerns, as have water allocation and distribution.

The Rhode Island Department of Environmental Management (http://www.state.ri.us/dem) is the primary state agency responsible for water resources management in the state. It contains the Bureau of Environmental Protection, Bureau of Natural Resources, Division of Fish & Wildlife, and Office of Water Resources. These agencies manage surface and groundwater resources for pollution prevention and well drilling, and they provide financial assistance to local governments. The Department also includes the Forest Environment Program (FEP), which manages 40,000 acres (16,188 ha) of state-owned forest land. FEP is responsible for developing forest and wildlife management plans, promotes environmental conservation and education, and enforces rules on regulations on forest lands.

The Rhode Island Department of Health is responsible for ensuring safe drinking water supplies in the state. The Office of Drinking Water Quality (http://www.health.ri.us/environment/dwq.home.htm) works closely with local water suppliers to monitor the quality of drinking water and acts as a liaison with the U.S. Environmental Protection Agency.

Rhode Island has a unique water management program with its Water Resources Board (http://www.wrb.state.ri.us/index.html). The board was created in 1964 with the original purpose of acquiring property for a reservoir water supply project. In 1967, the agency was given broader powers over statewide water supply, and in 1970, the Rhode Island Water Resources Board Corporate was created to serve as the financing agency for the Water Resources Board's activities. The two agencies work very closely together to identify and finance water supply projects, watershed protection, policy development, and program coordination. In 1997, the Water Resources Board assumed water supply planning duties that had been previously carried out by the Department of Environmental Management. In 1999, the state legislature further defined water allocation duties for the Water Resources Board.

POLICY ISSUE

State water agencies have a unique relationship with federal counterparts. On the one hand, states are equal partners with federal agencies since both levels of government enforce the laws of respective state and federal governments. State agencies are often empowered through federal legislation to carry out the water quality laws of the USEPA.

On the other hand, state water allocation law typically takes precedence over federal law, although endangered species and pollution laws can counter these rulings. This dichotomy of laws creates conflict where state water allocation law is preempted by federal regulations.

MULTISTATE WATER AGENCIES

Scores of multistate water management agencies in the United States work with issues that cross state lines. These agencies usually service large watersheds, groundwater aquifers, or other hydrologic boundaries. Generally, they are created through agreements between states and the federal government to promote planning, communication, and coordination of activities between a wide variety of local, state, and federal agencies. On the one hand, this creates excellent opportunities for improved water management activities; on the other hand, it creates problems of jurisdiction, common goals and objectives, and establishment of overarching policies. Examples of the wide range of functions performed by such groups are the Chesapeake Bay Commission and the Missouri River Basin Association.

THE CHESAPEAKE BAY COMMISSION

http://www.chesapeakebay.net

The **Chesapeake Bay Commission** was formed in 1980 to advise the general assemblies (legislatures) of Virginia, Maryland, and Pennsylvania to cooperatively manage the Chesapeake Bay and its upstream watershed. The Chesapeake Bay Program was created upon the signing of the Chesapeake Bay Agreement by the governors of Virginia, Maryland, and Pennsylvania, as well as the mayor of the District of Columbia and the head administrator of the U.S. Environmental Protection Agency. The Chesapeake Bay Commission is involved in all legislative, policy, and implementation decisions of the program and provides interagency coordination, information dissemination, and arbitration between states. Its main office is located in Annapolis, Maryland.

The overall goal of the program is to improve water quality in the Chesapeake Bay by reducing or preventing upstream pollution that could be transported to the Bay (Figure 10.7). The Commission was created to emphasize state responsibility for cleanup and improved policy coordination between states. The 21-member Commission includes 15 legislators (five each from the member states), three directors of state agencies, and three citizen representatives. (20)

FIG. 10.7. These students are participating in a storm drain stenciling program sponsored by the Chesapeake Bay Commission to reduce surface water pollution in the watershed. "Don't Dump—Drains into the Chesapeake Bay" is the painted message and is intended to stop dumping of paint, motor oil, household cleaning supplies, weed spray, fertilizer, and other common chemicals that could harm downstream aquatic wildlife.

MRBA

Missouri River Basin Association

MISSOURI RIVER BASIN ASSOCIATION

http://www.mrba-missouri-river.com

The **Missouri River Basin Association (MRBA)** was formed in 1981 by the governors of Missouri, Nebraska, South Dakota, North Dakota, Montana, Iowa, Kansas, and Wyoming. In 1990, the voting membership of the MRBA Board was expanded to include a representative of the basin's 28 Indian tribes. MRBA's purpose is to coordinate Missouri River planning activities. Representatives from eight federal agencies serve as advisors to the nine-member MRBA Board of Directors. The Association works with member states, tribes, federal agencies, river users, and others to improve the basin's economic and environmental resources. Its main office is located in Lewiston, Montana.

The Missouri is one of the country's most engineered rivers. Six large dams have been constructed across the river in North and South Dakota, and can store up to 73 million acre-feet (23.8 trillion gal, or 90.0 billion m^3) of water. Channelization in the lower river by the U.S. Army Corps of Engineers has reduced the floodplain to only 10 percent of its original width. This has improved navigation but has drastically altered adjacent wetlands and aquatic habitat.

Representatives to the MRBA Board of Directors are appointed by the governors of the eight member states. The executive director of the Mni Sose Tribal Water Rights Coalition serves as the tribal representative to the Board. (21)

OUR ENVIRONMENT

The Missouri River was an important highway of western expansion in the United States during the 1800s. The epic journey of Meriwether Lewis and William Clark (1804–1806) increased our knowledge of the region between its confluence with the Mississippi River near St. Louis, Missouri, and its source in the Rocky Mountains of Montana. Along the way, the Corps of Discovery (as the Lewis and Clark Expedition was called) discovered broad river valleys, pristine wilderness, and high mountain streams supporting a wide range of plant and animal life. The celebration of the two hundredth anniversary of the expedition is occurring as these words are being written. An excellent work on this topic is *Undaunted Courage* by Stephen E. Ambrose, published by Simon & Schuster in 1996.

By contrast, author William Least Heat-Moon retraced the journey of Lewis and Clark in 1995 and found a much different, and at times, bleaker picture of the Missouri River. In his excellent work *River Horse* (published in 1999 by Houghton Mifflin Co.), Least Heat-Moon finds a river that has been dammed and channelized throughout much of its course, resulting in a far different riparian habitat and riverine system. Choked passageways, pollution, and drastically reduced river flows below dams characterized a changed Missouri River system. Competing water uses were also evident as irrigation, barge traffic, and environmental needs all were vying for a share of limited water flows.

The Missouri River Basin Association faces many challenges as it develops plans for the twenty-first century. Issues of endangered species, minimum streamflow requirements for barge traffic, and variable water releases from on-channel dams—some of which impound and back-up river flows for hundreds of miles (hundreds of km)—create difficult water management situations. Should the barge transportation industry be shut down at times so that water releases from upstream reservoirs can be reduced to protect nesting shorebirds in Nebraska and Iowa? What level of flood protection should be provided to communities that have historically been devastated by high water flows? Should flood protection be enhanced at the expense of lost riparian habitat and associated wetlands? What role should a multistate water planning agency have in making these decisions?

POLICY ISSUE

Regional water agencies play a unique role in water resources management. Since these groups attempt to work with a variety of constituencies, including local, state, and federal agencies, their role is typically one of facilitation and cooperation. Regional agencies work to solve regional water conflicts, and they generally have a close working relationship with federal agencies. Jurisdictional issues and lack of defined authorities too often limit the overall effectiveness of regional water agencies. However, they do play a critical role in bringing parties together to negotiate or implement coordinated policy decisions.

WATER MANAGEMENT IN MEXICO AND CANADA

OVERVIEW

Canada and Mexico take approaches to water resources management that are very different from those found in the United States. Although the federal government of Canada does perform management functions in navigation and fisheries, the provinces hold most regulatory power regarding water allocation and use. Under the Constitution Act of 1867, provinces have proprietary rights regarding the use of surface and groundwater resources. Possessing proprietary rights means that provinces have the constitutional right to make laws and to regulate flows, water use and development, pollution control, and thermal and hydropower development. A unique partnership exists between the Canadian government and provinces for water management and planning—often in a complex and shared system. These shared issues include interprovincial water issues, agriculture, and national water policy. (22)

Mexico's system of water management contrasts sharply with Canada's. Water has been declared a national asset in Mexico and as such is managed by the federal government. Local, state, and federal water agencies, as well as individuals, can use water only through authorization from the Mexican government. As you read the following two guest essays on Mexico and Canada, consider the relationships between local, state, and federal water agencies in the United States. Which of the water management systems of these three countries appears to provide the most efficient, successful, and appropriate approaches? Do you agree that the U.S. system is a moderate version of federal water management in Mexico? Does the Canadian system appear to be the most reasonable? What role do climate and population play in the adoption of a national water management system, and are those two factors evident in the methods developed in the United States, Mexico, and Canada? Would citizens of Canada or the United States adopt Mexico's strong federal regulatory system? Why or why not?

GUEST ESSAY

Water Management in Mexico

by **Dr. Alvaro A. Aldama**
Director General of the Mexican Institute of Water Technology, Jiutepec, Morelos, Mexico

Dr. Alvaro Aldama

Alvaro Aldama was born on August 27, 1954 in Mexico City, Mexico. He received his Bachelor of Science in Civil Engineering from the National Autonomous University of Mexico in 1978, his Master of Engineering in Hydraulics from the National Autonomous University of Mexico in 1979, and his Doctor of Philosophy in Fluid Mechanics from the Massachusetts Institute of Technology in 1986. Dr. Aldama is currently professor of the Graduate Division at the School of Engineering of the National Autonomous University of Mexico and director general of the Mexican Institute of Water Technology. His research interests include computational fluid dynamics, problems of scale in fluid dynamics, and flood hydrology. He has written over 100 publications, including papers in refereed journals, conference proceedings and two books. He is the editor-in-chief of the journal *Hydraulic Engineering* in Mexico and has served on the editorial board of the journal *Advances in Water Resources*. He has received several prizes and awards, including the 1998 Enzo Levi Prize granted by the Mexican Association of Hydraulics to an outstanding career in research and teaching.

Introduction Water is the most important natural resource. Without water, it is not possible to

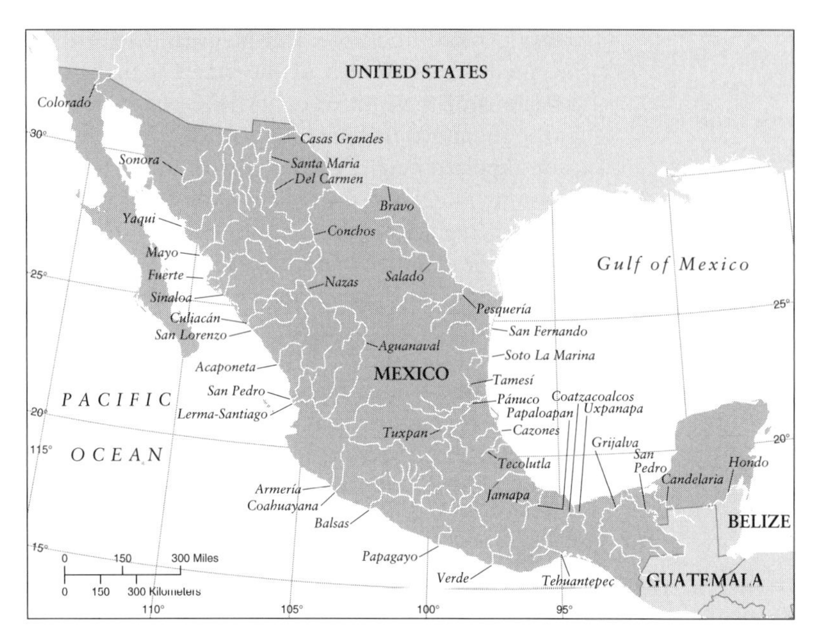

FIG. 10.8. Principal rivers in Mexico. Much of Mexico's northern and western regions are arid or semiarid climates. The discharge of rivers in these areas provides the life-blood for municipal and irrigation water demands.

have agriculture. Without water, industry would not exist. Without water, there would be no life. Water shortage brings suffering and desolation. In contrast, the excessive and uncontrolled occurrence of water causes destruction (see Figure 10.8).

The importance of water has long been recognized in Mexico, and it is an intrinsic feature of Mexican history. In fact, there is a longstanding hydraulic tradition dating back to the Pre-Hispanic civilization. Water played a significant role in indigeneous myths and religious beliefs, and influenced different aspects of social and economic development. Vestiges of irrigation systems and aqueducts can be found, illustrating the importance of water. Most notable among them are the works to control the hydrological system in the Valley of Mexico.[1]

Waterworks continued to be important throughout the period of Spanish dominance and during the first two decades following Independence. Some of the aqueducts serving colonial cities are still in use, and the irrigation works built in the large haciendas[2] are a close antecedent to the country's modern irrigation systems. After the Revolution of 1910 and for the past 75 years, the expansion of water in irrigation, cities, and industries has been based on the development of hydraulic infrastructure supported by major federal government involvement.

Water in Mexico Mexico is a country of approximately 2 million square kilometers (772,200 mi^2) and almost 100 million inhabitants. With a mean annual rainfall of 77 centimeters (30 in.), about 27 percent turns into runoff of 410,000 million cubic meters (332.3 million acre-ft, or 108.3 trillion gal) per year. Renewable groundwater is estimated at 63,000 million cubic meters (51.1 million acre-ft, or 16.6 trillion gal), and nonrenewable groundwater is close to 110,000 million cubic meters (89.2 million acre-ft, or 29.1 trillion gal), which gives a total of 473,000 million cubic meters (383.4 million acre-ft, or 125.0 trillion gal) of renewable freshwater per year.

Rainfall distribution is very irregular, as is reflected in the following statistics: only 20 percent of the total rainfall occurs in 40 percent of

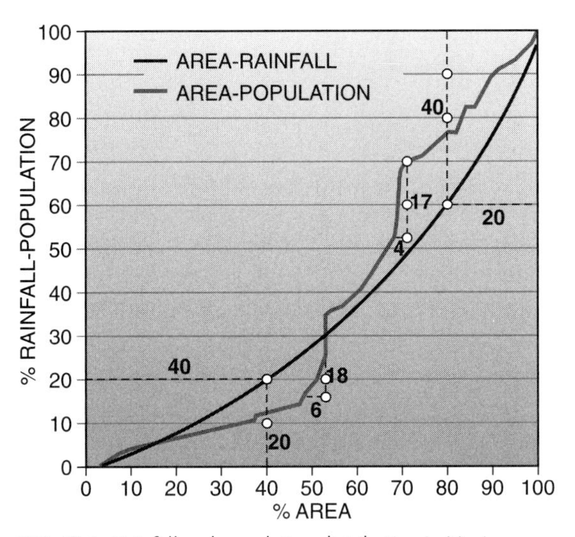

FIG. 10.9. Rainfall and population distribution in Mexico.

the total area, and 40 percent of the total rainfall falls over 20 percent of the area (see Figure 10.9). The irregular rainfall distribution is also observed in the seasons and interannually; rainfall is concentrated in a rainy season—from June to September—but it occurs in a torrential manner, making the resulting runoff difficult to regulate.

The average annual water availability per capita in Latin America is 38,560 cubic meters (31.3 acre-ft, or 10.2 million gal); in North America 15,370 cubic meters (12.5 acre-ft, or 4.1 million gal); in Europe 8580 cubic meters (7.0 acre-ft, or 2.3 million gal); and in Africa 5490 cubic meters (4.5 acre-ft, or 1.5 million gal).[3] In Mexico the average annual availability per capita is 4990 cubic meters (4.0 acre-ft, or 1.3 million gal).[4] This makes Mexico a low-water-availability country (see Figure 10.10).

The location of population and economic activity is inversely related to the availability of water. Only 28 percent of the runoff occurs where 77 percent of the population lives and 84 percent of gross domestic product is generated. Consequently, surface runoff and groundwater are increasingly insufficient to support population growth and economic activity, resulting in the overpumping (mining) of aquifers (Figure 10.11) and a need to implement water transfers between river basins. In addition, water pollution has reduced the potential beneficial use of certain rivers and water bodies (see Figure 10.12).

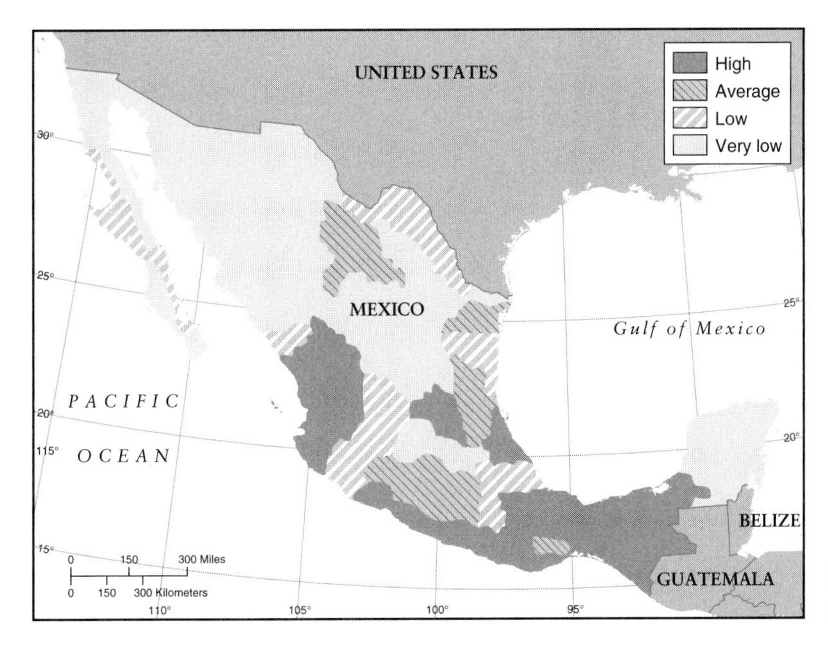

FIG. 10.10. Relative runoff availability in Mexico.

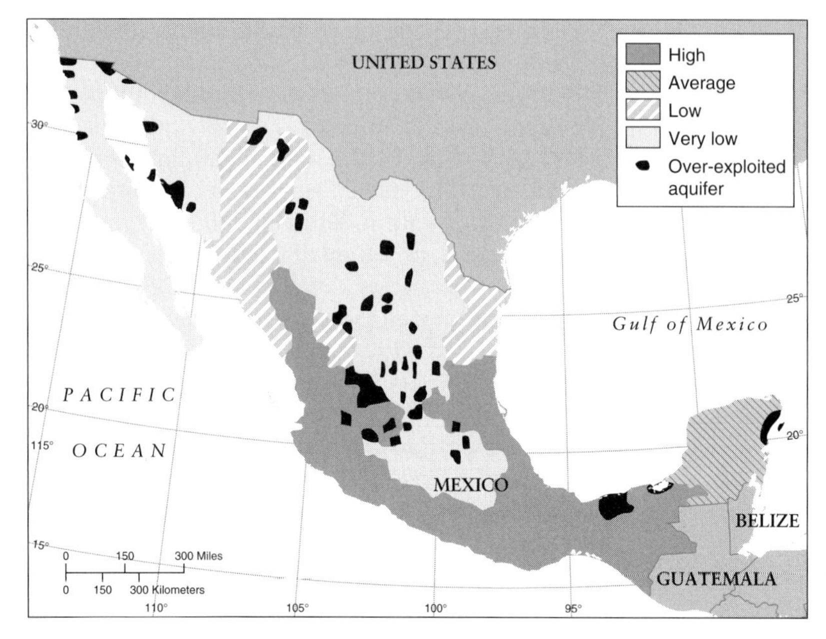

FIG. 10.11. Relative groundwater availability in Mexico.

On a national level, total water withdrawal has increased to about 161,000 million cubic meters (130.5 million acre-ft, or 42.5 trillion gal) per year, 34 percent of the country's renewable water, while total consumption represents only 9 percent of total renewable water (Table 10.1). Hydropower generation accounts for the largest volume withdrawn (60 percent); of the volume of water consumed in Mexico, 76.3 percent goes to agriculture, 17 percent to public use,

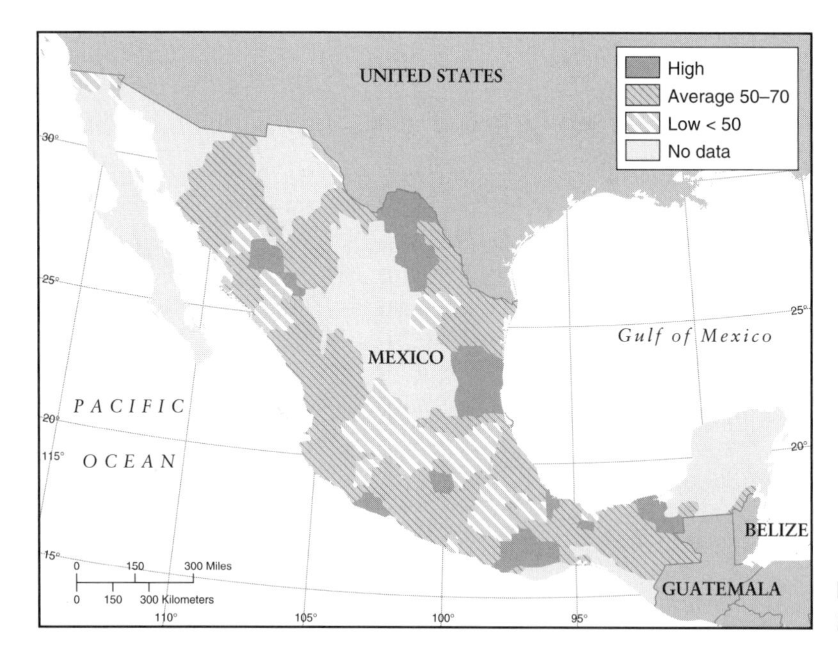

FIG. 10.12. Surface water quality in Mexico.

TABLE 10.1 Zonal Water Balance[a]

Zone	Area	Pop.*	Mean Rainfall	W/A*	Withdrawal				Consumption
					Water Use				
	%	%	mm	%	W/A %	IRR %*	HYDRO %*	WS %*	C/A %*
North	34	14	420	3	88	69	22	9	52
North and central Pacific	27	12	560	13	49	64	32	4	27
Central	15	50	940	16	80	22	71	7	18
Gulf and Southeast	24	24	1402	68	18	6	87	7	2
National	100	100	772	100	34	28	65	7	9
National	1.967 million km^2	97.483 million inhabitants	1,518,665 million m^3	472,540 million m^3	161,303 million m^3	45,726 million m^3	104,905 million m^3	10,671 million m^3	44,852 million m^3

[a]Comisión Nacional del Agua, Informe 1989–1993. Secretaría de Agricultura y Recursos Hidráulicos. México. 1994.

*Pop: Population; W/A: Withdrawal/Water availability; IRR: Irrigation; HYDRO: Hydropower generation; WS: Water supply; C/A: Consumption/Water availability.

5.1 percent to industry, 1.4 percent to aquaculture, and 0.2 percent to cooling processes in thermoelectric plants. However, the national water balance does not reflect the pressing problems that are affecting a large proportion of the country's aquifers and river basins. Regional water balances in over half of the territory show considerable deficits, demonstrating the degree of groundwater overdrafting, as well as an increasing problem of water pollution.

In addition, the country is subject to tropical cyclones on both littorals, as well as other major meteorological events that may produce severe floods, resulting in large loss of life and sometimes catastrophic economic damage.[5] Unfortunately, with the increasing deforestation and expansion of urban centers, flood damages will likely be even greater in the future. Extreme hydrometeorological events also include droughts. The northern and northeastern regions of the country are frequently overwhelmed by droughts, which of course affect water supplies to cities, agriculture, and industry.

During the past 20 years, natural disasters in Mexico have cost an estimated average of about $500 million per year. Except for the effects of the 1985 earthquake, the greatest losses have been associated with extreme hydrometeorological phenomena.[6]

Legal and Institutional Framework Based on the constitutional principle that water is a national asset, Mexico's legal and institutional framework for water management has evolved over the last 75 years, beginning with the creation of the National Irrigation Commission and the Irrigation Law of 1926. The integration of water development and management within the Secretariat of Water Resources (1946), the impulse toward regional development based on water resource development through River Basin Commissions (1940s and 1950s), the implementation of national sectoral water plans and schemes for the development of large regions like the northwest (1960s), and the integration of the first National Water Plan (1975) represent some milestones in Mexico's dynamic development of hydraulic infrastructure and management of water resources. Upon the complete reorganization of the federal government in 1976, the Secretariats of Water Resources and of Agriculture were transformed into the Secretariat of Agriculture and Water Resources (SARH), mainly to solidify the federal government's actions to solve agriculture's pressing problems.

Urban water supply and sewerage services provided by the federal government were first relocated within the Secretariat of Human Settlements and Public Works and later decentralized to municipalities, with SARH responsible for planning, designing, and constructing major aqueducts to supply bulk water to large cities and industrial areas. This arrangement of the federal government's role into sectors in relation to water caused serious problems of coordination, thus aggravating the already critical problems of scarcity, conflicting uses, and pollution in several river basins.

In response to the existing problems, highest priority was given to revising the institutional framework for water management and development. In 1989, the National Water Commission (NWC) was created as an autonomous agency attached to SARH, becoming the sole federal authority dealing with water management. In 1994, the NWC was attached to the Secretariat of the Environment and Natural Resources.

The Commission is charged with coordinating investment programs in the water sector and with setting priorities and constraints vis-à-vis the actual situation in each river basin. The basic responsibilities of the Commission are as follows:

- Define the country's water policies, and formulate, update, and monitor the implementation of the National Water Plan.
- Measure water quantity and quality.
- Regulate water use.
- Allocate water to users and grant the corresponding licenses and permits for both water withdrawal and water discharges.

- Plan, design, and construct the hydraulic infrastructure, totally or partly financed by the federal government (with some exceptions, such as hydropower development).
- Regulate and control river flows, as well as improve the safety of major hydraulic infrastructure.
- Provide technical assistance to water users.
- Define and, if necessary, implement financial mechanisms to support water development and the provision of water services for irrigation and urban water supply and for sewerage and sanitation.

In 1986, the Mexican Institute of Water Technology was created as a federal research institution to develop, adapt, and transfer technology, as well as to train qualified personnel for the management, conservation, and remediation of water. The Institute has provided scientific and technological support to federal, state, and municipal agencies related to water management and to water users.

The major achievements of Mexico's water policy include the following:

- Through the construction of 840 large dams, storage capacity totals 150,000 million cubic meters (121.6 million acre-ft, or 39.6 trillion gal).
- Mexico is now the seventh country in the world in irrigated area, with 6.3 million hectares (15.6 million acres); 95 percent of the total surface of 78 irrigation districts and all of the 16 rainfed agricultural districts have been transferred to the users.
- The current coverages for drinking water and sewage service are 87 percent and 73 percent, respectively.

The new National Water Law, enacted in 1992, provides a modern regulatory framework for water management. The new law was designed to achieve integrated water management, with the objective of achieving sustainable development to guide water development and management. River basins and aquifers are considered the hydrologic units for water planning and management. This legislation reinforces the role of the National Water Commission and makes water planning mandatory.

At the same time, the National Water Law of 1992 has expanded the mechanisms for public participation in planning and management of the nation's water resources. It has led the way to the creation of River Basin Councils as a mechanism for joint planning and management among the three levels of government and water users. The new law also gives the water services presently managed by the National Water Commission greater administrative autonomy and reinforces the notion that water services in the cities are essentially the responsibility of local government. Participation of the private sector is also emphasized, including the provision of water services.

Mexico has established a system of permits, very similar to those enforced in the United States under the amended Clean Water Act. This system has been coupled to a fiscal policy based on the "polluter pays" principle. Some general provisions account for the control of nonpoint pollution, as well as other increasingly important environmental concerns. Current challenges in Mexico's water resource management are to increase the effectiveness of the legal and institutional framework, the physical yield of water infrastructure, and the efficiency of water use.

Conclusions The adoption of a comprehensive policy framework treating water as an economic good, combined with decentralized management, greater reliance on pricing, and fuller participation of stakeholders, has become a necessity. That is why Mexico, like most countries, is striving to strike a delicate balance between governmental regulation and market mechanisms with regard to water resources management.

REFERENCES

[1]Subsecretaría de Infraestructura Hidráulica, Agua y Sociedad, *Una historia de la Obras Hidráulicas en México* (Secretaría de Agricultura y Recursos Hidráulicos, México, 1987).

[2]In Spanish America, a large, landed estate, one of the traditional institutions of rural life.

[3]E. J. Mestre, *Water Data in Emerging Economies: The Latinamerican Case* (RELOC-CEPAL, 1998).

[4]Compendio Básico del Agua, Comisión Nacional del Agua (México, 1999).

[5]J. Aparicio, "Inundaciones: la otra cara de la moneda." *Tláloc* 5, no. 11 (1998): 15–20.

[6]Comisión Nacional del Agua, *Presente y Futuro del Agua en México* (Secretaría de Medio Ambiente, Recursos Naturales y Pesca, México, 2000).

GUEST ESSAY

Water Management in Canada: The Inter-Jurisdictional Context

by **Ralph L. Pentland**
Former Director of Water Planning and Management
Canadian Federal Government,
Ottawa, Quebec

Ralph Pentland

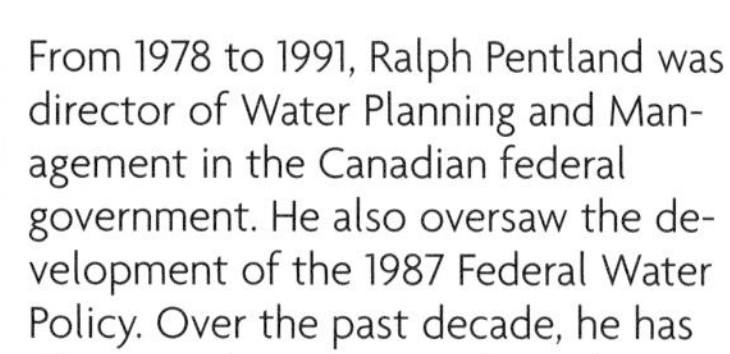

From 1978 to 1991, Ralph Pentland was director of Water Planning and Management in the Canadian federal government. He also oversaw the development of the 1987 Federal Water Policy. Over the past decade, he has served as a water policy consultant in a number of countries, including Canada, China, India, Indonesia, Poland, and Venezuela. In 1999–2000, he was the Canadian co-director of the International Joint Commission study of consumptive uses, diversions, and other potential removals of water from the Great Lakes.

In Canada, federal-provincial water relations have been determined more by the evolution of issues and the relative technical strength of the two levels of government in the various regions than by a strict interpretation of constitutional law. For that reason, the presentation that follows focuses on pragmatic as opposed to legal considerations.

Water resource development was very modest in Canada prior to Confederation in 1867. The British North America Act of 1867 (now the Constitution Act) did not mention water, but it did allocate exclusive legislative jurisdiction over fisheries and navigation to the federal government. Treaty-making, extraprovincial trade and undertakings, and criminal law also fell to the federal government, as did the powers to conduct the census and to collect statistics. Provincial powers were based largely on property and civil rights and control over local works and undertakings. This meant that provinces could manage natural resources within their boundaries without restriction except in areas subject to federal legislative control.

By the turn of the century, a number of disputes concerning water flowing along or across their common border had arisen between Canada and the United States. As a result, the Boundary Waters Treaty was signed in 1909. It was a far-reaching document that, among other things, created the International Joint Commission with equal representation from both countries. It also provided for joint studies, established rules for approving works and uses, and prohibited trans-boundary pollution that would result in injury to health or property.

Before about 1945, water development was aimed at addressing very specific problems and opportunities. For example, the Department of Public Works built a number of structures to facilitate the transport of logs and to improve navigation. In the drought-prone Prairie Provinces, the Prairie Farm Rehabilitation Administration (PFRA) built thousands of small dams, dugouts, community water supplies, and some large-scale irrigation projects. Hydroelectric generation expanded quickly in the early years of the twentieth century, fostered by provincial governments that saw immediate advantages in stimulating industrial growth.

By 1945, the pace of water development and emerging controversy regarding roles had reached the stage where national water policy began to take shape, if only slowly. At the Dominion-Provincial Conference on Reconstruction in 1945, the federal government formally adopted a policy on the division of responsibilities with the provinces. In the water resources area, the federal government undertook responsibility for research on a national scale, general and basic census surveys relating to resource development and public investment, protection of regional watersheds, and the integrated development of interprovincial river systems (see Figure 10.13).

During the 1945–1965 period, there was a broad consensus on rapid water development in support of economic growth. If economic growth was the theme, multiple use and coordination were the means most often mentioned in Canada. The merging of economic development and water management goals was evident in a number of joint federal–provincial programs such as the construction of dikes and breakwaters, a number of joint flood-control projects to protect regional economies, and federal contributions to

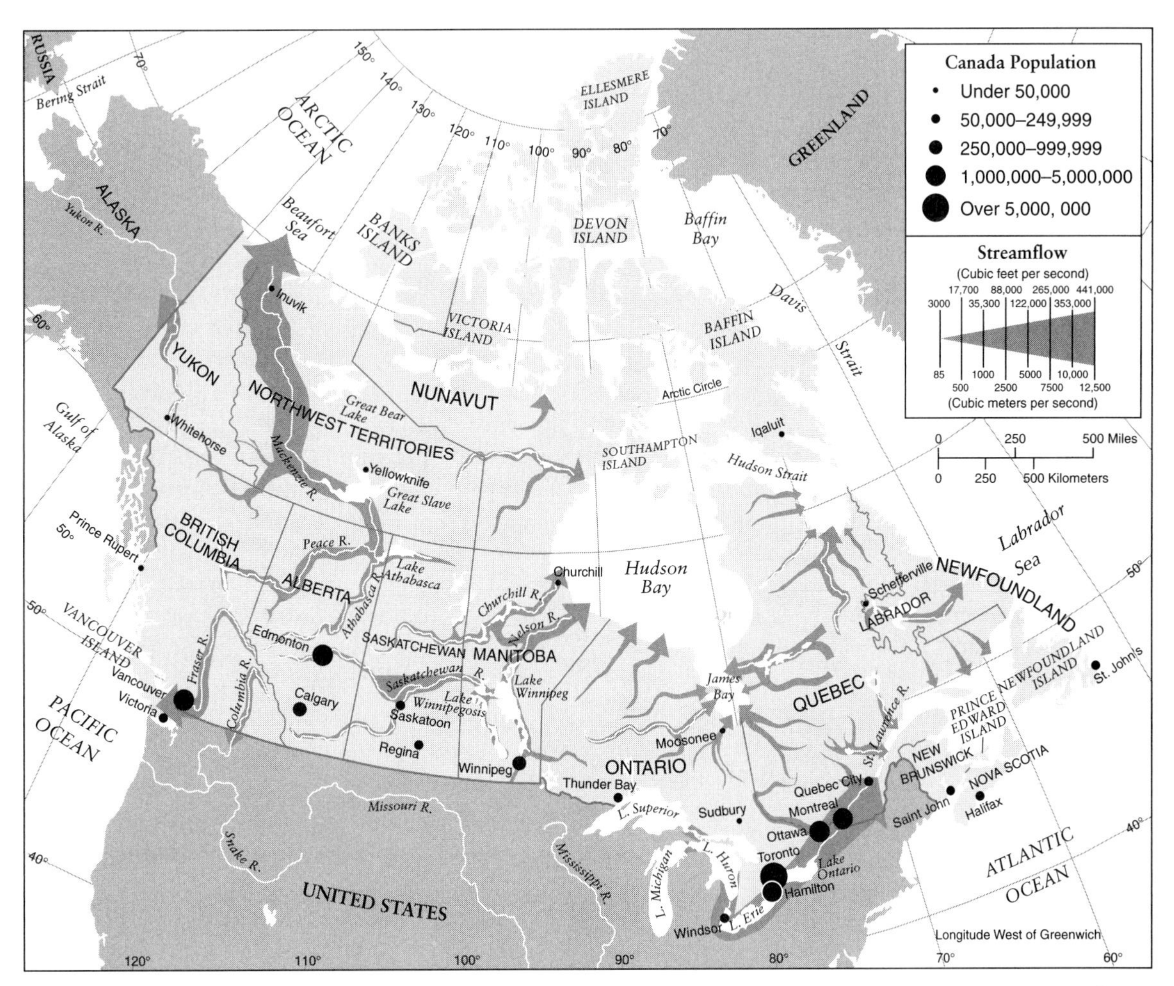

FIG. 10.13. Principal rivers of Canada. *Geography: Realms, Regions, and Concepts,* 10th Edition, by Harm de Blij and Peter O. Muller. Copyright © 2002 H. J. de Blij and John Wiley & Sons, Inc. This map was originally produced in color. Adapted and reprinted by permission of John Wiley & Sons, Inc.

municipal water and sewerage infrastructure. Cooperative development on an international scale also flourished, with major projects, such as the St. Lawrence Seaway, allowing large seagoing vessels access to the Great Lakes. A new federal policy announced in 1963 encouraged construction of major hydropower projects in Canada well before their output would be needed domestically, provided that markets could be found in the United States that would help pay project costs.

Early in the 1965–1985 period, it was recognized that a prerequisite to dealing with conflicts between uses in specific water-deficient areas was the establishment of limits on use within each jurisdiction sharing the same river basin. In Canada, that condition exists mainly in the three Prairie Provinces of Alberta, Manitoba, and Saskatchewan. In 1969, an apportionment agreement was reached between the federal government and the three Prairie Provinces on how the rivers flowing eastward between Alberta and Saskatchewan and between Saskatchewan and Manitoba would be shared. Additional apportionment agreements were negotiated with the United States for a few small transboundary streams.

Further policy changes came in two waves. The first made water management more comprehensive, and the second moved water management into a more environmental framework. Recognizing the increasing scale of potential water developments and the inevitability of conflicts among users and jurisdictions, the federal government opted to play a more proactive role in water planning and management. It passed the Canada Water Act, enabling it to enter into federal-provincial comprehensive planning agreements. Subsequently, it entered into numerous such agreements which both improved water management techniques and built capacity at the provincial level to continue these types of activity on their own.

Similarly, the federal government entered into numerous federal-provincial agreements to encourage more comprehensive flood management. Under these agreements, the emphasis shifted from the earlier focus on physical flood-control works to a new emphasis on floodplain management. A first generation of flood risk maps was prepared, and federal and provincial spending powers were used in a variety of ways to discourage future flood-vulnerable development. Again, as with basin planning, the outcome was both rapid progress and a stronger capacity at the provincial level to continue on their own.

The second wave of policy changes was brought about through profound changes in Canadians' attitudes about the environment. As in most other countries, these changes led to rapid strengthening of legislation and activities related to pollution and environmental assessment. It also led to the amalgamation of most natural resources and environmental responsibilities under an Environment Department. An increased emphasis on environmental matters was also reflected in intergovernmental arrangements, such as the milestone Canada-U.S. and Canada-Ontario Great Lakes Water Quality Agreements.

By the mid-1980s, with water development and most water management largely devolved to the provincial and lower levels, and increasing pressures for action in the environmental area, the federal government undertook a major public review of water policy. In the first step, it asked three commissioners to conduct extensive public hearings across the country. The Final Report of the Inquiry on Federal Water Policy (*Currents of Change,* 1985) made a number of specific recommendations but also proposed a more catalytic role for the federal government.

A CLOSER LOOK

The relationship between Canadian provinces and the federal government is quite unique, and can be compared, somewhat, to the symbolic union between Canada and Great Britain. Interestingly, all three—Canada, its provinces, and Great Britain—utilize the "Mace" to publicly display that union.

FIG. 10.14. Canada is a constitutional monarchy, which means that Canada's constitution is the supreme law and the official head of state is the Queen of England. The Mace (shown at right) is a ceremonial staff that represents the power of the House of Assembly (the provincial legislature) to make laws. No legislative business can be conducted unless the Mace is present in the Legislative Assembly. Each legislative day begins with the procession of the Speaker and the Sergeant-at-Arms carrying the Mace and placing it on a special stand located in the center of the Chamber.

The history of the Mace goes back to medieval England where soldiers carried swords into battle, while the bishops who went along carried a Mace engraved with the royal coat of arms. These early Maces were symbols of the power of the Crown, but they were also lethal weapons with spikes and blades that could penetrate a coat of armor. The Mace was used in the British Parliament beginning in about the fourteenth century to symbolically represent the power of the British Crown as well as the Parliament. It has always been unthinkable for members of Parliament to conduct business without the Mace present. In the nineteenth century, the British House of Commons had to delay legislative business when the key to the cupboard holding the Mace was missing. The ties between Britain and Canada have always been strong, and the tradition of the Mace in Canada continues to be a symbol of respect for the Crown, the Speaker, and the provincial government. The Mace in Figure 10.14 is used in the Nova Scotia House of Assembly in Halifax.

Two years following the release of the Inquiry Report, in 1987, the Minister of the Environment tabled a new Federal Water Policy in Parliament. Since 1987, many of the objectives laid out in the Inquiry Report and the 1987 Federal Water Policy have in fact come to fruition. The federal effort is now focused much more on direct federal responsibilities and on enabling others to do their jobs better through improved science and information efforts. The federal government continues to play a strong role in interjurisdictional matters, for example, by housing several technical secretariats that support interprovincial or international water allocation or reservoir operations. It also continues to lead in the scientific area and to provide a catalyst for addressing problems of national dimensions. Although it continues to participate in some regional planning studies, these studies now tend to be of a broader ecological nature rather than water specific.

Legislative and other initiatives have become more comprehensive, anticipatory, and preventative. Toxic substances legislation deals with the entire life cycle of chemicals, and environmental assessment legislation focuses more comprehensively on matters such as cumulative impacts. With many of the necessary frameworks well established, the Environment Department has shifted some of its focus from top-down imposition of regulations to a more enabling role. Although flood risk maps were early examples of enabling instruments, more recent ones include providing opportunities and incentives for voluntary pollution agreements, and a bottom-up pollution prevention strategy supported by strong public awareness efforts.

Even as conventional water issues in Canada appear to be well in hand, new forces of change are once again requiring a reevaluation of federal freshwater strategy. Some of the factors forcing this new evaluation include the information and communications revolution, which represents both a challenge and an opportunity; global environmental issues such as climate change and the long-range transport of air pollutants and their impacts on water; increasing capabilities in the provinces, the private sector, and the voluntary sector; and the internationalization of water issues.

With regard to the internationalization of water issues, the matter of bulk water export has recently come to the forefront. In Canada, there has always been overwhelming public opposition to the notion of moving large amounts of water between natural river basins, except for the purpose of hydroelectric generation in largely uninhabited regions. That opposition has recently focused on potential ecological impacts and has intensified as environmental interests have become concerned that international trade agreements may oblige nations to export water against their wishes.

The International Joint Commission, in its review of potential bulk water transfer from the Great Lakes, has concluded that "the provisions of NAFTA and the WTO agreements do not prevent Canada and the U.S. from taking measures to protect their water resources and preserve the integrity of the Great Lakes basin ecosystem where there is no discrimination by decision-makers against individuals from other countries in the application of those measures." The Commission has further concluded that "NAFTA and the WTO agreements do not constrain or affect the sovereign right of a government to decide whether or not it will allow natural resources within its jurisdiction to be exploited and, if a natural resource is allowed to be exploited, the pace and manner of such exploitation."

At a more pragmatic level, it would appear that the era of major diversions and water transfers is largely over. As we move further into the twenty-first century, such projects are being viewed as largely infeasible from economic and environmental perspectives. Even bulk water transfers by marine tanker are proving to be uneconomical, except in very short-haul situations. Instead, concerns are turning to ecosystem restoration, and sustainable management has begun to guide regional planning principles. It is now widely recognized that the difficulty and expense of moving water in bulk will force water managers around the globe to find local solutions to local problems, primarily by placing a greater emphasis on the efficient use of existing water sources.

REFERENCES

International Joint Commission, *Protection of the Waters of the Great Lakes,* Final Report to the Governments of Canada and the United States, February 2000.

J. Owen Saunders, *The Legal Context for Water Uses in the Great Lakes Basin,* Working Paper prepared for the International Joint Commission, 1999.

R. L. Pentland, *Institutional Reform at the Federal Level: Canada,* Contribution to a Report on Strategic Options for the Water Sector in China, prepared by Hydrosult, Inc., and the China Institute of Water Resources and Hydropower Research, 1999.

CAREERS

The agencies described in this chapter provide a wealth of opportunities for careers in the field of water resources management. Almost every local water agency employs a manager in charge of day-to-day operations, including coordination of activities with other local agencies, budget preparation, short- and long-range planning, public relations, and communication with the local board of directors. A manager may have a degree in engineering, community and regional planning, earth sciences, geography, business, environmental studies, biology, or chemistry.

Other employment opportunities exist, such as field technician positions involving the collection of water samples, pipeline inspection, performance of field studies, management and repair of water facilities, and monitoring of water flows. Opportunities also exist for accountants, public information and education specialists, engineers, geographic information systems (GIS) specialists, and water quality technicians. Degrees for these positions can include, but are not limited to, watershed science, biology, earth sciences, geology, geography, forestry, chemistry, English, and education as well as other related majors. For a listing of current water resources employment opportunities and salary ranges, go to http://www.jobreservoir.com or http://www.waterjobs.com.

CHAPTER SUMMARY

Chapters 9 and 10 have shown the large number and diversity of local, state, regional, and federal agencies that provide services ranging from drinking water supply to wastewater treatment, irrigation water delivery, flood control, and water quality protection. Generally, local water groups provide the grassroots level of service that so many residents desire. On the other hand, regional, state, provincial, and federal water agency providers allow economies of scale to reduce costs. Federal water agencies have varying relationships with local, state, and regional water groups. In some situations, the federal government may regulate and dominate local water use practices; in other instances, the federal role may be solely advisory. These relationships are changing rapidly, creating uncertainty in many regions of the United States over water management issues.

Mexico has eliminated the multilayered approach of the United States by vesting all water management authority in the federal government. Local, state, and federal water agencies, as well as individuals, can only use water through authorization from the federal government. In Canada, the relationship between the federal water agencies and provinces differs radically from that in either Mexico or the United States. Provinces have primary responsibility for all water management policies and regulations, and the federal government generally serves in an advisory capacity. Many water users in the United States would welcome this arrangement, whereas others would see it as an effort to thwart federal environmental protection programs.

QUESTIONS FOR DISCUSSION

1. How did municipal water departments originate in the United States?

2. What role did water quality play in the early creation of municipal water departments?

3. Name seven states that have local governmental agencies called water districts.

4. What is the difference between a mutual ditch company and an irrigation district? What advantages does one have over the other?

5. Do you believe John Wesley Powell would have supported the concept of natural resources districts in Nebraska and water management districts in Florida? Why or why not?

6. Groundwater management districts do not cover the entire State of Kansas. What arguments could be used to promote the creation of districts to serve the entire state? What reasons could you use to argue against that concept?

7. Name two water agencies in your community. Name two water agencies in your state.

8. Contrast water management in the United States to the systems used in Mexico and Canada.

9. Discuss job opportunities available in local, state, or regional water districts.

10. Contact a local water provider and ask about its relationship with federal water agencies. Which federal agencies are considered to be cooperative? Which ones are considered adversarial?

KEY WORDS TO REMEMBER

Chesapeake Bay Commission p. 303

conservancy districts p. 296

flood-control districts p. 289

groundwater management districts (GMD) p. 298

irrigation districts p. 294

levee districts p. 289

Missouri River Basin Association (MRBA) p. 304

municipal water departments p. 285

mutual ditch and irrigation company p. 293

natural resources districts (NRD) p. 297

open meeting laws p. 299

open records laws p. 299

state water agencies p. 301

water and sewer districts p. 288

water management districts (WMD) p. 299

SUGGESTED RESOURCES FOR FURTHER STUDY

READINGS

Dzurik, Andrew A. *Water Resources Planning*. Savage, MD: Rowman & Littlefield, 1990.

Koeppel, Gerard T. *Water for Gotham—A History*. Princeton, NJ: Princeton University Press, 2000.

Nash, Gerald D., and Richard W. Etulain. *The Twentieth-Century West*. Albuquerque: University of New Mexico Press, 1989.

WEBSITES

"Flood Control and Water Management in the Yazoo-Mississippi Delta," Social Science Research Center, Mississippi State University, Mississippi State, Mississippi, July 2003. http://www.leveeboard.org

"Managing Government Records: An Introduction to Kentucky's Public Record's Management Law," Public Records Division, Kentucky Department for Libraries and Archives, Education, Arts and Humanities Cabinet, Frankfort, Kentucky, July 2003. http://www.kdla.state.ky.us/pubrec/MANWRIT.HTM

Mexican Institute of Water Technology (IMTA—Instituto Mexicano de Technología del Aqua), July 2003. http://www.imta.mx

Environment Canada Homepage, Ottawa, Canada, July 2003. http://www.ec.gc.ca/envhome.html

REFERENCES

1. Koeppel, Gerard T. *Water for Gotham—A History*. Princeton, NJ: Princeton University Press, 2000, 17–18.

2. Ibid., 4–15.

3. Weidner, Charles H. *Water for a City—A History of New York City's Problem from the Beginning to the Delaware River System*. New Brunswick, NJ: Rutgers University Press, 1974, 18.

4. Highline Water District Homepage, Kent, Washington, http://www.highlinewater.org, September 2001; and personal communication with Laurie Van Leuven, Public Information Officer, Highline Water District, Kent, Washington, February 6, 2001.

5. Harrison, Robert W., and Joseph F. Mooney, Jr., "Flood Control and Water Management in the Yazoo-Mississippi Delta," Social Science Research Center Mississippi State, MS: Mississippi State University, July 1993, http://www.leveeboard.org, February 2001.

6. Grant, Ulysses S. *Grant—Personal Memoirs of U.S. Grant*. New York: Library of America, Literary Classics of the United States, 1990, 223.

7. Beman, Lamar T. *The Reference Shelf—Flood Control*. Volume V, No. 7. New York: H.W. Wilson Co., 1928, 116.

8. Ibid., 10–12.

9. Personal communication with Gary G. Peterson, Principal Planner, Pima County Flood Control District, Tucson, Arizona, February 17, 2001.

10. *Fourteenth Census of the United States, 1920—Irrigation and Drainage,* Volume VII. Washington, DC: U.S. Government Printing Office, 1923.

11. Ibid.

12. Nash, Gerald D., and Richard W. Etulain. *The Twentieth-Century West*. Albuquerque: University of New Mexico Press, 1989, 257–292.

13. Pick-Sloan Missouri Basin Program—Farwell Unit, Middle Loup Division, http://dataweb.usbr.gov/html/farwell.html, September 2001; and personal communication with Tom Knutson, General Manager, Farwell Irrigation District, Farwell, Nebraska, March 19, 2001.

14. Ohio Department of Natural Resources, Homepage, Columbus, Ohio, http://www.dnr.state.oh.us/odnr/water, February 2001.

15. Miami Conservancy District Homepage, Dayton, Ohio, http//:www.miamiconservancy.org, September 2001; and personal communication with Kelly Fackel, Community & Public Relations Manager, Miami Conservancy District, Dayton, Ohio, April 20, 2001.

16. North Dakota Legislative Council Staff, "Nebraska Natural Resources Districts—History,

Organization, and Powers," prepared for the Garrison Diversion Overview Committee, December 1997, http://www.state.nd.us/lr/memos/99205.html, February 2001.

17. Papio-Missouri River Natural Resources District Homepage, http://www.papionrd.org, September 2001; and personal communication with Emmett J. Egr, Information/Education Coordinator, Papio-Missouri River Natural Resources District, Omaha, Nebraska, April 24, 2001.

18. Northwest Kansas Groundwater Management District No. 4 Homepage, Colby, Kansas, http://www.gmd4.org, August 2003; and personal communication with Wayne Bossert, Manager, Northwest Kansas Groundwater Management District No. 4, Colby, Kansas, June 22, 2001.

19. Dzurik, Andrew A. *Water Resources Planning*. Savage, MD: Rowman & Littlefield, 1990, 69–77.

20. Chesapeake Bay Commission Homepage, Annapolis, Maryland, http://www.chesapeakebay.net, September 2001; and personal communication with Ann Pesiri Swanson, Executive Director, Chesapeake Bay Commission, June 12, 2001.

21. Missouri River Basin Association Homepage, Lewiston, Montana, http://www.mrba-missouri-river.com, September 2001; and personal communication with Richard Opper, Executive Director, Missouri River Basin Association, Lewiston, Montana, June 5, 2001.

22. Personal communication with Ralph Pentland, Director of Water Planning and Management, Environment Canada, Ottawa, Canada (1978–1991), Canadian Co-Chairman, Upper Great Lakes Plan of Study Team (2000–2001), and currently president of Ralbet Enterprises Incorporated, September 1, 2001.

CHAPTER 11

DRINKING WATER AND WASTEWATER TREATMENT

A fundamental promise we must make to our people is that the food they eat and the water they drink is safe.

President Bill Clinton,
Safe Drinking Water Act Reauthorization
August 6, 1996

Drinking water and wastewater treatment processes have evolved over centuries of experimentation and accidental discovery. In ancient times, drinking water was often allowed to "rest" to allow sediments to settle to the bottom of earthen vessels. This same concept was used in later years, but with sand filtering systems. Today, many drinking water treatment methods are elaborate variations of simple ideas derived thousands of years ago.

Wastewater treatment, on the other hand, is a relatively recent phenomenon. Historically, human waste was simply dumped into rivers near settlements or was flushed down city gutters. Paris was one of the first cities in the world to construct extensive sewer systems to carry away waste, thanks to Napoléon III. Later, in the twentieth century, treatment plants were constructed to clean wastewater before returning it to water bodies.

The processes of drinking water treatment and wastewater treatment will be described in detail later in this chapter. Keep in mind the physical processes of water quality discussed in Chapter 5, and note the role that turbidity, the nitrogen cycle, and waterborne diseases play in water treatment processes.

HISTORICAL PERSPECTIVE ON DRINKING WATER TREATMENT

In ancient times, the greatest concerns regarding drinkable water were its taste, temperature, and appearance. If the source of water had large amounts of suspended sediments, drinking water was sometimes allowed to stand in jugs to settle impurities. Later, the ancient Egyptians used alum (a white mineral salt) to remove suspended solids to improve taste and appearance. This may have been the first chemical process used for water treatment. Other civilizations, such as the Hindu in India, boiled foul water to improve taste and clarity.

The Greek physician Hippocrates (460 B.C.–377 B.C.) promoted the concept of "healthy" drinking water. The Romans followed his advice and transported pristine water great distances to their cities. Sand and other sediments were allowed to settle out of the water into depressions located at regular intervals along the delivery aqueducts. In addition, the open aqueducts allowed ultraviolet (UV) rays of the Sun to provide some disinfection of water flowing in the uncovered channels. The Romans also used sand filters, similar to those used in modern water treatment plants, to improve taste and appearance.

Roman military leaders were keenly aware of the need to provide clean drinking water to soldiers in the field. In A.D. 375, the Roman military writer Flavius Vegetius Renatus observed that bad drinking water was like the plague and should be avoided. He also noted that, if an army remained camped at the same location for an extended period of time, the drinking water in the area became unhealthy. However, if a new campsite was found, drinking water was cleaner and healthier. The Romans also observed that marshes and swamps could cause disease. The Roman government encouraged draining such wetland areas to reduce the chance of contracting the disease that caused high fever, chills, and a large spleen—the modern diagnosis of malaria. (1) This same policy of wetlands destruction was repeated by the U.S. government in the mid-1800s through the various Swamp Land Acts approved by Congress (see Chapter 9).

Roman cities also had extensive systems for removing wastewater and stormwater runoff, particularly in the capital city, Rome. Gutters and streets were routinely cleaned by unused water from aqueducts that regularly flowed down streets. Waste and debris from the city were simply flushed into neighboring streams.

In the Middle Ages, the concepts of cleanliness and healthy drinking water were largely ignored. Filth was common, and personal and public sanitation were basically nonexistent. Bathing and laundering were not common practices, and it was not unusual for individuals to bathe only once or twice in a year. A large part of the history of the Middle Ages revolves around disease, plague, and death.

In the Renaissance, scientists began investigating methods to improve the quality of drinking water. Britain's Sir Francis Bacon (1561–1626) conducted numerous experiments regarding water percolation, boiling, distillation, and coagulation. Later, Dutch scientist Anthony van Leeuwenhoek (1632–1723) built a microscope to view and document common forms of bacteria found in water. Drinking water treatment was slowly becoming a science.

During the eighteenth century, scientists continued to improve the level of knowledge regarding water quality and water treatment. In 1703, Philippe de la Hire (1640–1718) of Paris encouraged all households to install sand filters on their home water systems (which usually consisted of captured rainwater from rooftops stored in cisterns or barrels) before drinking. In 1746, Frenchman Joseph Amy was granted the first patent for a water filter, which was composed of charcoal, sponge, and wood. (2)

Living conditions improved very slowly. Communities were originally settled in locations with adequate water supplies, such as near springs, shallow groundwater, or clean surface water. However, human activities generally polluted local drinking water supplies, particularly groundwater from percolation of wastes. Lakes and rivers became conduits for sewage and industrial waste from expanding urban areas, effectively destroying these waters as sources of safe drinking water. Firefighting also became a major safety issue in cities. Generally, crude water supply and delivery systems were inadequate to stop major fires in downtown business districts.

During the eighteenth and nineteenth centuries, private aqueduct companies were established to provide water supplies in metropolitan areas. Private industry took advantage of inadequate public water supply systems, and developed and delivered drinking water to residents willing to pay for the service. (Recall the water company, established by Aaron Burr in New York City in 1800, discussed in Chapter 6.) However, outbreaks of cholera, typhoid, and other highly infectious diseases eventually forced most private companies out of business.

According to the American Water Works Association, in 1754 the town of Bethlehem, Pennsylvania, became the first American city to develop a municipal waterworks system. For years, residents enjoyed adequate water supplies from the crude system of spring water, pumped through bored hemlock logs and stored in a wooden reservoir. (3)

After years of yellow fever and typhus epidemics from contaminated public and private groundwater wells, the City of Philadelphia developed a system in 1801 to pump water from the Schuylkill River by using steam engines. Water was pumped from the Schuylkill to a reservoir on the site of Philadelphia's present City Hall. From there, water was distributed through 4.5-inch (11.4-cm) inside-diameter water mains made from hollowed logs. (4) A portion of these water system expenses was paid out from proceeds of the will of Benjamin Franklin (1706–1790), who left funds for his beloved city of Philadelphia for a new public water supply system (see Figure 11.1).

In 1804, Paisley, Scotland, became the first city in the world to provide treated filtered drinking water to an entire town. In 1806, Paris completed a water treatment plant for water taken from the Seine River. Water was allowed to settle for several hours before moving through sand filters. In 1834, the first slow sand filter system was developed in the United States in Richmond, Virginia.

In 1861, the Board of Inspection of the Pollution of the Passaic River and Its Tributaries was organized by the water departments of Newark and Jersey City, New Jersey. A sanitary patrol was immediately organized to remove carcasses of dead animals and refuse from the area. After the end of the Civil War in the late 1860s, the Board of Inspection, concerned about the return of cholera from polluted drinking water supplies, hired a chemist to test the quality of water in the Passaic. The chemist assured them that water from the Passaic River was as pure as the waters found in the Croton River (New York City's upstate supply). His opinion was based on the mistaken "law of purification," which held that every stream would purify itself through oxidation after only a few miles of flow.

Unfortunately, the Passaic River was woefully polluted upstream of Newark and Jersey City by sewage, rotting animal carcasses and human bodies, and industrial wastes from upstream cities in the mid-1800s. Cholera and typhoid became epidemic in spite of the chemist's assurances. Later, between 1890–1891 near Lowell, Massachusetts (the center of the U.S. Industrial

FIG. 11.1. Well-dressed people tour the inside of the Fairmount Water Works and examine its water wheel and pump system, Philadelphia, Pennsylvania, 1853. Powered by the Schuylkill River, pumps raised water to a reservoir on a nearby hill, Faire Mount. See http://www.fairmountwaterworks.org for more information.

Revolution during the late nineteenth century; see Chapter 8), a typhoid epidemic claimed 132 lives. The source of contamination was traced to several outhouses overhanging the banks of a tributary of the Merrimac River upstream of Lowell. (5) Disease continued as factory owners and municipalities continued to dump wastes into overloaded local rivers. Eventually, the lower Passaic River had to be abandoned as a source of municipal water supply. (6)

Municipal water suppliers worked to upgrade public water systems during the late nineteenth and early twentieth centuries. In 1870, the first filtration plant was constructed in Poughkeepsie, New York, for treatment of water from the Hudson River. (7) Filtration reduced the incidence of typhoid fever dramatically, and other cities followed Poughkeepsie's lead. Chlorination was first used in the United States at Jersey City, New Jersey, in 1909. The experiment was a huge success, bringing most bacterial counts down to negligible levels. (8)

POLICY ISSUE

It is unfortunate to note the relative lack of water quality protection prior to the 1900s. After all, the ancients had promoted the need for safe drinking water, and even in more recent times (1854) the connection between polluted drinking water and disease had been demonstrated by Dr. John Snow; see Chapter 5. In seeming disregard of all that knowledge, industrialists during the Industrial Revolution of the late 1800s helped create pitifully polluted rivers and groundwater, particularly in the eastern United States and in Western Europe.

The real beginning of serious environmental protection in the United States was a long time in coming—the U.S. Environmental Protection Agency was not created until 1970. Why did federal protection of water quality take so long? What role did the politics of economic development play in the protection of water quality in the United States in the early 1900s? What role did ignorance of water quality parameters also play?

Today **potable water** (supplies that can be used safely for drinking, cooking, and washing) is obtained from lakes, reservoirs, rivers, and groundwater aquifers. Small water suppliers and individual homeowners use groundwater almost exclusively because it can often be used with little or no treatment. However, the presence of nitrates, synthetic organic compounds, heavy metals, and other pollutants is becoming a problem in some locations.

Drinking water sources vary greatly around the world (Table 11.1). In the United States, over 50 percent of the population relies on groundwater. In Florida, New Mexico, Mississippi, Nebraska, and Idaho, the figure is over 90 percent. (9) **Desalination** is a relatively new technology used in some areas, such as California, Florida,

TABLE 11.1 Drinking Water Sources for Selected Cities

City	Primary Drinking Water Sources
Al-Jubail, Saudi Arabia	Salt water from the Arabian Sea
Chicago, Illinois	Lake Michigan
Edmonton, Alberta	North Saskatchewan River (glacial meltwater)
Montreal, Quebec	St. Lawrence River
New York City, New York	Delaware and Croton river watersheds
New Orleans, Louisiana	Mississippi River
Paris, France	Groundwater and surface water from the Seine and Marne River watersheds
Reykjavik, Iceland	Groundwater (untreated due to its high quality)
Washington, D.C.	Potomac River

Saudi Arabia, and the United Arab Emirates, to remove excess salts from ocean or sea water. (Salt water generally has total dissolved solids [TDS] of 35,000 ppm, while the Safe Drinking Water Standard for the United States has a TDS standard of 500 ppm.) Costs of desalination can be quite high, with ranges of over \$1000 per acre-foot (325,850 gal, or 1233 m^3) common, versus \$200 per acre-foot from freshwater supplies. However, Tampa, Florida, is now desalinating water at a cost of only \$650 per acre-foot (\$0.002/gal, or \$0.53/m^3), in part owing to the involvement of private developers in the process. (10) This plant came online in 2003 and is expected to provide 10 percent of the region's overall water supply by 2008. It will produce 25 mgd (77 acre-ft, or 94,635 m^3) and is the largest desalination plant in North America. It was built adjacent to Tampa Electric Company's Big Bend Power Station at a cost of \$110 million. (11)

The City of Abu Dhabi, capital of the United Arab Emirates, is located in a large desert area on the Persian Gulf. With a population of 800,000 and average annual rainfall of only 4 inches (10 cm) per year, water supplies are scarce. In the late 1990s, a desalination plant was constructed at a cost of \$1.7 billion and now produces 91.4 million gallons per day (281 acre-ft, or 345,987 m^3). (See Figure 11.2.) Excess freshwater is piped to an oasis 87 miles (140 km) away for irrigation. (12)

A relatively new drinking water treatment

FIG. 11.2. Historically, Kuwaitis have had a limited supply of freshwater, and its small population was served by a few artesian wells. Supplemental water was brought in on *dhows* (common local boats) manned by Kuwaiti seamen to obtain freshwater from the Shatt Al-Arab, the waterway created by the joining of the Tigris and Euphrates rivers between modern-day Iran and Iraq. Rapid population growth in the mid-1900s led to the construction of a desalination plant in Kuwait City in 1953. Later, two additional desalination plants and a reverse osmosis plant were constructed near Doha but were damaged or destroyed during the Iraqi invasion in 1990. Today, Kuwait is one of the world's leaders in the production of potable water from the sea. This desalination plant is located at Doha, Qatar.

process is **reverse osmosis (RO)**—a membrane separation process that removes salts and other contaminants through filtration. Many smaller communities on groundwater wells are installing RO systems to remove excess nitrates. Costs are higher for filtration than for normal drinking water system processes, but filtration does provide a means of using saline or otherwise contaminated water supplies for potable purposes.

GUEST ESSAY

Dr. Fares M. Howari

Water Desalination in the Middle East: One of the Realistic Options

by **Dr. Fares M. Howari**
United Arab Emirates University, Al Ain, UAE

Dr. Fares Howari has more than ten years of professional experience in water and environmental related issues. He worked with the Center for Environmental Resources Management at the University of Texas at El Paso on United States–Mexico environmental issues (1997–2001). Dr. Howari then worked at the Texas A&M Agricultural Research Station on salinity issues of the Rio Grande Valley (2001–2002). He has also been involved in investigations of salinity and sediments of the Jordan and Yarmouk rivers in the Middle East. He is currently on the Science Faculty of the United Arab Emirates University in Al Ain and is conducting research on water desalination using green technology.

Water scarcity is a growing worldwide phenomenon. The amount of renewable water resources per capita has declined dramatically over a single generation and in little more than 30 years will reach dangerously low levels. By the year 2025, the average net freshwater resources in the Middle East are expected to be less than half of today's supplies. Fortunately, water providers along the Arabian Gulf are developing successful examples of dealing with water scarcity through desalination.

Before we discuss this emerging technology, here are some brief facts about the six Arabian Gulf Countries (GC)—the Sultanate of Oman, United Arab Emirates (UAE), State of Bahrain, Emirate of Qatar, State of Kuwait, and the Kingdom of Saudi Arabia. These countries have similar physiographic, social, and economic characteristics, including extremely arid climates, sparse natural vegetation, and fragile soil conditions. The natural water resources consist of limited quantities of surface water runoff from floods, groundwater in alluvial aquifers, and extensive groundwater reserves in deep sedimentary aquifers. All six GC countries suffer from a scarcity of freshwater.

Desalination in the Middle East A wide range of options—water conservation, wastewater treatment, and desalination—have been used to meet freshwater demands. Of these methods, desalination has offered the best solution to water shortages in the Arab Gulf Countries. Of the world's 7500 desalination plants currently in operation, two-thirds of them are located in countries of the Middle East, particularly in the Arabian Peninsula. Groundwater supplies continue to be mined in the region and as a result, desalination technology is finding new outlets in locations where it had not previously been considered a viable long-term solution. Table 11A gives a summary of the desalination plants in the Middle East. Arabian Gulf Countries such as Saudi Arabia, the UAE, and Kuwait use dual-purpose power and desalination plants on a major scale; this practice has helped finance the expensive desalination projects and will be discussed later.

Kuwait was the first state to adopt sea-water desalination, linking electricity generation from the desalination process to generators to provide energy for the desalination plant. As a result, both energy and costs are minimized. Kuwait began desalinated water production in 1957, when 3.1 million m^3 (819 million gal, or 2513 acre-ft) were produced per year. By 1987, this figure had risen to 184 million m^3 per year (48.6 billion gallons, or 149,171 acre-ft), and the trend is still on the rise.

Saudi Arabia (the world's largest producer of desalinated water) represents another successful

TABLE 11A Desalination Units in the Middle East, 2003

Location	Number of Units	Total Capacity (m^3 per day)	(gal per day)
UAE	382	5,465,784	1.4 billion
Bahrain	156	1,151,204	304 million
Saudi Arabia	2074	11,656,043	3.1 billion
Oman	102	845,507	223 million
Qatar	94	1,223,000	323 million
Kuwait	178	3,129,588	827 million
TOTAL	2986	23,471,126	6.2 billion
Libya	431	1,620,652	428 million
Iraq	207	418,102	110 million
Egypt	230	236,865	63 million
Algeria	174	301,363	80 million
Tunisia	64	148,822	39 million
Yemen	66	132,897	35 million
Israel	n/a	149,594	40 million
TOTAL	1172	3,008,295	795 million

Source: Cooperation Council for the Arab States of the Gulf.

example. It also has the world's largest plant, which produces 484,533 m^3 per day (128 mgd, or 393 acre-ft) of desalted water. Saudi Arabia has networked 27 of its 2074 desalination plants to provide drinking water to major urban and industrial centers through water pipelines that extend for more than 3700 kilometers (2300 mi). Desalination meets 70 percent of this arid country's drinking water requirement.

The United Arab Emirates (UAE) also relies heavily on desalinated sea water for most of its needs (see Figure 11.3). Indeed, the UAE is the world's second largest producer of desalinated water, with production of more than 1.7 million m^3 (440 mgd, or 1350 acre-ft). A large portion of this water comes from the Al Taweelah Desalination Plant, which now produces freshwater for 600,000 people a day. Additional freshwater is produced at the Abu Dhabi solar plant, which cost more than $500 million to construct.

Bahrain currently has 156 desalination units producing 276,335 m^3 (73 mgd, or 224 acre-ft) of freshwater. Bahrain's largest plant at Hidd is to go through a 227,125 m^3 (60 mgd, or 184 acre-ft) capacity increase, planned for 2006. The estimated capital cost for this upgrade is approximately $300 million.

Desalination Problems Although desalination options are realistic and effective, we should not be overoptimistic about this relatively new technology, for it creates new problems. First, the economic viability of desalination is questionable for some poor countries in the Middle East. Although it has become less expensive in recent years as processes have matured, it remains an enormous investment. Most desalination plants in the region are subsidized by governments, gifts of land, free natural gas, or tax breaks to import necessary equipment. Poor countries will need to utilize international funds for construction and operation costs.

The second problem is that current desalination processes use enormous amounts of energy. For instance, with the multiflash method, a vacuum system is applied to reduce the boiling point of the sea water. A spray or thin film of water is exposed to high heat, causing flash evaporation. The sea water is flashed

FIG. 11.3. This main boiler feeds water pumps and auxiliary pumps (42 pumps in total) in the Fujairah Power and Desalination Plant in the United Arab Emirates.

repeatedly, yielding fresh distilled water which is collected for use. If energy prices increase, the multiflash method will become extremely expensive. This may not be a serious problem for most countries in the Arabian Gulf inasmuch as energy resources are plentiful, but it is a serious problem for energy-poor countries like Jordan. Multistage flash accounts for 86.7 percent of all desalting capacity, while reverse osmosis accounts for only 10.7 percent.

Third, desalination raises yet another environmental problem: the extremely saline waste product left behind after freshwater has been removed. Some desalination plants produce an estimated one to two truckloads of solid waste per week. This waste must be safely disposed or recycled. Fortunately, solar pond technology can offer a solution to this problem. The rejected salt can be used in a saline water pond to stop thermal water circulation. This condition occurs when large quantities of salt are dissolved in the hot, bottom layer of a body of water, making it too dense to rise to the surface and cool. Usually, the deep water has a higher salt content and is heavier, permitting sunlight to be trapped in the hot, bottom layer. A pilot study is being seriously considered at the United Arab Emirates University to investigate natural salinity gradients in solar ponds near salt-flat areas (*sabkhas*) of Abu Dhabi which can trap sunlight. This could allow useful energy to be withdrawn or stored for later use in water desalination processes.

Future Development The future of desalination technology will depend on reducing energy costs, improvements in reverse osmosis (RO) and distillation processes, the increased use of solar pond technology, and increased financing through public funds and privatization. In the short term, new solutions are required to address the trend of water demand growing at a greater rate than electricity and the dramatic seasonal variation of power. Some analysts claim that seasonal surpluses of power could be used by electrically driven technologies such as RO. Such solutions demand continuous research and development to achieve the goal of low-cost desalination to produce water at less than $0.50/m^3 ($0.002/gal, or $617/acre-ft).

FEDERAL PROTECTION OF DRINKING WATER IN THE UNITED STATES

In 1914, the U.S. Department of the Treasury set the nation's first bacteriological standard of 2 coliforms per 100 milliliters for drinking water. This standard marked a major shift in public policy: for the first time the federal government became responsible for the safety of public drinking water supplies. To meet the new standard, most large cities were required to add filtration and chlorination processes in their water treatment facilities. As a result, by the 1930s, most waterborne diseases such as typhoid and cholera were widely reduced in the United States.

In the late 1930s, the Potomac River Interstate Compact was approved by the legislatures of Maryland, Pennsylvania, Virginia, and West Virginia, the District of Columbia, and the federal government. It was ratified by the U.S. Congress in 1940 to protect water quality in the 14,679 square mile (38,019 km^2) Potomac River watershed. The Potomac begins as a small spring in West Virginia, approximately 500 feet (152 m) from the Maryland border. After 383 river miles (616 km), it empties into the Chesapeake Bay at Point Lookout, Maryland, where the river is more than 11 miles (18 km) wide. The river had been polluted for centuries, and the Compact was one of the first attempts in the United States to improve water quality through the implementation of an interstate agreement.

In 1942, the U.S. Public Health Service (USPHS) set the first comprehensive drinking water standards in the country. In 1948, the Federal Pollution Control Act was passed to reduce pollution entering rivers and lakes. However, these laws had no enforcement mechanisms and were often ignored until passage of the **Safe Drinking Water Act (SDWA)** in 1974. This Act is the most comprehensive law in U.S. history to protect drinking water.

The SDWA requires that public water systems, which provide water for human consumption (through at least 15 service connections or regular service to at least 25 individuals), must routinely monitor for and comply with certain contaminant levels. The U.S. Environmental Protection Agency was designated to develop standards and regulations regarding contaminant levels and provided funding to assist water providers to meet these standards. The USEPA was also given enforcement authority. (Individuals who obtain drinking water from domestic wells are not regulated or monitored by this Act.)

To enforce the SDWA standards, the USEPA set up mandatory, enforceable maximum contaminant levels (MCLs). These levels were set as close as possible to recommended health goals that would not pose significant health risks over a human lifetime. The USEPA set standards on microbiological contaminants (bacteria, viruses, and parasites); metals and inorganic chemicals (lead, copper, mercury, and aluminum); carbon-based chemicals such as sulfates, nitrates, nitrites, and asbestos; and volatile organic chemicals (VOCs) such as compounds from solvents, insecticides, and industrial waste including benzene, carbon tetrachloride, and heptachlor. Organic compounds (herbicides, pesticides, and radionuclides) are also regulated.

POLICY ISSUE

What political forces and world events in the 1940s would have led Congress to pass legislation (concerning drinking water standards and the Federal Pollution Control Act) without any federal enforcement mechanisms? What social/political changes occurred in the United States between the 1940s and the creation of the U.S. Environmental Protection Agency in 1970? In your opinion, does the USEPA, and the federal government in general, currently have too much or not enough regulatory authority regarding drinking water issues?

The **Safe Drinking Water Act Amendments of 1996** include a requirement that water providers inform customers of the quality of their drinking water. Brochures and other public information methods are used to meet this law. In addition, providers must inform the public within 24 hours if drinking water has become contaminated with constituents (such as microorganisms) that could cause immediate illness.

To ensure proper testing of drinking water, the USEPA has established testing schedules for various constituents:

- Bacteria, monthly or quarterly, depending on the size of the system
- Nitrates, yearly
- VOCs, twice every three years in large systems, once every three years in small systems
- Metals, once every three years for groundwater and surface water
- Lead and copper, annually
- Radionuclides, once every four years

The 1996 Safe Drinking Water Act Amendments also require states to develop programs to protect water supply areas (called *source water*) from pollution. The Act provides $10 billion for improving drinking water infrastructure in the country, and it requires that water utility operators be trained and licensed by states. See Table 11.2 for USEPA standards for safe drinking water.

Recently, perchlorate has come under close scrutiny by the USEPA but has not yet been given a maximum contaminant level for drinking water. Perchlorate is both a naturally occurring and human-made chemical. Most of the perchlorate manufactured in the United States is used in solid rocket propellant. Many municipal supply systems are finding trace elements of perchlorate in drinking water, and in 1998 it was placed on the USEPA's Contaminant Candidate List, since high levels can lead to thyroid tumor formation. See http://www.epa.gov/safewater/ccl/perchlorate/perchlorate.html for more information.

TABLE 11.2 USEPA Safe Drinking Water Standards of Selected Constituents, 2004

Constituents	Milligrams/ Liter	Health Risks	Sources of Contamination
Inorganic Chemicals			
Copper	1.3	Liver and kidney damage	Corrosion from plumbing; leaching from wood preservatives
Fluoride	4.0	Bone disease	Water additives; surface runoff
Nitrate (measured as Nitrogen)	10	Blue Baby Syndrome	Runoff from fertilizer use, septic tanks, sewage, runoff from natural deposits
Nitrite	1	Blue Baby Syndrome	Runoff from fertilizer use from lawns, farms, golf courses; septic tanks, sewage, runoff from natural deposits
Organic Chemicals			
Alachlor	zero	Risk of cancer	Runoff from herbicide use on row crops
Atrazine	0.003	Cardiovascular system problems	Runoff from herbicide use on row crops
2,4-D	0.07	Kidney, liver, or adrenal gland problems	Runoff from herbicide use on row crops
Dioxin	zero	Risk of cancer	Emissions from waste incineration
Xylenes (total)	10	Nervous system damage	Discharge from petroleum or chemical factories
Microorganisms			
Giardia lamblia	zero	Giardiasis	Human and animal waste
Cryptosporidium	zero	Gastrointestinal illness	Human and animal waste
Viruses	zero	Gastroenteric disease	Human and animal waste

Source: U.S. Environmental Protection Agency, Office of Water, Washington, DC, http://www.epa.gov/safewater, April 2004.

THE DRINKING WATER TREATMENT PROCESS

PROTECTION OF WATER QUALITY

The first step in acquiring safe drinking water is to protect raw water at its source. Watersheds used for municipal water sources often have restricted land uses, recreational activities, and development controls. Water providers must limit erosion of sediments, body contact sports such as swimming and water skiing, and waste disposal in such areas. Groundwater sources are often protected through Wellhead Protection Programs, discussed in Chapter 5.

INTAKES FOR RAW WATER

The second step in acquiring drinking water is to divert it from a river, reservoir, or groundwater. **Intakes** are the permanent connecting structures (pipes, concrete conveyance structures, etc.) that capture raw water and transport it to a drinking water treatment facility. The intake of a groundwater well includes the screened well casing in the aquifer and a piping system that delivers groundwater to the treatment plant or end user. The intake for surface water sources can include a diversion dam and headgate on a river (Figure 11.4), or pipes that divert water from a reservoir. Water intakes at reservoirs are usually located at different depths to obtain varying water temperatures and suspended sediments. Intakes are generally not placed near the water surface to avoid floating debris, or at locations that could collect bottom deposits. Drinking water sources can be intermingled to improve the temperature and water quality of raw water provided to the treatment plant from multiple sources.

FIG. 11.4. This intake structure for the Mount Werner Water Filtration Plant is located near the mouth of Fish Creek Canyon above Steamboat Springs, Colorado. Average diversions are 3 million gallons per day (9.2 acre-ft, or 11,356 m^3 per day). Flows are obtained directly from Fish Creek or from releases out of Fish Creek Reservoir or Long Lake. The Mount Werner Water District and the City of Steamboat Springs jointly own and operate the water treatment facility located downstream of this site.

A CLOSER LOOK

Zebra mussels (*Dreissena polymorpha*) are a new problem for water treatment facilities in parts of the United States and Canada, particularly along the western shores of Lake Erie. Mussels are a class of mollusk, similar to oysters, clams, and scallops, which originated in the Black and Caspian seas. Inadvertently, Zebra mussels have been transported to the Great Lakes by cargo ships. The mussels grow to approximately 1 inch (2.5 cm) in size and produce approximately 35,000 eggs per season per female. This proliferation of Zebra mussels is causing serious problems for raw water intakes and discharge lines since mussels attach to hard surfaces and clog water intakes as well as increase corrosion problems (see Figure 11.5). Water providers along the Caspian Sea did not experience these problems, perhaps owing to colder and deeper waters.

FIG. 11.5. Water intake clogged with Zebra mussels.

CASE STUDY

Department of Water Management, City of Chicago, Illinois

http://www.cityofchicago.org/watermanagement

The early settlers at the south end of Lake Michigan used the Chicago River as a primary source of water supply. By 1830, however, water in the Chicago River was becoming polluted from the growing population, and so a groundwater well was constructed (today in the vicinity of Michigan Avenue and Wacker Drive in downtown Chicago). Residents used buckets to carry water to their homes and businesses, and water peddlers sold well water door-to-door for 10 cents a barrel.

Chicago's first waterworks was established in 1842 by the privately owned Chicago Hydraulic Company. It operated a small pumping plant and distribution system off Lake Michigan and distributed water through wooden pipes to a relatively small number of Chicago residents. The water intake pipe extended 150 feet (46 m) into Lake Michigan near the outlet of the Chicago River. A steam engine pumped water from the lake into elevated wooden tanks that then flowed by gravity into the wooden pipes. Operating records tell of problems with turbid water, ice buildup, and fish-clogged water intakes. A cholera epidemic put the Chicago Hydraulic Company out of business, and in 1852 the City of Chicago took over the system. (13)

The original water supply intake pipe was located in Lake Michigan but not far from the outlet of the Chicago River into the lake. A 6-inch (15-cm) rainstorm in 1885 exacerbated the problem with this arrangement. By now, the Chicago River carried foul, polluted sewage and stormwater from city streets, sidewalks, and detention ponds (stormwater retention impoundments) to Lake Michigan. Water intake pipes delivered this polluted lake water directly back to the city for drinking water consumption. Disease and death followed.

The City of Chicago upgraded the facility after the 1885 plague and extended the intake pipes further into the lake, approximately 600 feet (183 m) from the shoreline. (14) Even so, pollution from the Chicago River continued to enter Lake Michigan, and water quality remained a serious public health threat. In 1892, the Metropolitan Sanitation District of Greater Chicago was created. One of the Sanitation District's first projects was to reverse the flow of the Chicago River by completing the 28-mile (45-km) Chicago Sanitary and Ship Canal. (15) Completion of the canal would prevent pollution from entering Lake Michigan by diverting the flow of the Chicago River into the Mississippi River Drainage. This was possible because Chicago is located at a low point between Lake Michigan and the Mississippi River watershed. The Illinois and Michigan Canal, built in 1848, allowed water from the Chicago River to be linked with the Mississippi via the Des Plaines and Illinois rivers. It took 8500 workers eight years to dig the sewage channel, which was 160 feet (49 m) wide and 24 feet (7 m) deep. The canal project, the largest earth-moving project in the history of municipal public works at the time, was completed in 1900. (16)

Residents in downstream states were horrified by the project, although the completion of the new sewage canal was widely hailed as a victory for wastewater treatment, since the natural cleaning processes of the Mississippi River would "dilute and assimilate" Chicago's wastewater. The downstream City of St. Louis and the State of Missouri thought otherwise. Together, they brought a lawsuit against the State of Illinois and the Chicago Sanitary District in 1900 to stop the dumping of raw sewage into the Mississippi. This contentious legal proceeding was eventually heard by the U.S. Supreme Court. Unfortunately for residents along the Mississippi River, the Supreme Court ruled in favor of Chicago and the State of Illinois. The Court reasoned that sewage could lawfully be delivered into the Mississippi River Basin since the Illinois River, which would receive waste from the newly constructed sewage canal, was less polluted than the Mississippi River at St. Louis. (17)

Chicago began chlorinating its water supply in 1912 and by 1915 provided chlorinated water to all residents, eliminating many health problems caused by contaminated water. To solve the problem of pollution in Lake Michigan, the flow of the Calumet River was also reversed in 1922. This did not completely solve the city's water quality problems, however, for excessive turbidity persisted in the lake owing to the mixing action of wind and waves. (18)

In 1928, city engineers began experimenting with filtration systems. The South Water Purification Plant was opened by 1947, and, with a capacity of 480 mgd (1473 acre-ft, or 1.8 million m^3 per day), was the largest plant in the world until surpassed in 1964 by the James W. Jardine Water Purification Plant (located just north of Navy Pier) with a capacity of 1440 mgd (4419 acre-ft, or 5.4 million m^3 per day). In 1967, the South Water Purification Plant was expanded to 720 mgd (2210 acre-ft, or 2.7 million m^3 per day). Today, Chicago provides excellent drinking water to its customers. (19)

PRETREATMENT OF DRINKING WATER

Once raw water is delivered to a drinking water treatment plant through an intake pipe, pretreatment usually occurs in large tanks or small reservoirs where a variety of water treatment steps begin. Screens are first used to remove large floating items, fish, fine solids, and other objects. Next, water is allowed to stand in tanks or reservoirs to promote sedimentation whereby larger silts, fines, and clay particles settle out of suspension. Pretreatment is particularly useful if water is diverted from a river that has high amounts of suspended sediments.

Flocculation/Coagulation The next step in drinking water treatment, **flocculation/coagulation,** is the process of adding chemicals to water to cause very fine suspended matter to settle out. Chemicals such as alum (aluminum potassium sulfate), activated silica, clay, and soda ash have been found to assist in this process when agitated (called *flash mixing*) into raw water. A precipitate of almost gelatinous particles will coagulate (form) usually within 10 to 30 minutes after the chemicals are added. This coagulation of fine suspended matter is called *floc* and will gain enough mass and weight to settle out of the water as a sludge.

The main objective of flocculation and coagulation is the formation of clear water that has floc visible and in suspension. This process can remove approximately 90 to 99 percent of all viruses present in water, although prechlorination and preozonation may be necessary if excessive organic material is present. Viruses are not actually killed during this process; instead, they are contained within the settled floc and sediments and are later removed. (20)

Concrete or steel sediment basins, generally 8 to 20 feet (2.4 to 6.1 m) deep, are used to hold water during the flocculation/coagulation process. Pretreated water continuously flows through these tanks. Sludge, created by settling floc, is generally removed from the bottom of sediment basins every six months.

Filtration The process of passing water through layers of sand and gravel to eliminate turbidity, odor, and color is called **filtration,** and follows flocculation and coagulation. The most common filtration method is the gravity rapid sand filter in which water is passed through beds of sand. Some drinking water plants can eliminate the need for flocculation and coagulation if raw water is obtained from protected (clean) sources of surface water or groundwater.

Filtration is a relatively simple process in which raw or pretreated water is slowly sprayed or sprinkled onto filtering media (sand). Gravity forces water through the sand particles until it exits the bottom of the media. Then the filtered water is conveyed to a storage area for additional treatment (pH, fluoridation, and disinfection).

If filters become clogged with sediments and other particulates, they must be backwashed (flushed) occasionally to remove unwanted materials. The backwashed material is considered waste and must be drained into the sewer system for treatment before being released back to rivers, lakes, or other water bodies. Improper backwash techniques can allow pollutants to contaminate a drinking water system.

CASE STUDY

Sewerage and Water Board, City of New Orleans, Louisiana

http://www.swbnola.org

New Orleans has a unique drinking water and wastewater system because of its location near the mouth of the Mississippi River. The city is located in a topographic bowl between the meanders of the Mississippi River and Lake Pontchartrain. Its high point is the French Quarter at 14 feet (4.3 m) above mean sea level, and its lowest point is 6 feet (1.8 m) below mean sea level. With an average annual precipitation of over 60 inches (152 cm), flooding is common in New Orleans and groundwater levels are quite high.

The French settled the swampy area in 1718. Most of New Orleans burned to the ground in 1788, and again in 1794, due to the lack of an adequate municipal water supply system. During the 1800s, drinking water was obtained either by capturing rainwater from rooftops or by filling buckets from the Mississippi River and allowing the turbid water to settle in large earthenware jars. Cholera, typhoid, and yellow fever were common diseases caused by polluted surface water supplies.

Finally, in 1899, the Louisiana Legislature created the Sewerage and Water Board of New Orleans to develop and operate a drinking water and wastewater system for the city. Today, New Orleans has a state-of-the-art drinking water treatment system that uses water directly from the Mississippi River (Figure 11.6). Raw water is purified, treated with ferric sulfate to promote coagulation, and combined with lime (calcium oxide) to adjust pH. Flocculation is used to remove suspended sediments. Chloramine (a combination of chlorine and ammonia)

FIG. 11.6. Raw water from the Mississippi River is pumped to the Carrollton Water Purification Plant (pictured at right), one of two such plants that serve the City of New Orleans. Settling basins allow coagulated particles to settle and form a sludge layer on the bottom of each basin, which is then removed by mechanical rakes, valves, and pumps. Further treatment includes disinfection with chlorine and ammonia, additional settling basins, fluoridation, and filtration through 44 dual-media rapid sand filters. The Carrollton Plant normally produces about 115 million gallons per day (353 acre-ft, or 435,322 m^3 per day) and removes almost 20,000 tons (18,144 metric tons) of solid materials from raw Mississippi River water annually.

is added for disinfection. Pretreated water then travels to secondary settling basins to allow additional contact time for disinfection and for smaller suspended particles to settle. Finally, water flows through 44 rapid sand filters for additional cleaning.

Laboratory personnel test drinking water for pH, alkalinity, hardness, fluoride, microorganisms, volatile organic compounds, and herbicides before it is transported by pumps and gravity through more than 1600 miles (2575 km) of water mains to over 140,000 service connections. (21)

FINAL DRINKING WATER TREATMENT

After filtration is completed, water is placed in holding basins where fluoridation and disinfection can occur. **Fluoridation** is considered a preventive medicine program to improve the health of teeth. Sodium fluoride (NaF) is generally used in this process. In a common method of disinfection using chlorine gas, called **chlorination,** the gas is mixed with water to kill remaining bacteria and some viruses. Chlorine odor and taste are common complaints of drinking water customers; activated carbon treatment can help reduce these complaints. Since chlorine is very toxic, safe storage and handling of the chemical at water treatment plants are extremely important. Chlorine compounds were used as early as the 1830s to eliminate foul smells. (22)

Ozone gas and **ultraviolet (UV) systems** may also be used in this final stage of water treatment to eliminate remaining bacteria and viruses. Ultraviolet treatment kills almost 100 percent of all microbiological organisms in water, but it is a very slow and expensive process. pH and corrosion control also occur at this stage of drinking water treatment by adding chemicals such as lime and soda ash.

DISTRIBUTION SYSTEM

The final step in the drinking water treatment process is the delivery network to consumers. Treated water is stored in covered concrete or steel reservoirs. Delivery pipes vary in size and can range from 24 to 60 inch (61 to 152 cm) main lines found beneath city streets down to 0.5 to 0.75 inch (1.3 to 1.9 cm) lines into a house. All waterlines are interconnected (looped) and generally run beneath roads or other accessible rights-of-way. A water delivery system is connected in a grid pattern to create multiple loops that eliminate dead ends where standing water could be trapped.

Why would standing water in a city's delivery system be a concern? Stagnant water allows bacteria to grow and corrosion of pipes to increase. Fire hydrants are located at set intervals and are periodically flushed by firefighters or city water department employees to prevent buildup of deposits and to cleanse the distribution system. However, flushing distribution pipes often stirs sediments or removes buildup from water lines. This causes a short-term increase in turbidity, color, and bad taste in water, followed, of course, by customer complaints.

Water pressure is also a major concern of water providers. Storage tanks are usually located at topographic high points of an area to allow treated water to be delivered to customers by gravity. Treated water may also be stored in water towers (large, elevated metal storage tanks), or it may be transported to customers at higher elevations by pumps. Adequate water pressure is critical to properly move treated water through elaborate systems of delivery pipes and to satisfy fire protection purposes.

The town of Clarkson, for example, located in northeast Nebraska approximately 85 miles (137 km) northwest of Omaha, has a municipal water system serviced by two wells with an average depth of 100 feet (30 m). The overhead storage system, shown in Figure 11.7, is a 200,000 gallon (0.6 acre-ft, or 757 m^3) water tower. Average daily demands are 250,000 gallons (0.8 acre-ft, or 946 m^3) with summertime peaks as high as 735,000 gallons (2.3 acre-ft, or 2782 m^3). Maximum system capacity is 1,337,600 gallons per day (4.1 acre-ft, or 5063 m^3). The quality of groundwater is so high that no treatment is

FIG. 11.7. The water tower at Clarkson, Nebraska, is located on a hill in the farming community of 700 residents in eastern Nebraska. Groundwater is pumped into the elevated storage tank to ensure adequate water supplies and water pressure for the entire town. Water towers are a source of pride in small towns in the Midwest, and Clarkson's steel tank is painted in the red and white colors of the local high school.

necessary. However, the community is developing a Wellhead Protection Program (see Chapter 5) with the assistance of the Nebraska Department of Environmental Quality. (23)

A CLOSER LOOK

How safe is our drinking water supply from terrorism? This is a question few considered in the past, but it is a very real concern today. One month after the terrorist attacks on the World Trade Center in New York City, the Pentagon in Washington, D.C., and in Pennsylvania, Congress held hearings to consider the level of risk to the United States' water resources. Speaking before the House Committee on Transportation and Infrastructure, Ronald L. Dick, deputy assistant director of the Counter Terrorism Division and Director of the National Infrastructure Protection Center (NIPC) at the Federal Bureau of Investigation (FBI), testified on the current safety of the nation's infrastructure. He stated that, when a terrorist attack occurs, the FBI is the lead federal agency for crisis management, whereas the Federal Emergency Management Agency (FEMA) is responsible for consequence (aftermath) management. If a water/wastewater facility is involved, the U.S. Environmental Protection Agency will assume lead responsibility. The NIPC currently provides water providers with "timely, substantive, and actionable information on specific threats." (24) A threat to a specific water/wastewater treatment plant would involve appropriate local, state, and federal security actions. Each FBI field office has a Weapons of Mass Destruction (WMD) coordinator whose primary responsibility is to coordinate the assessment and response to incidents that could involve the use of chemical, biological, and radiological/nuclear materials.

Dick reassuringly states that biological targeting of the nation's water supply could prove very difficult, owing to the properties of dilution. A terrorist would require large quantities (truckloads) of an agent, specific placement within the water supply network, and access to critical, and generally restricted, areas within the network. (25)

A statement before the Congress in October 2001 regarding terrorism and the nation's water supply system made the following four points:

1. Biological contamination of a water supply that causes illness or death of victims is possible but not probable.
2. Contamination of a water reservoir with a biological agent would probably not produce a large public health risk because of dilution, filtration, and disinfection of the water. In addition, only 1 to 2 percent of total water consumption is used for drinking water and food preparation.
3. A successful attack would require knowledge of, and access to, critical locations within the water supply network.
4. A successful attack would most likely involve disruption of the water treatment process (e.g., destruction of plumbing or release of disinfectants) or post-treatment contamination near the target. However, contaminated sources could be isolated from the distribution infrastructure. In addition, dilution, evaporation, and chemical and biological degradation would reduce the impact of a pretreatment assault. (26)

In cooperation with FEMA, a Technical Support Working Group has developed GIS software to assess the vulnerability of the nation's waterways to terrorist threats, as well as other natural and manmade disasters. The stream gaging program of the U.S. Geological Survey is used to provide flow data for the software program. Information on raw water intakes and water supply plants is obtained from the USEPA Safe Drinking Water Information System national database. (27)

Since 2000, the USEPA, the American Water Works Association, the Association of Metropolitan Water Agencies, the Centers for Disease Control and Prevention, the FBI, water utilities, and others have been meeting to assess the methods needed to protect the nation's water supply. These efforts were partially in response to President Clinton's Presidential Directives in 1998 to create an integrated government structure to combat terrorist attacks involving chemical, biological, or sabotage against infrastructure. (28)

HISTORICAL PERSPECTIVE ON WASTEWATER TREATMENT

In the Pulitzer Prize-winning book, *Angela's Ashes* (1999), Frank McCourt tells of his boyhood years in Ireland during the 1940s—a heartwrenching tale of poverty, disease, hope, and death. Pervasive throughout the book are the incredible unsanitary conditions of his home. Human waste, rats, and putrid smells were a part of everyday life for the poor in McCourt's neighborhood. Outhouses were standard in his community, and indoor plumbing was a luxury. (29)

Human waste disposal has been a problem for settlements since ancient times. Ever since civilizations developed along lakes and waterways, people have disposed of wastewater into the same rivers, lakes, and groundwater later used as sources of drinking water. Cholera, typhoid, and other waterborne diseases were common, greatly reducing the human life span. It was through this deadly process that nature controlled population growth.

Some sociologists contend that a civilization's sophistication can be judged by its sewage disposal practices. Sewage (wastewater or effluent), the culmination of human wastes from homes and workplaces, is composed of about 99.9 percent water and 0.1 percent solid or dissolved wastes. (30) It is usually dealt with in an "out-of-sight-out-of-mind" manner, but if not handled properly, it can cause severe environmental and health issues.

Sanitation laws, passed as early as 200 B.C. in Babylonia, mandated that waste not be thrown into wells or cisterns, and it was forbidden to locate a cemetery, furnace, tannery, or animal slaughterhouse within 80 feet (24 m) of a groundwater well. The ancient Greeks were particularly aware of the need for sanitation and clean drinking water. Hippocrates wrote that spring water or rainwater was preferable to stagnant water. The Greeks' strong emphasis on personal hygiene, especially bathing, made a great contribution to civilization. (31)

By the Middle Ages, the ancients' views on sanitation had seemingly been forgotten. In European cities, sewage was routinely dumped into the streets, adding to the "stench" of death during plague years. Around 1200 the streets of Paris (and other major cities) were paved with cobblestones, and open sewers ran down the center of these roadways. Rain, barefoot travelers, and roaming animals carried sewage throughout the city, most of which ultimately made it to the Seine River. In 1370, a vaulted sewer was constructed under rue Montmartre (a Parisian street) to drain wastes directly to the Seine. However, sanitary conditions in Paris still remained intolerable.

By the mid-nineteenth century, awareness of the importance of sanitation had finally improved. Under the reign of Napoléon III in the 1860s, for example, 181 miles (291 km) of additional underground sewers were constructed and later expanded. These sewage ways or *egouts* were constructed around four principal tunnels 18 feet (5.5 m) wide and 15 feet (4.6 m) high. The tunnels formed an underground city with street names clearly marked within the sewer tunnels. In contrast, during the U.S. Civil War, issues of water contamination and wastewater disposal were not widely understood. Union and Confederate soldiers alike often disposed of wastes in a river upstream while drawing drinking water from downstream sources. Dysentery was widespread and deadly. (32)

In 1867, the English civil engineer Baldwin Latham gave "A Lecture on the Sewage Difficulty" and discussed the problems of water pollution caused by animal and human waste. As a result, committees and commissions were established all over Europe to address the ongoing "sewage difficulty." One solution devised was to apply sewage onto farms where the natural system of the nitrogen cycle would decompose some pollutants. Thus, municipal sewage was piped directly to farms for the irrigation of crops, especially around London, Berlin, Paris, and Sydney in the late 1800s. This system was tried later in the United States. (33) By the early 1900s, most major cities in the United States began filtering their drinking water supplies.

SIDEBAR

The Musee des Egouts (Paris Sewer Museum) is located beneath the streets of Paris near the Eiffel Tower. The museum is an elaborate collection of photos and artifacts, and included boat rides in the sewer tunnels until they were discontinued in the 1970s. The museum shows the history of the sewage system, created under Napoléon III, which is still used today. However, wastewater is now transported beneath the Seine to modern wastewater treatment plants before it is returned to the river.

A CLOSER LOOK

Soil has long been known to cleanse water through the filtering action of the earth. Nutrients, microbes, disease-causing microbes, and other wastes are physically strained as surface water percolates through the soil. This process has been used for centuries and was carried to an extreme in the late 1800s in Germany. The City of Berlin purchased 19,000 acres (7689 ha) of adjacent farmland for disposal of municipal sewage from the city's 1.5 million residents. Raw sewage was pumped into tall stand-pipes (metal reservoirs) on the highest point of the farms and then distributed by gravity to various fields. Irrigation was either by the "wild flood irrigation" method, in which wastewater was simply allowed to flow across gently sloping pastures, or by "furrow" irrigation where wastewater flowed down crop rows (both discussed in Chapter 6). Excess wastewater at the end of the fields was allowed to drain off into nearby streams.

Fields had to be managed carefully, and a limit of 2000- or 3000-gallon (7.6 to 11.4 m^3) doses of sewage was applied per day per acre (0.4 ha) to give the soil time to "rest." A crew of 134 "sewage men" oversaw application rates to fields averaging 59 acres (24 ha) in size.

The Germans applied sewage to fields in the late 1800s with military efficiency. Parades were held at 6 A.M. and 6 P.M. to perform roll call, to examine tools, and then to march to assigned fields. Each sewage man carried a book with detailed instructions on irrigation and a list of punishments that would be imposed if the directions were not followed. Headgates to fields were numbered, and each worker wrote down the times when gates were opened and how many revolutions a valve was turned. This military style of management of the sewage men had come about in response to previous poor management which had produced numerous public complaints. Crop production now became prodigious. (34) During many winter months, the warmer sewage water kept the soil from freezing, and "irrigation" continued. When temperatures dropped to lower levels, sewage water was stored in temporary ponds of 5 to 22 acres (2 to 9 ha) in size.

Around 1870, the City of Paris began pumping sewage to the plain of Gennevilliers north of the city. At first a great public outcry arose over the idea of spreading sewage across farm ground and selling the resulting crops for human consumption. By 1880, however, criticism had waned. Other French farmers soon began to demand wastewater, and sewage produce became popular with Parisians. The city continued to operate a 15-acre (6 ha) plot, and closely monitored rainfall, weight and yields of produce, and sewage application rates. This information was shared with private landowners in the area who were also utilizing sewage effluent. (35)

Herman Roechling, an English engineer, visited the Plain of Gennevilliers when it served as a wastewater treatment site and noted the complete lack of offensive odors from the sewage farm. To prove the success of the filtration process, he even went so far as to dip a cup of water from the effluent channel and drank it. Roechling claimed that he subsequently suffered no ill effects. (36) Bacteria counts by a laboratory in Paris confirmed that the raw sewage had 29,454,000 bacteria per cubic centimeter, whereas the cleansed effluent at Gennevilliers had counts as low as 5380 per cubic centimeter. Later, Roechling visited the sewage farms of London (he survived his drink of effluent in France) and declared that they, too, provided absolutely no odor of any kind. (37)

One of the first formal land treatment projects for sewage in the United States started in 1881 on a municipal farm near Pullman, Illinois, just south of Chicago, modeled after the sewage farms in England, France, Germany, and Australia. Unfortunately, poor management led to excessive sewage applications, and soon the farm was abandoned as Chicago's urban growth engulfed the area. Other American cities that attempted this waste disposal practice, such as Brockton, Massachusetts, and Vineland, New Jersey, had only slightly more success. (38)

In 1928, the small town of Fessenden, North Dakota, completed a sewage collection system and was ready to install a wastewater treatment plant, but ran out of money. As a temporary solution, town officials decided to pipe the raw sewage to a natural pothole outside of town and to complete the wastewater treatment plant when they got more money. In the meantime, local officials hoped the smell from the raw sewage wouldn't become too bad before the project could be completed. The money never came, but neither did the smell. Local engineers found that evaporation, seepage through the pond bottom, and natural bacterial processes biodegraded solid waste and eliminated odors. The quantity of wastes generated by the town was small enough that it could be adequately treated by this low-tech wastewater treatment system (see Figure 11.8).

Those who heard of the Fessenden project soon replicated similar wastewater plants in other North Dakota communities. When, the U.S. Public Health Service finally learned about the new method, it tried it out in other states. Research confirmed that the process (called **sewage lagoons** or stabilization ponds) was a relatively inexpensive and effective method of wastewater treatment. Bacteria, oxygen, and sunlight combined to decompose sewage nutrients for their own metabolism. Wind on the surface of the standing wastewater also stimulated oxygen reactions. (39)

This method of treating sewage is quite common today in smaller communities around the world. If wastewater quantities are closely monitored, a sewage lagoon can handle a predetermined amount of incoming effluent indefinitely. However, the main disadvantage is time and space. A lagoon can only treat a certain amount of sewage in a given period of time, and adding too much effluent will overload a system and result in inadequate treatment.

FIG. 11.8. The Fessenden, North Dakota, sewage lagoon is famous in the realm of wastewater treatment around the world.

THE WASTEWATER TREATMENT PROCESS

Large-scale treatment of wastewater became commonplace in the United States, Canada, and many other regions of the world during the twentieth century. Wastewater facilities are generally located at a geographic low point in the topography of a city so that most wastes flow by gravity to the wastewater treatment facility. Areas that cannot be served by gravity require pump stations to lift wastewater to the treatment plant. (Drinking water treatment facilities are generally located at the highest topographic location in a city to allow treated drinking water to be delivered to customers by gravity.)

Complete sewage treatment consists of three steps: primary, secondary, and tertiary. Primary treatment involves little more than removing suspended solid materials from wastewater and then returning liquids to a stream. Secondary treatment removes suspended solids and a larger percentage of organic matter (see Figure 11.9). More elaborate systems include a third cleansing step called tertiary treatment.

PRIMARY TREATMENT

Primary treatment involves screening, grit removal, and primary settling. Raw sewage that arrives at treatment plants contains floating materials (wood, paper, grit, oils, etc.) that must be removed early in the treatment process both to protect mechanical equipment such as pumps and aerators and to prevent blockage of pipes.

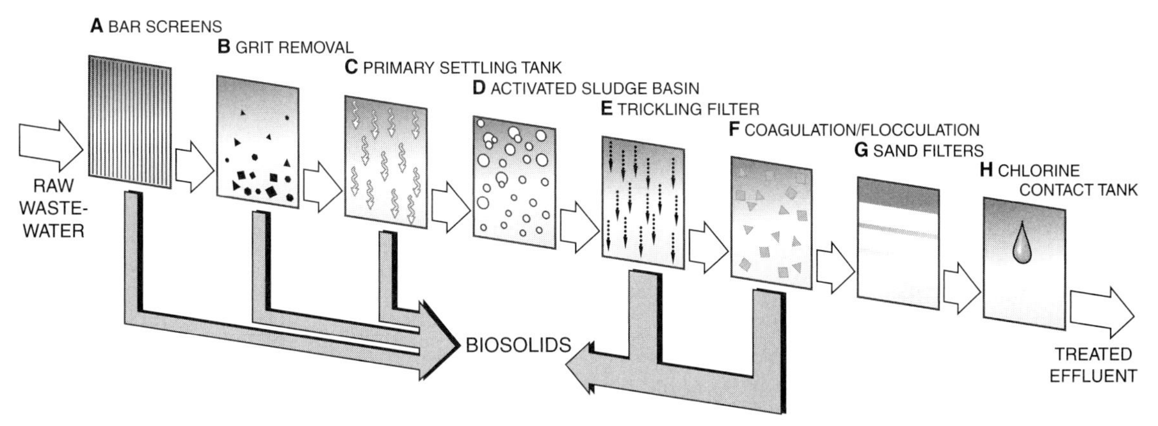

FIG. 11.9. The wastewater treatment process at larger facilities includes numerous steps such as bar screens, grit removal, primary and secondary settling (or clarification) tanks, aeration, flocculation and coagulation, sand filters, and chlorination. All combine to remove biosolids and, ultimately, to discharge treated effluent back to a river, lake, or other water body.

These materials are extracted with mechanical rakes or revolving screens. Since these contaminants contain potentially hazardous materials, they are discharged into containers and are disposed of by incineration or are transported by truck to a landfill site. Efficient screening is extremely important in the wastewater treatment process.

Water then moves into a grit chamber to allow cinders, sand, and small stones to settle to the bottom. Grit chambers are particularly important in communities with combined storm drainage and sewer systems where sand and gravel can wash into sewers after storms. Next, the wastewater enters primary settling tanks that are used to remove suspended solids that settle as sludge. This process is very similar to drinking water sedimentation. Time is allowed for the remaining grit and other suspended particles to settle from the wastewater; chemicals can be added to accelerate this process. Wastewater flows into the tanks at a constant rate so that heavier solids are deposited at the inlet end and lighter solids settle out at the outlet end. The bottom of settling tanks is usually a V-notch design (an inverted pyramid) which allows sludge to be mechanically scraped off the bottom of the tank. Radial (circular) settling tanks have floors that slope toward the center of the tank (an inverted cone) for easier sludge removal.

SECONDARY TREATMENT

Secondary treatment reduces the demand for dissolved oxygen (biological oxygen demand, or BOD) that wastewater will place on a waterway after discharge. This reduction in BOD takes place through the aerobic oxidation of nutrients in the water. Microorganisms are used in this denitrification process to consume nutrients that would act as food for dissolved oxygen in rivers and other water bodies. (See Chapter 5 for a review of the denitrification process.)

There are two types of secondary treatment processes: trickling filter and activated sludge systems. Aerobic microorganisms are used in both to decompose organic materials. BOD levels are high in these processes so that aeration of effluent is continuous.

Trickling filters are rectangular or circular beds filled with coarse media of rock and gravel with diameters of 2 to 4 inches (5 to 10 cm). Wastewater is sprayed on the surface and trickles down the filtering rocks until it reaches a drain system at the base of the filter system. A microbial film will develop on the surface of the coarse media and remove BOD as sewage trickles through the bed. Since no straining of particles occurs, this filtering treatment is strictly biological.

Air must also be distributed through the filtration system to promote aerobic oxidation. Air circulation is encouraged by the temperature differences between the air and wastewater, and causes upward air movement in tubes located at the sides of the filtering systems. Circular trickle systems have a rotating pipe that sprays effluent onto the surface of the coarse rock media. Rectangular trickle systems have a distributor (pipe) that is driven forward and backwards across the media surface to spray effluent for treatment. The depth of these filtering systems is usually 4 to 12 feet (1 to 4 m). Trickling filters are relatively simple and inexpensive to operate and are widely used. Disadvantages are the substantial land areas required, fly and odor nuisance (40), and its ability to remove only about 80 percent of organic matter from water. (41)

In another secondary treatment method, the **activated sludge** process, effluent is constantly agitated and aerated to assist in bacterial activity. The sludge contains large numbers of aerobic organisms that digest organic material. Microorganisms that grow in flocs transform the organic material into new bacteria, carbon dioxide, and water.

The primary function of the activated sludge process is to remove material that requires dissolved oxygen, or biological oxygen demand. Sludge flocs also promote good settlement in secondary sedimentation tanks. Flocs are continually flushed from the tank to make room for influent. Some floc returns to the activated sludge tank to provide sufficient bacterial growth to reduce BOD.

The remainder of the floc is removed as sludge. Oxygen is required by microorganisms and is provided by mechanical aeration (agitation or stirring) or by the diffusion (release) of air at the bottom of the tank. Air bubbles automatically form and create currents in the wastewater as the bubbles rise to the surface of the tank. This air/liquid interface is an efficient method of transferring oxygen to water.

Microorganisms play a large role in wastewater treatment, consuming nutrients and pathogens when the plant operator closely monitors pH, temperature, flow rates, and dissolved oxygen. Bacteria grows best in a narrow range of pH near neutrality, about 6.5 to 7.5. The other requirement for bacterial growth is the availability of carbon.

TERTIARY TREATMENT

Effluent from secondary treatment contains only 5 to 20 percent of the original quantity of organic matter and is generally discharged safely into rivers or lakes. However, nitrates and phosphates may still remain and can require **tertiary treatment**, a very expensive method involving physical and chemical methods such as flocculating chemicals, denitrifying bacteria in sand filters, and chlorine to remove additional contaminants. Some wastewater treatment plants use ultraviolet (UV) lights instead of chlorine because under some conditions chlorination can combine with methane gases to form carcinogens. A dense network of UV lights placed across the effluent will further disinfect wastewater.

Nutrient Removal Wastewater effluent is a major source of nutrient pollution in waterways around the world. (42) Municipal effluent can cause problems with algal blooms and eutrophication if ammonia or nitrates are present. Nutrients can be removed during the wastewater treatment process through ammonia stripping, additional chlorination, and selective ion exchange. However, these processes can be very expensive; they are also unreliable and unpopular.

The most common nutrient treatment is the use of natural nitrogen removal processes of the nitrogen cycle. The bacterial oxidation processes described in Chapter 5 can be enhanced by increasing the time wastewater resides in filtering tanks and by adding oxygen. However, these processes create capacity problems that require more tanks, additional wastewater treatment personnel, and other expenses. Economics always plays a major part in wastewater treatment, so that additional treatment requirements will sometimes be resisted.

POLICY ISSUE

Municipal effluent is a major nutrient pollutant of water. Although technology currently exists to remove nutrients from wastewater, cost is a limiting factor. If you were a member of the local water and sewer district board of directors, what criteria would you use to determine the amount of money to be allocated to wastewater treatment?

For example, suppose nutrient discharges from the local wastewater treatment plant could be reduced by an additional 50 percent if sewer service charges to customers were doubled. Would it be reasonable to increase customer sewer charges if you already met state or federal minimum water quality standards? How would you, as a board member, justify this rate increase to your constituents? Should degradation of downstream water supplies be avoided at any cost, or should compliance with existing water quality standards be considered adequate?

A **National Pollutant Discharge Elimination System (NPDES) Permit** is required to discharge wastewater into a navigable stream. The USEPA establishes water quality standards and can require implementation of additional technology-based standards based on five-day BOD standards and total suspended solids removal. (43) When developing the criteria for an NPDES Permit, the age of wastewater treatment facilities, quality of accepting waters, and volume of discharges are all considered. Permits are issued by the USEPA, although states may take over this responsibility except for federal facilities. (44) For more detailed information on the NPDES Permitting Program, see the USEPA NPDES Homepage at http://cfpub.epa.gov/npdes.

SEPTIC TANKS AND LEACH FIELDS

Septic tanks and leach fields are commonly used to dispose of sewage wastes from homes that are not connected to municipal wastewater systems. First used in the United States in 1884, septic tanks and leach fields are widely used today where sewer systems are not available. (45) A septic tank includes a buried tank that captures sewage directly from a home, settling out solids and breaking them down by bacteria. A drain or leach field may be used to spread out this partially treated water across a subsurface area. Other systems require that the waste be pumped out and hauled to a municipal wastewater treatment plant. Leach fields use the natural filtration capacity of soil to cleanse wastewater, where the waste is biodegraded by microorganisms before it reaches groundwater or exits into surface waterways. Septic tanks require low maintenance and have low operating costs. A septic tank with a leach field may require pumping only once every five years.

Leach (drainage) fields can be developed by burying perforated pipes from the septic system and allowing sewage seepage through a gravel bed and into the soil. Soil filters bacteria and some viruses, binds phosphates for vegetation uptake, and decomposes organic materials. Leach fields must be carefully located so that drainage will not percolate into the groundwater sources of a drinking water well. After about 10 years, most leach fields will clog and will have to be replaced. (46)

A common problem associated with septic tanks has to do with population density: Too many septic systems in one area can overwhelm the ability of soils to filter wastewater. In addition, disposing of synthetic organic compounds (SOCs) down sink drains is a major cause of pollution. Among the worst of these compounds are pesticides, cleaners, auto wastes such as oil, antifreeze and other fluids, paints, and solvents.

WETLANDS AND WATER TREATMENT

Wetlands, including swamps, marshes, bogs, and bayous, play a vital role in the natural treatment of water resources, serving as both regulators and incubators in the processes of water quality treatment. Floods, high groundwater tables, and drainage ways are regulated by impoundment in wetlands. This natural water storage can be useful in regulating flows in nearby streams or percolation rates into groundwater aquifers. Because water quality can be greatly affected by water quantity, wetlands provide a vital role in naturally regulating flows and diluting contaminants in adjacent water bodies.

Wetlands serve as incubators by providing the time needed for natural chemical and biological reactions to occur. The nutrient cycle is enhanced by stagnant waters that contain organic matter from decaying trees, leaves, and other plants. Floods and other high-water events, as well as wind action, provide the dissolved oxygen needed for the nutrient cycle to decompose organic matter.

Microorganisms in wetlands can modify nutrients and immobilize heavy metals. Wetlands also promote sedimentation and can remove up to 95 percent of sediments in a column of water. (47) Pesticides such as Atrazine can actually be reduced by wetlands. (48) Wetlands can be used for small-scale sewage treatment as well as stormwater retention and treatment. (Wetlands will be discussed in more detail in Chapter 12.)

OUR ENVIRONMENT

Proper drinking water treatment has produced dramatic improvements in human health over the past 100 years. Many potentially fatal diseases of the 1800s, especially cholera and typhoid fever, are almost unknown in countries such as Canada, the United States, Japan, Australia, and Western Europe. Similarly, proper wastewater treatment processes in more industrialized countries have greatly improved the health of river and lake habitats by prohibiting raw human waste from being dumped into such water bodies. However, these improvements have occurred relatively recently (only since the 1970s in the United States, for example) and have required huge federal expenditures.

Is there an existing drinking water quality or wastewater treatment issue in your area that needs to be addressed? Do you know if the water quality problem is being caused by a point source or nonpoint source pollutant? How much might it cost to correct this problem, and what local, regional, state, provincial, or federal agency would most likely take the lead in addressing this issue?

CHAPTER SUMMARY

This chapter shows the remarkable improvement of drinking water and wastewater treatment during the past century. Historically, drinking water quality was extremely poor and was often polluted by upstream sewage disposal. Unsanitary waste disposal and waterborne diseases devastated human health. The U.S. government began regulating drinking water quality in the early 1900s but didn't establish modern safety standards until the 1970s. Wastewater treatment standards were also adopted in the 1970s and have significantly improved the nation's water bodies. See Table 11.3 for a listing of the important water quality events discussed in this chapter.

TABLE 11.3 Selected History of Stages in Water Quality Planning, 2000 B.C.–A.D. 2004

Date	Location	Process
2000 B.C.	India	Water is boiled to improve taste and clarity.
1500 B.C.	Egypt	Alum is used to treat drinking water to improve taste and appearance.
400 B.C.	Rome	Hippocrates promotes the use of "healthy water."
200 B.C.	Babylonia	Laws are passed to prohibit waste from being thrown into wells or cisterns, and it is forbidden to locate a cemetery, furnace, tannery, or animal slaughterhouse within 80 feet (24 m) of a groundwater well.
A.D. 300	Rome	Sand filtration system is used to improve the taste and appearance of drinking water.
A.D. 375	Rome	Vegetius recommends movement of Roman field camps to avoid disease caused by water polluted from large groups of soldiers.
1200s	Europe	Period of plagues and the Black Death occurs.
1370	Paris	Vaulted sewers are constructed beneath rue Montmartre to drain wastes to the Seine.
1600s	England	Sir Francis Bacon conducts numerous experiments regarding water percolation, boiling, distillation, and coagulation.
1600s	Holland	Anthony van Leeuwenhoek discovers common forms of bacteria in water with microscope.
1703	Paris	Phillippe La Hire encourages all households to install sand filters on their water supply (usually captured rainwater from rooftops) before drinking.
1746	France	Joseph Amy is granted the first patent for a water filter.
1700–1800s	United States	Private companies develop and manage drinking water systems for larger eastern cities.
1754	Bethlehem, PA	First municipal drinking water system in the United States is developed.
1801	Philadelphia, PA	First large municipal government develops drinking water pumping plant and delivery system in the United States.
1804	Paisley, Scotland	First city to provide treated filtered drinking water to an entire town.
1806	Paris	Drinking water treatment plant is constructed for water taken from the Seine River. Water was allowed to settle for several hours before moving through sand filters.
1834	Richmond, VA	First municipal slow sand filter system is developed in the United States.
1842	Chicago	First waterworks is established by the privately owned Chicago Hydraulic Company. It operated a small pumping plant off Lake Michigan.
1860s	Newark, NJ	Aqueduct Board hires chemist to test the waters of the Passaic River for pollution.
1860s	Paris	181 miles (291 km) of underground sewers are constructed.
1861	Newark and Jersey City, NJ	Board of Inspection of the Pollution of the Passaic River and Its Tributaries is organized.
1867	England	Civil engineer Baldwin Latham gives "A Lecture on the Sewage Difficulty" and discusses the problems of pollution in rivers from animal and human waste.
1870	Poughkeepsie, NY	First drinking water filtration plant is constructed. Incidence of typhoid fever drops dramatically.
1870–1890s	London, Berlin, Paris, and Sydney	Farms are used extensively for irrigation with treated sewage water to grow vegetables.

(*continued*)

TABLE 11.3 Selected History of Stages in Water Quality Planning *(continued)*

Date	Location	Process
1899	New Orleans	Sewerage and Water Board is created by the Louisiana Legislature.
1899	Washington, DC	U.S. Congress passes the Rivers and Harbors Act prohibiting the dumping of refuse into navigable waters.
1900	Chicago	Flow of the Chicago River is reversed to prevent sewage from contaminating drinking water supplies of Lake Michigan.
1900	State of Missouri	Lawsuit is filed against the State of Illinois and the Chicago Sanitary District to stop the dumping of raw sewage into the Mississippi. U.S. Supreme Court rules in favor of Chicago since the Illinois River was less contaminated than the Mississippi River.
1909	Jersey City, NJ	Chlorination is first used by a U.S. city.
1912–1915	Chicago	Chlorination of the city's entire water supply began in 1912 and is used systemwide by 1915. It solves many health problems caused by contaminated water.
1914	Washington, DC	U.S. Department of the Treasury sets the nation's first bacteriological standard of 2 coliforms per 100 milliliters.
1928	Fessenden, ND	Sewage lagoons are discovered for wastewater treatment.
1930s	United States	Filtration and chlorination almost eliminates waterborne diseases such as typhoid and cholera in the United States.
1940	Washington, DC	First interstate compact for water quality is established on the Potomac River.
1942	Washington, DC	U.S. Public Health Service (USPHS) sets the first drinking water standards in the country.
1948	Washington, DC	Federal Pollution Control Act is passed but contains no enforcement mechanisms.
1974	Washington, DC	Safe Drinking Water Act is approved by Congress.
1996	Washington, DC	Safe Drinking Water Act Amendments approved by Congress.
1998	Abu Dhabi, UAE	Desalination plant is constructed.
1998	Washington, DC	USEPA places perchlorate on the Contaminanted Candidate List to consider maximum MCL for drinking water.
2004	United States	Threat of terrorism causes concern regarding safety of municipal drinking water supplies and wastewater treatment facilities.

QUESTIONS FOR DISCUSSION

1. Describe treatment methods for drinking water prior to 1800.

2. Describe wastewater treatment methods prior to 1850.

3. What role has population growth played in the need to improve drinking and wastewater treatment practices?

4. Describe some of the requirements of the Safe Drinking Water Act of 1996.

5. Soil is often used as a wastewater treatment medium. Discuss the concept of "sewage farms" in the late 1800s and the use of septic tanks today. Is there a marked difference between these two wastewater treatment methods?

KEY WORDS TO REMEMBER

activated sludge p. 340
chlorination p. 334
coagulation p. 332
desalination p. 323
filtration p. 332
flocculation p. 332
fluoridation p. 334
intakes p. 330
National Pollutant Discharge Elimination System (NPDES) Permit p. 341
ozone gas p. 334
potable water p. 323
primary treatment p. 338
reverse osmosis (RO) p. 325
Safe Drinking Water Act (SDWA) of 1974 p. 328
Safe Drinking Water Act Amendments of 1996 p. 328
secondary treatment p. 339
sewage lagoons p. 338
tertiary treatment p. 340
trickling filters p. 339
ultraviolet (UV) systems p. 334

SUGGESTED RESOURCES FOR FURTHER STUDY

READINGS

American Public Works Association. *History of Public Works in the United States: 1776–1976*. Chicago: American Public Works Association, 1976.

Barzilay, Joshua I., Winkler G. Weinberg, and J. William Eley. *The Water We Drink—Water Quality and Its Effects on Health*. New Brunswick, NJ: Rutgers University Press, 1999.

Csuros, Maria, and Csaba Csuros. *Microbiological Examination of Water and Wastewater*. Boca Raton, FL: Lewis Publishers, 1999.

De Zuane, John. *Handbook of Drinking Water Quality*. 2nd ed. New York: Van Nostrand Reinhold, 1997.

Hey, Donald L., and Nancy S. Philippi. *A Case for Wetland Restoration*. New York: John Wiley & Sons, 1999.

Horan, N. J. *Biological Wastewater Treatment Systems—Theory and Operation*. West Sussex, England: John Wiley & Sons, 1990.

Kent, Donald M. *Applied Wetlands Science and Technology*. Boca Raton, FL: Lewis Publishers, 1994.

Krenkel, Peter A., and Vladimir Novotny. *Water Quality Management*. Orlando, FL: Academic Press, 1980.

McCourt, Frank. *Angela's Ashes*. 2nd ed. New York: Simon & Schuster, 1999.

Melosi, Martin V. *Pollution and Reform in American Cities, 1870–1930*. Austin: University of Texas Press, 1980.

Stevens, Leonard A. *Clean Water—Nature's Way to Stop Pollution*. New York: E. P. Dutton, 1974.

Stewart, John Cary. *Drinking Water Hazards*. Hiram, Ohio: Envirographics, 1990.

Warshall, Peter. *Septic Tank Practices*. Garden City, NY: Anchor Press, 1979.

WEBSITES

American Water Works Association, Homepage, July 2003. http://www.awwa.org

Centers for Disease Control and Prevention, Homepage, Atlanta, Georgia, July 2003. http://www.cdc.gov

Environment Canada, "Water Quality," July 2003. http://www.ec.gc.ca/water/en/manage/qual/e_qual.htm

Health Canada, "Water Quality and Health," July 2003. http://www.hc-sc.gc.ca/hecs-sesc/water/publications.htm

Tampa Bay Water Homepage, Tampa, Florida, August 2003. http://www.tampabaywater.org

U.S. Environmental Protection Agency, Office of Wastewater Management, "National Pollutant Discharge Elimination System Permit Program," Washington, DC, July 2003. http://cfpub1.epa.gov/npdes

U.S. Environmental Protection Agency, Office of Ground Water and Drinking Water, "The Safe Drinking Water Act," Washington, DC, July 2003. http://www..epa.gov/OGWDW/sdwa/sdwa.html

U.S. Geological Survey, "Periodic Water Fact: Water Desalination," Washington, DC, July 2003. http://water.usgs.gov/watuse/wuweeklyfact.html

REFERENCES

1. Barzilay, Joshua I., Winkler G. Weinberg, and J. William Eley. *The Water We Drink—Water Quality and Its Effects on Health*. New Brunswick, NJ: Rutgers University Press, 1999, 10.

2. State of Rhode Island, Water Resources Board, "A Brief History of Drinking Water," by Ellen L. Hall and Andrea M. Dietrich, http://webster.wrb.state.ri.us/programs/eo/historydrinkingwater.htm, September 2001.

3. American Public Works Association. *History of Public Works in the United States: 1776–1976*. Chicago: American Public Works Association, 1976, 217.

4. Ibid., 218.

5. Melosi, Martin V. *Pollution and Reform in American Cities, 1870–1930*. Austin: University of Texas Press, 1980, 39.

6. Ibid., 46–47.

7. De Zuane, John. *Handbook of Drinking Water Quality*. 2nd ed. New York: Van Nostrand Reinhold, 1997, 7.

8. Barzilay et al., 17.

9. Stewart, John Cary. *Drinking Water Hazards*. Hiram, Ohio: Envirographics, 1990, 35.

10. U.S. Geological Survey, "Periodic Water Fact: Water Desalination," http://water.usgs.gov/watuse/wuweeklyfact.html, September 2001.

11. Tampa Bay Water, "Seawater Desalination Project," http://www.tampabaywater.org, August 2003.

12. De Zuane, 456.

13. Melosi, 40.

14. Ibid., 40.

15. American Public Works Association, 227–228.

16. Stevens, Leonard A. *Clean Water—Nature's Way to Stop Pollution*. New York: E. P. Dutton, 1974, 61–62.

17. American Public Works Association, 416.

18. Department of Water, Chicago, Illinois, http://www.cityofchicago.org/water, September 2001; and personal communication with Gary Litherland, Chicago Department of Water, Chicago, Illinois, May 5, 2001.

19. Ibid.

20. De Zuane, 440.

21. Sewerage and Water Board of New Orleans, Homepage, http://www.swbnola.org/water, New Orleans, Louisiana, September 2001; and personal communication with Harold Gorman, Executive Director, Sewerage and Water Board of New Orleans, New Orleans, Louisiana, July 9, 2001.

22. American Public Works Association, 238.

23. Personal communication with Lorraine Smith, Clerk-Treasurer, Town of Clarkson, Clarkson, Nebraska, July 12, 2001.

24. "Congressional Statement—Federal Bureau of Investigation," House Committee on Transportation and Infrastructure, Subcommittee on Water Resources and Environment, Washington, DC, http://www.fbi.gov/congress/congress01/rondick101001.html, October 2001.

25. Ibid.

26. Ibid.

27. "Technical Brief—RiverSpill Infrastructure Protection and Consequence Management Tool," Technical Support Working Group, http://www.tswg.gov/tswg/NewTech/RiverSpillTB.htm, October 2001.

28. Sandia National Laboratories, "Sandia Develops Program to Assess Water Infrastructure Vulnerabililties," October 3, 2001 Press Release, http://www.sandia.gov/media/NewsRel/NR2001/watinfr.htm, October 2001.

29. McCourt, Frank. *Angela's Ashes*, 2nd ed. New York: Simon & Schuster, 1999, 262–263.

30. Csuros, Maria, and Csaba Csuros. *Microbiological Examination of Water and Wastewater*. Boca Raton, FL: Lewis Publishers, 1999, 75.

31. Barzilay et al., 9.

32. Stewart, 76.

33. Stevens, 25–28.

34. Ibid., 29–36.

35. Ibid., 36–40.

36. Ibid., 37.

37. Ibid., 39.

38. Ibid., 41–45.

39. Ibid., 71–73.

40. Horan, N. J. *Biological Wastewater Treatment Systems—Theory and Operation*. West Sussex, England: John Wiley & Sons, 1990, 52–55.

41. Csuros and Csuros, 76.

42. Horan, 217.

43. U.S. Environmental Protection Agency, Office of Wastewater Management, "National Pollutant Discharge Elimination System Permit Program," http://cfpub1.epa.gov/npdes, September 2001.

44. Krenkel, Peter A. and Vladimir Novotny. *Water Quality Management*. Orlando, FL: Academic Press, 1980, 34–35.

45. Stewart, 175.

46. Csuros and Csuros, 77.

47. Kent, Donald M. *Applied Wetlands Science and Technology*. Boca Raton, FL: Lewis Publishers, 1994, 63.

48. Hey, Donald L., and Nancy S. Philippi. *A Case for Wetland Restoration*. New York: John Wiley & Sons, 1999, 25.

CHAPTER 12

WATER, FISH, AND WILDLIFE

We can not always distinguish between the results of man's actions and the effects of purely geological or cosmical causes. The destruction of the forests, the drainage of lakes and marshes, and the operations of rural husbandry and industrial art, have unquestionably tended to produce great changes in the hygrometric, thermometric, electric, and chemical condition of the atmosphere, though we are not yet able to measure the force of the different elements of disturbance, or to say how far they have been neutralised by each other or by still obscurer influences; and it is equally certain that the myriad forms of animal and vegetable life, which covered the earth when man first entered upon the theatre of a nature whose harmonies he was destined to derange, have been, through his interference, greatly changed in numerical proportion, sometimes much modified in form and product, and sometimes entirely extirpated. (1)

George Perkins Marsh (1864)

In his classic book *Man and Nature* (1864), George Perkins Marsh (1801–1882) of Vermont wrote of the interaction of forests, soil, and water. Marsh grew up in the wooded Northeast of the United States and saw at firsthand the devastation created by improper harvesting of timber. He later traveled extensively as minister to Turkey under President Zachary Taylor and as ambassador to Italy under President Lincoln. His global perspective on the impact of humans on fish and wildlife and the natural environment gained international attention.

At an early time in our history, Marsh alerted the public to the assault on nature caused by canal building, deforestation, and poor water quality. In his own native Northeast, he was alarmed by the degradation of lands cleared by Vermont farmers for crop production, as well as the loss of wetlands. He is known as the father of the environmental movement.

EARLY FISH AND WILDLIFE PROTECTION

Although few laws existed in ancient times to protect the natural environment, several were enacted to protect fish and wildlife, particularly in the Roman Empire. Wild animals, or *ferae naturae*, were classified in the same manner as the oceans and the air—they were owned by no one, but they represented a common resource. There was one exception, however; individuals had the right to hunt for food on their own property. Later, this hunting right was restricted, particularly under Anglo-Saxon law in feudal Europe around A.D. 450. Only nobility were allowed to hunt wild animals in "royal forests," with hunting limits imposed on surrounding private lands to increase royalty's chances of a successful hunt.

The Magna Carta (1215) reformed the taking of fish and wildlife, and the government was given control of managing animals for the public good in England. Mill weirs were removed from rivers to allow for fish passage, but the English king and Parliament still retained hunting privileges. The tradition of governmental control over fish and wildlife was taken to America, although individuals had unrestricted hunting privileges on their own private property until the mid-

FIG. 12.1. This image of Tower Falls and Sulphur Mountain in Yellowstone National Park in Wyoming was painted by Thomas Moran. He was one of two artists on a historic expedition to the area in 1871 led by Ferdinand Vandiveer. Paintings such as this captured the wonderment of the public and led to increased protection of unique natural settings in future years around the world.

1800s. Fish and wildlife were considered a common resource, but little management was needed because populations were abundant. Unfortunately, the destruction of habitat was widespread during settlement, particularly following the draining of wetlands authorized under the Louisiana Swamp Land Act of 1849 and the Swamp Land Acts of 1850 and 1860 (see Chapter 9).

By 1890, all U.S. states had adopted laws to regulate hunting and fishing. However, these efforts were often implemented too late and with no effect on water development projects of the era. During earlier westward expansion and settlement, wildlife preservation was generally disregarded except in special situations (see Figure 12.1).

FISH AND WILDLIFE PROTECTION IN THE TWENTIETH CENTURY

Gifford Pinchot (1865–1946), first director of the Division of Forestry (later renamed the U.S. Forest Service) in 1898, joined the growing movement to protect forests and other public lands from destruction. He criticized excessive harvesting of forests and the damage done to fish and wildlife habitat, and as noted earlier, he promoted the concept of wise use of natural resources. Pinchot also fought against the destruction of land wreaked by hydraulic mining; this process used high-pressure water hoses to wash away entire hillsides while prospecting for gold and other precious minerals. Pinchot observed sadly that a horse thief could be hung for stealing one animal, but that public lands could be destroyed in defiance of current laws with little or no concern from the public.

During the early 1900s, President Theodore Roosevelt (1858–1919) joined the wise use movement; the philosophy behind this movement was to use natural resources for the greatest good of the greatest number of people. Although Roosevelt supported the protection of animals through the creation of national wildlife refuges in Alaska, Oregon, California, and Montana, his concern did not extend to predators such as the grizzly bear, mountain lion, or wolves.

Fish and wildlife protection in the United States remained at a somewhat low level during the first half of the twentieth century. Rapid urban development, as well as the use of new chemicals that polluted waterways across the country, caused the rapid decline of many species, particularly the American bald eagle. By 1940 it was rarely seen in the lower 48 states. Finally, Congress was forced into action to prevent further destruction of the national symbol. In 1940, the Bald Eagle Protection Act was approved, incorporating a preamble almost as majestic as the national bird itself:

Whereas the Continental Congress of 1782 adopted the bald eagle as a national symbol; and Whereas the bald eagle thus became the symbolic representation of a new nation under a new government in a new world; and Whereas by that act of Congress and by tradition and custom during the life of this nation, the bald eagle is no longer a mere bird of biological interest but a symbol of the American ideals of freedom; and Whereas the bald eagle is now threatened with extinction: Therefore be it enacted . . . (2)

Clearly, the most influential environmentalist of the twentieth century was Rachel Carson (1907–1964), who jolted the world with her book *Silent Spring* in 1962. An avid birdwatcher, she noted the reduced number of bird species and attributed their declining numbers to DDT and other chemical pesticides (see Chapter 5). She warned that with the large numbers of birds that were dying, other members of the food chain would soon follow, including humans, unless drastic measures were taken immediately.

A CLOSER LOOK

Rachel Carson was born in Springfield, Pennsylvania, attended Pennsylvania College for Women and Johns Hopkins University, and did graduate work at Woods Hole Marine Biological Laboratory in Massachusetts. Professor Carson was on the zoology staff at the University of Maryland and later worked as an aquatic biologist with the Bureau of Fisheries in Washington, D.C. She was chief editor for the U.S. Fish & Wildlife Service before she left to work full time on her own writing. She derived the material for *Silent Spring* largely from research conducted at the U.S. Geological Survey's Patuxent Wildlife Research Center in Laurel, Maryland. (3)

THE WILD AND SCENIC RIVERS ACT

http://www.nps.gov/rivers/wildriverslist.html

The sign is neatly painted. It stands in a creek-bottom pasture so short you could play golf on it. Near by is the graceful loop of an old dry creek bed. The new creek bed is ditched straight as a ruler; it has been "uncurled" by the county engineer to hurry the run-off. On the hill in the background are contoured strip-crops; they have been "curled" by the erosion engineer to retard the run-off. The water must be confused by so much advice. (4)

A Sand County Almanac: And Sketches Here and There,
by Aldo Leopold, © 1949, 1977 by Oxford University Press, Inc.
Used by Permission of Oxford University Press, Inc.

Historically, water development has meant the construction of dams, alteration of rivers, and reengineering of watersheds. Too often, fish and wildlife preservation have not been a great consideration. However, this attitude changed in the United States in the 1960s with congressional actions to implement a national policy of environmental protection. One of the first federal laws to be passed was the **Wild and Scenic Rivers Act** in 1968. The Act protects "free-flowing" rivers from development (primarily from dam construction). Although strongly opposed by many legislators, particularly those from the arid West, President Lyndon Johnson of Texas signed the bill into law in the Rose Garden of the White House. The Act states:

It is hereby declared to be the policy of the United States that certain selected rivers of the Nation which, with their immediate environments, possess outstandingly remarkable scenic, recreational, geologic, fish and wildlife, historic, cultural, or other similar values, shall be preserved in free-flowing condition, and that they and their immediate environments shall be protected for the benefit and enjoyment of present and future generations. The Congress declares that the established national policy of dams and other construction at appropriate sections [reaches] *of the rivers of the United States needs to be complemented by a policy that would preserve other selected rivers or sections thereof in their free-flowing condition to protect the water quality of such rivers and to fulfill other vital national conservation purposes.* (5)

Wild and Scenic Rivers Act, 1968

Wild and scenic rivers rank with national parks, wildlife refuges, forests, and wilderness areas in terms of importance in protecting fish and wildlife and associated habitat. Currently, the Wild and Scenic Rivers program protects reaches of 160 rivers in 39 states, with a total of 11,292 river miles (18,173 km). The three types of river designations under this program are Wild Rivers, Scenic Rivers, and Recreational Rivers. **Wild Rivers** are untouched vestiges of primitive America generally not accessible by roads; **Scenic Rivers** are primarily primitive areas with shorelines generally not accessible by roads; and **Recreational Rivers** are readily accessible by roads and may have some development located along shorelines. Each designation carries a different level of protection, although no federal construction activities are allowed that would damage a designated river.

SIDEBAR

The *Federal Register* is the official document of the U.S. Congress and is published Monday through Friday except for federal holidays. It is a legal newspaper published by the National Archives and Records Administration (NARA), and it includes federal agency regulations, proposed federal rules and notices, executive orders, proclamations, and other presidential documents. A primary purpose of the publication is to inform citizens of official activities of the federal government. For additional information on the *Federal Register*, see http://www.archives.gov/federal_register/index.html.

Applications for wild and scenic designation are made to the secretary of the Department of Interior. The secretary, in turn, notifies the Federal Energy Regulatory Commission to publish the application in the *Federal Register*, the official document of the U.S. Congress. Comprehensive land-use studies, fish and wildlife surveys, and habitat analysis are conducted, followed by public meetings held by the National Park Service, the U.S. Forest Service, or the Bureau of Land Management. Areas designated under the Act are administered by an agency or political subdivision of the state or states concerned, unless located on federal lands. The federal government cannot regulate private lands under the Wild and Scenic Rivers Act.

POLICY ISSUE

The Wild and Scenic Rivers Act has created much conflict between those who support the need to preserve segments of rivers and those who support continued development of water projects, particularly dam construction. Irrigators and municipal representatives often argue that new dams are needed to conserve water for periods of drought. Intensifying this controversy is the fact that numerous remaining locations for dam construction are located along rivers with attributes of wild and scenic rivers. As mentioned earlier, the Wild and Scenic Rivers Act has successfully blocked the construction of dams on numerous occasions. Those who wish to protect free-flowing rivers maintain that someone must protect unique natural areas from inundation. They also argue that fish and wildlife needs must be met even if it means less reliable water supplies or the adoption of more expensive water conservation methods by municipal, industrial, and agricultural water users.

Do you agree that a federal law should be used to stop dam construction on wild and scenic rivers? Or would it be more appropriate for the federal government to develop a national water policy that would identify feasible locations for future water development, while designating other areas for protection? Would it be appropriate for the federal government to develop such a plan, and if so, which federal agency should be designated to create such a national water resources plan? Should it fall to state or local government to create regional water management plans, with little or no input from the federal government? What role should individuals have in such a water planning process?

THE NATIONAL ENVIRONMENTAL POLICY ACT

http://www.epa.gov/compliance/nepa/index.html

The political and social climate in the United States during the 1960s was one of general unrest. The Vietnam War raged, racial relations in the nation were bitter (the very violent Watts race riots in Los Angeles occurred during this period), President Richard Nixon was extremely preoccupied about the participation of college students in Washington peace rallies, and rock groups like the Doors, Rolling Stones, Beatles, and The Who were calling for radical social change to the general acclaim of their fans. The stage was set for a significant change in environmental protection around the world.

A CLOSER LOOK

Earth Day exploded on the American culture scene on April 22, 1970. In New York City, Mayor John Lindsay commemorated the day by closing traffic for two hours along Fifth Avenue, from Fourteenth Street to Central Park. Teach-ins, lectures, and games of frisbee, as well as a good deal of theatricality, ruled the afternoon. A college student, dressed as the "Grim Reaper," taunted shareholders at a General Electric Company meeting. Demonstrators pulled a net full of dead fish down Fifth Avenue and cried, "This could be you!" Oil-coated ducks were dumped on the front steps of the Department of Interior in Washington, D.C. President Nixon kept his regular schedule at the White House and refused to issue a national proclamation in support of Earth Day; instead he issued proclamations for National Archery Week and National Boating Week. (6)

Wisconsin Senator Gaylord Nelson was a leader in Congress during this period of unrest and was deeply concerned about environmental degradation. Senator Nelson used his national stature to form a nonpartisan organization called the "Environmental Teach-In." He pledged $15,000 of personal funds to start the effort, and in December 1969, he asked former Stanford University student body president Denis Hayes to serve as the national coordinator for the first Earth Day the following spring. (7)

The **National Environmental Policy Act (NEPA),** signed into law by President Nixon on January 1, 1970, sought to foster improved protection of the environment, particularly fish and wildlife, and associated habitat. A Council of Environmental Quality (CEQ) was established to administer the Act. No other law had a greater effect on water resources development, management, or policy than NEPA.

The Act declared in part:

> *(a) The Congress, recognizing the profound impact of man's activity on the interrelations of all components of the natural environment, particularly the profound influences of population growth, high-density urbanization, industrial expansion, resource exploitation, and new and expanding technological advances and recognizing further the critical importance of restoring and maintaining environmental quality to the overall welfare and development of man, declares that it is the continuing policy of the Federal Government, in cooperation with State and local governments, and other concerned public and private organizations, to use all practicable means and measures, including financial and technical assistance, in a manner calculated to foster and promote the general welfare, to create and maintain conditions under which man and nature can exist in productive harmony, and fulfill the social, economic, and other requirements of present and future generations of Americans.* (8)

> [NEPA is] *the most important piece of environmental legislation in our history.* (9)
>
> Senator Gaylord Nelson
> Wisconsin—1980 (Founder of Earth Day)

SIDEBAR

Under the National Environmental Policy Act, any federal action that could negatively affect the environment requires the completion of an environmental impact statement. The NEPA process is as follows: Identify a Need; Make a Proposal; Inform the Public; Identify Concerns; Develop Alternatives; Prepare an Environmental Assessment or Environmental Impact Statement; Conduct a Public Review of Analysis; Make a Decision.

The President's **Council on Environmental Quality (CEQ)** was also created on January 1,

1970, to assist with administration of NEPA. (The U.S. Environmental Protection Agency was not created until later that summer and was not operational until December 2, 1970.) The CEQ, made up of three members appointed by the president, has the authority to carry out policies of the Act and provides oversight of individual federal agencies. The CEQ sends an annual report to Congress, although in recent years much of this work has been completed by the U.S. Environmental Protection Agency. The report covers the status of major natural, manmade, or altered environmental aspects of the nation, trends in the quality of the nation's environment, the adequacy of existing natural resources to fulfill future human and economic requirements, a review of ongoing environmental programs, and recommendations to remedy the deficiencies of existing programs and activities, including potential federal legislation.

A CLOSER LOOK

Some people do not agree with the methods used to implement the National Environmental Policy Act. The process of obtaining permits and other approvals can often take years, and it can be extremely expensive. The frustration of an industry member, required to comply with NEPA, is evident in the following exchange between Congressman Richard W. Pombo of California and Tim J. Leftwich, senior environmental scientist and principal of GL Environmental, Inc., of Rio Rancho, New Mexico, at a congressional hearing in 1998:

MR. POMBO: I think that's an important point and it is something that . . . other people [testified] to during this hearing, is that we end up spending the money, the time, the energy on the process and begin to forget that the reason we're doing this is for the environment. What's the typical cost of going through a NEPA process on a new mine that you're involved with?

MR. LEFTWICH: Of the four that we went through, the least expensive one was about $1.2 million. Another one was $6 million and the company—

MR. POMBO: That's just process.

MR. LEFTWICH: That's to get to a record of decision.

MR. POMBO: You're not talking about doing anything for the environment. You're talking about $6 million of paper.

MR. LEFTWICH: That's correct. And what's deceiving about an EIS is if I were to bring in a typical mining EIS document, it may be two or three inches thick. But what most people don't understand is the huge amount of research, baseline data work that goes into compiling—it's a summary document. If you look at your handout of the process and the steps there, each one of those baseline studies may be another pile of paper, depending on the issue, ground water modeling, wildlife, all of those resources that are studied. And so those become a huge pile of paper there that back up the document that is really written to summarize the studies and to make some decision. (10)

Hearing of the Committee on Resources,
U.S. House of Representatives,
March 18, 1998

Others would argue that implementation of NEPA is sometimes slow and costly but necessary to protect and restore fish and wildlife habitat. The federal government is the appropriate instrument for protecting the nation's environment. Do you agree or disagree? Why?

POLICY ISSUE

The President's Council on Environmental Quality (CEQ) creates new and interprets existing NEPA regulations that are binding on all federal agencies. The CEQ conducts informal consultations with federal agencies to ensure appropriate implementation of NEPA, informally consults with state and local governments and private citizens, comments and testifies on proposed federal legislation, designates lead federal agencies for the preparation of environmental impact statements, participates in international environmental activities, and provides information about the NEPA process to interested parties.

The CEQ was the primary agency responsible for implementation of NEPA. As the role of the USEPA evolved, and as other federal agencies gave it more environmental responsibilities, the CEQ also assigned several NEPA duties to the USEPA. President

Richard Nixon started this process with a Special Message to Congress on July 9, 1970.

To the Congress of the United States:

As concern for the condition of our physical environment has intensified, it has become increasingly clear that we need to know more about the total environment—land, water, and air. It also has become increasingly clear that only by reorganizing our Federal efforts can we develop that knowledge, and effectively ensure the protection, development and enhancement of the total environment itself. (11)

Why would duties under NEPA be divided between the executive branch (CEQ) and an independent agency (the USEPA)? Why didn't the federal government give overall responsibility for NEPA administration to an existing agency, such as the Department of Interior, the U.S. Army Corps of Engineers, or the U.S. Fish & Wildlife Service, rather than creating a new federal agency? In 1968, the administration of the Wild and Scenic Rivers Act was given to individual states. Would this have been an appropriate component of NEPA? Why or why not?

Do you believe Congress would have created the CEQ if the USEPA had been in existence and functioning for several years prior to the adoption of NEPA? Do you suspect the Nixon Administration insisted upon strong involvement from the executive branch of government (vis-à-vis formation of the CEQ) in order to obtain its political support for the passage of NEPA?

THE ENDANGERED SPECIES ACT

http://endangered.fws.gov

In 1973, Congress approved the **Endangered Species Act (ESA)** to protect plants, vertebrates and invertebrates believed to be on the brink of extinction. The U.S. Fish & Wildlife Service, located in the Department of Interior, normally takes the federal lead on the ESA's coordination and implementation provisions. However, the National Marine Fisheries Service (NMFS) in the Department of Commerce has responsibility for protecting certain marine animals (all whales, porpoises, sea lions, and seals) and anadromous fish species (such as salmon).

When the ESA was approved in 1973, 109 species were listed for protection, including the California condor (*Gymnogyps californianus*), southern bald eagle (*Haliaeetus leucocephalus*), black-footed ferret (*Mustela nigripes*), grizzly bear (*Ursus horribilis*), and the whooping crane (*Grus americana*). As of April 2, 2004, there were 1263 U.S. species on the list, with an additional 287 species considered as candidates for listing. Only 39 species have been delisted ("removed") from the endangered species list since 1973. (12)

The goals of the Endangered Species Act are to increase the populations of threatened and endangered wildlife, as well as to secure suitable habitat to promote the recovery of these species to acceptable levels. ESA programs often involve the acquistion of additional water supplies in rivers and lakes, and can greatly interfere with longstanding water use practices. Since recovery of a species is typically a gradual process, it may take several generations of successful reproduction to reach acceptable population goals. (13) Recovery plans are either prepared by a panel of recognized experts under the direction of the U.S. Fish & Wildlife Service or are contracted to an appropriate consulting expert on the species.

What is the difference between an endangered and a threatened species? A plant or an animal species or subspecies is considered **endangered** when plants or animals (both vertebrate and invertebrate) are in danger of extinction, within the foreseeable future, throughout all or a significant portion of its range. A **threatened** classification is used when a species is likely to become endangered within the foreseeable future. A species can be considered a **candidate species** for listing as threatened or endangered if petitions are filled by the general public or wildlife protection groups, or if reports are filed by federal or state agencies. Anyone may request the U.S. Fish & Wildlife to have a species listed or reclassified. First, findings are required, and then the request must be published in the *Federal Register*. The U.S. Fish & Wildlife Service then determines whether the listing is warranted.

SIDEBAR

The bald eagle once nested throughout the United States, but by the mid-1960s much of its habitat had been significantly altered. Declining numbers were also the result of the pesticide DDT, which is toxic to eagles as well as other species of birds. In 1967, only 417 nesting pairs of bald eagles remained in the United States. Through the Endangered Species Act, as well as the banning of DDT and other private efforts, there are currently over 4000 nesting pairs in the United States. In 1995, the U.S. Fish & Wildlife Service upgraded the status of the eagle from endangered to threatened. (14)

Under Section 7(a)(2) of the ESA, all federal agencies must ensure that no federal action authorized, funded, or carried out will jeopardize the existence of a listed species, or destroy or adversely modify habitat needed by the species. The Bureau of Reclamation and the U.S. Army Corps of Engineers are particularly affected by this requirement since the Bureau of Reclamation develops and manages irrigation projects, and the USACE controls flood regulation dams, navigational improvements, and issues Section 404 Dredge and Fill Permits under provisions of the Clean Water Act (discussed in Chapter 5).

Section 9 of the ESA prohibits the taking (killing, harming, or harassment) of any federally listed threatened or endangered species by anyone. The federal government can use this provision of the Act to prohibit or stop private activities that could result in a taking. A federal action is required to trigger a Section 7 proceeding, but a private action can cause implementation of Section 9 of the Act. An "incidental take" provides a means for nonfederal projects to be permitted under federal ESA laws under closely prescribed conditions. Incidental take permits are intended to provide a balance between orderly economic development and species conservation.

Some of the language used in Endangered Species Act proceedings is as follows:

Jeopardy Opinion—Official federal opinion that states a proposed action (such as building a dam) would be reasonably expected, directly or indirectly, to appreciably reduce the likelihood of both the survival and recovery of a listed species in the wild by reducing reproduction, numbers, and distribution of that species, or adversely changing its critical habitat.

Reasonable and Prudent Alternatives—Alternative actions that are economically and technologically feasible but do not adversely jeopardize listed species or modify critical habitat for the species.

Safe Harbor—Incentives to private and other nonfederal property owners to restore, enhance, or maintain habitats for listed species. Since most habitat is found on private property, the federal government can therefore provide technical assistance to property owners without fear of future land-use restrictions or practices on the land involved.

Scoping Process—Informing the public of a proposal and asking for public input toward the proposal.

POLICY ISSUE

The Endangered Species Act has provided protection to a wide range of plants and animals since its implementation in 1973. Also feeling the effects of the ESA have been developers, irrigators, municipal water providers, highway construction officials, and many, many others. In Oregon, as discussed in Chapter 9, irrigators were forced to watch crops wither and die during the drought of 2001 (see Figure 12.2). This was due to limited irrigation water from a Bureau

FIG. 12.2. Bureau of Land Management rangers stand guard, at right, as the headgates of the A Canal of the Klamath Project slowly close, Thursday, Aug. 23, 2001, in Klamath Falls, Oregon. Federal officials shut down irrigation headgates to hold back water for endangered fish after a limited amount of federal irrigation water was ordered released for Klamath Basin farmers in July.

of Reclamation Project, held in Upper Klamath Lake by the federal government, for the benefit of endangered salmon. (This situation will be discussed further later in this chapter.)

In Texas, groundwater withdrawals near San Antonio are restricted because of the presence of six endangered species in the karst limestone formations of the Edwards Aquifer (discussed in Chapter 4). The threatened species are the Texas Blind Salamander (*Typhlomolge rathbuni*); Fountain Darter (*Etheostoma fonticola*); San Marcos Gambusia (*Zizania texana*); Comal Springs Riffle Beetle (*Heterelmis comalensis*); Comal Springs Dryopid Beetle (*Stygoparnus comalensis*); and Peck's Cave Amphipod (*Stygobromus pecki*).

Should the costs and benefits of protecting species such as the Texas Blind Salamander or the San Marcos Gambusia be weighed against the costs and benefits to local economies and society in general? How would these costs and benefits of species protection be calculated?

Is there an endangered species listed in your area? Has it affected the use or development of water resources in the region? What methods are being used in the recovery efforts for that species, and are the numbers increasing? What is the target population goal for the species, and how long is it projected to reach those goals?

WETLANDS AND WILDLIFE

Hope and the future for me are not in lawns and cultivated fields, not in towns and cities, but in the impervious and quaking swamps. (15)

—Henry David Thoreau

Wetlands have been misused, degraded, and destroyed for centuries, and they are too often viewed as mosquito-infested, disease-ridden wasteland filled with snakes and other undesirable wildlife. It has been estimated that close to 612,500 square miles (1.59 million km^2), or more than 11 percent of the total U.S. landmass, was covered with wetlands before Europeans and others arrived. (16)

George Washington (1732–1799) was one of the first landowners in America to attempt to convert a large wetland into productive farmland. In 1763, Washington, having formed the Dismal Swamp Land Company and the Adventurers for Drainage of the Great Dismal Swamp, purchased 40,000 acres (16,188 ha) of "swamp ground" along the border of Virginia and North Carolina. Washington planned to construct a canal to drain a portion of the swamp, and so directed the surveying and digging of a 5-mile (8 km) ditch (called the Washington Ditch). Unfortunately for Washington and his investors, the drainage project was not successful. Today the area is federally protected within the Great Dismal Swamp National Wildlife Refuge and is administered by the U.S. Fish & Wildlife Service.

Almost a century after Washington's efforts in the Great Dismal Swamp, the U.S. government established the first federal policy on wetlands. In 1849, the Louisiana Swamp Land Act was passed, followed by the Swamp Land Acts of 1850 and 1860 (discussed in Chapter 9). These federal laws encouraged the draining of marshy lands to promote development and reduce the spread of disease, particularly in Alabama, Arkansas, California, Florida, Illinois, Indiana, Iowa, Kentucky, Louisiana, Michigan, Minnesota, Mississippi, Missouri, Ohio, and Wisconsin. In part due to the Swamp Land Acts of the mid-1800s, it has been estimated that 53 percent of the original wetlands in the United States have been drained. California, Illinois, Indiana, Iowa, Kentucky, Missouri, and Ohio have lost more than 80 percent and were specifically targeted in the Swamp Land Acts for removal. (17)

SIDEBAR

Many Native American tribes understood the important role of wetlands in the natural environment through their protection of the beaver. Beaver dams created extensive ecosystems that provide habitat for fish, deer, muskrat, antelope, and plant life. These areas were often prime sources of food for tribes. Beaver were highly regarded in religious beliefs, and hunting the animal was strictly forbidden in many tribes.

Organized drainage projects were also widespread in the United States between 1870 and 1920, draining more than 75,000 square miles (194,250 km^2). (18) In 1906, the U.S. Department of Agriculture surveyed the states west of the 115th Meridian (generally California, Oregon, and Washington) to identify swamp and overflow lands that could be drained and reclaimed for agricultural purposes. The federal government offered technical assistance to landowners interested in converting such "wasteland" into cropland.

Section 404 of the Clean Water Act, as well as the National Environmental Policy Act and the Endangered Species Act, drastically changed federal wetland policy. The USEPA describes wetlands as "biological supermarkets" where decaying plants and animals provide food sources for aquatic insects, shellfish, and small fish. These, in turn, are food sources for larger predatory fish, reptiles, amphibians, birds, and mammals. Wetlands also serve as breeding pools, resting zones during migration, flood catchment basins, and water quality improvement vessels, depending on the time of year the wetland contains water. Wetlands are now understood to be extremely important components of ecosystems.

Loss of a wetland can cause serious food shortages for local species, leading to reduced populations and, in extreme cases, extinction. (19) More than one-third of the threatened and endangered species found in the United States live only in wetlands, and nearly one-half use wetlands at some stage in their life cycle. (20)

Fortunately, the societal value of wetlands has increased significantly since the mid-1900s. The U.S. Department of Agriculture no longer provides funds to drain wetlands to improve farm ground productivity. The U.S. Fish & Wildlife Service has expanded the number of national wildlife refuges, typically in wetlands complexes, to protect critical areas from development. Wetlands that are hydraulically connected to a waterway are protected by the U.S. Army Corps of Engineers and require federal permits (Section 404 Dredge and Fill Permits) or federal exemptions before construction can occur in those locations. The USEPA also reviews Section 404 permit applications to determine whether dredging or filling of a wetland will have an adverse effect on municipal water supplies, shellfish beds, fisheries, or wildlife or recreational areas. (21)

SIDEBAR

Major Causes of Wetland Loss and Degradation:

Human Actions

Drainage
Dredging and stream channelization
Deposition of fill material
Diking and damming
Tilling for crop production
Levees
Logging
Mining
Construction
Runoff
Air and water pollution
Changing nutrient levels
Releasing toxic chemicals
Introducing nonnative species
Grazing by domestic animals

Natural Threats

Erosion
Subsidence
Sea-level rise
Droughts
Hurricanes and other storms

Source: U.S. Environmental Protection Agency, *America's Wetlands*, Office of Wetlands, Oceans and Watersheds, EPA843-K-95-001, Washington, DC, December 1995, p. 10.

A CLOSER LOOK

Walden Pond, located just outside Boston near Concord, Massachusetts, is one of the most famous lake and wetland complexes in the world. It was here, between July 1845 and September 1847, that Henry David Thoreau (1817–1862) wrote *Walden: or, Life in the Woods*, first published in 1854. Thoreau combined philosophy with views toward the environment and observed the changes that occurred in the natural systems at Walden. He noticed changes in vegetation, the color of lake and marsh water, and the movement of animals. For many, Thoreau's Walden Pond is the birthplace of the American environmental movement.

Today, Walden Pond is a national historic landmark managed by the National Park Service. The pond has no surface water inflow or outflow but is hydraulically connected to the local groundwater system. Changes continue at the pond, but unfortunately, not all of the changes are positive. Surface water runoff from a nearby landfill and adjacent trailer park, nutrient loading from adjacent septic tank leach fields, and bacterial contamination from large numbers of visitors who use a swimming beach in the summer are contributing to declining water quality in the pond. Algae species have also invaded its waters, with eutrophication and aquatic species degradation now serious concerns. (22)

A national wetlands forum sponsored by the U.S. Environmental Protection Agency in 1987 focused on wetlands protection in the United States. "No net loss of wetlands" is the current national policy for federally funded programs for highway projects and construction, although approximately 100,000 acres (40,470 ha) are still lost annually owing to other development activities in the United States. (23) The U.S. Army Corps of Engineers encourages the use of wetland mitigation banks, where wetlands are restored, created, or enhanced to replace wetlands lost due to development. (24) Under the criteria of Section 404 Dredge and Fill Permits, a developer may be required to replace 1, 1.5, and up to 2 acres (0.4, 0.6, and up to 0.8 ha) of new wetlands for every acre (0.4 ha) of existing wetlands destroyed.

New wetlands must be created within the general vicinity of the destroyed area and must function in perpetuity. Wetland mitigation banks have been created in many states and are utilized as replacement wetlands for areas destroyed by highway projects or other development. Skeptics argue that replacement wetlands can never replicate historic sites. Poor design, improper chemical and water balances, and sloppy construction have proved skeptics correct in some situations. See Table 12.1 for a list of important dates of

TABLE 12.1 Timeline of Selected Events: Fish and Wildlife Management

Year	Event
450	Creation of "Royal Forests" for exclusive hunting by the king in England.
1849	Louisiana Swamp Land Act approved.
1864	*Man and Nature* written by the father of the environmental movement, George Perkins Marsh.
1872	Yellowstone National Park created.
1890	Washington Legislature requires fish passage devices at dams.
1892	Sierra Club formed.*
1894	Prohibition against all hunting in Yellowstone National Park.
1905	National Audubon Society formed.*
1936	National Wildlife Federation formed.*
1937	Ducks Unlimited formed.*
1940	Bald eagles rarely seen in the continental United States.
1947	Everglades National Park created.
1949	Aldo Leopold promotes protection of wild and scenic river areas.
1951	Nature Conservancy formed.*
1962	Rachel Carson's *Silent Spring* becomes best-seller.
1967	Environmental Defense formed.*
1968	National Wild and Scenic Rivers Act becomes law.
1970	First Earth Day held.
1970	National Environmental Policy Act (NEPA) becomes law.
1970	Natural Resources Defense Council formed.*
1971	Construction of the Tellico Dam in Tennessee halted by environmental lawsuit.
1973	Endangered Species Act becomes law.
1977	Hit List approved by President Jimmy Carter to stop construction of 18 dams.
1987	National Wetlands Forum held—"No net loss of wetlands in the U.S."
1994	Only 2½ million salmon return to spawning grounds along the Columbia River Basin.
1997	Three-States Cooperative Agreement signed by Nebraska, Wyoming, and Colorado to protect endangered species along the Platte River.
2001	Klamath River controversy arises between irrigators and the federal government in Oregon.

*History and additional information of this organization can be found in the Appendix.

wetlands protection as well as for other issues of concern to fish and wildlife.

A CLOSER LOOK

The Everglades ecosystem in Florida is probably one of the most endangered wetlands complexes in the United States. The Everglades region is actually a slow-moving river over 50 miles (80 km) wide and only 6 inches (15 cm) deep. This wide sheet of water, which empties into an estuary at Florida Bay located at the southern tip of Florida, creates the largest freshwater wetlands in the world along its path. Although at one time the Everglades was one of the most diverse and productive wetland systems on Earth, today over one-half of the original marshes have been drained. The remaining area is dissected by 1400 miles (2253 km) of canals loaded with nutrients from fertilizer and waste runoff from urban and agricultural lands in the region (Figure 12.3).

The Florida Everglades include 4 national parks and preserves, 13 national wildlife refuges, 2 national marine sanctuaries, 17 state parks, 10 state aquatic preserves, and 5 wildlife management areas. The Everglades National Park was created in 1947 to protect approximately 20 percent of the remaining wetlands, although about 50 percent had already been lost to development from residential areas, roads, cropland, and dairy farms. The Everglades provides recharge water for aquifers across the state and is home to 14 endangered or threatened species, including the Florida Panther *(Felis concolor coryi)*, wood stork *(Mycteria americana)*, and the Florida Everglade Kite (Florida Snail Kite, or *Rostrhamus sociabilis plumbeus*).

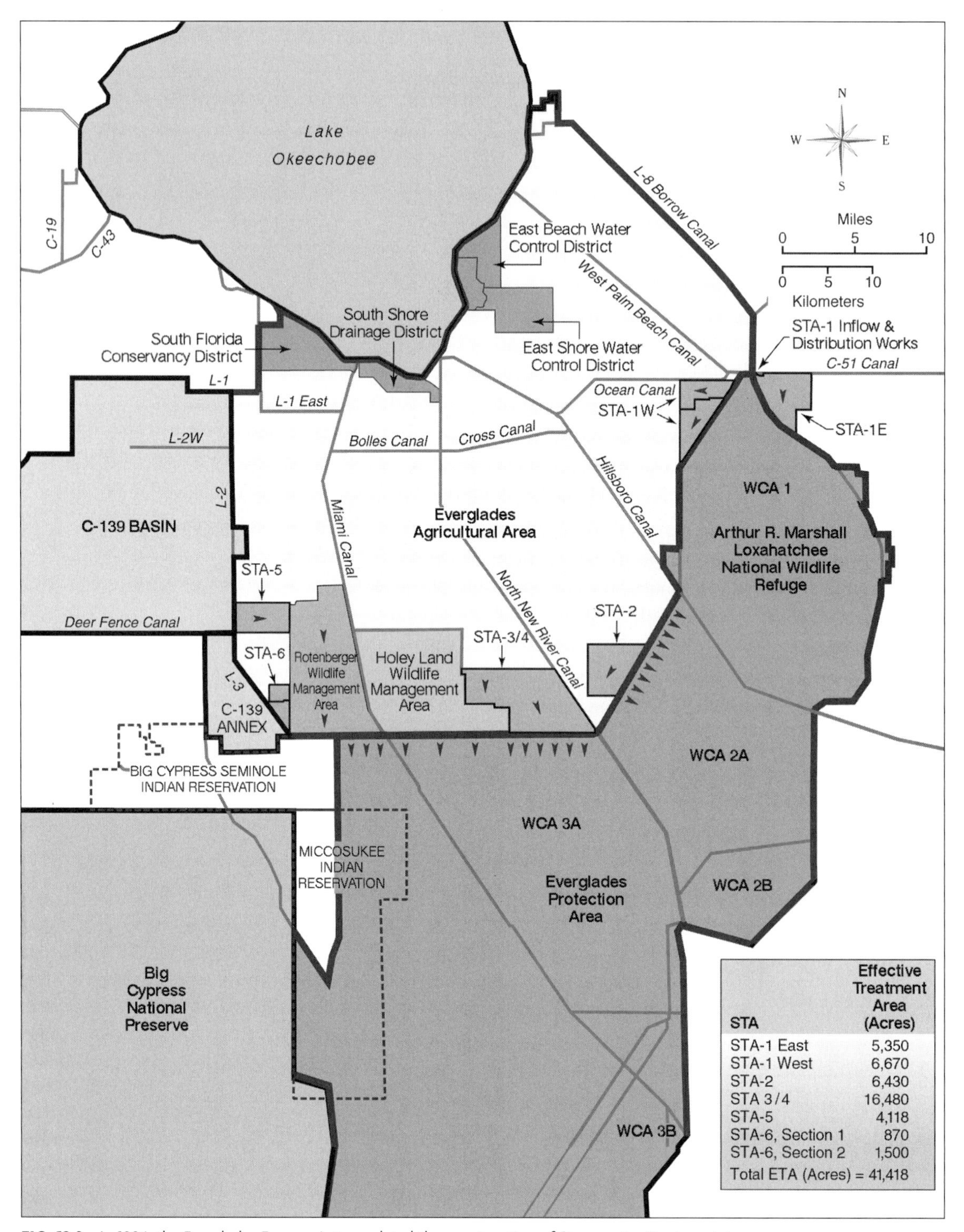

STA	Effective Treatment Area (Acres)
STA-1 East	5,350
STA-1 West	6,670
STA-2	6,430
STA 3/4	16,480
STA-5	4,118
STA-6, Section 1	870
STA-6, Section 2	1,500
Total ETA (Acres) = 41,418	

FIG. 12.3. In 1994, the Everglades Forever Act mandated the construction of Stormwater Treatment Areas (STAs) to improve water quality in the Florida Everglades. Activities within an STA can include the construction of levees and culverts to capture stormwater runoff, and creation of artificial wetlands to naturally remove nutrients such as nitrates and phosphorus. Several STAs are shown in this illustration.

In the 1960s, the U.S. Army Corps of Engineers straightened a 103-mile (166 km) reach of the Kissimmee River to provide flood protection. This construction drained wetlands and allowed cattle grazing to occur. Manure from the animals, as well as fertilizer from sugar cane and vegetable fields, washed into Lake Okeechobee and increased the rate of eutrophication. Algae blooms caused drastic reductions in dissolved oxygen and led to the death of many aquatic species. In the past 60 years, nesting bird populations have decreased by 90 percent. (25)

A number of efforts to restore the Everglades have been made, such as increasing water flows through the area to mimic historic flow patterns, cleaning up polluted waters, and purchasing private lands to protect them from development. In 1993, Congress authorized the U.S. Army Corps of Engineers to restore the channelized Kissimmee River to its more natural, meandering condition in the hope of replacing lost wetlands. Later, in November 1996, Florida voters defeated a proposed tax of one penny on each pound of raw cane sugar produced in the Everglades. The tax monies would have been used for the conservation and protection of the Everglades ecosystem. Sugar industry officials suggested that Floridians were not only against higher taxes but were also concerned about potential lost jobs in the sugar industry if the tax proposal had passed. (26)

C A S E   S T U D Y

Snail Darters and the Little Tennessee River

http://www.tva.gov

In 1967, the Tennessee Valley Authority (TVA) (see Figure 12.4) began work on the Tellico Development Project, which included plans to construct the $120 million Tellico Dam and Lake. The federally funded project was designed to dam the Little Tennessee River in eastern Tennessee. Extensive industrial and other development was planned around the 30-mile-long (48 km) reservoir created by the dam, and the stored water would be used to help operate an adjacent hydroelectric plant to provide power to the region. (27)

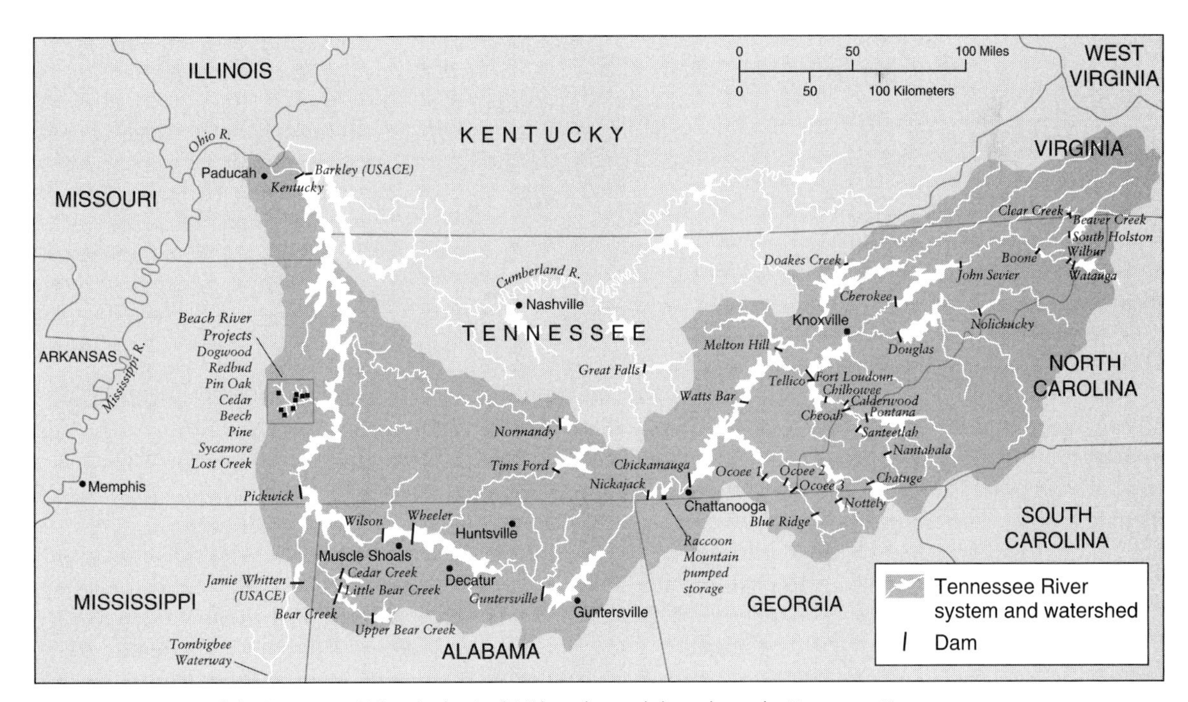

FIG. 12.4. Projects of the Tennessee Valley Authority (TVA) are located throughout the Tennessee River watershed.

The TVA had spent $15 million on the concrete dam prior to passage of the NEPA in 1970. However, in 1971 several environmental groups, led by the Environmental Defense Fund (EDF, discussed on page 437 in the Appendix), filed an injunction against the project, stating that an environmental impact statement (EIS) had not been completed. Construction had to stop until the TVA completed and filed an EIS. The report mentioned a rare species of fish, the previously unknown 3-inch (7.6-cm) long, tan-colored snail darter (eventually described as *Percina tanasi*), which was found just upstream of the dam site in the Little Tennessee River. Litigation followed, but in late 1973 the Federal District Court in Nashville ruled that the EIS was adequate and so construction on the dam resumed.

On December 28, 1973, President Nixon signed the Endangered Species Act into law. (It passed the Senate on a vote of 92–0 and the House of Representatives by 355–4.) On October 9, 1975, the snail darter was officially classified in the *Federal Register* as endangered, and another lawsuit was filed. Plaintiffs in the suit argued that completion of the Tellico Dam would destroy the snail darter's habitat, which was a clear violation of the Endangered Species Act. By now, $116 million had been spent on dam construction, but the project was again stopped, this time by the Sixth Circuit Court of Appeals in Cincinnati, Ohio.

The TVA and the U.S. Fish & Wildlife Service reintroduced the endangered fish into other rivers in the area, but the success of those populations was not assured. Eventually, the U.S. Supreme Court heard the lawsuit and finally issued a ruling in 1978 in favor of the environmental groups. The following year, however, an amendment was attached to the Public Works Appropriations Bill in Congress to exempt the Tellico Dam Project from the Endangered Species Act. The bill passed and was signed by President Jimmy Carter in 1979. (28)

During this process, the Cherokee Indian Tribe also filed a lawsuit to stop construction of the dam since Tellico Lake would inundate tribal burial grounds. They were unsuccessful, and the reservoir was filled in 1979. On July 5, 1984, the U.S. Fish & Wildlife Service reclassified the snail darter and downgraded it as threatened. (29)

Although circumvented in the particular case of Tellico Dam, citizen lawsuits are usually an effective way of forcing implementation of the Endangered Species Act. A provision of the ESA gives individuals the right to sue the federal government for violating provisions of the law. Environmental groups commonly file lawsuits, such as the 1971 suit by the Environmental Defense Fund on Tellico Dam, or a 1976 suit by the Sierra Club regarding a proposed U.S. Army Corps of Engineers dam near St. Louis, Missouri. The impounded water of the Corps project in eastern Missouri would have flooded caves that provided habitat for the endangered Indiana bat (*Myotis sodalis*). (30) Also in 1976, the National Wildlife Federation filed suit against an interstate highway project in Mississippi that would have destroyed habitat for the endangered Mississippi sandhill crane (*Grus canadensis pulla*). (31) Because of the high cost of litigation, national environmental organizations generally undertake such lawsuits rather than individuals. (Numerous additional environmental organizations are highlighted in the Appendix.)

Conversely, during the 1990s landowners filed suit against the federal government claiming the loss of property rights due to enforcement of the Endangered Species Act. The "takings issue" was hotly debated in legislatures and courts across the country, with private landowners protesting the loss of the right to use their property for farming, development, mining, and other commercial uses.

In 1975, the U.S. Fish & Wildlife Service issued a ruling that provided clear definitions of the term *harm*. Section 9 of the Endangered Species Act makes it unlawful for a person to "take" a listed species. According to the Act, "The term 'take' means to harass, harm, pursue, hunt, shoot, wound, kill, trap, capture, or collect or attempt to engage in any such conduct." (32) The Secretary of Interior, through the implementing regulations, defined the term *harm* in this passage as "an act which actually kills or injures wildlife. Such an act may include significant habitat modification or degradation where it actually kills or injures wildlife significantly impairing essential behavioral patterns, including breeding, feeding, or sheltering." (33)

Critical habitat is a specific legal designation defined in Section 4 of the Endangered Species Act as geographic areas "on which are found those physical or biological features essential to the conservation of the species and which may require special management considerations or protection." Critical habitat may include areas not occupied by the species at the time of listing, but essential to the conservation of the species. Critical habitat designations affect federal agency actions and federally funded or permitted activities. Harm is harm, regardless of who does it, who funds it, or where it actually occurs. (34)

A CLOSER LOOK

In 1992, Oregon irrigators sued the U.S. Fish & Wildlife Service to prevent the loss of irrigation water during a drought. Three endangered fish—the Lost River sucker *(Deltistes luxatus)*, the short-nosed sucker *(Chamistes brevirostris)*, and the Southern Oregon/Northern California coasts' coho salmon *(Oncorhynchus kisutch)*—live in the Klamath River and were negatively impacted by irrigation diversions. Because of severe water shortages, the Bureau of Reclamation ordered the USFWS to seal off its refuge wetlands and release water directly to the Klamath River. As a result, water supplies to irrigators dwindled, with crops wilting and dying in the fields. Water shortages also caused loss of vegetation in local parks. Although government officials admitted they were unsure whether the minimum water levels would adequately protect the endangered fish, irrigators were still forced to drill groundwater wells and to pay the Bureau of Reclamation for irrigation water that was never received during the summer.

During the subsequent court battle, irrigators argued that the USFWS had ignored the economic consequences of their water restrictions, resulting in a $75 million loss to the local economy. The U.S. Supreme Court unanimously agreed with the ranchers' argument and ruled that an individual can sue the federal government on the basis that the ESA may have been carried out excessively. Justice Antonin Scalia wrote for the Supreme Court that an individual could sue the government to reduce enforcement of the ESA as well as to require enforcement. The level of enforcement ultimately rests on the principles of prudence and reasonableness. The Court ruled that minimum water-level requirements in the lakes had to be eliminated. (35)

CASE STUDY

Whooping Cranes and the Platte River

http://www.platteriver.org

The Platte River has been called "a mile wide and an inch deep" and "too thin to plow and too thick to drink." It has a total length (without tributaries) of 625 miles (1006 km) and includes a drainage area of 86,000 square miles (222,740 km^2). The Platte is a braided stream with numerous channels, plentiful sandbars, and wide-open corridors. It has two major tributaries: the North Platte, which begins in north-central Colorado, flows through Wyoming, and then into the panhandle of Nebraska; and the South Platte River, which has its headwaters southwest of Denver, Colorado. The two tributaries join to form the Platte River near North Platte in western Nebraska, and both empty into the Missouri River near Omaha on the eastern side of the state. There are 15 major dams and over 200 diversions on the Platte, North Platte, and South Platte rivers in Colorado, Wyoming, and Nebraska. (36)

Historically, spring floods from melting snow off the Rocky Mountains provided peaking flows that altered the Platte River channel. This, in turn, developed an ideal habitat for the endangered whooping crane, least tern, and piping plover, as well as hundreds of thousands of ducks, geese, and sandhill cranes.

Every spring nearly 500,000 of the world's lesser sandhill cranes (*Grus canadensis canadensis*), a subspecies of sandhill cranes, use the Platte River as a critical staging area (Figure 12.5). The cranes winter in the southern regions of Texas and in northern Mexico. The

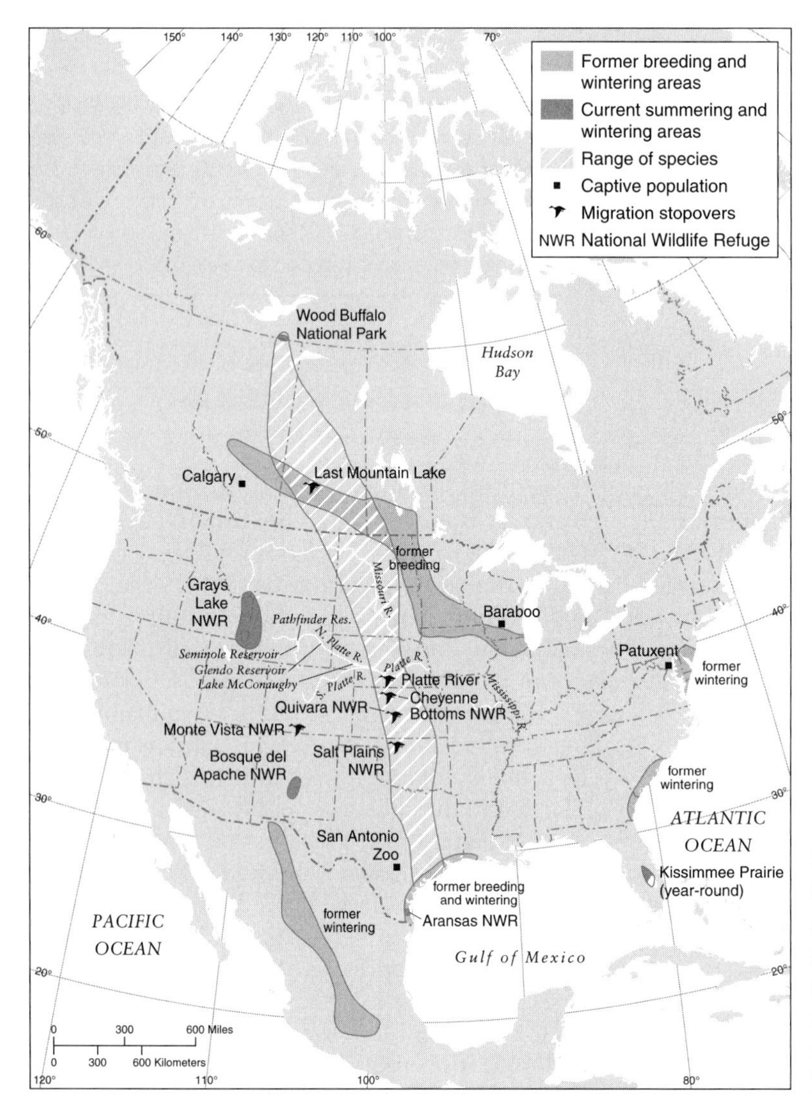

FIG. 12.5. The migration corridor of sandhill cranes (*Grus canadensis canadensis*) and whooping cranes (*Grus americana*) extends from the south shores of Texas north to the Arctic Circle. Notice the constriction of the flyway along the Platte River in central Nebraska.

migrating birds then fly some 3000 miles (4828 km) to the north to breeding grounds in the arctic regions of Canada. Every March and early April, 80 percent of the world's lesser sandhill cranes, and hundreds of thousands of ducks and geese, converge on a 150-mile (241 km) reach of the Platte River and its adjacent farmland in central Nebraska between Kearney and Grand Island. The crane's primary diet during this period is wasted corn from the fall harvest. During the daytime, flocks of cranes stand in cornfields feeding, while at night they roost on the shallow sandbars of the Platte River. (37) (See Figure 12.6.)

The cranes arrived at the roost at sundown the night before, first one or two, nervous and alert to danger, and then the others. They rained out of the sky like stilt-legged parachutists. As darkness enveloped the sandbars along the Platte, the cranes walked into shallow water, preening and rearranging their feathers. New cranes

FIG. 12.6. Sandhill cranes roosting on the Platte River near Grand Island, Nebraska.

swarmed to a vacant sandbar behind them. It was the early hours of the morning before the air was clear of the call of homing cranes. (38)

"Sandhill Cranes—Wings over the Platte"
NEBRASKAland Magazine

The sandhill crane is an archaic bird, with fossil remains found from the Eocene epoch some 55 million years ago. Other fossil evidence shows that cranes have been part of the Nebraska landscape for over 10 million years. A mature sandhill crane stands 3.5 feet (1.1 m) tall and has a wing span of 6 feet (1.8 m). (39)

Sandhill cranes are fastidious in their selection of a nighttime river roost. During its month's stay along the critical habitat area of the Platte River, cranes require shallow water not more than 6 inches (15 cm) deep, with a broad channel width of approximately 500 feet (150 m), and no vegetation on sandbars. Why? A primary concern is protection from predators. River corridors with a clear view and deep, braided channels provide unobstructed views of the shoreline. In addition, the deep channels form a barrier of deep flowing water between the shoreline and the sandbar roosts. Today these conditions are being lost due to changing water uses and encroaching vegetation. In some locations channel width has degraded in the past 100 years to less than 10 to 20 percent of its original width, and in other areas the reduction is approximately 30 to 40 percent. Sandbars choked with willows and other bushy vegetation have reduced usable habitat in other areas. (40)

Approximately 400 endangered whooping cranes remain in the world, and they are closely monitored to determine migration patterns. Cranes are tracked using satellite telemetry through a radio transmitter fused to leg bands placed on each crane. Whooping cranes are numbered, and movements of individuals are followed carefully.

In the late 1970s, the National Wildlife Federation and the State of Nebraska sued the Basin Electric Power Cooperative over construction of the Grayrocks Dam in Wyoming. At issue was the potential negative impact of diverting 19,000 acre-feet (6.2 billion gal, or 23.4 million m^3) each year from the Laramie River (a tributary of the North Platte) on the habitat of the endangered whooping crane in central Nebraska, approximately 500 miles (805 km) downstream. Water stored behind the dam, in 104,000 acre-foot (33.9 billion gal, or 128.3 million m^3) Grayrocks Reservoir, would be used for steam generation and cooling of the Laramie River Station, a 1650-megawatt coal-fired power plant near Wheatland, Wyoming. Ultimately, a settlement was reached, and Basin Electric paid $7.5 million to the newly created Platte River Whooping Crane Maintenance Trust for the improvement of whooping crane habitat in central Nebraska. The Trust is guided by a board of trustees of the National Wildlife Federation, the State of Nebraska, and the Basin Electric Power Cooperative.

In the late 1980s, numerous Front Range cities in Colorado, such as Greeley, Fort Collins, Loveland, Boulder, and Denver, were preparing to renew land use permits with the U.S. Forest

Service. The permits would allow the growing cities to store municipal water in high mountain reservoirs on tributaries of the South Platte River. The permits were issued decades earlier and allowed the cities to build and utilize small reservoirs on Forest Service land for a 30-year period before renewal was required. Historically, the permitting process was simple—the permit was reissued upon completion of the required paperwork and payment of a few hundred dollars.

This time the federal permit renewal process changed dramatically. Since storage of water on federal lands now triggered the National Environmental Policy Act, the U.S. Fish & Wildlife Service required a review of the action. Later study showed that storage and consumption of water in Colorado negatively affected the habitat of downstream endangered species along the Platte River in central Nebraska. The federal government asserted that permits would not be renewed without a one-for-one replacement of water diversions (called *bypass flows*). This meant that for every acre-foot (325,851 gal, or 1233 m^3) of water stored in a high mountain reservoir in the South Platte Basin of Colorado (that required a federal permit), the federal government was going to require the release of 1 acre-foot of water for delivery to critical habitat in central Nebraska, approximately 500 miles (805 km) downstream.

Municipal water users in Colorado were shocked. Since the state had used the Doctrine of Prior Appropriation since 1876, the requirement of bypass flows would cause the loss of hundreds of millions of dollars in senior water rights. Water rationing, nonissuance of building permits, and investment in new sources of water supplies would be required to meet these new federal requirements. Water users argued that the economic loss to the area would be devastating.

Years of negotiations followed during the 1990s and continues today. The federal government, led by the U.S. Fish & Wildlife Service but assisted by the USEPA, Bureau of Reclamation, FERC, and the U.S. Army Corps of Engineers, agreed to consider other alternatives to the one-for-one water replacement requirement. Finally, after nearly 10 years of meetings, the Platte River Basin states of Wyoming, Colorado, and Nebraska agreed to work together to reengineer the broad and shallow Platte River. It was agreed that a basinwide approach was warranted to solve this difficult issue.

In 1997, the governors of Wyoming, Colorado, and Nebraska, and the U.S. Department of Interior signed the Three-States Cooperative Agreement, which was designed to develop and implement a recovery program to improve and conserve the habitat of four threatened and endangered species: the whooping crane, piping plover, interior least tern, and pallid sturgeon. (The pallid sturgeon is found east of the whooping crane's critical habitat area near the confluence of the Platte and the Missouri rivers near Omaha, Nebraska.) Implementation of adequate practices to protect these "target species" or "species of concern" will allow limited water development upstream.

A draft EIS was prepared to provide analysis of the proposed Platte River program, and public review provides feedback. The NEPA process requires agencies that propose a major federal action to evaluate and report the effects of those proposed actions and reasonable alternatives to both decision makers and the public. Some of the reports completed as part of the Platte River NEPA process include:

Physical History of the Platte River in Nebraska: Focusing upon Flow, Sediment, Transport, Geomorphology, and Vegetation, Simons and Associates, August 2000.

Identification and Evaluation of Potential Third Party Impacts Related to the Habitat Component of the Proposed Platte River Recovery, Hazen and Sawyer, June 2000.

Third Party Impacts Executive Summary, Hazen and Sawyer, June 2000.

The Feasibility of Operational Cloud Seeding in the North Platte River Basin Headwaters to Increase Mountain Snowfall, Medina, May 2000.

Estimated Additional Water Yield from Changes in Management of National Forests in the North Platte Basin, Troendle and Nankervis, May 2000.

Final Summary of Scoping Input—Platte River Programmatic Environmental Impact Statement, July 1998.

A phased management approach is being used to protect the critical habitat. Phase one requires improvement of 29,000 acres (11,736 ha) of habitat; "target flows" (flows in the Platte River at the critical habitat area) need to increase annually by 130,000 to 150,000 acre-feet—42.4 billion gal, or 160.4 million m^3, to 48.9 billion gal, or 185.1 million m^3—over the next 10 years; enlargement of the storage capacity in Pathfinder Reservoir (a Bureau of Reclamation facility in Wyoming) for the benefit of downstream endangered species; establishment of a 200,000 acre-foot (65.2 billion gal, or 246.7 million m^3) environmental account in Lake McConaughy, a nonfederal facility in western Nebraska, to be released by the U.S. Fish & Wildlife Service as needed; and development of a groundwater recharge project at the Tamarack State Wildlife Area in eastern Colorado. Ultimately, 417,000 acre-feet (135.9 billion gal, or 514.4 million m^3) of new water flows (consumptive use water) will be required to meet the needs of the endangered species in the critical habitat area of central Nebraska.

SIDEBAR

Private organizations have become actively involved in preservation of critical habitat for the endangered whooping crane. The Audubon Society, through a bequest of Lillian Annette Rowe of New Jersey, purchased the first land to be protected as wildlife habitat along the Platte River near Kearney, Nebraska. The Rowe Sanctuary uses mechanical clearing of encroaching willows and trees (bulldozing and tree cutting) to provide improved habitat as well as improved crane watching. During the spring migration north, it is not unusual for viewers to witness 15,000 cranes per mile (every 1.6 km).

POLICY ISSUE

Why save threatened and endangered species? The U.S. Fish & Wildlife Service cites many reasons:

Environmental—It is important to maintain the food web of plants, and animals. Loss of critical elements of this web, sometimes called *keystone species,* could mean the total disruption of our natural environment.

Economic—Plant and animal species, such as chinook salmon, provide economic benefits to businesses such as commercial fisheries.

Medicine—Approximately 50 percent of all prescribed medicines are derived from substances found in plants and animals.

Agriculture—Wild strains of grain can be used to develop hybrids that are more resistant to crop diseases, pests, and marginal climatic conditions.

Recreation—Ecotourism, such as bird watching, whale watching, and plant investigation, is important.

Religious/spiritual/aesthetic—Many feel that the stewardship of nature's resources greatly enhances our quality of life. Many people have religious beliefs that stewardship respects their God's creation.

Critics of the Endangered Species Act argue that despite all its efforts the federal government removes only one species per year, on average, from the endangered species list. With so few species recovering from the ESA's efforts, they maintain that the law is not working. For example, the peregrine falcon was taken off the list in 1999, but numbers increased largely because of a ban on pesticides enforced by the USEPA. Others argue that efforts by Ducks Unlimited and other private conservation efforts are having a greater beneficial effect on wildlife habitat and, in turn, have helped restore endangered species. These nonfederal programs are generally much less expensive than federal efforts and are often more productive as well. Opponents to this argument state that the ESA is implemented only when species numbers are at critical levels and that it takes decades to turn negative trends around.

At what costs should society protect endangered species? Is there a point where the value of saving a species is offset by the enormous expense required? Should cost-benefit analyses (discussed in Chapter 7) dictate policy decisions regarding wildlife protection?

CASE STUDY

Salmon and the Columbia River

http://www.salmonrecovery.gov

The Columbia River Basin is North America's fourth largest watershed, including 250,000 square miles (647,500 km^2) of land as far north as British Columbia and as far south as Nevada. The basin contains over 250 reservoirs and 150 hydroelectric projects. Eighteen mainstem dams (dams constructed across the main river channel) exist along the 735-mile (1183 km) length of the Columbia River and its main tributary, the Snake River (see Figure 12.7). The U.S. Army Corps of Engineers operates nine of the ten major federal projects in the basin and provides flood control, navigation, recreation, fish and wildlife, municipal and industrial water supplies, and irrigation water to residents of the region. Grand Coulee Dam, the largest hydropower producer in North America, is a major dam on the system and is operated by the U.S. Bureau of Reclamation.

Five species of salmon live along the Columbia River—the chinook (king salmon), coho, chum, sockeye, and pink—in addition to steelhead trout, shad, smelt, and lamprey. Salmon hatch in the freshwater of the Columbia and its tributaries where they rear for up to two years. Later, the young salmon migrate to the Pacific Ocean where they grow in size and mature for two to five years. (Pink salmon are an exception and begin migrating downstream almost as soon as they leave their gravel beds.) The adult then returns up the Columbia Basin—sometimes a journey of over 2000 miles (3219 km)—to its place of origin to spawn.

Overharvesting in the late 1800s and early 1900s caused a serious decline in salmon numbers. In addition, the construction of dams prevented salmon from migrating past these locations. In 1890, the Washington Legislature approved a law requiring devices for fish passage, such as fish ladders, to allow salmon and other species to migrate past dams. Federal laws also required fish ladders on selected dams, but by 1940 eight unladdered dams had been built across the Yakima River in the Columbia River Basin of Washington. As a result, the annual salmon run decreased from 6 million to 9000 fish into the Yakima River. When the Grand Coulee Dam was constructed, 1000 miles (1609 km) of natural spawning grounds in the upper Columbia River were essentially lost forever. (41)

The life of a wild salmon along the Columbia River today goes something like this:

1. Fry are hatched in the sediments of the Columbia River and its tributaries at the same location as the birth of their parents and grandparents. These beds of gravel and other coarse sediments are called *spawning grounds*.
2. Fingerlings hide behind logs, large rocks, or anything else that provides cover. Food is grabbed quickly to keep exposure to predators brief. During a freshwater residence period of as much as two years, the fingerling grows in size and becomes a smolt. Soon, the juvenile salmon begin to swim downstream toward the Pacific Ocean off the coasts of Oregon and Washington.
3. Smolts encounter up to 10 dams and reservoirs along the mainstem of the Columbia

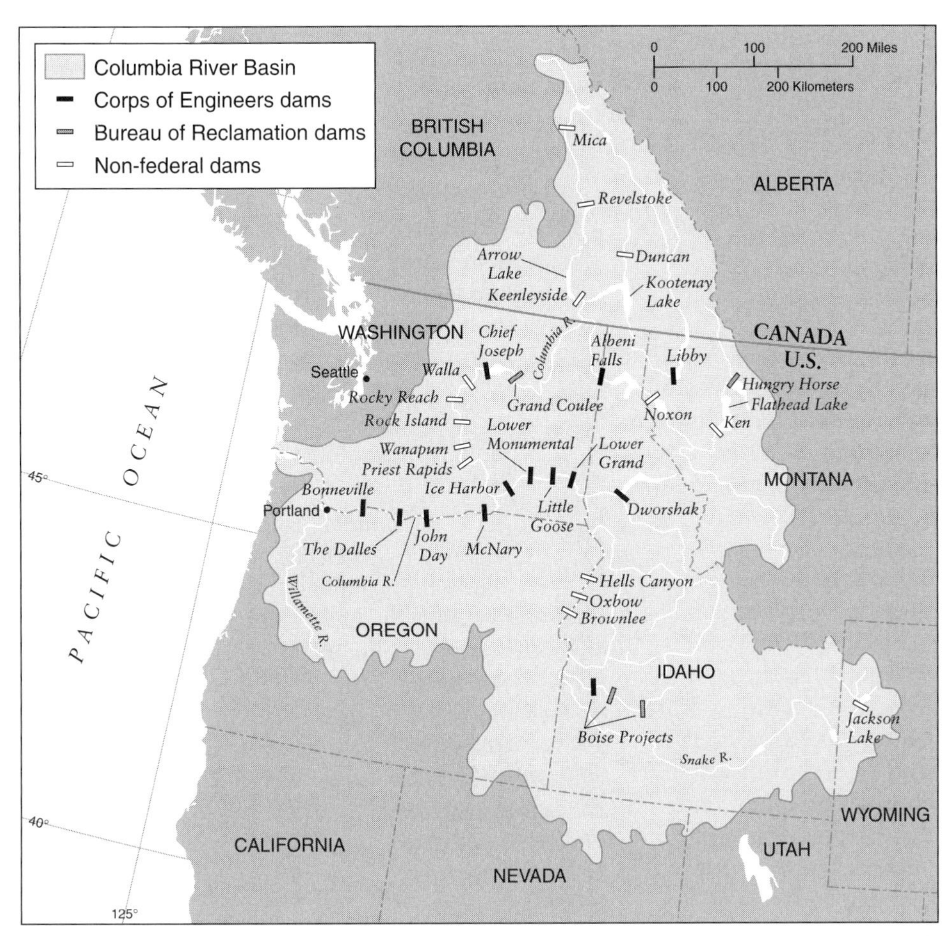

FIG. 12.7. The Columbia River Basin drains lands in Washington, Oregon, Idaho, Montana, Wyoming, Nevada, Utah, and British Columbia. What coordination problems might exist since many dams in the basin are operated by nonfederal agencies, and 10 of the major dams in the Columbia River Basin are operated by the federal government?

River before they reach the Pacific. Reservoir temperatures can be too warm near the surface for smolts to survive but too low in dissolved oxygen at lower levels to provide safe passage. Smolts must continually seek the appropriate layer of water in a reservoir to survive while avoiding predatory lake fish. Lake trout, sturgeon, and pike can actually create a barrier of predators in a reservoir for migrating smolts, and generally consume large numbers of the juvenile salmon. The worst predators of juvenile salmon are Northern pikeminnow, catfish, and birds such as gulls and terns. (42)

4. If a smolt survives the gauntlet of reservoir fish, it will then have four options to get past a dam that blocks its downstream path. One risky route is to enter an intake pipe and then pass through the hydroelectric turbines, where the fish may strike a structure or be exposed to severe pressures

and fluid forces. Federal dam operators have placed submerged screens in front of turbine intakes to direct juvenile fish away from these dangerous flow paths through a dam.

A fortunate smolt may find the entrance to bypass pipelines (called *juvenile fish bypass systems*) which allow it to avoid hydroelectric turbines (see Figure 12.8). Unfortunately, downstream predatory fish, such as sturgeon and pike, have learned that these downstream release locations of bypass pipes are smolt buffets. Large numbers of the unsuspecting juveniles are eaten as they emerge from the safety of the bypass tunnel.

If it's a high-water year, some of the river water may be spilled over the dam (rather than passed through the turbines), and smolt may be able to simply ride over a dam's spillway, dropping hundreds of feet (hundreds of meters) to the swirling waters below. Unfortunately, this journey has its drawbacks because the turbulent water at the base of a spillway is generally supersaturated with air. This harmful water can cause the smolt to get the bends as a result of nitrogen bubbles forming under their skin. Their eyes may bleed, and internal organs

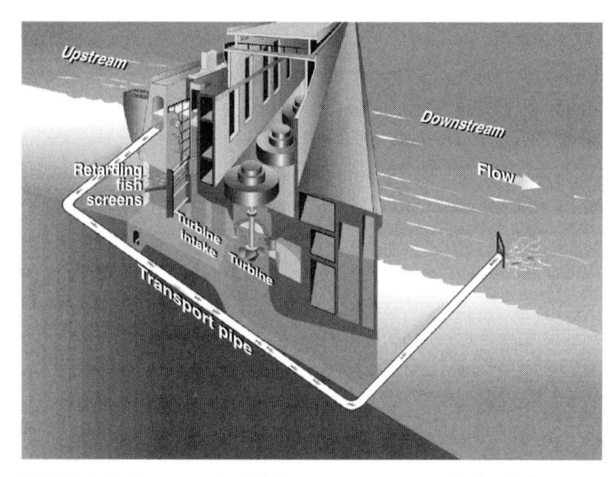

FIGURE 12.8. Juvenile fish bypass system at a federal dam along the Columbia River.

can actually explode. In 1970, nitrogen levels in the Columbia and Snake rivers were unusually high, and the federal government estimated that 70 percent of the salmon and steelhead trout were killed by nitrogen supersaturation. (43)

A fourth method for a smolt to get past a dam is to hitch a ride. At some dams, juvenile fish are captured in holding and loading facilities, operated by the U.S. Army Corps of Engineers, and placed into barges or trucks for transport downstream to a tidal estuary (Figures 12.9 and 12.10). Although entirely unnatural, it provides a higher level of safety for the small fish.

5. If a smolt survives the ride through a whirling hydroelectric turbine, or the fall over a dam spillway, and is quick enough to avoid the mouth of a predator fish at the bypass pipeline, its journey to the Pacific Ocean may still be in jeopardy since nine more downstream reservoirs and dams are still ahead.
6. If the smolt is successful, it will arrive at the Pacific Ocean in relatively good health and swim around for about four to five years (depending on the species) to put on weight. Salmon will follow prevailing North Pacific currents and sweep a large area following the ocean bed, currents, and regional climate to feeding grounds. A salmon will log about 10,000 miles (16,093 km) during this period.
7. When spawning time arrives, the adult salmon will head back inland, up the Columbia River and its tributaries, to find its unique spawning grounds.
8. The first Columbia River dam that creates a barrier to the adult salmon migrating upstream is the Bonneville Dam. Here, salmon try to enter the fish ladders that were installed at the time of dam construction. The fish generally enter a holding pattern with others at the base of the ladder since a bottleneck forms along its course. Fish ladders

FIG. 12.9. The Walla Walla District of the U.S. Army Corps of Engineers operates an extensive fish transportation program (called Operation Fish Run) to carry juvenile salmon downstream past the gauntlet of dams along the Columbia River system. Here, a tanker truck carries juvenile salmon that were gathered at Little Goose Lock and Dam on the lower Snake River for release below Bonneville Dam, the last dam on the Columbia River.

FIG. 12.10. Where possible, the U.S. Army Corps of Engineers uses barges in Operation Fish Run to transport juvenile salmon downstream past dams along the Columbia River.

are constructed in a series of steps and pools that provide a gradual uphill climb to the reservoir behind the dam (see Figure 12.11). "Attraction" flows are created at the base of the fish ladders to resemble the conditions at the base of natural waterfalls. This encourages migrating salmon toward these locations.

9. At the top of the fish ladder, the salmon enter strange territory. The water behind the dam is placid and has varying water temperatures and dissolved oxygen levels depending on depth. The salmon will seek a water level that most closely equals the waters of the lower Columbia River, but it can be difficult to find. Because a salmon relies on smell (i.e., taking cues from chemicals

FIG. 12.11. A fish ladder on each shore of the Snake River allows migratory fish to swim upstream past Ice Harbor Dam and Lake Sacajawea in western Washington.

dissolved in the water) to follow its path to historic spawning grounds, deep reservoir water can create additional obstacles.

10. After conquering the Bonneville Dam and Reservoir, the next structures are the Dalles Dam, then John Day Dam, and finally McNary Dam. Each structure has fish ladders, bottlenecks, varying temperature and dissolved oxygen levels, and deep water to confuse the sense of smell. Then come the Priest Rapids Dam, the Beverly Dam, and the Rock Island Dam (although some salmon may continue up the Snake River and encounter other dams).
11. After migrating upstream for up to 2000 miles (3219 km), the adult salmon stops at its natal stream. Once in the river, the entire journey is determined by memory of dissolved chemicals in the water (e.g., leached from vegetation and soils in the local watershed. Adult salmon in the Pacific Ocean find their way to the mouth of the Columbia River in other ways, primarily through sensitivity to magnetic fields). If the salmon is lucky, its original streambed will still be intact and will not be smothered in silt from soil erosion in the watershed. Salmon create a redd (nest of eggs) in the gravel, which contains some 5000 eggs each. The incubation period for the eggs in the gravel nest is two months or more.

A CLOSER LOOK

A Passive Integrated Transponder, or PIT tag, is a radio-monitored fish tagging system used to track selected salmon. An antenna and computer chip are placed in a small glass tube and then inserted into the body cavity of a juvenile fish where it remains the rest of its life. The PIT tag provides data whenever the fish is near a dam that contains a tag-monitoring facility. At all other times the tag is inactive. Each carries a code so that the data gathered are unique to each fish. Unfortunately, a present limitation of the PIT tag system is its short range; a fish must pass within 7 inches (18 cm) of the tag-monitoring facility at each dam to be detected.

Another method uses balloon tags that are inserted into a salmon. The tag can be inflated to bring the study fish to the water's surface to be recaptured and examined. This method of tracking is used strictly for turbine passage and spillway passage studies. As soon as the tagged fish is released, the balloon starts inflating, so the fish is recaptured within 10 minutes or so. Radio tags are a third monitoring method that can also be used on adult salmon, but this method requires insertion of a battery into the fish. Such batteries have a limited energy lifetime. One advantage, however, is that a radio tag signal can be read up to 0.5 mile (0.8 km) away, so it has a distinct advantage over a PIT tag. (44)

Biologists have been studying fish migration along the Columbia River for many years (see Figure 12.12 and Table 12.2). Mortality rates have been very alarming, and fish ladders, bypass pipelines, and other features at dams have been inadequate to restore the quantities of native fish that were present prior to dam construction. An alternative to this complex problem has been to build fish hatcheries. The first salmon hatchery in the United States was built in 1871 on the Columbia River. Between 1900 and 1930, the number of fry released rose from 25 million to 90 million annually. Believing this approach was more efficient and economical than fish ladders, officials did not include fish ladders in dams constructed on the White Salmon, Chehalis, or Elwha rivers.

By the 1930s, fish numbers were declining rapidly. During this period, Canadian biologists proved that fry released from fish hatcheries would not survive if natal spawning grounds were cut off by dams. If hatchery fry return to spawn, the adult salmon will stack up near the discharge pipe of a hatchery in an attempt to reenter the facility to spawn. The Canadian government stopped its hatchery program around 1940, after which many hatcheries in the northwest United States were also closed. Later it was discovered that hatchery fry did not learn how to hide from predators since it was a skill not needed in the concrete hatchery tanks. In addition, more recent research is showing that genetics plays a huge role in a fry's survival skills. As genetic strains are being lost, hatchery-raised fish appear to be less dynamic than their wild

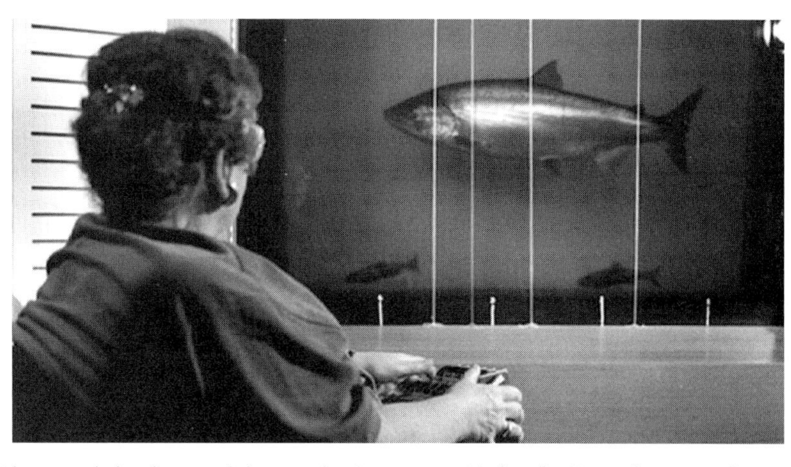

FIG. 12.12. Every fish that uses a fish ladder at the Bonneville Dam on the Columbia River is counted. Vertical lines on the 5-foot-wide (1.5 m) window allow fish counters to judge the length of each fish, as well as age and species. On a slow day, only a few hundred fish may pass by the window. However, on April 18, 2001, 12,020 chinook salmon (known as Kings) swam past the two counting windows at Bonneville Dam. Fish counters work 8-hour shifts with a mandatory 10-minute break every hour. Salmon that pass by the counting room at night are video recorded and counted the next day. Data are provided to the National Marine Fisheries Service to help produce the official federal year-end migration total.

TABLE 12.2 Daily Fish Passage Report, U.S. Army Corps of Engineers—The Dalles Project on the Columbia River (Oregon/Washington), October 2, 2001

Water Transparency: 5.5 Secchi Units
Water Temperature: 67.0°F
Location: Fish ladders

Species	Right Ladder	Left Ladder	Total
Chinook Adults	82	1319	1401
Chinook Jacks[a]	48	968	1016
Chinook Total	130	2287	2417
Coho Adults	37	366	403
Coho Jacks	0	16	16
Coho Total	37	382	419
Steelhead	109	3392	3501
Wild Steelhead	37	570	607
Sockeye	0	0	0
Lamprey	0	7	7
Chums	0	0	0
Pinks	0	0	0
Shad	0	0	0
Other	0	0	0
TOTAL ALL FISH	313	6638	6951

[a]"Jacks" are subadult salmon that are reproductively mature. Jacks usually return from the ocean two to three years before typical adult salmon; this can be an indicator of adult returns two to three years in the future.

Source: U.S. Army Corps of Engineers, "Project Daily Passage Report," Portland Office, Portland, Oregon, http://www.nwd-wc.usace.army.mil/ftppub/cafe/r091.txt, October 2001.

counterparts. Competition for food reduces the number of salmon found in the wild, which further reduces the wild gene pool available. The result is a less active salmon that is more prone to predators. (45)

POLICY ISSUE

The future administration of the Endangered Species Act (ESA) may be exemplified by an agreement made with the federal government and the largest developer in southern California. In exchange for setting aside 21,000 acres (8499 ha) of prime development land for a nature preserve, the Irvine Company was given a "no surprises" guarantee that it could construct housing tracts, shopping malls, and industrial parks in other locations without fear of legal battles regarding the ESA. This compromise provided the company with the freedom to develop major urban projects without the fear of being stopped in midproject. The cost of losing 21,000 acres (8499 ha) of potential development land was offset by the "no surprises" guarantee from the federal government. In contrast, the federal government obtained a substantial reserve of property for the protection of habitat. (46)

Extortion, sellout by the federal government, or great compromise? What do you think?

GUEST ESSAY

Larry Rogstad

Careers in Fish and Wildlife Management

by **Larry Rogstad**, District Wildlife Manager, Greeley North District for the Colorado Division of Wildlife Greeley, Colorado

Larry Rogstad received the John D. Hart Award as the Outstanding Field Officer in the Colorado Division of Wildlife in 1995. His district includes the rapidly urbanizing Front Range and eastern plains of northern Colorado. He received a Bachelor of Science degree in Zoology from the University of Oklahoma at Norman and has been with the Colorado Division of Wildlife since 1981.

I have had the great good fortune to work in the job I have wanted since I was in the third grade. I grew up in central Oklahoma in an area where the tall grass prairies give way to the cross timbers, the westernmost extension of eastern hardwood forests. It was an area that received ample, if not plentiful, moisture in most years. I spent my childhood fishing, hunting and trapping, and exploring the many creek bottoms that wind their way eventually to the Cimarron River. My Uncle Bud, a childhood hero, had been a game warden in Minnesota, and on a camping trip to the Grand Canyon one summer I met a ranger naturalist for the National Park Service who had a huge influence on me. Therefore, I have never had a desire to be anything other than a game warden.

As a district wildlife manager, I am the initial field contact between the Colorado Division of Wildlife and our customers. I devote about 25 percent of my time to wildlife and general law enforcement. I am also a field biologist responsible for doing counts on many diverse species, setting harvest quotas on game species to balance populations within carrying capacity of the habitat, assisting with compiling and implementing species management plans, working with landowners to resolve game damage issues (Colorado statutes require the Division to compensate landowners for financial losses caused by Big Game), handling nuisance and/or injured wildlife, disease monitoring, and so on.

I also have duties as a habitat biologist developing and implementing management plans on state wildlife areas and other public lands, as well as working with landowners on private lands. We are the Division's contact with county and local governments on land-use planning and a host of other issues. The district wildlife manager is also a primary contact with local newspaper and electronic media concerning issues of local interest. We are educators, too, providing a variety of learning opportunities concerning natural resource topics for all age groups. The variety in my job is what makes it so exciting. There is no typical day. I may start a day counting Greater Prairie Chickens on booming grounds, spend the middle of the day in meetings or checking anglers at a pond, and finish the day with a frog count.

By statute, the Colorado Division of Wildlife has two jobs to perform. First and foremost we are supposed to preserve, protect, and enhance Colorado's invaluable wildlife resource. This is a huge task. With the variety of habitats and altitudinal variation in our state, Colorado is home to over 960 different species of terrestrial and aquatic wildlife. The Division's second charge is to manage wildlife for the use and benefit of the people of the state and visitors as well. In my years as a wildlife manager I have discovered that people management is the most interesting, most challenging, and at times the most vexing part of my job.

When I took the job I naively thought, "This is great! I'll be working the backcountry, working with animals and not people." Nothing could have been further from the truth. My reality as a modern-day game warden is that I spend 90 percent of time working with people in myriad situations, sometimes friendly, sometimes antagonistic, and at times in very dangerous circumstances. In school I had always emphasized the biology and ignored development of the people skills. My salvation in this job has been my social skills which allow me to handle whatever situation I encounter. In short, my perceptions of what the job should be have at times differed markedly from reality. What follows are some of the assumptions I had when I first began in this field and the realities I have enjoyed working with.

ASSUMPTION 1: Decisions are made based strictly on biological information. The Division of Wildlife, like any governmental agency, exists in a political realm. My agency is within the executive branch of state government under the leadership of the governor. Our budget and the laws we work with are developed by the state legislature. A politically appointed Wildlife Commission oversees the Division's planning and long-range goals. Our human customers, if dissatisfied, can and will appeal directly to the Division or any of the above groups. To survive in this world, the Division must work as a biopolitical organization. However, as a field biologist required to provide information on a controversial topic, my first duty is to protect wildlife. Consequently, my comments should be biologically oriented with the understanding that the final decision may involve some compromise.

ASSUMPTION 2: Just because I wear the uniform or represent the agency, the customers will/should do what I say. I learned the reality of this misconception during the first Pronghorn Antelope season I worked as a wildlife officer. I was standing alongside a trail in a pasture talking to some hunters when I saw a pickup truck approaching. I could tell the men in the truck were hunters, so I decided to stop them. When the truck got close, I stood in the trail and, fully expecting them to obey my request as an officer, signaled with my hand for them to stop. When they got about 10 feet (3 m) away from me, the driver sped up and aimed the truck at me. I rolled up over the hood and into sagebrush and cactus. Lesson learned. I assumed they would do what I wanted them to do, but they had different plans. In working with people, it is important to remember that they may have different ideas, may not understand what you want, or may have different ways of communicating and different expectations. It is essential to be ready to adapt as necessary to get the job done.

ASSUMPTION 3: We (the agency) are the only experts on wildlife issues. This may have been more true in the past than it is today. When I joined the Division of Wildlife in 1981, it was a traditional governmental agency of the mid-twentieth century. Information for decision making came almost entirely from within the organization. Our customers expected us to be the decision maker. Our constituency tended to be from the ranks of sportsmen who seldom questioned our authority or decisions. During the last 20 years, society has become less trusting of regulatory agencies, more vocal in expressing opposition, more willing to participate in decision-making processes, and more prone to seek redress through litigation.

ASSUMPTION 4: Wildlife/habitat preservation plays a key role in local land-use planning decisions. Like many western states, Colorado tends toward a conservative land-use ethic. Many westerners feel that landowners should be free to use property as they see fit, without governmental interference. Our state statutes reflect the traditional western view toward land use by placing the primary decision-making authority in land-use planning onto local governments.

As a representative of the Colorado Division of Wildlife, I become involved in the review of development plans only at the request of a county or municipal planning agency. My comments are limited to biological impacts, with recommendations of a variety of alternative treatments that hopefully will preserve quality wildlife habitat or mitigate its loss. The local planning department is responsible for making the ultimate decision concerning the development plan. In addition, recent court cases involving "takings" issues have placed financial responsibility on local governments that have imposed development restrictions on landowners. Consequently, planners may be reticent in placing restrictions on development.

The Front Range of Colorado, including my district, is experiencing a tremendous growth in human population and therefore development. This is the biggest threat to our wildlife resource—alteration of prime wildlife habitat. The bottom line in land use is money. Arguments for habitat preservation for its intrinsic value and aesthetics will fall on deaf ears in a market-driven economy. It is our challenge to show that habitat preservation is of economic value to the developer and community.

ASSUMPTION 5: I have adequate information or data on which to make decisions. In the face of our rapidly developing landscape and management responsibility for over 900 species, the Colorado Division of Wildlife constantly faces new questions and challenges. Other entities, both public and private, also have their data sets, which may be in conflict with Division information. Consequently, the Division is in constant need of new and/or updated information. For example, human-induced changes in the hydrology and limnology of the South Platte River have altered aquatic habitat and consequently species composition of the river, prompting concern for several small fish species.

In response, the Division undertook a small fish survey of the South Platte Basin to develop data on our native fish fauna and establish population baselines for management. As a result, the Division has developed management plans, including construction of hatchery facilities and habitat protection for several species of native small fish. Hopefully, these efforts will enable the Division to restore these species to healthy population levels and avoid listing them as threatened or endangered. Currently, we are beginning amphibian and reptile surveys of northeast Colorado to establish baselines for our herptile fauna.

ASSUMPTION 6: Species can be managed separately. In the past, wildlife managers tended to manage for individual species. Ecosystem management has replaced this outmoded concept. Obviously, species coexist with all other species in their shared habitat. Changes affecting one species will influence others in the system. Therefore, it is essential to consider other associated groups when making management decisions for a particular species. For example, the Colorado Pikeminnow (Squawfish), Bonytail and Humpback Chubs, and Razorback Sucker, all native fish of the Colorado River drainage, have been placed on the federal and Colorado threatened/endangered (T&E) species lists. The population of these species has declined primarily because of changes in the hydrology of the Colorado River Basin in western Colorado. These species traditionally make a long migration to spawning habitat. Dams along the rivers in this basin have prevented migration and have altered critical spawning habitat.

ASSUMPTION 7: If it's not of concern now, it won't be a concern later. After a problem has been analyzed and a decision has been made, one usually forgets about it and goes on to other matters. Situations change, and new and better information may appear. As a manager, it is essential to keep on top of things. Whirling Disease, a parasitic disease of salmonids caused by an amoeba that infects larval trout, made its appearance in Colorado in the 1980s. The parasite lives in trout as the primary host with a tubifex worm, which lives in mud along the bottoms of lakes and streams as a secondary host. The parasite is native to Europe and was first discovered in North America in the 1950s.

Whirling Disease was brought into Colorado on a load of trout from Idaho by a privately owned hatchery. The private hatchery, on the Arkansas River, was upstream from state hatcheries, and soon the Division's hatcheries on the Arkansas were also infected. The organism quickly spread to almost all of our state hatcheries and to 13 of Colorado's 15 major river basins. The protozoan infects young fish, causing deformation of the spine and killing heavily infected fry. Young fish inoculated with the parasite swim in circles, hence the name.

We have spent millions of dollars making changes to our hatcheries, which hopefully will break the disease cycle. We have implemented a new stocking program for cold-water fish that will protect pristine, uninfected waters. We are still involved in an exhaustive research effort to find a permanent solution for this problem, which at first seemed to have a minimal impact on our fishery resources.

To get my job done, I spend a great deal of my time in my state pickup truck. I put on about 35,000 miles (56,326 km) a year and jokingly call it my field office. I carry spare clothes and a sleeping bag in it, and when an assignment goes overtime, it serves as a convenient bivouac. Since I spend a lot of time in my truck, it also serves me as a place where I sort out my thoughts. As I write this essay, it is mid-April, and we are experiencing a much needed, wet spring storm that started with rain but has wound up a blizzard. It is turkey season, so I began my morning at sunrise by driving out east along the lower South Platte River. I wanted to see if I could find anyone foolhardy enough to be out turkey hunting on such a miserable day. I wound up on the Riverside Irrigation Ditch access road, which had dissolved into an almost impassable quagmire. On one side of the road the Riverside Canal was running full, waiting to swallow my truck and me too if I slid a little to the right. On the other side there was a big dropoff to swallow my truck if I miscalculated in that direction. I had shifted into four-wheel drive and was driving gingerly down the road trying to get to where I needed to get to when it came to me. Navigating down this sodden, poor excuse of a trail kind of represents what I do in my job on a daily basis.

Each of my tasks has a beginning and a destination. Sometimes the journey is a pleasant one with sunny skies and smooth pavement all the way. Occasionally I hit a pothole or other minor inconvenience. However, every once in a while I wind up on a messy road with mudholes and problems on either side just waiting to suck me in. Survival in these situations requires a few minor changes in attitude.

First, recognize and admit that it is getting a little slick out there. Make full use of the equipment that is available. Slow down to a pace appropriate for the situation you are in. Avoid the tendency to oversteer because that may cause a spinout and land you where you don't want to wind up. Have patience and try not to panic. Let the truck do its work, and eventually you will get to where you really need to go to. No matter what situation you find yourself in, enjoy the ride. It really is an amazing journey, and getting there is only half the fun.

CHAPTER SUMMARY

Environmental protection was extremely limited before the late 1800s in the United States and was not greatly enforced by the federal government until the 1970s. Conflict, costly studies, and prodigious meeting schedules are typical byproducts of the process to protect endangered species and critical habitat. The success rate of recovering species has been extremely low—only a few per year on average—and has led many to question the process being used. Others argue that species restoration will take decades, even centuries, before reasonable populations return. The debate continues.

QUESTIONS FOR DISCUSSION

1. Describe the level of fish and wildlife protection that occurred in the United States prior to the mid-1800s and contrast it with the federal laws enacted in the past 40 years.

2. Why is it important to protect fish and wildlife? Should this protection be provided at any cost, or should a benefit-cost analysis be required? Why or why not?

3. Is there a Wild and Scenic River designated in your region? If so, when was it established? What types of uses are permitted? What agency administers the program?

4. Should fish and wildlife management be an entirely local, state, or federal obligation? Why?

5. Contrast the water resources philosophies of George Washington and Henry David Thoreau as described in this chapter.

6. The snail darter was a landmark case regarding endangered species protection and dam construction. Have any similar conflicts occurred in your area regarding a dam and the Endangered Species Act? What was the outcome?

7. Whooping cranes have been migrating between Canada and the Gulf Coast for tens of thousands of years, yet their numbers were quite low in the mid-1900s. Salmon on the Columbia River also faced declining populations during the 1900s. Some salmon are currently transported by barge and truck to assist during migration to the Pacific Ocean. What if the federal government proposed to transport (by plane or truck) whooping cranes, between spring nesting sites in the north and wintering grounds in the south, to avoid human-caused obstacles along the route? How unusual is that concept given the current methods used to transport salmon in the Pacific Northwest? Should any assistance be provided to fish and wildlife that cannot migrate because of human development?

KEY WORDS TO REMEMBER

candidate species p. 354
Council on Environmental Quality (CEQ) p. 352
Endangered Species Act (ESA) of 1973 p. 354
endangered species p. 354
Jeopardy Opinion p. 355
National Environmental Policy Act (NEPA) of 1970 p. 352
Reasonable and Prudent Alternatives p. 355
Recreational Rivers p. 351
Safe Harbor p. 355
Scenic Rivers p. 351
Scoping Process p. 355
threatened species p. 354
Wild and Scenic Rivers Act (1968) p. 350
Wild Rivers p. 351

SUGGESTED RESOURCES FOR FURTHER STUDY

READINGS

Crenshaw, Larry. *Earthbook*. Asheville: North Carolina Outward Bound School, 1995.

Farrar, John, and Ken Bouc. "Sandhill Cranes—Wings over the Platte," *NEBRASKAland Magazine,* Lincoln, Nebraska, no date.

Hey, Donald L., and Nancy S. Philippi. *A Case for Wetland Restoration*. New York: John Wiley & Sons, 1999.

Kent, Donald M. *Applied Wetlands Science and Technology*. Boca Raton, FL: Lewis Publishers, 1994.

Leopold, Aldo. *A Sand County Almanac*. London, England: Oxford University Press, 1949.

Marsh, George P. *The Earth as Modified by Human Action (A Last Revision of "Man and Nature")*. New York: Charles Scribner's Sons, 1898.

Matthiessen, Peter. *Wildlife in America*. 3rd ed. New York: Viking Press, 1967.

Outwater, Alice. *Water—A Natural History*. New York: Basic Books, 1996.

Reich, Charles A. *The Greening of America*. New York: Random House, 1970.

Switzer, Jacqueline Vaughn, and Gary Bryner. *Environmental Politics—Domestic and Global Dimensions*. 2nd ed. New York: St. Martin's Press, 1998.

U.S. Environmental Protection Agency. *America's Wetlands: Our Vital Link Between Land and Water*. Washington, DC: Office of Wetlands, Oceans and Watersheds, EPA843-K-95-001, 1995.

WEBSITES

"Fish Passage Center Reporting Sites," Fish Passage Center, Portland, Oregon, July 2003. http://www.fpc.org/fishway/map.html

Great Outdoors Recreation Pages (GORP), "Destinations: Wild and Scenic Rivers," July 2003. http://www.gorp.com/gorp/resource/usriver/main.html

Nebraska Game and Parks Commission, "Sandhill Cranes," July 2003. http://www.ngpc.state.ne.us/wildlife/cranes/sandhill.htm

U.S. Department of Interior, National Park Service, "Endangered, Threatened, and Rare Species Management," July 2003. http://www.aqd.nps.gov/nps77/t&e.new.html

U.S. Environmental Protection Agency Homepage, July 2003. http://www.epa.gov

VIDEO

Winged Migration, Directed by Jacques Perrin, Sony Pictures Classics, New York, NY, 2003, 85 minutes.

REFERENCES

1. Marsh, George P. *The Earth as Modified by Human Action (A Last Revision of "Man and Nature")*. New York: Charles Scribner's Sons, 1898, 9.

2. 16 U.S.C. 668-668d, 54 Stat, 250. See U.S. Department of the Interior, Bureau of Indian Affairs, "Eagle Protection Act," http://www.doi.gov/bia/information/eagleproact.htm, September 2001.

3. Personal communication with Mitch Snow, Chief of Media Services, External Affairs Office, U.S. Fish & Wildlife Service, Washington, DC, May 29, 2001.

4. Leopold, Aldo. *A Sand County Almanac: And Sketches Here and There*. London, England: Oxford University Press, 1949, 118–119.

5. Wild and Scenic Rivers Act of 1968 (U.S. Code Title 16, 1271–1287).

6. U.S. Environmental Protection Agency Homepage, http://www.epa.gov/history/topics/epa/15c.htm, April 2001; and Switzer, Jacqueline Vaughn, and Gary Bryner, *Environmental Politics: Domestic and Global Dimensions,* 2nd ed. New York: St. Martin's Press, 1998, 10.

7. U.S. Environmental Protection Agency, History Office, "The Spirit of the First Earth Day," Washington, DC, http://www.epa.gov/history/topics/earthday/01.htm, September 2001.

8. National Environmental Policy Act of 1970, Section 101 [42 USC § 4331].

9. U.S. Environmental Protection Agency Homepage, http://www.epa.gov/history/topics/epa/15c.htm, April 2001.

10. U.S. House of Representatives, Hearing of the Committee on Resources, U.S. House of Representatives, Washington, DC, March 18, 1998, "Problems and Issues with the National Environmental Policy Act of 1969," http://commdocs.house.gov/committees/resources/hii47866.000/hii47866, September 2001.

11. U.S. Environmental Protection Agency, History Office, "Reorganization Plan No. 3 of 1970," July 9, 1970, http://www.epa.gov/history/org/origins/reorg.htm, September 2001.

12. U.S. Fish & Wildlife Service, "Species Information," http://endangered.fws.gov/wildlife.htm#Species, April 2004.

13. U.S. Fish & Wildlife Service, "The Endangered Species Act of 1973," http://endangered.fws.gov/esasum.html, September 2001.

14. U.S. Department of Interior, Fish & Wildlife Service. *The Road Back—Endangered Species Recovery*. Washington, DC, no date. p. 1.

15. Thoreau, Henry David. "Walking." *The Atlantic Monthly* 9 (June 1862): 657–674.

16. Hey, Donald L., and Nancy S. Philippi. *A Case for Wetland Restoration*. New York: John Wiley & Sons, 1999, 2.

17. Ibid., and U.S. Environmental Protection Agency Homepage, http://www.epa.gov, September 2001.

18. Hey and Philippi, 36.

19. U.S. Environmental Protection Agency. *America's Wetlands: Our Vital Link Between Land and Water*. Washington, DC: Office of Wetlands, Oceans and Watersheds, EPA843-K-95-001, December 1995, 5.

20. Ibid., 8.

21. Kent, Donald M. *Applied Wetlands Science and Technology*. Boca Raton, FL: Lewis Publishers, 1994, 10.

22. U.S. Department of the Interior, U.S. Geological Survey. "Walden Pond, Massachusetts: Environmental Setting and Current Investigations," USGS Fact Sheet FS-064-98, June, 1998.

23. Hey and Philippi, 3.

24. U.S. Environmental Protection Agency, "Four Years of Progress: Meeting Our Commitment for Wetlands Reform," August 1993–August 1997, http://www.epa.gov/OWOW/wetlands/plan/4years.html, September 2001.

25. Switzer, Jacqueline Vaughn, and Gary Bryner. *Environmental Politics—Domestic and Global Dimensions*. 2nd ed. New York: St. Martin's Press, 1998, 154.

26. Ibid., 154–155.

27. Switzer and Bryner, 249; and personal communication with Dr. Ted Nelson, Manager, Navigation Development, Tennessee Valley Authority, Knoxville, Tennessee, September 18, 2001.

28. Switzer and Bryner, 250.

29. U.S. Fish & Wildlife Service, "Snail Darter," http://endangered.fws.gov/i/e/sae15.html, September 2001.

30. *Sierra Club v. Froehlke*, 534 F. 2d 1289 (8th Circuit Court, 1976).

31. *National Wildlife Federation v. Coleman,* 529 F. 2d 359 (5th Circuit Court, 1976).

32. U.S. Fish & Wildlife Service, "The Endangered Species Act of 1973," http://endangered.fws.gov/esa.html, September 2001.

33. Ibid.

34. Personal communication with Mitch Snow, U.S. Fish & Wildlife Service, May 29, 2001.

35. Switzer and Bryner, 254.

36. Nebraska Game and Parks Commission, "Sandhill Cranes," http://www.ngpc.state.ne.us/wildlife/cranes/sandhill.htm, September 2001.

37. Farrar, John and Ken Bouc. "Sandhill Cranes—Wings over the Platte." *NEBRASKAland Magazine,* Lincoln, Nebraska, no date, N-2 to N-4.

38. Ibid., N-5.

39. Ibid., N-2.

40. Ibid., N-7.

41. Outwater, Alice. *Water—A Natural History*. New York: Basic Books, 1996, 109–110; and personal communication with Dr. Glenn F. Cada, Environmental Sciences Division, Oak Ridge National Laboratory, Oak Ridge, Tennessee, August 24, 2001.

42. Personal communication with Dr. Glenn Cada, August 24, 2001.

43. Outwater, 112; and personal communication with Dr. Glenn Cada, August 24, 2001.

44. Personal communication with Dr. Glenn Cada, August 24, 2001.

45. Outwater, 113–114; and personal communication with Dr. Glenn Cada, August 24, 2001.

46. Switzer and Bryner, 235.

CHAPTER 13

THE ECONOMICS OF WATER

"When the well's dry, we know the worth of water."

Benjamin Franklin
Poor Richard's Almanac, 1746

INTRODUCTION

The economics of water is based on the allocation of the resource among different uses, based on its value. The value of water is determined by two elements: 1) demand—the utility to humans and their willingness to pay for that utility, and 2) supply—the cost of providing the resource in a certain quality, quantity, and location—which varies in different parts of the world. At some times and locations, the value of water can be very high due to scarcity, but at other times and locations it can be very low or even free due to relatively plentiful supplies. For example, have you ever had to pay to take a drink from a public water fountain? By contrast, have you ever paid for bottled water when it was readily available at no cost from a nearby water faucet? Would you ever pay more for clean drinking water than you would for a diamond ring? Leading economists are finding the economics of water to be a fascinating (and growing) field of study (Figure 13.1).

According to Falkenmark and Lindh (1), human cultures generally go through three stages of using and developing water supplies. During early cultural development phases, water is generally adequate to meet the basic needs of residents—drinking, sanitation, and other necessities such as irrigation (in dry climates) or for navigation (if in humid climates). Conflicts are minimal, and water is generally affordable to residents.

In the second phase of urban/agricultural development, water projects are constructed to divert water from lakes and rivers for use in cities or for irrigation, flood control, hydropower generation, or other economic purposes. Interbasin transfers are pursued to encourage economic

FIG. 13.1.

"I'll begin today's proceedings by saying that we have enough food and water to last us until some sort of eventual turnaround."

growth in one watershed but generally at the expense of the basin of origin. Unevenly distributed water resources are reallocated, and water markets may be implemented to ensure the marketability of developed water supplies.

In the third phase of urban/agricultural development, population growth continues, as does the need for additional supplies of water. Although water supplies have been developed for maximum beneficial economic use, environmental concerns become more pronounced. Improving the efficiency of water distribution networks becomes very expensive, and nonconventional methods may be implemented. The economic viability of such a region has become extremely dependent upon safe, reliable sources of water for a variety of uses.

Government generally plays a large role in the management of water resources. It must determine the proper allocation methods for efficient use of the finite resource, and must protect water resources from unacceptable depletion rates, environmental degradation, and pollution. Government also serves an important role in preventing monopolization of water supplies by private individuals or companies. As we'll see later in this chapter, the role of government can be that of both water provider and water regulator.

THE VALUE OF WATER

Aristotle (384–322 B.C.) held that the source of an item's value was based on its need. Without the need for an item, no exchange of goods would occur. Aristotle also distinguished between the value of an item for *use* and the value of an item for *exchange:*

> *Of everything which we possess, there are two uses; For example, a shoe is used for wear and it is also used for exchange.*
>
> Aristotle, *Politics*, Book I

The paradox of this concept was explored by Adam Smith (1723–1790) in his *Wealth of Nations* in 1776. (2) Smith's groundbreaking text discussed the famous **"water–diamond" paradox**—the fact that water is very necessary

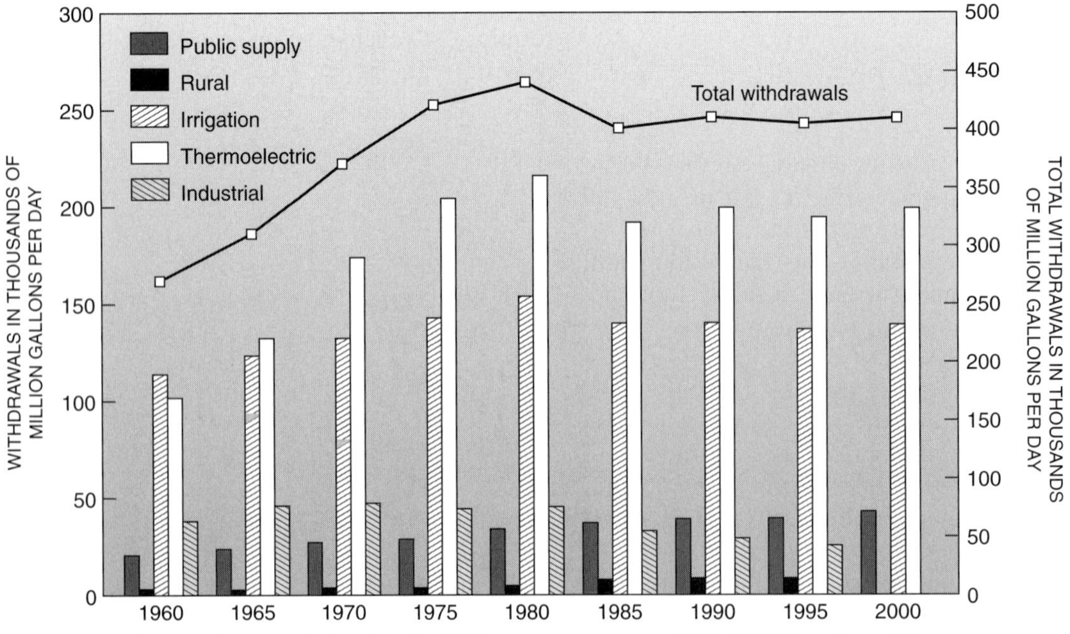

FIG. 13.2. Trends in total water withdrawals in the U.S. by water-use category, 1960–2000. Note that public supply and rural uses are slowly increasing while industrial and irrigation are lower than the peaks in 1980.

and essential to life, yet it is relatively inexpensive (and in many cases, free). Paradoxically, diamonds are relatively inessential to life, yet are extremely expensive. Smith argued that the cost of producing a commodity explained the value placed on it (a theory he called the Labor Theory of Value). Water was quite plentiful during the early stages of urban/agricultural development in the United States and Canada (and in most other regions of the world) and, in Smith's view, explained its relatively low economic value.

Today, the value of water is generally determined by its location, quantity, characteristics (physical, chemical, biological, and thermal), and use (Figure 13.2). The value does not usually include the cost of health (loss of wages and economic output due to illness caused by contaminated water), the cost of recreation (how much people will spend to travel, recreate, and vacation near a body of water), the impacts on third parties, or ecological values.

Many of the world's water shortages occur because we do not place proper economic value on water supplies. (3) Underpriced water resources lead to inefficient and wasteful practices. For example, irrigation water is sometimes subsidized by government agencies and delivered to growers at a fraction of its real value—especially if the government owns and operates the collection and conveyance system. In Mexico, irrigators pay on the average just 11 percent of the water's full cost. In Indonesia and Pakistan, the cost is about 13 percent, while in Egypt, farmers are not charged directly for any costs for irrigation water. In the United States, the Reclamation Act of 1902 authorized the Bureau of Reclamation to subsidize the development and operation of surface water irrigation projects. Many times, the repayment period for irrigators has been lengthened, and payments have been limited to the irrigators' ability to pay. (4) Groundwater irrigators generally rely on privately owned wells, so that subsidies for water are typically nonexistent. Irrigators under mutual ditch companies or irrigation districts are also not generally subsidized by local, state, or federal government agencies.

WATER AS A PUBLIC VERSUS A PRIVATE GOOD

Should water resources be viewed as a *public good* or a *private good?* A **public good** refers to the right of use of water for all people—for basic human consumption and sanitation needs, for aesthetic values, and for environmental protection. It's generally agreed that no human should be deprived of drinking water even if he or she has no ability to pay for its use. Government has the responsibility for ensuring that basic human needs are met in terms of water quantity and water quality (Figure 13.3). (5)

FIG. 13.3. Villagers at a community well use ropes tied to jugs to retrieve drinking water from the Jaspar village well in northern India. This photo was taken on June 13, 2003, and the temperature in the village was 107°F (42°C) with no rain forecast.

In the 1800s and even today, however, many view water in the United States as a **private good** to be developed, used, traded, and sold for economic productivity and financial gain. Water marketing (discussed later) is embraced in many water-short regions, and variations are being developed in more humid locations (see Chapter 14). The environmental movement of the 1960s took issue with the use of water as a private good. Through legislation and legal action, local, state, and federal water agencies were forced to reevaluate water development projects as more of a public resource issue. Environmental protection and stewardship became important components of water development projects. This inherent conflict of water as a private good and a public good continues today throughout the world.

PRIVATIZATION

Privatization is a growing practice by local governments whereby private firms are utilized to deliver public services. Whereas privatization reduces the direct costs of providing basic services such as drinking water treatment, it has certain drawbacks: it poses the potential for corruption in the pursuit of contracts with municipalities, it relies on smaller, less-experienced companies that are subject to bankruptcy, and it may involve higher user fees that could disproportionately burden lower income users. (6)

Several U.S. cities are currently served by privately owned water systems. Some of these include San Jose, California; Indianapolis, Indiana; Lexington, Kentucky; Baton Rouge, Louisiana; Chattanooga, Tennessee; Bridgeport, Connecticut; Hackensack, New Jersey; Charleston, West Virginia; St. Louis County, Missouri, and Peoria, Illinois. According to the National Association of Water Companies (NAWC), private water companies provide approximately 15 percent of all water services in the United States. This proportion has remained relatively constant since the mid-1940s. Worldwide, privatization of water supply systems includes all of the United Kingdom, Berlin, Buenos Aires, Johannesburg, Manila, and Mexico City. (7)

SIDEBAR

In Atlanta, United Water assumed operation of the city's water system in 1999 under a 20-year agreement. Excitement was high when the privatization agreement was signed with the North American subsidiary of Suez, a French multinational organization based in Paris. United Water intended to operate a 136 mgd (514,816 cmd) and a 56 mgd (211,983 cmd) water treatment plant along with 2400 miles (3862 km) of water distribution lines. It was also to provide maintenance, read water meters, and operate the monthly billing system. Operation of three wastewater treatment plants in the region was also part of the highly touted agreement that was closely watched by other municipalities and private water supply companies around the world. The agreement gave United Water $21.4 million in annual revenues, with a projected savings to the City of Atlanta of $400 million during the 20-year period. Two years of negotiations led to this innovative privatization arrangement. (8)

Unfortunately, on January 24, 2003, the City of Atlanta issued a news release from the Mayor's Office stating that the City and United Water had dissolved the 20-year agreement. (9) At issue were higher than expected operation and maintenance costs incurred by United Water and allegedly poor water service, broken water mains, and water quality issues. The failure of this agreement was a huge blow to future privatization plans in other cities.

Most water supply, treatment, and distribution systems in the United States were largely built over 100 years ago and many are now in need of repair. In addition, population growth in some areas is forcing the expansion of existing infrastructure. Expenditures for these system upgrades can be extremely high, taxing the local government's ability to respond appropriately. As a result, many water systems have experienced large backlogs of deferred maintenance and expansion. In the United States, there are over 54,000 community water systems (systems serving over 25 people). However, most of these systems serve small populations, with 85 percent of the community water systems in the country serving only 10 percent of the population using such systems. (10) In 1995, privatization of water supply utilities accounted for 14 percent of the total water revenues and about 11 percent of total water system assets in the United States. (11)

WATER AFFORDABILITY

At the 2002 World Summit on Sustainable Development in Johannesburg, South Africa, great concern was expressed about the 1.1 billion people in the world who do not have access to safe drinking water and the 2.4 billion who live without proper sanitation. The United Nations has called for a 50 percent reduction in the number of people who do not have safe drinking water by 2015. (12)

Safety, reliability, and affordability are the most important issues faced by a water service provider. However, the cost of meeting these criteria is often disproportionately paid by lower-income households. Water is used to meet economic, social, and environmental objectives, but water pricing does not always promote all three objectives.

In the early 1990s, France passed antipoverty legislation to guarantee access to water supplies, for reasons of dignity and public health and sanitation, to households at or below the poverty level. During this time and even after passage of the law, 130,000 water disconnections occurred every year in France, most generally for only one day. However, some 2000 of those cutoffs lasted more than a day. In 1998, legislation was passed to require government and private water providers to assist the poor with their payment of water bills. In 2000, further regulations held that no water disconnections could occur in France if infants or the elderly lived in the household that had not paid their water bill. (13) Similar programs are already in place in the United Kingdom, the United States, and many other countries around the world.

According to the Organization for Economic Co-operation and Development in France, the ratio of average household water charges to average household income ranges from 0.5 percent to 2.4 percent in countries such as the United States, Italy, Poland, Mexico, Austria, France, Scotland, Japan, and Portugal. The United States was at the low end at 0.5 percent, while Poland was highest with 2.4 percent. (14) What does this tell us? First, countries in need of infrastructure rehabilitation generally require higher water fees than do countries with established and relatively sound water supply, delivery, and treatment systems. Second, these figures may be a bit misleading because more-developed countries may be using water supplies for uses such as lawn irrigation, fountains, pools, and other nonessential purposes. Elimination of these costs from the average household water charges would increase the spread between the ratios of less-developed and more-developed countries.

Some water agencies have increased the cost of water to residential areas to reduce consumption. However, this approach can be quite controversial when issues of public health, low-income households, and equity over the use of a public good are considered. In water-short areas, large amounts of water are often used for irrigation. Water demand for irrigation water is generally inelastic at low water prices (little economic incentive to reduce water use) and elastic at high water prices (great economic incentive to reduce water use). This elasticity is greatly dependent on annual weather conditions and on industrial and urban development in a region. During times of drought, competition for limited water supplies can drive up the price of water significantly if urban areas or industry experience water shortages.

SIDEBAR

The price elasticity of demand is the responsiveness of the quantity of an item to the market price of an item. Generally, as the market price of an item increases, the demand for the item will usually decrease. This can be expressed as:

$$Ed = Pdem \div Pmkt$$

where Ed = price elasticity of demand
$Pdem$ = percentage change in quantity demanded
$Pmkt$ = percentage change in price

Approximately two-thirds of urban water use in the United States is attributable to in-house use, and most of that is utilized by toilets (36%), the bath and shower (28%), and the washing machine (20%). (15) For that reason, government water utilities and politicians generally oppose using water pricing as an incentive to conserve water. (Remember the public use value of water—that everyone has the "right" to use water for basic human consumption and sanitation needs discussed earlier.) Because of this basic human need for water and the acknowledgment that it is a basic human right, technology has been promoted as a means to conserve water through promotion of ultra-low-flow toilets, restricted shower heads, and water-efficient washing machines. Between 1989 and 1992, 17 states in the United States considered or passed laws to mandate the use of low-flow toilets. Municipalities such as Denver, San Diego, Los Angeles, San Francisco, and Washington, D.C. have also adopted such requirements. In 1992, the U.S. Congress passed the Energy Policy Act, which required all new and remodeled commercial buildings and residential homes to have low-flow toilets and restricted shower heads. (16) Communities in most western states and provinces also have ordinances limiting lawn and landscape watering during periods of drought, although some have adopted outdoor watering restrictions on an annual basis.

WATER MARKETING

Water marketing can be defined as the sale or lease of water rights in a market-based system. In water-short regions, such as the western United States and southeastern Australia, water markets are very active, with cities and industry paying large sums of money for water. Water values increase based on the location and quality of water. Competition for scarce water supplies can become very intense, particularly in urban/industrial markets. For example, some irrigators are confronted with the economic reality of temporarily leasing or permanently selling valuable water rights to a city. This might prevent foreclosure but would force a drastic change in lifestyle that had been passed on by generations of farmers. Leasing or selling water, though economically reasonable, destroys the agricultural economy and lifestyle of a rural area. Many residents in such areas place a higher value on the rural lifestyle and rural economic systems than on the high economic return of selling water to cities, industry, or developers.

Surface Water Marketing In the United States, water marketing is prominent in Colorado, Utah, Washington, California, Nevada, and New Mexico, particularly between individual irrigators, between irrigators and municipalities, and between irrigators and industry. In California, water resource administration is dominated by large irrigation companies and municipal providers, and has active intraorganizational markets. (17) Price is generally set by supply and demand market forces.

Two obstacles faced by water marketing in the western United States include the historical paradigm that water is a *public* resource to be equitably shared and allocated by society and that water is a *local* resource over which local communities should have a special claim. These two paradigms are influencing state water law and federal water regulations, private water ownership systems, and societal expectations. (18)

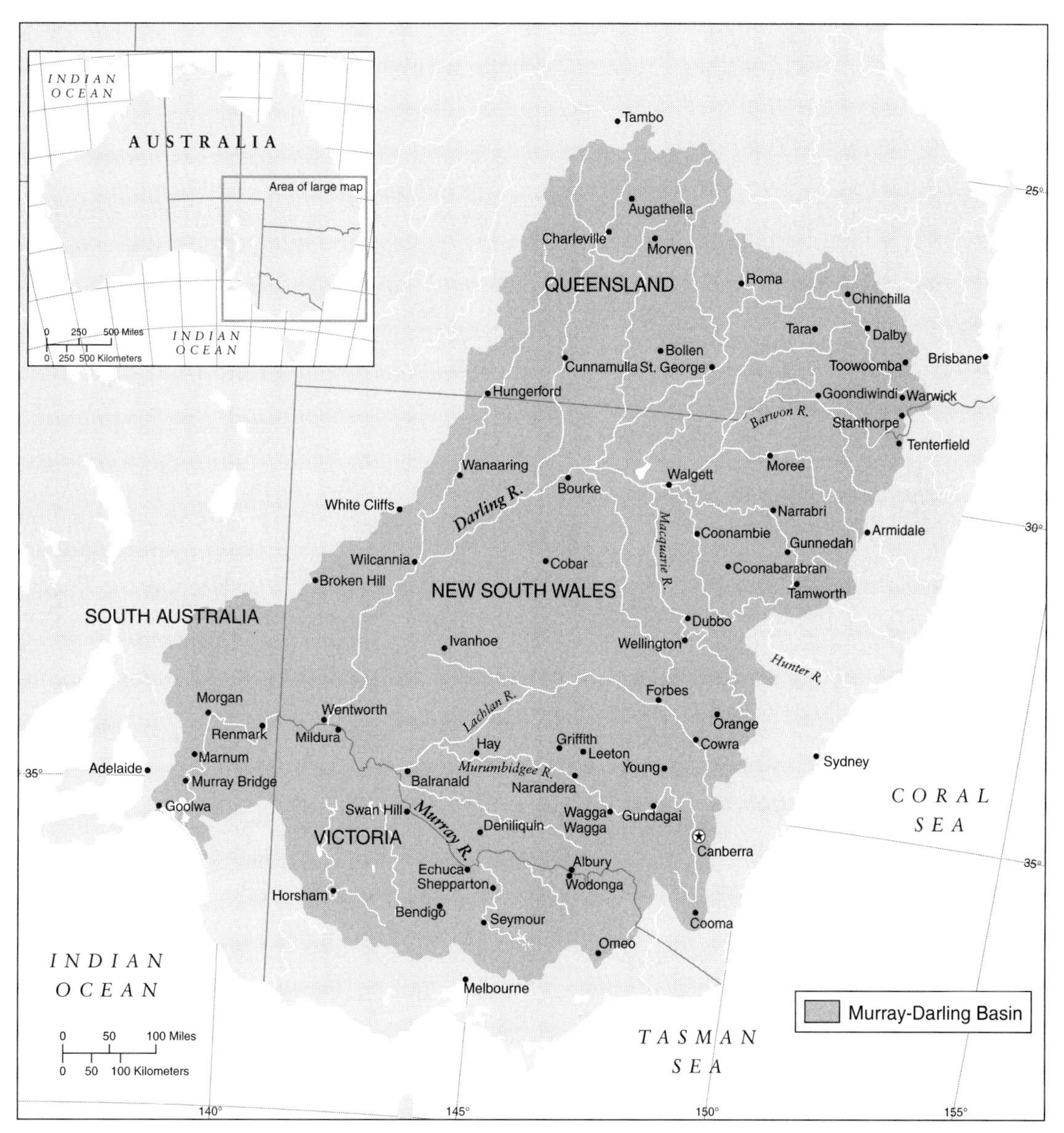

FIG. 13.4. The southeastern Murray-Darling River basin receives an average annual precipitation of 19 inches (48 cm). Water management discussions began as early as 1863. The River Murray Commission was formed in 1917.

A CLOSER LOOK

In Australia, the beginnings of the Murray-Darling Basin Commission were established in 1917 with the signing of a federal and state agreement to construct storage projects, and a series of locks and weirs, to manage the Murray River in southeastern Australia. The Murray-Darling River Basin drains the states of New South Wales and Victoria, as well as parts of South Australia and Queensland (Figure 13.4). The entire region includes approximately 50 percent of the country's cropland and about 1.8 million people, or roughly 10 percent of its population. Irrigation is extensive in the region and accounts for about 75 percent of Australia's total. Some small towns obtain their water from the

Murray River, and Adelaide, the capital of South Australia, depends on the river for over half of its water supply.

The agreement sets up a sharing structure between New South Wales, Victoria, and South Australia. Water is distributed by the commission based on the ownership and control of the three states. The states are both shareholders and customers of the water supplies. Historically, irrigation water costs were subsidized by the state to encourage settlement and cultivation of sparsely populated areas. However, in recent years state governments have phased in price increases to cover the cost of delivering water. The Victorian government has a policy of collecting full operation costs; previously, the government had a policy of subsidizing its water charges by 30 percent. In 1995 the New South Wales government implemented an additional "water management charge" of $1.35 per megaliter ($1.67 per acre-foot, or per 325,851 gallons per day) to preserve instream flows. (This greatly angered irrigators in the region.) (19)

Until the 1980s, water use for irrigation in the Murray-Darling River Basin was unlimited for irrigators, and water consumption increased dramatically. Shortages, salination of soils, and waterlogging became serious problems. In 1981, volumetric water allocations were introduced on the Murray River, and the temporary trading of water rights within irrigation districts was started. By the mid-1990s, water rights were trading for as much as $70 per megaliter ($86 per acre-foot, or per 325,851 gallons per day) on an annual basis. Water marketing evolved incrementally in response to demands by irrigators, and reflected existing drought conditions in much of the area. (20)

In Chapter 10 we discussed the formation of private mutual ditch companies and irrigation districts. The district concept was promoted to allow a local government entity to operate and manage a federally funded irrigation project. Local communities did not want the federal government to be involved in the daily operation of a local irrigation project. Although government subsidies helped with the development and operation of these districts, some question the equity of individual irrigators profiting from the sale of water rights in irrigation systems developed through federal subsidies. Irrigation districts (and member irrigators) control a vast amount of marketable water in the West, particularly in Washington and California.

Transferring water rights between watersheds continues to be of grave concern to many local communities. John Wesley Powell argued in the late 1800s that removing water supplies from one watershed for the benefit of another would lead to increased conflict between regions over a common water management regime. In reality, transbasin water diversions have led to disagreements, political controversy, and lawsuits. In these situations, water marketability often meets insurmountable legal and political challenges waged by local communities, water agencies, and politicians.

Federalization of water rights was a growing movement in the United States during the late 1900s. Those opposed to federal control of water saw federal laws, such as the Wild & Scenic Rivers Act of 1968, the National Environmental Policy Act of 1970, and the Endangered Species Act of 1973, as attempts to gain control of private property rights. Elwood Mead, the first state engineer of Wyoming, argued during the early 1900s that water marketing encouraged appropriators to speculate in water—claiming much more water than was realistically needed—thereby created water monopolies. He helped draft the original Wyoming Water Code and warned against treating water as a property right; Mead stated that water in a democracy

> *"belong(s) to the people, and ought forever to be kept as a public property, for the benefit of all who use them, and for them alone, such use to be under public supervision and control."* (21)

Groundwater Marketing Groundwater marketing is not a common feature of groundwater allocation in most states in the United States. In the wetter climate of the east coast, the riparian doctrine generally applies and allows a landowner to withdraw groundwater with few limitations. However, when well-to-well interference occurs, the Reasonable Use Principle generally applies to prohibit the unreasonable use of groundwater if it negatively affects a neighboring well.

In the arid West, some states follow the strict Doctrine of Prior Appropriation for groundwater. In this case, the "first in time, first in right" concept is followed, and well-to-well interference may not be prohibited. However, some states have well-spacing requirements that reduce the possibility of such well-pumping interference. In Texas, the Rule of Capture is followed and allows a well owner to sell groundwater from beneath his or her land. This is similar to Spanish law which allowed a landowner to dig a well and remove groundwater with total disregard as to the effect on neighboring wells.

In Arizona, some cities have turned to "water ranching." State law can make it difficult to buy groundwater rights separately from the land in the Grand Canyon state, so entire farms and ranches are acquired for the use of the attached water rights. The land is then "dried up" by permanently removing irrigation water and returning the land to dryland cropping or native vegetation. With loss of tax revenues a major problem in Arizona, a state law was passed in 1991 that requires cities to pay sums of money equal to the property taxes (payments in lieu of taxes) that would have been assessed on the property. In some areas where groundwater mining has occurred, a "replenishment tax" may be required from those who have overdrawn their groundwater accounts. (22)

Under the Doctrine of Prior Appropriation, it would seem logical that a senior groundwater right should have value. In some states, however, such as in Colorado, use of some types of groundwater (particularly tributary groundwater) is legally limited to a prescribed location and acreage limit. This requirement greatly limits the marketability of tributary groundwater in Colorado and limits its economic value. On the other hand, nontributary groundwater in Colorado is not restricted to a location of use and can be sold and moved to a willing buyer at an agreed-upon price.

Nebraska has a quota system on groundwater use in limited locations, such as in the Upper Republican Natural Resources District in southwest Nebraska, but generally a quota, or allocation of groundwater, cannot be sold. An exception to this restriction is in Arizona, where the state directly manages most overdrafted aquifers and allows the marketing of groundwater between well pumpers of a use class or across uses. In addition, California allows the marketing of groundwater rights among pumpers in an adjudicated groundwater basin. Restrictions are imposed on the transferability of use rights or quotas in both California and Arizona. (23)

Many states prohibit the sale and export of water, despite a 1982 U.S. Supreme Court ruling that invalidated Nebraska's prohibition of exporting groundwater. In *Sporhase* v. *Nebraska*, the Court ruled that the Commerce Clause allows the export of water resources between states. (24) This ruling caused concerns that water could be transferred from the Great Lakes to the western United States. To date, however, no such schemes have occurred.

In Georgia, controversy over surface and groundwater depletions led to the first-of-its-kind water auction where irrigators were paid not to pump. The first auction was held in 2001 (described in Chapter 14), and irrigators were paid $136 per acre (0.4 ha) to not irrigate their land. Under this innovative program, 33,000 acres (13,355 ha) of land were not irrigated, and surface water depletions in the Flint River were reduced by approximately 200 cfs (129.3 million gallons per day, or 489,315 cmd) at a cost of $4.5 million. In 2002, approximately 40,000 acres (16,187 ha) of land were taken out of irrigation at a cost of $127 per acre (0.4 ha) and a total cost of $5.2 million (Figure 13.5).

In Texas, entrepreneur T. Boone Pickens is planning to sell groundwater from beneath land that he owns, overlying the Ogallala Aquifer in the Texas Panhandle, to growing cities to the south. Prices would be in the $800–$1400 per acre-foot (325,851 gal/day, or 1233 cmd) range to deliver groundwater to Dallas, El Paso, and San Antonio.

In a statement to Congress on June 4, 2003, Pickens stated that neither the Canadian River Municipal Water Authority nor the city of Amarillo—the only two major water markets in

FIG. 13.5. Potential bidders in Dawson, Georgia (above, left) receive water auction rules and instructions from state officials. Dawson is located in the heart of the peanut belt in southwest Georgia. Officials at Auction Central monitor offers coming in (above, right). Success was achieved as irrigators were paid $135.70 per acre (325,851 gal/day, or 1233 cmd) not to irrigate in 2002. This allowed more surface water to remain in the Flint River system in Georgia and Florida for other uses.

the Texas Panhandle—needed the groundwater beneath his property (held by the Mesa Water Group). He stated that Mesa was prepared to deliver 150,000 acre-feet (48.9 billion gal or 185.0 million m^3) of water per year to the Dallas-Fort Worth Metroplex, San Antonio, or El Paso by January 1, 2009 at competitive costs. He referred to his panhandle groundwater sources as "low-hanging fruit for the purchasers" and a huge benefit to other regions of the state. (25)

WATER BANKING

Water banking utilizes a local clearinghouse to hold water rights available for sale or lease under a set of locally defined or state-mandated rules. Water banking is being developed to diminish the impacts of permanently removing agricultural water from farms and ranches.

During the drought of the 1990s in California, a water bank was created through which the California Department of Water Resources leased water for $125 per acre-foot from irrigators. At the time, the price was less than $50 per acre-foot per year. Irrigators responded very favorably to this offer, and the water bank acquired 820,000 acre-feet (267.2 billion gallons, or 1.0 billion cubic meters) of water for other uses. (26) This unique economic water conservation strategy required the careful balance of protecting property rights and maintaining political balance with local communities.

In 2001, the Arizona Water Banking Agreement was approved allowing the State of Nevada to store unused and surplus Colorado River water in aquifers in the neighboring state of Arizona for later use. Nevada must pay for any undiverted river water stored underground in Arizona, but will then receive a "credit" in the Arizona Water Bank. Under this agreement, up to 1.2 million acre-feet (391.0 billion gallons, or 1.5 billion cubic meters) of Nevada's Colorado River water can be stored underground in Arizona. The State of Nevada anticipates that it will begin using some of these credits after 2015. Cost to store this water underground is in the range of $200 per acre-foot (per 325,851 gal, or 1233 cubic meters). (27)

Water demands will continue to increase if water prices remain low, particularly under government subsidy programs. Waste of water, or inefficient use of the resource, is simply a response to low water prices and not necessarily due to a lack of education or unwillingness to protect the resource. Water running off the end of an irrigated field, or down a gutter from excess or inefficient lawn irrigation, is an economic response to plentiful and cheap water supplies. A study in 1992 estimated that the urban demands of the western United States would be met for the next 25 years if only 5 percent of the water used by agriculture was transferred to municipal uses. (28) This transfer could include the construction of pipelines and the use of nonrenewable groundwater supplies. While this prospect is appealing to some, it does not take into account the loss of agricultural jobs, open space, wildlife habitat, and the destruction of rural communities. Water banking is a new concept being developed to diminish the impact of permanently removing agricultural water from farms and ranches.

POLLUTION FEES AND CREDITS

Pollution fees and credits are a common method of pollution control in Europe. In Germany, the Federal Water Act and Effluent Charge Law was adopted in 1976 to maintain or improve the water quality of the Ruhr River. This was not a new concept in Germany; as early as 1904, effluent charge systems were in place in some German states. During the debate over the 1976 legislation, local governments expressed concern over the level of control by the federal and state (*länder*) governments. Nonetheless, the effluent fees were set high enough to create market-like incentives to develop pollution abatement measures. (29)

In 1989, the Tar-Pamlico Association was formed in North Carolina and became the first community of point source dischargers to create a water pollution brokerage association in North America. The group facilitates trade among dischargers and nonmember farmers to reduce phosphorus and nitrogen discharges into the Pamlico estuary and the Atlantic Ocean. The Association is given a discharge allowance and members are required to make a financial contribution into a nonpoint source pollution fund. These monies are used to implement land use practices (such as improved farming practices) that reduce polluted surface water runoff and groundwater percolation. Any member of the association can reduce nutrients on its own and obtain a credit, trade credits within the group, or pay a $56 per kilogram ($25 per pound) effluent fee. Membership in the Tar-Pamlico Association is voluntary, whereas the German method noted previously is mandatory. (30)

In response to growing concerns over increased salinity levels in the Murray River in Australia as discussed earlier, salinity credits were introduced to allow trading of such credits for salinity mitigation practices. In 1994, a privately owned paper mill leased salinity credits from the state to offset the impact of a new manufacturing process at its pulp factory. The Murray-Darling Basin Commission has evolved into an active manager of the Murray River and its environment and as an agency that establishes a common water market in the watershed. This role comes at the expense of the local governments' reduced autonomy or sovereignty in water resources issues, and has created a great deal of controversy.

In some situations, subsidies are used to allow municipalities, irrigation organizations, and other water utilities to provide low-cost water supplies. This encourages equity among water users, but it can also lead to little or no incentives to conserve water resources. Full-cost pricing of water supplies would greatly encourage conservation but could be devastating to local economies and lower-income groups.

ENVIRONMENTAL VALUES

In recent times, water use and development have led to concerns over environmental damage. Natural cycles of groundwater recharge, surface water runoff, and the preservation of ecologically sensitive areas are altered and even destroyed by new, and existing, water resources projects. In addition, the development and consumption of limited water supplies has an intergenerational component. Water use today should not unfairly deprive future generations of the use of adequate and nonpolluted water supplies—both consumptive and nonconsumptive uses—for human needs as well as ecological protection and enhancement.

The nonconsumptive use of water can take many forms. Recreational activities such as boating, fishing, and swimming, nutrient cycling, pollution dilution, habitat for wildlife, and the presence of water for aesthetic purposes, all have benefits and economic value. Placing an economic value on these nonconsumptive uses remains a difficult and at times controversial duty of society. Adding to this difficulty is the determination of appropriate values when volume, depth, flow, and quality vary. Regardless of these problems, to be properly used and protected, the economic value of water must adequately reflect its true worth to society.

CHAPTER SUMMARY

The economics of water means that water will be allocated among competing uses according to the highest valued use, measured in willingness to pay. Economists know that the first units of water consumed are precious, and greater than diamonds. Fortunately, most of us live in circumstances where the cost of supplying a great many units (far in excess of primary needs) means that we don't have to pay the full utility value and, as a result, reap huge consumer surpluses. (31)

QUESTIONS FOR DISCUSSION

1. What are three roles that government plays in the management of water resources?
2. What are some of the properties or characteristics that determine the value of water?
3. Explain the difference between viewing water as a public good or as a private good.
4. Do you believe that privatization of water delivery systems is a positive or a negative trend?
5. Do you believe that water marketing is a positive or a negative economic system?
6. Discuss Adam Smith's water-diamond paradox.

KEY WORDS TO REMEMBER

private good p. 384
privatization p. 384
public good p. 383
water banking p. 390
water marketing p. 386
water-diamond paradox p. 382

SUGGESTED RESOURCES FOR FURTHER STUDY

READINGS

Green, Colin, *Handbook of Water Economics: Principles and Practice*. Hoboken, NJ: John Wiley & Sons, 2003.

Postel, Sandra, *Last Oasis*. New York: W.W. Norton & Company, 1997.

Renzetti, Steven, *The Economics of Water Demands*. Boston: Kluwer Academic Publishers, 2002.

Privatization of Water Services in the United States. Washington, DC: Joint Center for Political Studies Press, 2002.

WEBSITES

"Murray-Darling Basin Initiative," Murray-Darling Basin Commission, http://www.mdbc.gov.au/, April 2004.

"Water Use in the United States," U.S. Geological Survey, http://water.usgs.gov/watuse/, April 2004.

REFERENCES

1. Falkenmark, Malin, and Gunnar Lindh, "Water and Economic Development," Chapter 7 in *Water in Crisis: A Guide to the World's Fresh Water Resources,* edited by Peter H. Gleick, Pacific Institute for Studies in Development, Environment, and Security, Stockholm Environment Institute, Oxford University Press, Oxford, United Kingdom, 1993, 80.

2. The full title of the work is *An Inquiry into the Nature and Causes of the Wealth of Nations.*

3. Postel, Sandra, *Last Oasis,* W.W. Norton, New York, 1997, 166.

4. Ibid., 167–168.

5. *Social Issues in the Provision and Pricing of Water Services,* Organization for Economic Co-operation and Development, Paris, France, 2003, 19.

6. Suggs, Robert E., *Minorities and Privatization,* Joint Center for Political Studies Press, Washington, DC, 1989, 16–17.

7. *Privatization of Water Services in the United States,* National Research Council, National Academy Press, Washington, DC, 2002, 3.

8. Ibid., 64.

9. "City of Atlanta and United Water Announce Amicable Dissolution of Twenty-Year Water Contract," Mayor's Office of Communications, City of Atlanta, January 24, 2003, http://www.ci.atlanta.ga.us/homepage/PressReleases/UnitedWater1-24-03.htm, October 2003.

10. *Privatization of Water Services in the United States,* 2–3.

11. Ibid., 3.

12. *Social Issues in the Provision and Pricing of Water Services,* 17.

13. Ibid., 62–63.

14. Ibid., 34.

15. *Current Issues in the Economics of Water Resources Management,* edited by Panos Pashardes et al., Kluwer Academic Publishers, Dordrecht, The Netherlands, 2002, 125.

16. Renzetti, Steven, *The Economics of Water Demands,* Kluwer Academic Publishers, Boston, MA, 2002, 10.

17. Thompson, Barton H., Jr., "Institutional Perspectives on Water Policy and Markets," *California Law Review* 81(3), 1993: 671–764.

18. *Water Marketing—The Next Generation,* edited by Terry L. Anderson and Peter J. Hill, Rowman & Littlefield, Lanham, MD, 1997, 2.

19. *Water Marketing,* 130–133.

20. Ibid., 135–137.

21. Mead, Elwood, *Irrigation Institutions: A Discussion of the Economic and Legal Questions Created by the Growth of Irrigated Acreage in the West,* Macmillan Co., New York, 1903.

22. Postel, 174.

23. Commission on Geosciences, Environment and Resources, *Sharing the Fish: Toward a National Policy on Individual Fishing Quotas,* National Academies Press, Washington, DC, 1999, 53.

24. 458 U.S. 941 (1982). *Sporhase v. Nebraska.*

25. House Transportation and Infrastructure, Subcommittee on Water and Resources and Environment, U.S. Congress, June 4, 2003, http://www.house.gov/transportation/water/06-04-03/pickens.html, October 2003.

26. *Water Marketing,* xiii.

27. Southern Nevada Water Authority, "Arizona Groundwater Bank," http://www.snwa.com/html/wr_az_banking.html, October 2003, and Keith Rogers, "Report Outlines LV Valley's Sources of Water after 2016," *Las Vegas Review-Journal,* Las Vegas, NV, March 22, 2002.

28. Spencer, Leslie, "Water: The West's Most Misallocated Resource," *Forbes* (April 27, 1992): 68–74.

29. Brown, Gardner M., and Ralph W. Johnson, "Pollution Control by Effluent Charges: It Works in the Federal Republic of Germany, Why Not in the U.S.?" *Natural Resources Journal* 24 (October 1984): 929–966.

30. *Water Marketing,* 154–156.

31. Personal communication with Dr. Marie Livingston, Department of Economics, University of Northern Colorado, Greeley, Colorado, January 20, 2004.

CHAPTER 14

WATER USE CONFLICTS

Isaac left . . . and made the Wadi Gerar his regular campsite. But when Isaac's servants dug in the wadi and reached spring water in their well, the shepherds of Gerar quarreled with Isaac's servants, saying, "The water belongs to us!" . . . Then they dug another well, and they quarreled over that one, too. . . . When he had moved on from there, he dug still another well; but over this one they did not quarrel.

Genesis 26:17–22

(This first documented water confrontation was settled when Isaac agreed to dig his well at a different location.)

Water and *conflict* are two words that have often been linked in recent years. Unfortunately, fighting over water is nothing new and has been a cause of dissension since biblical times. Water shortages, drought, polluted water supplies, and environmental degradation continue to be major problems during the twentieth century. The next 100 years will inevitably witness continued conflict over water issues between states, provinces, territories, and foreign countries. The Middle East will most likely be at the epicenter of such disputes, although population growth and declining water supplies around the world will create serious water shortages in other regions as well.

REASONS FOR WATER USE CONFLICTS

Population growth will be a root cause of water conflict in the future (see Figure 14.1). The demand for water around the world increased 900 percent during the twentieth century (1), though it is expected to moderate substantially by 2025 as a result of increases in water efficiency, improvement and more widespread adoption of recycled water, and shifts in energy technologies. (2) Drinking water quality, on the other hand, is already inadequate in many parts of Africa, the Middle East, Mexico, and India (usually due to abject poverty), whereas shortages of irrigation water for food production will become more common in India, China, the Middle East, and in some regions of the United States in the twenty-first century.

Exploding urban growth in China, particularly around Beijing, is taking scarce water supplies from rural villages and farms downstream of the city. This same scenario currently exists along rivers downstream of Atlanta, Georgia; Denver, Colorado; and many other U.S. cities. The potential for increased conflict over inadequate water supplies is very real around the world.

Globally, 47 percent of all land lies within river basins that encompass multiple countries (Table 14.1). A total of 214 river basins are multinational, with over 2 billion people dependent on international cooperation to share water supplies and to maintain adequate water quality. Some of these rivers are governed by international water treaties, but many remain open to dispute. (3)

Poor water quality, deriving from inadequate infrastructure in less-industrialized countries,

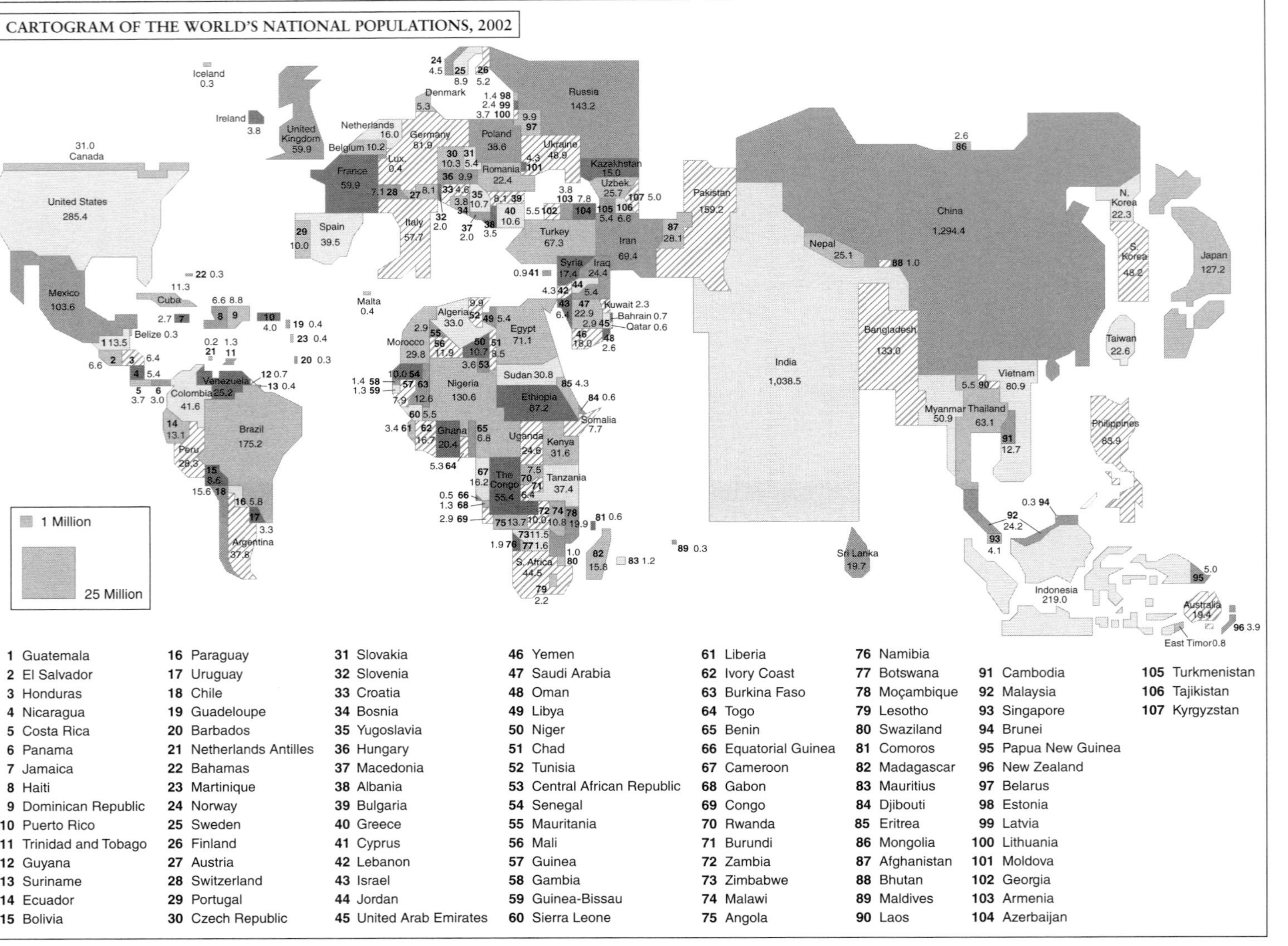

1 Guatemala	**16** Paraguay	**31** Slovakia	**46** Yemen	**61** Liberia	**76** Namibia	**91** Cambodia
2 El Salvador	**17** Uruguay	**32** Slovenia	**47** Saudi Arabia	**62** Ivory Coast	**77** Botswana	**92** Malaysia
3 Honduras	**18** Chile	**33** Croatia	**48** Oman	**63** Burkina Faso	**78** Moçambique	**93** Singapore
4 Nicaragua	**19** Guadeloupe	**34** Bosnia	**49** Libya	**64** Togo	**79** Lesotho	**94** Brunei
5 Costa Rica	**20** Barbados	**35** Yugoslavia	**50** Niger	**65** Benin	**80** Swaziland	**95** Papua New Guinea
6 Panama	**21** Netherlands Antilles	**36** Hungary	**51** Chad	**66** Equatorial Guinea	**81** Comoros	**96** New Zealand
7 Jamaica	**22** Bahamas	**37** Macedonia	**52** Tunisia	**67** Cameroon	**82** Madagascar	**97** Belarus
8 Haiti	**23** Martinique	**38** Albania	**53** Central African Republic	**68** Gabon	**83** Mauritius	**98** Estonia
9 Dominican Republic	**24** Norway	**39** Bulgaria	**54** Senegal	**69** Congo	**84** Djibouti	**99** Latvia
10 Puerto Rico	**25** Sweden	**40** Greece	**55** Mauritania	**70** Rwanda	**85** Eritrea	**100** Lithuania
11 Trinidad and Tobago	**26** Finland	**41** Cyprus	**56** Mali	**71** Burundi	**86** Mongolia	**101** Moldova
12 Guyana	**27** Austria	**42** Lebanon	**57** Guinea	**72** Zambia	**87** Afghanistan	**102** Georgia
13 Suriname	**28** Switzerland	**43** Israel	**58** Gambia	**73** Zimbabwe	**88** Bhutan	**103** Armenia
14 Ecuador	**29** Portugal	**44** Jordan	**59** Guinea-Bissau	**74** Malawi	**89** Maldives	**104** Azerbaijan
15 Bolivia	**30** Czech Republic	**45** United Arab Emirates	**60** Sierra Leone	**75** Angola	**90** Laos	**105** Turkmenistan
						106 Tajikistan
						107 Kyrgyzstan

FIG. 14.1. Cartogram of the world's national populations, 2002. Note how India and China dominate this figure. *Geography: Realms, Regions, and Concepts,* 10th Edition, by Harm de Blij and Peter O. Muller. Copyright © 2002 H. J. de Blij and John Wiley & Sons, Inc. This map was originally produced in color. Adapted and reprinted by permission of John Wiley & Sons, Inc.

TABLE 14.1 Rivers and Lakes with Five or More Nations Forming Part of the Drainage Basin

Danube	17	Volta	6
Congo	11	Aral Sea	6
Niger	11	Ganges/ Brahmaputra/ Meghna	6
Nile	10		
Rhine	9		
Zambezi	9	Jordan	6
Amazon	8	Mekong	6
Lake Chad	8	Tigris/Euphrates/ Shatt al Arab	6
Tarim	7		

Source: Adapted from A. T. Wolf, J. A. Natharius, J. J. Danielson, B. S. Ward, and J. Pender, "International River Basins of the World," *International Journal of Water Resources Development* 15, No. 4 (December 1999), pp. 387–427. Reprinted by permission of Taylor & Francis Ltd., http://www.tandf.co.uk/journals.

drought, and population growth, will create intense pressure on usable water supplies around the world. The conflict that results may be in the form of personal confrontation, litigation, legislative changes to longstanding water allocation law, or international hostility. The form of the battle will vary by location, existing water law, the degree of the water shortage, and the desire of water users and governments to reach consensus.

The four case studies that follow present a wide range of water conflict that is presently occurring in Texas, the southeastern United States, California, and the Middle East. As you read through the examples, notice the similarities and differences of situations and geographic settings. What issues are at the center of the water conflict? Is population growth a major issue in all four case studies? Why is cooperation so difficult? Are there any long-term solutions to these conflicts?

THE TEXAS PANHANDLE

In the Panhandle region of north Texas, irrigated agriculture has relied on groundwater pumped from the Ogallala Aquifer (also called the High Plains Aquifer) for over half a century. This extensive groundwater resource extends from South Dakota in the north to Texas in the south and provides supplies for almost 11,000 square miles (28,490 km^2) of irrigated ground. (4) Because natural recharge is less than 1 inch (2.5 cm) per year across much of the region, extraction of groundwater, for the most part, permanently reduces the amount available for future users. This process of removing groundwater from an aquifer at a faster rate than can be naturally or artificially recharged is called *groundwater mining,* or groundwater overdraft (see Chapter 4).

The development of the center pivot irrigation system in the 1940s led to heavy pumping of the Ogallala, and by 1990 almost 25 percent of the aquifer had been economically depleted in Texas. This mining of groundwater led to the abandonment of nearly one million acres (404,700 ha) of irrigated farm ground in the state between 1974 and 1989. (5) Fortunately, rates of groundwater declines slowed regionally somewhat during the 1990s, although the decline in Texas was over 38 feet (12 m) between 1980 and 1995, according to the U.S. Geological Survey. (6)

In response to the statewide drought of 1996 and growing concerns regarding other water resource issues, the Texas Legislature passed Senate Bill 1 (SB1) in 1997. SB1 is a statewide water planning law to determine the best use of water in the state, including the Ogallala Aquifer. The new law established regional planning groups to develop long-range plans that were reviewed extensively by the public. In addition, the legislation directed the Texas Natural Resource Conservation Commission (TNRCC) to develop computerized water availability models (WAMs) for the six major river basins in the state. The purpose of the models is to provide long-range planning tools to evaluate various water management options across the state. (7)

Within this setting of groundwater depletions, drought, and statewide water planning, entrepreneur T. Boone Pickens developed plans with

OMB No. 1545-0520

Form-10318 (11-96)

Department of the Treasury - Internal Revenue Service

DEDUCTION FOR DEPLETION ON GROUND WATER USED FOR IRRIGATION

Taxpayer's Name | Tax Year Ended

Address | (County) | (State) | (ZIP Code)

IMPORTANT: If your farm was acquired in more than one acquisition, prepare a separate depletion schedule for each acquisition (tract) making up your farm.

1. (a) Number of acres ________ : Show only number of acres in this acquisition (tract).
   (b) Give compete legal description of this tract and locate it in the spaces provided on the back of this form.
2. Is this farm within and part of a water conservation district? ☐ Yes ☐ No
   If Yes, please give your water conservation district number ____________ .
3. (a) Date of acquisition: Month ______________ Year __________ .
   (b) How did you acquire this tract? ☐ Purchase ☐ Inheritance ☐ Gift ☐ Exchange
   (c) If acquired by gift, show date acquired by donor ______________________ .
4. Basis at time of acquisition .... $ __________
   (a) If by purchase, your purchase price.
   (b) If by inheritance, the fair market value on that date as shown by Federal estate tax return, if filed; inheritance tax return; or value shown if valued under Section 2032A.
   (c) If by gift, donor's basis plus gift tax paid, but total not to exceed fair market value at time of gift.
   (d) If by exchange, give details on back.
5. Value of improvements, including residence, at time of acquisition .... $ __________
6. Basis attributable to land and water (line 4 less line 5) .... $ __________
7. Basis attributable to land and water per acre .... $ __________
8. Portion of basis attributable to ground water per acre .... $ __________
   (a) You should use the percent of value attributable to ground water as shown in the guideline table for that area approved by the District Director. However, that value should not exceed the upper limit for water, nor should the balance attributable to land be less than the lower limit for land as shown in the guideline table. (Show percentage used __________ %; upper limit for water $ __________ ; lower limit for land $ __________ .)
   (b) If you computed the amount on line 8 by some other method, check here and explain the basis of your computation on back. (See Note 1 on back of form.)
9. Allowable decline in water table under this tract for this year ________ ft. Show source of information: ______________________ . (If there was any increase in the thickness of the water table in a previous year, see Note 2 on back of form.)
10. (a) Saturated thickness of water formation under this tract at time of acquisition __________ ft.
    Show how this was determined: ______________________ .
    (b) Number of feet of ground water used since this tract was acquired, as shown on saturated thickness and water decline maps: __________ ft.
11. Water depletion allowance per acre in this year. To determine this, divide decline in water table (line 9) by saturated thickness (line 10(a)) and multiply result by basis of ground water per acre (line 8).
    (Line 9) / (Line 10(a)) ________ ft. X $ __________ (line 8) = depletion per acre .... $ __________
12. (a) Your water depletion allowance for this year .... $ __________
    To determine this, multiply acres of this tract __________ line 1(a) by water depletion per acre $ __________ line 11.
    (b) Did you acquire or dispose of this farm in this tax year? ☐ Yes ☐ No. If Yes, show date __________ and see Note 3 on back of form.

This form is for computing cost depletion deductions by taxpayers who extract ground water from the Ogallala formation for irrigation purposes. (See Revenue Procedure 66-11, Revenue Ruling 65-296 and Revenue Ruling 82-214.)

(over)

Form -10318 (11-96)
Previously issued as SWR E-665 (3-94) Catalog Number 23167Y

FIG. 14.2. Irrigators that rely on nonrenewable groundwater supplies in some regions of the United States, particularly from the Ogallala Aquifer, are entitled to claim a deduction for groundwater depletion on their annual income tax payments.

Mesa Water, Inc., to transport groundwater from the Ogallala Aquifer in the Texas Panhandle to water-short cities to the south. Since Texas utilizes the "Law of Capture" for groundwater (discussed in Chapter 8), a landowner can capture and use all groundwater found directly beneath his or her land for any purpose. Mesa Water, Inc., developed plans to acquire 200,000 acre-feet (65.2 billion gal, or 246.7 million m^3) of groundwater from the Ogallala Aquifer for delivery to Texas cities such as Dallas, Fort Worth, El Paso, San Antonio, Midland, Odessa, and Lubbock. Water would be delivered in a 108-inch (274 cm) diameter pipeline, with that cost being paid by the participating cities. The cost of the pipeline to San Antonio alone is estimated at $2.5 billion. (8)

Opponents of the project argued that nonrenewable groundwater supplies would be lost forever and that local economies of north Texas would be devastated. A portion of the 200,000 acre-feet (65.2 billion gal, or 246.7 million m^3) of water supply developed already existed beneath a 26,000-acre (10,522 ha) ranch in

A CLOSER LOOK

Groundwater depletion has been a concern in north Texas for years. In 1965, several Panhandle farmers brought suit against the U.S. Treasury Department to obtain an **Internal Revenue Service (IRS) water depletion allowance** on their annual income tax (Figure 14.2). The plaintiffs had to convince the federal government that the Ogallala Aquifer was not an inexhaustible supply of groundwater and that irrigation would cause permanent depletions. Since groundwater levels declined a few feet (or several meters) some years and as much as 20 feet (6 m) in others, the cost of pumping eventually required too much horsepower and became uneconomical. The IRS agreed, and today it allows irrigators in many regions of the Ogallala Aquifer to declare an annual depletion allowance for groundwater use. (9)

Roberts County. Land was purchased for $350 per acre to obtain the water rights to pump underlying groundwater from the Ogallala Aquifer. (10) Texas legislators considered changes to the Rule of Capture water allocation law in order to prevent the redistribution of groundwater in the state, but others argued that water beneath an individual's land was his or hers to develop.

During the Seventy-seventh Session of the Texas Legislature in 2001, Senate Bill 2 authorized the creation of 30 new groundwater conservation districts, along with substantial new powers for all districts to regulate groundwater withdrawals, and approved the charge of export fees on water moving out of a district. (11) Do you believe this a proper role for local water districts, or should state government water agencies regulate exports of groundwater within its borders? Or should individual landowners be allowed to choose the best use of groundwater resources located beneath their property, even if it involves transport to another region?

ALABAMA AND FLORIDA V. GEORGIA

Water conflicts are not typically found in the wet and humid regions of the southeastern United States, but one conflict has been raging for more than a decade in Georgia, Alabama, and Florida (see Figure 14.3). The staggering urban growth

MIDWEEK EDITION

Moseley wins region title 7A

Miss Bainbridge crowned 6A

A night of honors at BC 12A

The Post-Searchlight

Vol. 95, No. 35 Wednesday, May 2, 2001 2 Sections

Welcome: M

...r Artsfest

'Water war' closer to end?

Negotiating commissioners from Alabama, Georgia and Florida met Monday, April 30, and agreed to extend their negotiating period for an additional 45 days beyond the deadline of May 1, 2001. They will continue the closed talks mediated by Florida State University President Sandy D'Alemberte that have gone on since November. Florida negotiators are working on a proposal that may lead to water allocation formulas, possibly ending a decade-long legal fight.

The negotiators are the ACF (Apalachicola, Chattahoochee and Flint) River Basin Compact Commission and ACT (Alabama, Coosa and Tallapoosa) River Basin Compact Commission members. They each have representatives from the three states who have been holding talks on how to share the water in the two river systems for

WATER WARS, ...

The Post is Cherokee County's No. 1 weekly paper!

The Post

This week

Every week

Water deal goes down the drain

Make 'Em, Break 'Em

Road Apples

State ranks ... bottom ...

FIG. 14.3. Water negotiations and disputes are front-page news in many regions of the United States and around the world. The articles shown here were in the Bainbridge, Georgia, *Post-Searchlight* and *The Post* of Cherokee County, Alabama, in 2001.

of Atlanta, unusual drought, and expanding irrigated lands in Georgia are creating heavy demands for water. The Chattahoochee River, and five other river systems that have their headwaters in Georgia and then flow into Alabama or Florida, are at the center of the controversy. At issue are economic growth and environmental protection within the three states.

Two major watersheds are in dispute. One is the ACF Basin (named after the first initials of the Apalachicola, Chattahoochee, and Flint rivers), which begins in Georgia, runs along the Alabama border, and then enters the Gulf of Mexico at Apalachicola Bay in Florida. The other is the ACT Basin (named after the Alabama, Coosa, and Tallapoosa rivers), which begins in Georgia, flows through Alabama, and then empties into Alabama's Mobile Bay on the Gulf of Mexico. Approximately 40 percent of Georgia's land area and 60 percent of the state's population are found within these river basins. (12) Water demand in the metro Atlanta area has increased dramatically since the 1970s, with a population that doubled to 3.6 million people between 1970 and 2000. (13) The ACF and ACT River Basins are shown in Figure 14.4.

The interstate water conflict came to a head in 1990 when the State of Alabama filed a lawsuit against the U.S. Army Corps of Engineers. The Corps was planning to divert more water from the ACF Basin for multiple uses in Atlanta and other northern Georgia cities. Florida, seeing the lawsuit as an opportunity to protect the Apalachicola Bay in the state from future depletions on the Flint River in Georgia, became a party to the negotiations. Since a long and costly legal battle was anticipated, the governors of the three states met and eventually signed two interstate water compacts to facilitate negotiations and place litigation on hold. The legislatures of Georgia, Alabama, and Florida passed identical legislation to create the ACT and ACF Compacts in 1997, laws ratified by Congress and signed by President Bill Clinton on November 20, 1997.

The **ACT and ACF Compacts** created two water commissions for:

> *the purposes of promoting interstate comity* [the ethic of recognizing the need to support legislation and judicial rulings in neighboring states], *removing causes of present and future controversies, equitable apportioning* [of] *surface waters of the ACT and ACF, engaging in water planning, developing and sharing common data bases.* (14)

The governors of the three states became the ACT and ACF compact commissioners but appointed alternates to develop a resolution to these issues. A federal compact commissioner oversaw the proceedings. It was agreed that the two compacts would dissolve if resolution was not reached by a specified period of time, but the three states could agree to extend the negotiating period upon mutual agreement. The negotiating period has been extended several times.

Georgia compact commissioners argued two primary issues: (1) Adequate water supplies must be available to support the long-term growth of the state's major city, Atlanta, as well as for the rest of Georgia's population in the ACT and ACF Basins, and (2) irrigation must be allowed to prosper in the southwestern regions of the state.

Alabama commissioners insisted that (1) adequate water supplies must flow out of Georgia to support the long-term growth of the state's hydropower industry, and (2) adequate flows must be maintained in Alabama for navigation needs. Florida compact commissioners, meanwhile, were concerned with maintaining adequate river flows out of Georgia for the Apalachicola Bay area since this estuary serves as a nursery for significant marine life.

Georgia farmers could be the primary group impacted if the three states adopt streamflow restrictions. Such restriction of pumping from rivers and alluvial aquifers would provide additional surface water flows to Florida and would also protect water supplies for future urban growth in Atlanta. (15) Since impervious surfaces such as parking lots, highways, subdivisions, and other urban development have re-

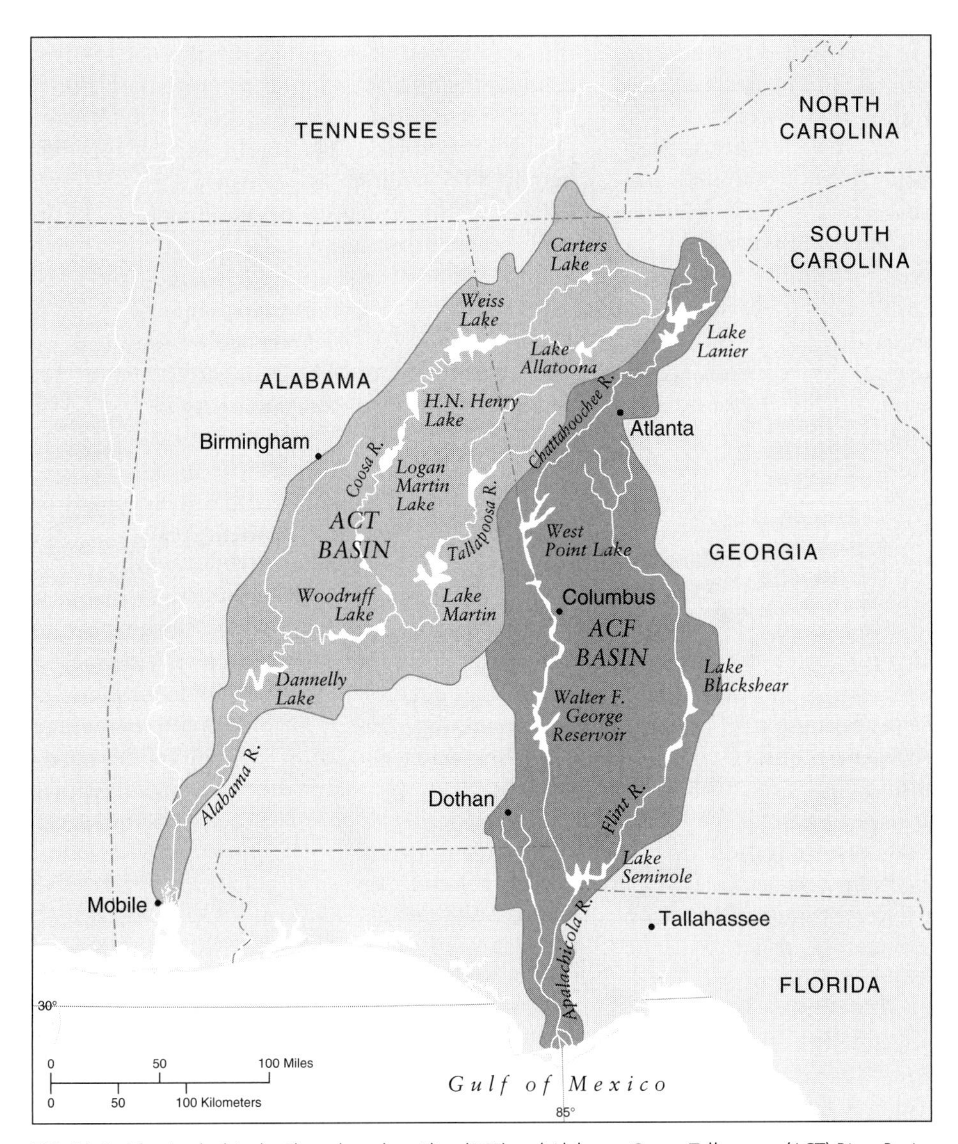

FIG. 14.4. The Apalachicola-Chattahoochee-Flint (ACF) and Alabama-Coosa-Tallapoosa (ACT) River Basins.

duced groundwater recharge in Georgia, particularly in the northwest region of the state, a limitation on irrigation could be a potential beneficial solution if the water was somehow transferred to the metropolitan Atlanta area. As expected, Georgia farmers are opposed to any significant reductions in pumping since 5400 permitted alluvial groundwater wells and 800,000 irrigated acres (323,750 ha) in southwest Georgia would be impacted. (16) In addition, over 2000 applications were filed in 2001 to drill new irrigation wells in southwest Georgia and would increase irrigated lands by 100,000 acres (40,470 ha). (17)

Negotiators from Florida and Alabama are trying to restrict the amount of water Georgia can divert from the Flint and Chattahoochee rivers and its alluvial aquifers. In response to that potential threat, Georgia farmers are adopting conservation methods, such as low-pressure center pivot systems, that would reduce evaporation. Irrigators understand that, politically, urban growth issues affect more voters than does concern for irrigation. However, irrigators are worried that state officials might restrict their use of water during the middle of the growing season. According to a grower quoted in a Georgia Peanut Producers Association newsletter:

Can you imagine how disastrous it would be to have someone walk into a farmer's field during the growing season and tell him he had to shut down his irrigation? (18)

Studies by the University of Georgia have placed the economic benefits of existing irrigation in that state at almost $3 billion. (19) In response, the State of Georgia established the Flint River Drought Protection Fund in the 2000 Legislative Session to reimburse farmers if they were required to cut back or cease irrigation due to compact restrictions in the future.

In 2001, the Flint River Drought Protection Act was implemented, and 575 irrigators in the basin were invited to participate in an auction of water rights. This was the first of its kind in Georgia and probably for the entire United States. Invited Flint River irrigators entered the silent auction to determine who would be compensated for not pumping from perennial streams in the basin between April and December of that year. In exchange, the State of Georgia compensated the selected irrigators from proceeds of the State's Tobacco Settlement with the U.S. Justice Department. At the silent auction, irrigators used secret ballots to list the payment they would require to cease irrigation for the year. After the first round, the Georgia Environmental Protection Division (EPD) listed the offers it would accept if the auction stopped at that point. After each additional round, the EPD posted additional offers that were acceptable. At the end of five rounds, the EPD purchased the irrigation rights of 209 farmers for one year from 33,000 acres (13,355 ha) of land. The total cost was approximately $4.5 million, or $136 per acre (0.4 ha). Surface water depletions of the Flint River Basin were reduced by approximately 200 cfs (129.3 million gal per day, or 489,315 cmd) and provided maintenance flows for aquatic habitat downstream. (20) The drought continued in 2002, and the Flint River Drought Protection Act was implemented once again. A total of 41,145 acres (16,651 ha) of land were removed from irrigation for that year at a cost to the State of Georgia of $5.3 million. In 2003, the drought finally subsided, and the Drought Protection Act was not implemented.

The Three-States Negotiating Commissioners continued talks under the leadership of Florida State University president Sandy D'Alemberte. Short-term deadlines were set, and negotiations continued. Any agreement by the commissioners regarding water allocation will require the signatures of the governors of the three states. At that point, the federal compact commissioner will have three options:

1. Veto the agreement on the basis that it violates federal law. The Three-States Negotiating Commissioners would have 45 days to resolve the issue, or the veto would void the agreement.

2. Do nothing. After 255 days, the compact agreement would automatically become law.

3. Concur with the Agreement.

If the three states are unable to reach a solution, the U.S. Supreme Court could appoint a federal water master to determine a forced solution. The allocation of water between the three states does not take into consideration water quality issues since the compacts are intended to address only water quantity issues. Weekly, monthly, yearly, and multiple-year minimum flows are contemplated and will be based on the river hydrology records of the 1939–2000 time period. Throughout this negotiating process, 11 federal agencies have been involved in the review of all options considered by the Three-States Negotiating Committee. How would you compare these negotiations to those of the Colorado River Compact in 1922 discussed in Chapter 8? What issues are similar, and what new concerns are being considered with the ACF and ACT Compacts?

FIG. 14.5. Artist Sandow Birk of Long Beach, California.

NORTHERN AND SOUTHERN CALIFORNIA

California's water conflicts are caused primarily by expansive population growth in very dry, desert locations. Like much of the world, California has large population centers in areas that do not have substantial water supplies. To readjust water availability, water agencies in southern California have constructed extensive aqueduct systems to obtain water supplies from northern California and from the Colorado River to the east.

Northern Californians have watched with concern as their resident neighbors to the south have developed expansive water projects across the state. In the 1920s, they observed construction of the Los Angeles Valley aqueduct/pipeline that dried up the Owens Valley. Radical opponents repeatedly bombed the pipeline system in an effort to stop water raids by Los Angeles. In later years, plans to construct pipelines to divert water from northern to southern California were met with strong political resistance, and state legislators engaged in heated battles.

Against this backdrop, Artist Sandow Birk (Figure 14.5) and the Laguna Art Museum in Laguna Beach, California, recently presented the ambitious series *In Smog and Thunder: Historical Works from the Great War of the Californias*. This extensive art and audio exhibit explored the concepts of popular culture, fictional military battles, and crushing satire to present a wildly unique view of California's water future. Paintings included scenes from a classical war that started when the fictional characters General James Walker, a Grateful Deadhead, and Commander Thomas Park, a former Los Angeles Raiders fan who was still upset over the return of his team to Oakland, led brigades of soldiers, out-of-work soap opera actors, migrant workers, and pizza delivery personnel north on Interstate 5 and Highway 1 toward San Francisco. The art exhibit showed a fictitious assault that was on schedule and with high spirits until a group of heavily armed farmers were encountered at Big Sur. The hemp farmers thought the soldiers were DEA agents, and a skirmish developed. This was the start of the futuristic Great War of the Californias. (21)

The first battle of the fictional war was short, and the mythical armies of Walker and Park continued north toward Fog City (San Francisco). However, the logistics of feeding and clothing their troops from southern California quickly escalated, and gas stations and mini-marts were soon overrun by squadrons of hungry soldiers. Food establishments quickly ran out of beer and Fritos, and ATMs ran out of twenties. (22)

The residents of San Francisco were totally surprised by the attack that came on three fronts. Los Angeles ships attacked from the Golden Gate, while Park's Army marched toward the city center from Golden Gate Park. The remaining brigade followed the coast and met no resistance. Artillery was strategically aimed with the use of encrypted cell phone technology that probably had unlimited access to the Internet. Two days later, the battle was over. (23)

Or was it? Artist Sandow Birk describes how the residents of Fog City, led by Commander Toma Aqua (Figure 14.6), regrouped and gained control of the California Aqueduct to prevent water supplies from heading south. Then they organized a counterattack with mobs in SUVs and minivans headed south toward the San Fernando Valley of Smog City (Los Angeles) and the high ground of the Getty Center. Angelenos strategically placed artillery the length of Mulholland Drive to protect northern approaches into the city, and a few hundred troops held 10 freeway lanes against 5000 invading San Franciscans. However, the battle was short-lived because looting and fires caused by disgruntled Los Angelean troops left little for the invaders from San Francisco to seize. Eventually, a peace agreement was drafted, and the Californias became one again. (24)

FIG. 14.6. Commander Toma Aqua rallies his troops in defense of the California Aqueduct during the Great War of the Californias.

Ultimately, the war had taken its toll on both sides. Water, if it could be found, was now $14 a gallon ($3.70 per liter), while three-day rentals at Blockbuster Video hit $36 for new releases. "We are all Californians," Commander Park said at the peace celebration. "And no one is the victor." (25)

THE MIDDLE EAST

In the late 1800s, John Wesley Powell firmly believed that state boundaries should coincide with watersheds. Author Marc Reisner has described Powell's philosophy:

States bothered Powell, too. Their borders were too often nonsensical. They followed rivers for convenience, then struck out in a straight line, bisecting mountain ranges, cutting watersheds in half. Boxing out landscapes, sneering at natural reality, they were wholly arbitrary and, therefore, stupid. In the West, where the one thing that really mattered was water, states should logically be formed around watersheds. Each major river, from glacial drip at its headwaters to the delta at its mouth, should be in a state or semistate. The great state of the Upper Platte River. Will the Senator from the state of Rio Grande yield? To divide the West any other way was to sow the future with rivalries, jealousies, and bitter squabbles whose fruits would contribute solely to the nourishment of lawyers. (26)

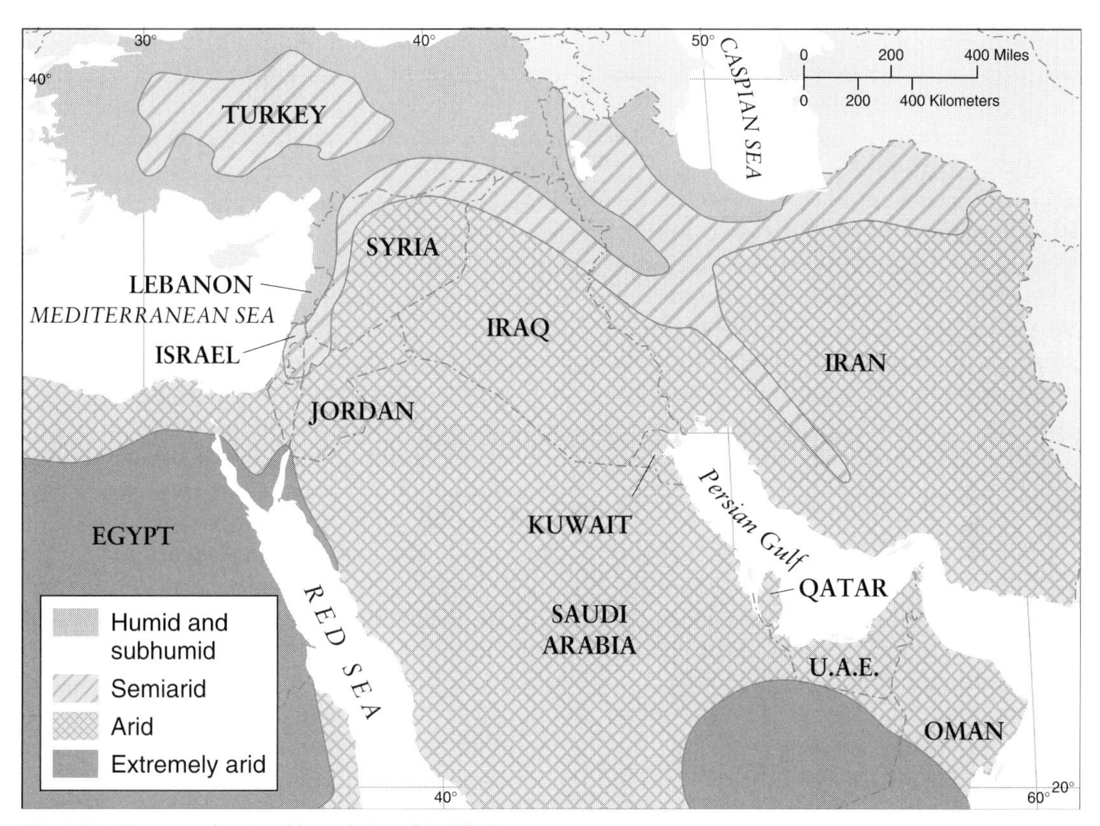

FIG. 14.7. Climate and national boundaries of Middle Eastern countries.

The national boundaries of the Middle East (Figure 14.7) would have served its citizens well if its organizers had heeded Powell's nineteenth century philosophy. Unfortunately, after World War I and World War II, the boundaries of Middle Eastern countries were delineated to divide rivers by straight lines that bisected watersheds and groundwater aquifers and to generally follow totally irrational international boundaries. This political disregard for the volatile nature of water management in a desert region has led to the use of political envoys, water commissions, and, in rare cases, military action to settle water disputes in the region.

RELIGIOUS/POLITICAL BACKGROUND

The country known as Palestine has been at the heart of the conflict in the Middle East because of its religious significance to Arabs, Jews, and Christians in the region and around the world. To Jews, the area is called *Eretz Yisrael* and is regarded as the origin of their nation. Christians venerate the region as *Terra Sancta,* the birthplace of Christ and the site of holy places such as Bethlehem, Jerusalem, and Galilee. Muslims refer to Jerusalem as *Al Quds* ("The Holy") and in particular honor the Temple Mount with its mosque. All three groups consider the area to be central to their religion.

In 1918, Palestine was taken away from Turkish control and administered by the British military. Plans were immediately developed to encourage Jewish settlements within the multinational Palestine. There was considerable disagreement within the Arab-speaking communities of the British mandate. An initial plan to divide the region into Arab and Jewish lands would have left several Jewish settlements without access to the Jordan River and so was rejected by Jewish negotiators. Arab nationalists disagreed with alternate proposals. In late 1919 and early 1920, Bedouin tribes and marauding Arabs attacked new Jewish settlements in the area, killing a number of defenders and forcing others to leave their homes. At the time, approximately 600,000 Arabs lived in the region, of which 90 percent were Muslim and 10 percent were Christian. Approximately 80,000 Jews also lived in the disputed area. In spite of the attacks, plans were developed to allow the relocation of three to four million Jews into unused and uncultivated British-controlled land in Palestine. (27)

The Settlement of 1922, between the British, French, and others, renegotiated the boundary between Palestine and Lebanon to give Palestinians access to the Sea of Galilee. In addition, and more importantly, access was also provided to the spring of Dan, the largest of the upper sources of the Jordan River. A few years later the Golan Heights was included within the borders of Syria, and British occupation of that area ended. Contrary to the philosophy of John Wesley Powell, these new international boundaries placed the upper reaches of the Jordan River in four countries: Dan Spring in Palestine (today in Israel); the Hasbani River which is the western tributary of the Jordan River (in Lebanon); the Banias River which is the eastern tributary of the Jordan River (in Syria); and the Yarmouk River (now the border between Syria and the present-day Kingdom of Jordan). (28)

After World War II, the United Nations approved the partition of Palestine into two separate regions—a Jewish state and an Arab state. Soon after, on May 14, 1948, the Jewish community of Palestine declared the creation of the State of Israel. Once again, Arabs totally disagreed with this reorganization of Palestine. Palestinian Arabs, aided by neighboring Arab states, attacked the new State of Israel shortly after its formation and seized the West Bank while the Egyptian Army gained hold of the Gaza Strip.

After months of fighting, a truce was declared and the United Nations mediated an armistice between the warring factions. The Kingdom of Jordan annexed several cities including East Jerusalem, Nablus, and Hebron, calling them the West Bank. Negotiators agreed that the British would relinquish all control over Palestine immediately (in 1948), over Jordan (in 1956–1957), and over Iraq (in 1958). France gave independence to Syria and Lebanon in 1958. Prior to this time, foreign military occupation of the region provided some control over international tensions. However, shortly after the departure of occupation forces, religious fundamentalism reawakened, populations increased dramatically, and demands for water resources, for both irrigation and urban uses, increased exponentially.

The religious wars between Arabs and Jews in the Middle East smoldered and erupted time and again for decades. Full-scale wars occurred in 1956, 1967, 1973, and 1982, while smaller military skirmishes were constant. The war in 1967 was preceded by Syria's attempt to divert water from the Jordan River. Israeli warplanes attacked a Syrian construction site to prevent the diversion of water from Israel. This conflict led to the Six-Day War a few months later. Victorious Israeli forces took control of the Sinai Peninsula and the Gaza Strip from Egypt, the West Bank from Jordan, and the Golan Heights from Syria.

In 1973, Egypt and Syria counterattacked Israel on Yom Kippur to take control of lands lost in the conflict in 1967. During the interim between the two wars of 1967 and 1973, the Israeli Army had constructed strong defenses on its side of the Suez Canal to prevent military reoccupation of the Sinai Peninsula by Egyptian troops. The main line of defense was a massive embankment made of desert soil and sand. It was constructed at great expense along the entire length of the Canal's east bank, and included concrete and stone bunkers at regular intervals along the dike. Israeli commanders considered the steep dike to be an impenetrable line of defense. (29)

They were wrong. Arab gunboats on the Suez Canal used water cannons to dissolve the embankments (not unlike the hydraulic engineering processes used in California during the 1800s for mining purposes) and punched through the Israeli defenses. Amphibious tanks entered through several holes in the dike and formed an attack behind Israeli military lines. Later, Israeli forces counter-attacked and encircled Egypt's Third Army on the East Bank. Egyptian President Anwar Sadat visited Jerusalem a few years later to initiate a peace agreement with Israel over this disputed region. In 1979, both countries signed a landmark agreement. Israel agreed to withdraw from the Sinai Peninsula (including its air bases, oil fields, farms, towns, and tourist attractions that had been developed).

A relatively stable peace followed between Israel and Egypt since they are separated by empty desert and do not share any water resources. However, conflict continues between Israel, Syria, Jordan, and Lebanon. Since the Israeli occupation of the West Bank and the Golan Heights in 1967, Yasser Arafat, leader of the Palestinian Liberation Organization (PLO), has led various efforts to regain control of the region for Palestinian Arabs. To this day, Palestinians refer to the creation of Israel as *al nakba* or "the catastrophe" since it has resulted in many of them living in neighboring Arab states.

WATER RESOURCES IN THE MIDDLE EAST

Water use and management are extremely contentious issues in the Middle East for a variety of reasons. First, the current political borders disregard hydrologic boundaries, thereby ensuring future conflict. Second, freshwater is very unevenly distributed in the Middle East. This requires cooperation between countries that have no desire to work together. Third, total water use in the region has increased nearly 1000 percent over the last 100 years. Fourth, inadequate water treatment and disposal have led to steadily deteriorating water quality in surface and groundwater sources. (30)

The West Bank Mountain Aquifer The West Bank Mountain Aquifer in Palestine provides about one-third of the water consumed in Israel. (31) Although the aquifer is recharged by rainfall in Palestine, the groundwater table gradient slopes toward Israel. Therefore, the underground movement of groundwater in this highly transmissive, limestone aquifer is to the west, toward Israel. Since Israeli occupation of the West Bank region in 1967, Israel has used military force to protect recharge of the West Bank Mountain Aquifer. (32)

The Jordan River In 1953, Jordan, Lebanon, Syria, and Israel agreed on a priority of use of Jordan River water and its tributaries. The Unified Plan (also called the Johnston Plan after President Dwight Eisenhower's special envoy, Eric Johnston) was developed to allow the Yarmouk River (a tributary of the Jordan River) to be used exclusively by Syria and Jordan. The Jordan River was divided 52 percent to Jordan, 32 percent to Israel, 13 percent to Syria, and 3 percent to Lebanon. (33)

Although technical experts from the four countries agreed to the plan, government officials rejected it for a variety of political reasons. When these negotiations failed, Israel and Jordan proceeded with water development projects in their own countries. However, the two neighboring countries generally followed the technical aspects of the Plan until 1967. (34) In 1958, Israel began work on the National Water Carrier (completed in 1964), which now diverts 90 percent or more of the flow of the Upper Jordan River to Israel's water-deficient southern regions. Jordan receives mostly urban wastewater and irrigation return flows from upstream water users in Israel. This water is virtually unusable, even for saline-tolerant crops. (35) Israel's Arab neighbors saw the completion of the National Water Carrier as a strategy to improve the economy of Israel. In response to that perceived economic threat, neighboring Arab countries developed plans to divert almost half of the water anticipated to be used by Israel from the National Water Carrier. Israel responded to this Arab threat in April 1967 by bombing water facility construction sites in Syria. These actions, as well as larger strategic, security, and political factors, led to the Six-Day War of 1967.

A CLOSER LOOK

On October 26, 1994, Jordan and Israel signed a peace treaty ending international water disputes between the two countries. Following are selected sections of the treaty. Notice the definitions used regarding water withdrawals and plans for water pollution prevention. Try to find several examples of words in the Treaty that could be interpreted in one way by Jordan but in an entirely different way by Israel. Multiple or vague definitions can easily lead to disagreement and, ultimately, to failure of an agreement or treaty. In the real world, a treaty is only as successful as the will of the parties to maintain mutual relations.

Peace Treaty Between Jordan and Israel 1994

[Due to the length of the treaty, only selected sections are presented.]

ARTICLE 6: WATER

With the view to achieving a comprehensive and lasting settlement of all the water problems between them:

1. *The Parties agree mutually to recognize the rightful allocations of both of them in Jordan River and Yarmouk River waters and Araba/Arava ground water . . .*
2. *The Parties, recognizing the necessity to find a practical, just and agreed solution to their water problems . . . jointly undertake to ensure that the management and development of their water resources do not, in any way, harm the water resources of the other Party . . .*

ANNEX II: WATER RELATED MATTERS

Pursuant to Article 6 of the Treaty, Israel and Jordan agreed on the following Articles on water related matters:

Article I: Allocation

1. *Water from the Yarmouk River*
   a. *Summer period—15th May to 15th October of each year. Israel pumps 12 MCM and Jordan gets the rest of the flow.*
   b. *Winter period—16th October to 14th May of each year. Israel pumps 13 MCM and Jordan is entitled to the rest of the flow subject to provisions outlined herein below . . .*
   c. *In order that waste of water will be minimized, Israel and Jordan may use, downstream of point 121/Adassiya Diversion, excess flood water that is not usable and will evidently go to waste unused.*
2. *Water from the Jordan River*

[A similar allocation follows . . .]

Article II: Storage

1. *Israel and Jordan shall cooperate to build a diversion/storage dam on the Yarmouk River . . .*
2. *Israel and Jordan shall cooperate to build a system of water storage on the Jordan River . . .*
3. *Other storage reservoirs can be discussed and agreed upon mutually.*

Article III: Water Quality and Protection

1. *Israel and Jordan each undertake to protect, within their own jurisdiction, the shared waters of the Jordan and Yarmouk Rivers, the Arava/Araba ground water, against any pollution, contamination, harm or unauthorized withdrawals of each other's allocation.*
2. *For this purpose, Israel and Jordan will jointly monitor the quality of water along their boundary, by use of jointly established monitoring stations to be operated under the guidance of the Joint Water Committee.*
3. *Israel and Jordan will each prohibit the disposal of municipal and industrial wastewater into the course of the Yarmouk or the Jordan Rivers before they are treated to standards allowing their unrestricted agricultural use. Implementation of this prohibition shall be completed within three years from the entry into force of the Treaty.*

[Terms of the treaty continue regarding water quality issues . . .]

Article IV: Groundwater in Emek Ha'arava/Wadi Araba

[Terms follow regarding well drilling and pumping . . .]

Article V. Notification and Agreement

1. *Artificial changes in or of the course of the Jordan and Yarmouk Rivers can only be made by mutual agreement.*
2. *Each country undertakes to notify the other, six months ahead of time, of any intended projects which are likely to change the flow of either of the above rivers . . .*

Article VI: Cooperation

1. *Israel and Jordan undertake to exchange relevant data on water resources through the Joint Water Committee.*
2. *Israel and Jordan shall cooperate in developing plans for purposes of increasing water supplies and improving water use efficiency, within the context of bilateral, regional or international cooperation.*

Article VII: Joint Water Committee

1. *For the purpose of implementation of this Annex, the Parties will establish a Joint Water Committee comprised of three members from each country . . .*

[The treaty continues with a lengthy explanation of the composition of the Joint Water Committee . . .] (36)

A similar agreement was signed on September 28, 1995, between Israel and the PLO. It set forth mutual terms for water use and sewage disposal along the West Bank and the Gaza Strip. Provisions for water purchases, a Joint Water Committee, and enforcement of terms were also included. (37)

THE TIGRIS AND EUPHRATES RIVERS

The mountains of eastern Turkey contain the sources of both the Tigris and Euphrates rivers, although much of the flow in the Tigris originates in Iran. The Tigris has a total length of 1180 miles (1899 km) and flows through Turkey and Iraq before emptying into the Persian Gulf. The Euphrates River flows for 1677 miles (2699 km) through Turkey, Syria, and Iraq before also reaching the Persian Gulf. The land between the two rivers is often referred to as Mesopotamia.

Turkey has launched the Southeast Anatolia Development Project, known by the Turkish acronym GAP (Guneydogu Anadolu Projesi), to develop hydropower and increase the irrigated land area by 50 percent. GAP would include construction of 68 irrigation systems to cover 7720 square miles (19,995 km^2) and 22 dams. Total cost of the project is estimated in excess of $20 billion. (38) Syria is concerned that the flow of the Euphrates could be reduced by 40 percent in normal years and by much more during droughts. It's estimated that Iraq could lose between 60 and 80 percent of its pre-GAP flow. (39) Already, the Syrian cities of Damascus and Aleppo have experienced shortages. (40) Turkey argues that it needs the new sources of water for irrigation to feed its growing population—65 million in 2000 and estimated to be as high as 77 million residents by the year 2010. Syrian growth estimates are as high as 23 million residents by the end of the decade. (41) Iraq has one of the highest growth rates in the region (3.5 percent annually) and will double in population in about 23 years. (42) Population growth in this region, as well as around the globe, will create critical water management situations in the future.

In 1990, Turkey completed construction of the Ataturk Dam on the Euphrates River and stopped the entire flow of the Euphrates River for one month to fill the reservoir. Turkey's downstream neighbors were extremely concerned, particularly Syria. Quite often, Syrian cities have no electricity owing to a lack of water to turn hydroelectric turbines at dams along the Euphrates River. Turkish officials argue that downstream countries have no claim on Turkish water. In 1991, Turkish president Suleyman Demiral declared that Syria and Iraq had no more right to Turkey's upstream waters from the Euphrates than Turkey had to oil supplies in those two countries. (43)

In spite of their rigid view on water management, Turkish officials proposed a "peace pipeline" to transfer water from the Tigris and Euphrates rivers to neighboring countries. Although not a viable option at the present time, two pipelines, each 10 to 13 feet (3 to 4 m) in diameter, were planned to divert freshwater for beneficial use that now empties into the Mediterranean Sea. The first pipeline would provide water to Jordan, Saudi Arabia, and Syria, and the second would deliver water to Kuwait, Saudi Arabia, the United Arab Emirates, Qatar, Oman, and Bahrain. The project cost was estimated at $21 billion. However, all parties would have to agree to a fair division of water and would also have to agree to Turkey's role as the ultimate "water broker" in the region. Agreement is also necessary before the World Bank will provide funding to complete GAP. (44) The prospects for this occurring are slim because other countries have a deep aversion to placing their fate in the hands of Turkish officials. In addition, Turkey's population growth may require the use of more water than would be available if supplies were provided to downstream neighboring countries.

TRAGEDY OF THE COMMONS

What is common to the greatest number gets the least amount of care. Men pay most attention to what is their own: they care less for what is common.

Aristotle (384–322 B.C.)

We want the maximum good per person; but what is good? To one person it is wilderness, to another it is ski lodges for thousands. To one it is estuaries to nourish ducks for hunters to shoot; to another it is factory land. Comparing one good with another is, we usually say, impossible because goods are incommensurable. Incommensurables cannot be compared. (45)

"The Tragedy of the Commons." Garrett Hardin. (Washington, DC: Science, 162, 1243. © 1968). Reprinted with permission from the American Association for the Advancement of Science.

Why do people, states, and countries fight about water? Why would one group develop water resources to the detriment of others? In 1968, author Garrett Hardin proposed the concept of the **Tragedy of the Commons** to describe the human perspective that often leads to resource mismanagement and conflict. In his essay, Hardin discussed the concept of competing demands on a resource, and the issue that personal gain often injures the common good. Society as a whole is not protected from the actions of the individual, state, or country.

For example, a public water supply system is often developed with little or no regard for downstream water users unless laws, agreements, compacts, or treaties are in place to provide such protection. Without these legal restrictions, there are no incentives to "protect" downstream water users. This lack of incentive is the tragedy of protecting the "public good" when using a "common" resource. Other common resources that are often defiled include air quality of a region or water quality in a stream. Without legal protection, these common resources are often misused, such as the salinity problems in the Jordan River downstream of Israeli irrigation projects, or the loss of downstream hydroelectric generating power when Turkey diverted the entire flow of the Euphrates River in 1990.

Other examples of the concept of the Tragedy of the Commons are college dormitory lounges, public hallways and restrooms, laundry rooms, or classrooms used by multiple teachers and students. Are there any incentives to keep these areas uncluttered and pristine? When is the last time you saw a spotless common area at your school, apartment, or neighborhood? Without economic or other value incentives, public areas are often cluttered, unkept, and degraded.

Fortunately, in some situations individuals and groups speak out against polluted streams, poor air quality, or dysfunctional public use areas. Volunteer groups may be organized to pick up trash along highways, painting crews may form to brighten public spaces, and protesters often gather to oppose degradation of our natural resources. The individual can reach a point where the "Tragedy of the Commons" will no longer be tolerated.

CHAPTER SUMMARY

Water conflicts in the twenty-first century will take the form of seemingly endless negotiations between divergent parties, prolonged courtroom battles, and, in extremely limited cases, armed confrontation. Conflict over water will occur because of limited surface and groundwater supplies, and also because of inadequate water quality. Drought, urban growth, and poverty will be the root cause of most conflicts.

In some situations, water shortages will create unique water management opportunities. The 2001-2002 water rights auctions in Georgia were an exchange of public funds for private assets. Funds from the State of Georgia were diverted to irrigators for the purpose of fish and wildlife protection. In contrast, money was reallocated between willing sellers of groundwater rights in the Texas Panhandle and growing urban centers to the south. A land development company, Mesa Water, Inc., was simply a facilitator of the exchange. The beneficiaries will be residents of the cities, individuals who sold water rights, and the development company.

The scenario presented for California was fictional, exaggerated, and highly unlikely. However, the Middle East presents a stark contrast in reality. A war in the region over water, combined with religious fanaticism and nationalism, is a possibility in the twenty-first century.

QUESTIONS FOR DISCUSSION

1. What will be the primary cause of water conflicts in the future?
2. How many river basins in the world are divided into multiple countries?
3. How many people around the world are dependent on international cooperation to share water supplies and to maintain adequate water quality?
4. What is the primary source of irrigation water in the Texas Panhandle?
5. Describe the conflict between water users in the Texas Panhandle and the plans of Mesa Water, Inc.
6. Explain concerns you might have regarding the ACT and ACF Compacts if you were a peanut grower in Georgia who relied on alluvial groundwater to irrigate your crop.
7. What countries are at the center of water conflicts in the Middle East?
8. How many countries share water from the upper Jordan River? How would John Wesley Powell have configured national boundaries in this region?
9. In the context of water resources, why did Israel and Egypt sign a peace agreement in 1979?
10. Explain the "Tragedy of the Commons."

KEY WORDS TO REMEMBER

ACT and ACF Compacts p. 400

Internal Revenue Service (IRS) water depletion allowance p. 399

Tragedy of the Commons p. 410

SUGGESTED RESOURCES FOR FURTHER STUDY

READINGS

Amery, Hussein A., and Aaron T. Wolf. *Water in the Middle East—A Geography of Peace*. Austin: University of Texas Press, 2000.

Baden, John A., and Douglas S. Noonan. *Managing the Commons*. 2nd ed. Bloomington: Indiana University Press, 1998.

Birk, Sandow. *In Smog and Thunder: Historical Works from the Great War of the Californias*. Laguna Beach, CA: Laguna Art Museum, 2000.

Biswas, Asit K. *International Waters of the Middle East—From Euphrates-Tigris to Nile*. Water Resources Management Series: 2. Oxford, England: Oxford University Press, 1994.

Clarke, Robin. *Water: The International Crisis*. Cambridge, MA: MIT Press, 1993.

Fromkin, David. *A Peace to End All Peace: Creating the Modern Middle East 1914–1922*. New York: Henry Holt & Co., 1989.

Gleick, Peter H. *The World's Water: 2000–2001*. Washington, DC: Island Press, 2000.

Hillel, Daniel. *Rivers of Eden: The Struggle for Water and the Quest for Peace in the Middle East*. New York: Oxford University Press, 1994.

Mazarr, Michael J. *Global Trends 2005: An Owner's Manual for the Next Decade*. New York: St. Martin's Press, 1999.

Opie, John. *Ogallala: Water for a Dry Land*. Lincoln: University of Nebraska Press, 1993.

Postel, Sandra. *Last Oasis: Facing Water Scarcity*. New York: W.W. Norton, 1992.

Reisner, Marc. *Cadillac Desert: The American West and Its Disappearing Water*. New York: Penguin Books, 1986.

Simon, Paul. *Tapped Out: The Coming World Crisis in Water and What We Can Do About It*. New York: Welcome Rain Publishers, 1998.

VIDEOS

Dune. David Lynch, Director, Universal Studios, 1984. 2 hrs., 17 min.

Lawrence of Arabia. David Lean, Director. Columbia/Tri-Star, 1962. 3 hrs., 41 min.

Milagro Beanfield War. Robert Redford, Director, Universal Pictures, 1988. 1 hr., 58 min.

REFERENCES

1. Mazarr, Michael J. *Global Trends 2005: An Owner's Manual for the Next Decade*. New York: St. Martin's Press, 1999, 53.

2. Gleick, Peter H. *The World's Water: 2000–2001*. Washington, DC: Island Press, 2000, 54–55.

3. Clarke, Robin. *Water: The International Crisis*. Cambridge, MA: MIT Press, 1993.

4. Opie, John. *Ogallala: Water for a Dry Land*. Lincoln: University of Nebraska Press, 1993, 3.

5. Postel, Sandra. *Last Oasis: Facing Water Scarcity*. New York: W.W. Norton, 1992, 33.

6. U.S. Geological Survey, Virginia L. McGuire, and Jennifer B. Sharpe, "Water-Level Changes in the High Plains Aquifer, 1980 to 1995," U.S. Geological Survey, Fact Sheet FS-068-97, Lincoln, Nebraska, 1997; http://ne.water.usgs.gov/highplains/hpfs95_txt.html, September 2001.

7. Texas Natural Resource Conservation Commission, "WAM: Water Availability Modeling," http://www.tnrcc.state.tx.us/permitting/waterperm/wrpa/wam.html, May 2001.

8. Personal communication with C. E. Williams, General Manager, Panhandle Groundwater Conservation District, White Deer, Texas, July 9, 2001.

9. U.S. Treasury, Internal Revenue Service, "Market Segment Specialization Program—Chapter 14," http://www.irs.ustreas.gov/plain/bus_info/mssp/grain-9.html, May 2001.

10. Copple, Brandon. "Ankle Deep in Trouble." *Forbes Magazine*, October 16, 2000, 52.

11. Personal communication with C. E. Williams, July 9, 2001.

12. The Georgia Public Policy Foundation, "Tri-State Water Compact," http://www.gppf.org/environment/tristate.htm, May, 2001.

13. Montaigne, Fen. "There Goes the Neighborhood!" *Audubon Magazine*, March–April 2000, 60–70.

14. Office of Water Resources, Interstate Water Compacts, State of Alabama, http://www.adeca.state.al.us/AOWR/compacts/ACF/acf-overview.html, May 2001; and Office of Water Resources, State of Alabama, http://www.adeca.state.al.us/ADECA/Pages/Pages_HTML/act-overview.html, May 2001. Reprinted with permission.

15. "Tri-State Water Compact Negotiations at Impasse." *Progressive Farmer,* Birmingham, Alabama, http://www.progressivefarmer.com/issue/0800/water/compact.asp, May 2001.

16. Lightsey, Ed. "Water WARS," Peanut Patriot, Georgia Peanut Producers Association, Albany, Georgia, 23; also found at http://www.georgiapeanuts.org/articles/h_a_9912_waterwars.html, May 2000.

17. Personal communication with Bob Kerr, Director of Pollution Prevention Assistance Division, Georgia Department of Natural Resources, Atlanta, Georgia, and Chief Negotiator for the Governor of Georgia on the ACT and ACF Compact Commissions, October 1, 2001.

18. Lightsey. Reprinted with permission.

19. Ibid.

20. Memorandum from Nolton G. Johnson—Chief, to Harold F. Reheis—Director, Georgia Department of Natural Resources, Atlanta, Georgia, April 12, 2001.

21. Birk, Sandow. *In Smog and Thunder: Historical Works from the Great War of the Californias*. Laguna Beach, CA: Laguna Art Museum, 2000.

22. Ibid., 21.

23. Ibid., 21–24.

24. Ibid., 33–38.

25. Ibid., 38. Reprinted with permission.

26. Reisner, Marc. *Cadillac Desert: The American West and Its Disappearing Water*. New York: Penguin Books, 1986, 47.

27. Fromkin, David. *A Peace to End All Peace: Creating the Modern Middle East 1914–1922*. New York: Henry Holt & Co., 1989, 515–522; and Hillel, Daniel. *Rivers of Eden: The Struggle for Water and the Quest for Peace in the Middle East*. New York: Oxford University Press, 1994, 84–85.

28. Fromkin, 558–567; and Hillel, 84–86.

29. Hillel, 89.

30. Biswas, Asit K. *International Waters of the Middle East—From Euphrates-Tigris to Nile*. Water Resources Management Series: 2. Oxford, England: Oxford University Press, 1994, 1.

31. Amery, Hussein A., and Aaron T. Wolf. *Water in the Middle East—A Geography of Peace*. Austin: University of Texas Press, 2000, 46.

32. Ibid., 34.

33. Biswas, 127.

34. Personal communication with Dr. Thomas Naff, Director, Middle East Research Institute, University of Pennsylvania, Philadelphia, Pennsylvania, October 11, 2001.

35. Amery and Wolf, 40.

36. U.S. Embassy—Israel, "Israel—Jordan Peace Treaty," October 26, 1994, http://www.usembassy-israel/org.il/publish/peace/annex2.htm, May 2001.

37. Facts on File, "Key Event: Israel, PLO Sign Breakthrough Peace Accord," http//www.facts.com/vooo11.htm, May 2001.

38. Hillel, 104–105.

39. Personal communication with Dr. Thomas Naff, June 10, 2001.

40. Postel, 81.

41. Simon, Paul. *Tapped Out: The Coming World Crisis in Water and What We Can Do about It*. New York: Welcome Rain Publishers, 1998, 51–52.

42. Personal communication with Dr. Thomas Naff, June 10, 2001.

43. Amery and Wolf, 228.

44. Clarke, Robin. *Water: The International Crisis*. Cambridge, MA: MIT Press, 1993, 105.

45. Baden, John A., and Douglas S. Noonan. *Managing the Commons*. 2nd ed. Bloomington: Indiana University Press, 1998, 5; cited from an article by Garret Hardin, "The Tragedy of the Commons," *Science* 162 (1968): 1243.

CHAPTER 15

EMERGING WATER ISSUES

FUTURE GLOBAL WATER MANAGEMENT ISSUES

FUTURE GLOBAL WATER MANAGEMENT SOLUTIONS

THE NEED FOR COOPERATION

CONCLUSIONS

Anyone who solves the problem of water deserves not one Nobel Prize but two—one for science and the other for peace.

President John F. Kennedy

Water resources management will continue to be a contentious issue in the twenty-first century. Population growth, polluted water supplies, food shortages, and climatic change will create water management situations that will be both challenging and desperate. Some governments will respond by constructing mega dams to harness large rivers for irrigation, water supply, and flood control. In other regions—because of economics, environmental concerns, or a lack of reliable water supplies—dam construction or other water development options are not a viable alternative and water conservation, groundwater recharge, and even desalination will become more common.

Local and regional water providers are faced with increasing demands from population growth as they work to solve worldwide water problems. From the beginning of human existence until the end of the 1700s, the world population grew to a total of 1 billion people. During the 1800s, human numbers increased to about 1.7 billion. On October 12, 1999, the United Nations celebrated "A Day of 6 Billion" humans on the planet. A reduction in birth rates will result in a slower growth rate in the future, but still the world's population is expected to reach 8 billion people by 2025 and approximately 9.2 billion by 2050 (over a 50 percent increase since 2000). See Figure 15.1. (1) In many

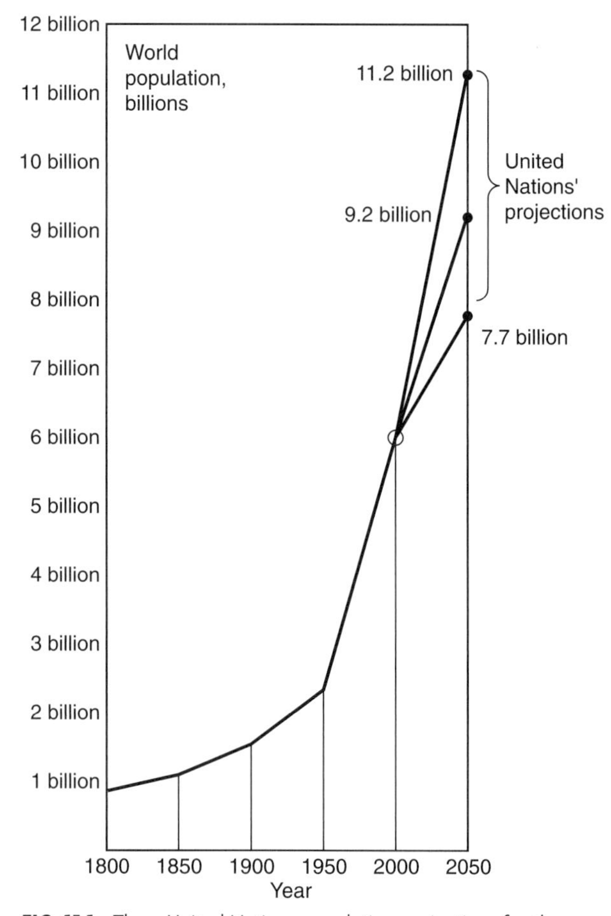

FIG. 15.1. Three United Nations population projections for the twenty-first century are presented in this graph. Variations are due to different estimates of fertility rates.

FIG. 15.2. The Caribbean is an area of great beauty but too often contains extreme poverty. Haiti is the Western Hemisphere's poorest country. In Port-au-Prince, the capital of Haiti, women carry drinking water through their polluted slum. These slums rank among the worst in the world and accounted for the desperation of Haitian boat people who search for a better life in America.

less-industrialized countries, women still walk long distances to obtain poor quality water for household uses. Disease and death are prevalent as more than a quarter of the world's population lack a clean water supply and the means for proper wastewater sanitation. (2)

FUTURE GLOBAL WATER MANAGEMENT ISSUES

POPULATION

Population growth around the world will constrain our ability to provide safe and adequate water supplies in the future. The Earth's population is expected to increase by approximately 3 billion people between 2000 and 2050, yet the world's freshwater supply will remain roughly constant (Figure 15.2). In addition, the World Meteorological Organization estimates that the world's water consumption will double between 1960 and 2010. (3)

LACK OF WASTEWATER TREATMENT

The quality of our freshwater will decrease in the future as the world's burgeoning population pollutes lakes and rivers with waste (see Table 15.1). Today, nearly 3 billion people (half the world's population) live without basic household sanitation such as sewage disposal, septic tanks, or even latrines (see Table 15.2). More than 5 million people die every year from waterborne diseases such as cholera. (4) It has been estimated that 1.2 billion people become sick annually as a result of poor quality drinking water; these illnesses contribute to the death of 15 million children every year under age five. (5)

In spite of sanitary improvements over the past several decades, efforts have not kept up with population growth. In Africa alone, over 400 million residents have no proper wastewater sanitation systems. Most African cities use rudimentary sewage treatment systems, which are often incomplete or exist only in the central urban areas. Septic tanks, though common, frequently pollute local groundwater supplies. An estimated 3 million Africans die each year as a result of waterborne illnesses. The workload of many African women, already great, is exacerbated by the need to spend hours each day walking to obtain water (often polluted) for domestic uses. (6) According to author and global water expert Peter Gleick, the sanitation systems of ancient Rome would be a significant improvement for almost 3 billion people in the world today. (7)

ENVIRONMENTAL DEGRADATION

The loss of species and habitat will continue to plague environmental protection efforts. As cities and rural areas experience population growth, agricultural lands and open space will be converted into industrial, commercial, and residential uses. In less-industrialized countries, the need for food will lead to the conversion of forested lands to agricultural uses. In more-industrialized countries, farms will be replaced by roads, subdivisions, and industrial parks. Overall, the effect on fish and wildlife habitat could be very negative.

In the future, the preservation of diverse ecosystems will be a challenge. Conflict will continue between the need to protect wetlands, riparian zones, and surface and groundwater quality, and the pressures to provide new housing, greater quantities of food, and sites for industrial and commercial expansion. No easy solutions are evident.

What predictions do you have for the future of water resources use, development, and conservation? Throughout the rest of the chapter, four scenarios will be presented for your consideration. Which one presents the most likely outlook? Can you develop other scenarios?

TABLE 15.1 Population Growth in Selected Metropolitan Regions Around the World, 1980–2000 (population in millions)

City	Country	1980	1990	2000	% Growth (1980–2000)
Beijing	China	9.0	10.8	14.0	56
Buenos Aires	South America	9.9	11.9	12.9	30
Calcutta	India	9.0	11.8	15.7	74
Los Angeles	United States	9.5	11.9	16.4	73
Mexico City	Mexico	14.5	20.2	25.6	77
Moscow	Russia	8.2	8.8	9.0	10
New York City	United States	15.6	16.2	21.2	36
Shanghai	China	11.7	13.4	17.0	45
Tokyo	Japan	16.9	18.1	19.0	12

Source: Phillip Guest, "The Impact of Population Change on the Growth of Mega-cities," *Asia-Pacific Population Journal,* Population Programme, United Nations, New York, 1994. http://www.unescap.org/pop/journal/1994/v09nla3.htm, May 2001; and U.S. Census, "U.S. Department of Commerce News," U.S. Department of Commerce, Washington, DC, April 2, 2001, found at http://www.census.gov/press-release/www/201/cb01cn64.html, October 2001.

TABLE 15.2 Indicators on Water Supply and Sanitation, 2000

	Percent of Population with Access to Improved Drinking Water Sources			Percent of Population with Access to Improved Sanitation Facilities		
Area	Total	Urban	Rural	Total	Urban	Rural
Australia	100	100	100	100	100	100
Austria	100	100	100	100	100	100
Bulgaria	100	100	100	100	100	100
Finland	100	100	100	100	100	100
Netherlands	100	100	100	100	100	100
Slovakia	100	100	100	100	100	100
Sweden	100	100	100	100	100	100
Switzerland	100	100	100	100	100	100
United Kingdom	100	100	100	100	100	100
United States	100	100	100	100	100	100
Canada	100	100	99	100	100	99
Jordan	96	100	84	99	100	98
Kazakhstan	91	98	82	99	100	98
Saudi Arabia	95	100	64	100	100	100
Zimbabwe	85	100	77	68	99	51
Argentina	79	85	30	85	89	48
Brazil	87	95	54	77	85	40
Chile	94	99	66	97	98	93
Costa Rica	98	98	98	96	98	95
Egypt	95	96	94	94	98	91
India	88	92	86	31	73	14
Iran	95	99	89	81	86	74
Iraq	85	96	48	79	93	31
Mexico	86	94	63	73	87	32
Nigeria	57	81	39	63	85	45
South Africa	86	92	80	86	99	73
Turkey	83	82	84	91	98	70
Afghanistan	13	19	11	12	25	8
Cambodia	30	53	25	18	58	10
Chad	27	31	26	29	81	13
Fiji	47	43	51	43	75	12
Haiti	46	46	45	28	50	16
Rwanda	41	60	40	8	12	8

Source: World Health Organization and UNICEF, from the United Nations Statistics Division, "Indicators on Water Supply and Sanitation," http://unstats.un.org/unsd/demographic/social/watsan.htm, February 2004.

SCENARIO 1: "BUSINESS AS USUAL"

World population growth continues at its current pace with no change in sight. Extended periods of flood and drought affect greater numbers of people as populations increase, particularly in the Middle East, Africa, and India. Thousands die during prolonged droughts in the Sudan and Ethiopia in Africa, while floods in Bangladesh kill tens of thousands annually. Water quality becomes so poor in Jordan that outbreaks of cholera are common. Women and children continue to carry the burden of poor water quality in Africa as millions of infants die annually from waterborne diseases. World leaders hold water summits to provide additional funds to less-industrialized countries for construction of drinking water plants and wastewater treatment facilities. However, these efforts cannot keep up with burgeoning population growth around the world. Even the populations of many major cities in the United States are expected to double by the year 2020.

On a scale of 1 to 10, with 1 being most likely, how would you rank the probability of this scenario occurring by the year 2020? 2050?

FUTURE GLOBAL WATER MANAGEMENT SOLUTIONS

PRIVATIZATION OF WATER TREATMENT AND DELIVERY

In Chapters 5 and 11, we read about private water suppliers of the 1800s. In New York City and London, these entrepreneurs provided potable water to homes on carts or via community water systems made of hollowed logs. Private water merchants satisfied a great service need since most local governments of the era were unable to serve all residents with a safe and reliable supply of water. The situation in less-industrialized countries is somewhat similar today.

In Africa, for example, an unrelenting population growth rate of 3 percent per year is a major factor in ongoing water problems. Cities such as Nairobi, Kenya, and Lagos, Nigeria, have experienced tremendous population increases in the past 40 years. To meet these increased water demands, private water companies are common. In fact, most residents in Nigeria obtain their water from private water supply companies. Water vendors typically charge much higher prices than public water suppliers.

Private water companies are also common in countries like Honduras. However, water service associations have been recently organized to assist residents to develop, operate, and maintain their own public water systems. The cost of water from these associations is much lower than that charged by unregulated water vendors but require up-front capital investments that are usually borrowed. Typically, a local water committee is responsible for overseeing its use.

The trend toward privatization of water systems is occurring in more-industrialized countries as well (see Chapter 13). The British water utility, Thames Water, has over 12 million domestic customers in England as well as 11.5 million customers in international projects from Chile to Australia and from Turkey to China. Thames Water recently merged with RWE Aqua and now supplies 10 million customers in Germany and Hungary. (8) However, the spread of private water suppliers in the United States will be slower as 55,000 local water departments currently provide most drinking water supplies in the country. (9)

The largest private water suppliers in the world are French giants Suez and Vivendi. E'Town, the seventh largest supplier in the United States, located in Westfield, New Jersey, was recently acquired by Thames Water. Suez Lyonnaise has now acquired the Hackensack Water Company of New Jersey. Private water providers are becoming a major player in water development, treatment, and distribution around the world.

SCENARIO 2: "TECHNOLOGY SAVES THE DAY"

In our second scenario of the future, private water providers discover improved desalination methods that generate massive amounts of affordable potable water along sea coasts around the world. These new technological breakthroughs significantly reduce the cost of water treatment operations, to the point that new lands are placed under irrigation in northern Africa, the Middle East, and the western United States to meet the growing food demands of the world's population.

By the year 2050, science moves us beyond the need for naturally occurring water from the hydrologic cycle. Households use replicators to create water as needed. Individuals are clothed in water-restorative body wear to absorb hydrogen and oxygen atoms from the air and inject the converted water molecules into the body as needed. Farmers irrigate using a similar innovative process of piping hydrogen and oxygen gases and converting them to water. The world's water problems are solved!

On a scale of 1 to 10, with 1 being most likely, how would you rank the probability of this scenario occurring by the year 2020? 2050?

GROUNDWATER RECHARGE

Replenishment of groundwater aquifers could extend water supplies in many locations of the world in the future. In California and other western states in the United States, water is currently diverted from streams for recharge into holding basins such as dry ponds, dry creek beds, or other appropriate sites (Figure 15.3). There, the water is impounded until it percolates into the groundwater aquifer. Feasible recharge sites are located on permeable soils with gravel or other porous material below ground. In addition, the aquifer must have adequate saturated thickness, and the top of the groundwater table must be deep enough to allow large volumes of water to migrate downward from the land surface. This method of groundwater recharge is relatively inexpensive.

A second method of groundwater recharge today is the use of injection wells. Treated water is pumped into an aquifer, usually several hundred feet deep, to recharge declining groundwater levels. The injected water must be of very high quality and of proper temperature, or "plugging" of the aquifer can occur due to impurities collecting within pore openings of a groundwater formation. This method is expensive since the water must be treated before injection, and then pumping costs are incurred for both injection and removal of water supplies.

WATER CONSERVATION

In some countries, water conservation will provide additional water supplies for future demands. Since just one leaking faucet can waste

FIG. 15.3. The Orange County Water District in Fountain Valley, California, is a leader in groundwater recharge. A staff of 20 employees manages the year-round operation that recharges 250,000 acre-feet (81.5 billion gal or 308.4 million m^3) annually. Recharge ponds vary in depth from 2 to 3 feet (0.6 to 0.9 m) to 150 feet (46 m) deep. The District is looking for additional land to develop recharge basins and is working on a Basin Cleaning Vehicle to remove silt and other sediments from the bottom of recharge areas without first draining the ponds.

3280 gallons (12,415 l) of water every year, simple household and system repairs can save enormous amounts of water. Low-flush toilets, restrictive shower heads, and water-efficient dishwashers and washing machines all help conserve water. In 1992, the U.S. Congress passed the National Energy Policy Act and mandated that all new residential toilets sold in the United States after January 1994 had to be efficient, low-flush models that required only 1.6 gallons (6.1 l) per flush.

In dry climates, xeriscape landscaping with drought-tolerant plants has been used to reduce the need for water for lawn irrigation. Xeriscape is the use of native plant species that have adapted to variable and low water availability. This process minimizes the amount of water used by as much as 50 percent by consumers, particularly during hot summer months. Apricot and plum trees, chokecherry bushes, grapes, pansies, penstemons, and irises all do well in dry climates. Municipal water departments in Denver, Colorado, and Albuquerque, New Mexico, as well as several in California, have extensive xeriscape programs.

DAM CONSTRUCTION

As population increases and water supplies run low, most countries have turned to dam construction as a viable alternative. In Chapter 7 we discussed the components of dam construction, and in Chapter 12 we discussed the impacts of these projects. At least 45,000 dams have been constructed around the world over the past several thousand years, and today nearly half of the world's rivers have at least one large dam. (10) In the twenty-first century, with the need for adequate water supplies, food production, flood control, and electricity production increasing, many less-industrialized countries will construct large dams to meet critical water needs.

According to the World Commission on Dams based in Switzerland, about one-third of the countries of the world currently rely on hydropower for more than half of their electricity supply, while half of the world's large dams are used for irrigation purposes. Inundated lands behind reservoirs have forced the relocation of 40 million to 80 million people around the world. Population growth may necessitate 15 to 20 percent more irrigation water by 2025; by that time, 3.5 billion people will be living in water-stressed countries. Presently, 2 billion people live without electricity, and the demand will only grow. Freshwater species, particularly fish, will continue to be threatened, and wetlands will be lost, which will reduce the capacity of ecosystems to serve extremely important functions within the environment. (11)

SCENARIO 3: "GLOBAL WARMING FLOODS THE WORLD"

In this scenario, the world's average temperature has increased several degrees. Higher temperatures mean increased crop water demands, and nonirrigated lands now require supplemental water supplies. Irrigated lands experience critical water shortages to the extent that land is taken out of production. Groundwater recharge projects and water conservation programs that were developed in the late 1900s cannot keep pace with population growth and increased water demands in the warming global climate. Food shortages follow.

Global warming causes significant changes in precipitation patterns in the middle and high latitudes in particular, including the United States, Western Europe, northern Canada, and Siberia. Coastal areas experience permanent inundation as ocean water levels increase. Mean sea level is now 10 feet (3.0 m) higher than historic levels in the twentieth century. New Orleans is entirely flooded, and 1 billion people around the world are forced to relocate from coastal areas due to inundation of their homes. (Sea levels have fluctuated throughout the Earth's history in response to climate change. Eighteen thousand years ago, during the last Ice Age, mean sea level was approximately 330 to 500 feet [100 to 152 m] lower than it is today.) (12)

The U.S. Congress appropriates $2 trillion to build dikes and other retarding structures to hold back the rising oceans, and relies on consultants from the Netherlands (experts in land reclamation in low-lying coastal areas) to provide planning and construction expertise. Social programs are gutted to provide funding for the new "Operation ZuiderZEE," a massive land reclamation program patterned after a similar effort in the

Netherlands. Flooding of wastewater plants and septic tanks in low-lying areas creates highly polluted zones for hundreds of miles or kilometers of newly created saltwater habitat. Cholera epidemics kill thousands along the Florida coast.

On a scale of 1 to 10, with 1 being most likely, how would you rank the probability of this scenario occurring by the year 2020? 2050?

WATER EDUCATION

Water education efforts have developed over the past 25 years in the United States and around the world in order to promote the wise use of water resources. Innovative and unbiased programs, such as those presented in this section, all serve to educate, inform, and promote the wise use of water across the globe.

The Watercourse and International Project WET

http://www.projectwet.org

The Watercourse was established in 1989 as a not-for-profit water science and education program at Montana State University in Bozeman. The Watercourse develops educational materials on water and water management issues, and distributes them through training workshops and institutes. The goal of the Watercourse is to "promote and facilitate public understanding of atmospheric, surface, and ground water resources and related management issues through publications, instruction, and networking."

The Watercourse has developed several programs: International Project WET (a broad-based water science and education program for educators of K–12 students); Healthy Water, Healthy People (water quality curriculum related to the environment and human health); Conserve Water (water conservation education); KIDs (Kids in Discovery activity booklets that cover diverse water topics such as rivers, wetlands, groundwater, and springs for fourth through seventh grade students); Discover a Watershed series (for educators of students in grades 6 through university about specific watersheds); WOW! Wonders Of Wetlands (wetland education); and Native Waters (water resources from the Native American perspective).

The mission of International Project WET is to reach children, parents, educators, and communities of the world with water education. The project was originally developed in 1984 by the North Dakota State Water Commission but moved to Montana State University in 1989. With funding from the U.S. Bureau of Reclamation, the Project WET Curriculum and Activity Guide was published and the national network was expanded. Today Project WET U.S.A. is active in all 50 states, the District of Columbia, and four U.S. islands. Project WET has also expanded internationally and has been adopted by Canada, Mexico, the Philippines, and Korea and is a partner with the Peace Corps. Through the many programs of the Watercourse, curricula and teacher training are provided for thousands of teachers each year in the areas of wetlands, groundwater, water conservation, watersheds, issue resolution, water history, and water monitoring. (13)

The Water Education Foundation

http://www.watereducation.org

The Water Education Foundation began in 1977 as the Western Water Education Foundation and serves as an impartial nonprofit organization with a mission to "create a better understanding of water issues and help resolve water resource problems through education programs." (14) The Sacramento, California-based Water Education Foundation has developed an extensive program of fact sheets, public television documentaries and videos, maps, classroom curricula, and other publications to provide students, teachers, journalists, water managers, and politicians with unbiased information regarding various water issues in the western United States.

Western Water magazine is the flagship of the Water Education Foundation efforts. It is published six times each year and covers a variety of water-related subjects in a nonbiased manner. Topics include water quality, water rights, transfers of water supplies between agriculture and municipal users, and endangered species.

The Groundwater Foundation

http://www.groundwater.org

The Groundwater Foundation was formed in 1985 as a nonprofit organization dedicated to informing the public about one of our greatest hidden resources, groundwater. The Foundation seeks to make learning about groundwater fun and understandable for kids and adults alike. (15) The Foundation office is located in Lincoln, Nebraska.

In 1989, the Foundation held the first-ever Children's Groundwater Festival in Grand Island, Nebraska. This one-day event provides hands-on water education activities that include water quality, water history, groundwater and art, dance, and music. Since its beginning in 1989, the idea of water festivals has spread across the United States and to numerous international settings, including Canada, Mexico, Great Britain, and India.

The Groundwater Foundation also sponsors activities such as Groundwater Guardian, a community-based program to protect and improve drinking water supplies across the country; Groundwater University, a summer science camp for students in eighth through tenth grades; and the Awesome Aquifer Club, which provides after-school activities. It also provides issues of *The Recharge Report*, and *The Aquifer* to members.

The American Ground Water Trust

http://www.agwt.org

The American Ground Water Trust is a national not-for-profit organization based in Concord, New Hampshire. The Trust promotes groundwater protection, sustainability, technology solutions, and citizen involvement. Since its formation in 1986, the Trust has organized many technical conferences regarding well drilling, groundwater recharge, and community water resources planning, and it has promoted awareness of the need to protect groundwater resources. It also serves as a resource for groundwater information through its toll free number (1-800-423-7748) regarding questions on groundwater, groundwater wells, and water systems.

The Trust is funded through donations, the sale of products, and conference fees. For several years it has operated a scholarship program for students entering the field of groundwater resources. It also operates a national media program with the nation's 10,000 newspapers to provide public information articles about groundwater. In addition, the Trust's Wheels for Water program (1-800-929-3507) is an innovative means of developing funds to provide education and training programs to develop and protect drinking water in developing countries. The Trust accepts donations of cars, boats, and trailers (anything with wheels), which are then sold for cash to be used in water protection programs. Vehicles are picked up at no cost to the owner.

Stockholm International Water Institute

http://www.siwi.org

In Sweden, the Stockholm International Water Institute was commissioned in 1997 to administer the annual Stockholm Water Symposium each August. The Symposium is designed to develop "practical solutions and strategies that will help to alleviate the world water crisis." (16) The Symposium is an event of global proportions and seeks to develop a comprehensive understanding of the complex relationships between society and global water resources for policymakers, scientists, practitioners, and decision makers. Previous symposia have included over 600 delegates from over 80 countries.

Each year, the Stockholm Water Prize is awarded at the Symposium to honor outstanding achievements in science, engineering, technology, education, or public policy that leads to the protection of the world's water resources. It is given to individuals, organizations, or companies and was first awarded in 1991.

The project-related work recommended by the Symposium is known as the Stockholm Water Initiative. This effort promotes cooperative meetings, negotiations, and facilitation of transboundary water-related problems. Initiative projects are carried out in cooperation with international organizations, governments, and other institutions.

THE NEED FOR COOPERATION

THE HUMAN FACTOR

No large group of people has ever lived together without rules, laws, or consequences from actions that were detrimental to the group. The reason for this is simple: diverse personalities do not coexist without conflict. Wars are the result of human actions created by diverse viewpoints, inability to compromise, and sometimes simply the egos of leaders.

Cooperation is at the center of solving future water conflicts, and yet it is human nature to distrust someone who is viewed as an adversary. Lack of knowledge of what others intend often leads to suspicion, skepticism, fear, and, ultimately, confrontation.

Individuals and countries often act in a similar manner. Road rage (aggressive driving which can lead to confrontation) is not so different from an altercation between two countries. An action by one party may lead to a counteraction by another. On the highway, tailgating may lead the driver in front to slow down in retaliation. This in turn may result in the tailgater flashing his or her lights in an attempt to distract the impeding driver. Too often, these confrontations can lead to accidents and bodily harm. States often act in the same way. State A diverts water from a stream to the detriment of downstream neighbors. Downstream State B files an injunction in court or, if the dispute is in the Middle East, a country may launch attack planes to stop the diversion. Bodily injury, or at the least ill will, is the result.

SIDEBAR

Dr. Aaron Wolf of the Oregon State University College of Science argues that "Water Wars" are a myth and that cooperation between countries is quite common. He recently compiled a database that examines water treaties around the world since approximately 3000 B.C. It shows, according to Dr. Wolf, that history is full of cooperation and compromise over water, not war. "The study of water disputes throughout recorded time reveals only one single war that was actually linked to a water resource conflict," Wolf said. The single fight occurred in about 2500 B.C. along the Tigris and Euphrates rivers. (17)

Dr. Wolf's database shows that some 3600 water treaties have been created around the world. (18) Many of these included serious debate and controversy but ultimately resulted in fair and equitable apportionment of the waters in dispute. "I study what has driven even hostile combatants to compromise," Wolf said. "It's really quite fascinating how people who hate each other's guts can end up working together." (19)

Dr. Wolf has assisted the U.S. State Department with negotiations of water disputes between Arab nations and Israel, and has helped develop options for solutions to the interstate compact between Georgia, Florida, and Alabama.

How can state governments, or individuals, stop this cycle of mistrust and conflict? Information, cooperation, and planning apply in both situations. In traffic, adequate signs, proper lane markings, and speed limits provide a basis for safe traveling. With regard to countries, knowledge of river operations, sound technical data, and historic streamflow data provide a basis for common agreement on existing conditions.

Cooperation is necessary in both situations. To maintain a smooth flow of traffic and to provide a safe drive, drivers need to refrain from weaving in and out of traffic, to maintain posted speeds, and to allow adequate distances between cars. Similarly, states and countries need to develop cooperation to ensure a fair apportionment of streamflows, to refrain from new water construction projects without prior agreement between parties, and to provide adequate time to develop such agreements.

EXPERT ANALYSIS

In his book, *Leadership Secrets of Attila the Hun,* author Dr. Wess Roberts describes one of the world's greatest, though most ruthless, leaders. Around A.D. 450, Attila, leader of the Huns, set himself the goal of conquering the world, and headed up a marauding band that was brutal, barbaric, and wild. Even so, the group followed unwritten laws, customs, and traits that provided a degree of cooperation which others came to envy and fear. Here is a sampling of the traits as Roberts lists them:

Do not consider all opponents to be enemies. You may have productive, friendly, confrontations with others inside and outside your tribe.

Time is your ally when you're negotiating. It calms temperaments and gives rise to less-spirited perspectives. Never rush into negotiations.

Honor all commitments you make during negotiations lest your enemy fail to trust your word in the future.

We must refrain from charging prematurely and furiously into unfamiliar situations.

Huns may enter war as the result of failed diplomacy; however, war may be necessary for diplomacy to begin.

Huns never take by force what can be gained by diplomacy. (20)

Consider the above statements and think about the conflicts discussed in Chapter 14. What types of negotiating tactics would work well in the various situations? How important is diplomacy in the discussions among the States of Alabama, Florida, and Georgia? How would negotiations differ between the conflicting groups in the Middle East and the conflicting groups in north Texas? Do water shortages create similar conflicts around the world, or do culture, climate, and historical precedence play a larger role in the extent and form of a conflict? What other philosophies would you add to Dr. Roberts' statements?

ETHICS

The word *ethics* comes from the Greek word *ethos,* which means "custom," "usage," or "character." To have ethics is to use fairness and integrity when dealing with others, including the relationship between people and countries, or with the environment. An ethical person or country is one whose word is trusted, whose actions are respected, and whose judgment is considered sound.

Ethical choices are usually not black and white decisions. Unfortunately, many choices are in the "gray" area and require decisions based on what's right or best for individuals and a community at a given time.

Mahatma Gandhi (1869–1948) of India stressed the importance of protecting the poorest of individuals, the *Antyodaya,* and the welfare of the entire human society, the *Sarvodaya.* (21) Aldo Leopold (1887–1948), who is considered the father of wildlife ecology, stated that ethics must constrain the individual to act as a member of a community, which includes other people and the environment. The decision-making process in water resources development, management, and policy is littered with influences from individuals, society, and personal satisfaction (pride, greed, fame, power). Small decisions develop into a body of broader actions, and even the smallest choices will lead to the "ethics" of an individual. From the broad perspective, it would be wise to ask yourself, "What if everyone made this choice?" (22)

Whenever urban growth occurs, the natural or developed environment is modified. It may involve the construction of homes, the need for additional water supplies, or the increased

FIG. 15.4. Chad Pregracke was the driving force behind the creation of the Mississippi River Beautification and Restoration Project, which has grown considerably since its inception in 1997.

pollution of streams. Public policy attempts to minimize the impact of such growth. In some situations, in-depth scientific analysis is used to protect society and the environment through careful mitigation of impacts. Unfortunately, in other situations, money, politics, or a lack of public funds for such scientific studies allow the development process to proceed largely on its own.

Public policy toward growth and water resources management could well follow Gandhi's philosophy that all individuals must be protected, but too often, in practice land and water developers do not follow this approach. This is where ethics enters. Inadequate funding can lead to limited implementation of government policy, and policymakers must determine the "best use" of funds for the "greatest good." What activity or project will provide the greatest public good? What activity or policy will provide the best return on the investment? Ethics plays a key role in selecting the "best" uses of funds.

THE POWER OF THE INDIVIDUAL

Never doubt that a small group of thoughtful, committed citizens can change the world. Indeed, it is the only thing that ever has.

Margaret Mead (1901–1978) Anthropologist

Chad Pregracke has been cleaning the Mississippi River since 1997 (Figures 15.4 and 15.5). Since that time, he has collected over 500 tons (453.6 metric tons) of garbage along 1000 miles (1610 km) of shoreline between Iowa, Illinois, and Missouri. He has pulled out bicycles, refrigerators, gas tanks, stoves, car batteries, a school bus top, and a 1970 Ford Econoline van among many other items. His goal is to remove more trash and contaminated metal drums from his beloved river. (23)

For six years, I like to say I crawled the bottom of the [Mississippi] *river from Fort Madison to Dubuque* [Iowa]. *I've camped on most of these islands, some of the most*

FIG. 15.5. Chad Pregracke and his crew have removed over 500 tons (453.6 metric tons) of trash from the Mississippi River and its tributaries since 1997. This photo shows what Chad and his crew collected between March and October 2000.

pristine areas you'll ever see, filled with old refrigerators and TV sets. I decided I was going to do something about it. (24)

Chad Pregracke

Here's some of the junk Chad Pregracke pulled from the Mississippi River:

5 engine blocks
12 bathtubs
15 air conditioners
6 motorcycles
8272 tires
675 refrigerators
1 grand piano
15 lawn mowers
32 freezers
1 van
26 dishwashers
178 coolers
83 water heaters
34 bicycles
49 sinks
380 chairs
75 televisions
70 washing machines
50 cooking grills
338 antifreeze containers
5 messages in a bottle (25)

Chad grew up just a few feet from the Mississippi River in East Moline, Illinois, and spent summers diving for freshwater mussels. To save money, he camped on islands in the braided river and fell in love with the Mississippi. But he noticed that trash was everywhere, and it bothered him. In 1997, with no financial backing, he began cleaning a 100-mile (160 km) reach. (26)

Since that time, major corporations such as Alcoa, Anheuser-Busch, and Cargill have provided some funding to help with expenses. Living Lands and Waters was formed to accept charitable donations to create a shoestring annual budget of $200,000. Chad is now working on the Illinois and the Ohio rivers under the Ohio River Beautification and Restoration Project. Work is also beginning on the Missouri River. (27) See http://www.cleanrivers.com for additional information.

SCENARIO 4: "SPACE IS THE ANSWER"

In our final scenario, humans have escaped the restrictions of the Earth. Water, as needed, is secured by means of new technology, space water mining, and replication. Water education programs have created support for government funding of water exploration programs in outer space.

Deimos, the outer moon of Mars, is explored for water to use as a propellant and life support for more distant space voyages. Water is already being transported to the Unified Earth Space Station and the orbiting hotel launched by Japan for the tourist industry. Robotic spacecraft land on Deimos, the smallest of the two moons of Mars, to drill approximately 330 feet (100 m) into the lunar rock and obtain and analyze drill cuttings for water ice.

On a scale of 1 to 10, with 1 being most likely, how would you rank the probability of this scenario occurring by the year 2020? 2050?

GUEST ESSAY

Where Do We Go from Here?

by **Susan S. Seacrest**, President
The Groundwater Foundation

Susan Seacrest

Susan Seacrest is president and founder of the Groundwater Foundation, located in Lincoln, Nebraska. She started the foundation at her kitchen table in 1985 to educate people about groundwater, and since then she has led the way in establishing a wide variety of water education activities, including the world's first Children's Groundwater Festival in 1989, the Groundwater Guardian and Groundwater University programs, and the Awesome Aquifer Club. In August 1999, *Time* magazine named Susan a "Hero for the Planet," and she was also featured in the March 1993 issue of *National Geographic* magazine.

A single, present moment contains the infinity of the past. Water exemplifies this truth. Our philosophy and beliefs about the future protection and preservation of our precious water resources rest on a constant process of learning from both failure and success.

Our direction and destination are constantly shifting based on what we know and what we need. As someone who has devoted a career to groundwater, my perspective on groundwater stewardship and on identifying the next steps in its protection is connected to past history, present conditions, and future realities such as technological advances and water scarcity.

In reflecting on a considered answer to the question, "Where do we go from here," I returned again and again to the idea that we need to move toward a closer connection with each other and to the water resources that nourish us. It is my belief that to become able water stewards in the twenty-first century we must become more aware of each other's needs, more receptive to the importance of vital communities and an expanded sense of place, and finally more committed to our common interest in a plentiful, safe, water supply.

Following is my personal perspective on the forces driving this need to discover common ground and connect with each other.

1. The explosion of scientific knowledge and its rapid application. The rate at which science is changing our world is difficult for me to grasp. Scientific research and data used to drive change—today the force of change drives science. The human genome project and cellular therapies, quantum physics, and new computing systems exemplify the leap from idea to reality. The future is unknown but here today.

Water connection: Too numerous to mention. One close to my home and heart is the use of computer calibration in center pivot irrigation systems. Thanks to innovations in hardware and a more complete knowledge of the water needs of particular crops, mechanized irrigation has become a powerful ally in our struggle to conserve water quantity and protect water quality. Irrigation systems are utilized in diverse geographic settings, and these various adaptations will yield rich sources of future data and further innovation.

2. The availability and use of technological tools. The Internet has made us more connected in every way and has led to huge increases in available information. Computer hardware and software also touch every part of our lives and are becoming more invisible all the time. As a result, we become dependent on computer technology without even knowing it. For example, my car engine is calibrated to make adjustments through an on-board computer completely invisible to the eye.

We routinely use technology that I know about through my days as a "Star Trek" fan. Remember Captain Kirk's talking computer? Recently, I rented a car, and my dashboard Global Positioning System (GPS), dubbed "Never Lost," spoke to me throughout my trip, alerting me to construction and the correct exit. I found the experience simultaneously helpful and scary!

Water connection: Powerful but difficult to predict. Communities and counties can now use GPS to locate sources of water contamination at relatively low cost. Geographic information systems (GIS) allow even the nonprofessional to map water supplies and monitor natural resources. The end result will be more grassroots leadership and greater numbers of people involved in water protection.

3. Globalization. Driven in part by the Internet and by strong political and economic pressures, we are truly becoming "one world, connected by water." The stewardship and conservation in one part of the world matter because our resources, economies, and knowledge are increasingly shared. The reverse of globalization will be the human need to focus on local action as an effective response to the larger, more complex world we live in.

Water connection: The trend toward privatization of water companies and water suppliers will continue. Private companies are able to respond with greater speed and flexibility in a global market. In a market-driven environment, it is difficult to embargo knowledge, practices, and in some cases even water itself. As a result, private companies will provide water to customers willing to pay, and water as a commodity will increase in value.

4. Resource scarcity. Competition for increasingly scarce water resources is a long-term and growing problem. Population pressures especially in parts of Asia and Africa, along with legal and environmental concerns, make it necessary to develop every available source of water. Thanks to the explosion of technology and science, our limited resources will stretch farther than ever before.

Water connection: Growing use of water reclamation—especially in arid and semiarid regions. This is happening now in the western part of the United States and desert regions in the Middle East. However, advances in technology will make water reuse more common and cost effective. As a result, localities facing temporary water shortages or needing supplemental supplies will use reclamation.

5. Watershed management. This trend has been gaining momentum in recent years and is being fueled in part by an expanded understanding of the interrelationships between water, soils, land-use practices, and development. A watershed approach to water management has become a template for integrated natural resources stewardship, and this approach is being adapted on almost every level—local, state, and federal. The growth in grassroots watershed groups may also be part of our need to "act locally, think globally."

Water connection: Watershed management and protection creates a focus on the importance of local action. Local water protection activities will no longer stop at the city limits but will extend to include counties, states, and large watershed areas. In some areas, this approach will create multistate management plans and basinwide water management strategies.

The most compelling aspect of my list is that each of these forces is connected to the other. For example, we are able to manage water in a watershed context because we have a more complete understanding of the interrelationships between groundwater and surface water, the role of wetlands, the importance of soils, and even the impact of air quality. Widespread access to scientific data and falling costs for technology help empower citizens, who in turn can take meaningful action to protect water locally and stretch increasingly limited resources. Communication technology makes it possible for us to quickly disseminate data and technologies, and to share successes and lessons learned around the world or down the hall.

The interrelated nature of water is one of the reasons that I've found groundwater protection so engaging. Groundwater is physically connected to almost every other natural resource because, thanks to gravity, a community or watershed protecting its groundwater is also protecting its surface water, disposing of its toxics and solid waste carefully, and managing its soils in a sustainable way.

We haven't always had the knowledge we need to protect water in an integrated manner. Many years ago, most people lived in rural areas and in small towns. People farmed for a living; there was limited irrigation, mostly from surface water sources, and water supply was probably a well or a nearby river. The connection between water and land uses existed, but there was little understanding and no best practices or zoning. The outhouse might have been too close to the well, the local landfill was almost always on the outskirts of town close to the water supply, and urban sewage was flushed into surface water bodies and forgotten.

When nineteenth-century scientists discovered that illness could be waterborne, water managers began to separate water management into its various uses. Infrastructures were built so that wastewater could be treated and discharged away from drinking water. Centralized drinking water distribution systems became more common, especially in growing urban areas. Drinking water was increasingly viewed as a consumer product discharged through miles of pipes into people's homes. Groundwater was pumped and consumed directly with little or no treatment.

As the twentieth century progressed, groundwater became an increasingly important part of the total water picture: a source of drinking water, irrigation, and surface water recharge. We also learned, much to our regret, that the Earth wasn't the perfect filter we had come to expect and that the result of human activity could and did end up in groundwater. We also learned that preventing pollution was the most viable strategy for groundwater protection because once contaminated, groundwater might be almost impossible to reclaim or restore.

Because of its function as the environmental "bottom line" and with the focus on pollution prevention as the management strategy of choice, groundwater emerged as a signpost toward the future. We now understand the connections between groundwater and surface water, and we know that in order to protect either one, we need to protect both. Groundwater protection also demands a high degree of commitment to the future and the courage to make difficult decisions, sometimes in the face of daunting political and economic challenges.

I am privileged because the people I work with every day demonstrate this commitment on a regular basis. The way I see it, it is because of people that we protect groundwater, and it is only through the efforts of people that we can do so. Following are some examples of this leadership, commitment, and connection from the Groundwater Foundation's files:

1. Central Platte Natural Resources District in Grand Island, Nebraska. Landowners have been participating in a groundwater education program since 1973. Beginning in 1986 and continuing to this day, a unique nitrogen management program that includes a multitude of voluntary, groundwater protective land-use practices have successfully stabilized nitrate levels throughout the central Platte River Basin.
2. Anaheim, California, and El Paso, Texas. Older citizens led programs to carefully locate and promote the sealing of abandoned wells in each of the communities. These voluntary initiatives have protected hundreds of wells from spills and thousands of people from potential harm.
3. Cape Girardeau, Missouri, and Desert Hot Springs, California. Leaders of the League of Women Voters in Cape Girardeau and of the Mission Springs Water District in Desert Hot Springs implemented community education campaigns to help citizens understand the importance of pollution prevention. Both communities have vulnerable groundwater because of local springs. The end results were sewer systems and ongoing pollution prevention activities.
4. Upper Republican Natural Resources District near Imperial, Nebraska. An innovative well metering and allocation program in place since the early 1980s has maintained groundwater supplies and agricultural production in an area that saw serious depletion during previous decades. Producers are given a five-year water account and use best practices to conserve water and maintain a constant supply. The practical result is less water use from the account in wet years so that more will be available in dry years.
5. East Lansing, Michigan. An innovative multijurisdictional approach to groundwater protection has helped the area around East Lansing protect groundwater quality and supply. A common water supply has created the ability to share a resource across jurisdictional boundaries in order to maintain a safe and ample water supply. This model will be increasingly common and will create an expanded sense of community in the process.

Each of these communities is led by people who understand the importance of connecting with their water and their community. These leaders are willing to think beyond immediate need and instead focus on future needs. These groups are taking care of water in their own community and in doing so are taking care of the water in everyone's community. Water is our common language, our common need, and our common future.

To achieve the sustainable future we all seek, it will be vitally important for each of us to:

1. Continue learning about water resources—bridging the gap between what we know and what we do.
2. Stay in touch with the needs and priorities of our communities—following the leadership and examples of water stewardship that surround us.
3. Understand that each of us can make a difference by becoming involved—while realizing that caring about and for water is an enormous and complex task.
4. Realize that we have come very far but that we still have far to go.
5. Rest assured that in connecting with water, each other, and our sense of place, we will finally become the stewards we were meant to be. And so we pause, reflect, and move forward, together.

CONCLUSIONS

In 1992, 500 participants, including government-designated experts, attended the International Conference on Water and the Environment (ICWE) in Dublin, Ireland. Later that same year, over 2500 delegates attended the United Nations Conference on Environment and Development (UNCED), a similar event in Rio de Janeiro, Brazil. Through these two conferences, basic principles were developed to protect water resources of the world for future generations. The main issues include the following:

- Freshwater is a finite and vulnerable resource, essential to sustain life, development, and the environment.
- Water development and management should be based on a participatory approach, involving users, planners, and policymakers at all levels.
- Women play a central part in the provision, management, and safeguarding of water.
- Water has an economic value in all its competing uses and should be recognized as an economic good. (28)

Also in 1992, the United Nations General Assembly declared March 22 of each year as World Day for Water. The idea is to devote one day each year for activities that will create public awareness regarding water resources through the creation of publications, documentaries, seminars, and expositions. The focus has been on problems with drinking water supplies and the need for conservation and protection of those water sources. Themes have included:

2004	"Water and Disasters"
2003	"Water for the Future"
2002	"Water for Development"
2001	"Water and Health"
2000	"Water for the Twenty-first Century"
1999	"Everyone Lives Downstream"
1998	"Groundwater—the Invisible Resource"
1997	"The World's Water: Is There Enough?"
1996	"Water for Thirsty Cities"
1995	"Women and Water"
1994	"Caring for Our Water Resources Is Everyone's Business"
1993	No theme

In 2000, the message of World Day for Water was straightforward:

- Water will be scarce.
- Water will be under increasing threat from pollution.
- We may suffer from increasingly severe periods of flood and drought.
- Safe and adequate water supplies should be the concern and responsibility of all.

What is your prediction for the future of our water resources? What are you going to do about it?

QUESTIONS FOR DISCUSSION

1. Do you agree that population growth is our most critical concern in future global water management issues?

2. In your opinion, should privatization of drinking water treatment and delivery be encouraged, or should it remain under government control?

3. Do you feel that dam construction will be widely encouraged in the future because of food shortages?

4. What water education organizations are active in your area? How can you become involved in such programs?

5. Do you agree that water will be the issue in international conflicts in the twenty-first century?

SUGGESTED RESOURCES FOR FURTHER STUDY

READINGS

Gleick, Peter. *The World's Water: 1998–1999*. Washington, DC: Island Press. 2000.

Leslie, Jacques. "Running Dry: What Happens When the World No Longer Has Enough Freshwater?" *Harper's Magazine,* July 2000.

Qing, Dai. *The River Dragon Has Come! The Three Gorges Dam and the Fate of China's Yangtze River and Its People*. Armonk, NY: M. E. Sharpe, 1998.

Roberts, Wess. *Leadership Secrets of Attila the Hun*. New York: Warner Books, 1985.

Silver, Cheryl Simon, and Ruth S. DeFries. *One Earth, One Future: Our Changing Global Environment*. Washington, DC: National Academy Press, 1990.

WEBSITES

Stauth, David, "Water Disputes: Cooperation, Not Conflict," News and Communication Service, Oregon State University, Corvallis, Oregon, July 2003. http://www.orst.edu/dept/science/srsp99wolf.html

United Nations Environment Programme, "World Water Day, 2001: Water for Health," July 2003. http://www.unep.org

REFERENCES

1. U.S. Census Bureau, "World Population Profile: 1998—Highlights," http:www.census.gov.ipc/www/wp98001.html, June 2001.

2. World Meteorological Organization, "The Dublin Statement on Water and Sustainable Development," http://www.wmo.ch/web/homs/icwedece.html, May 2001.

3. World Meteorological Organization, "Water for the Twenty-First Century," http://www.health.fgov.be/WH13/krant/krantarch2000/kranttekstmar/000322m05wmo.htm, May 2001.

4. Leslie, Jacques. "Running Dry: What Happens When the World No Longer Has Enough Freshwater?" *Harper's Magazine,* July 2000, 37–38.

5. United Nations Environment Programme, "World Water Day, 2001: Water for Health," http://www.unep.org, May 2001.

6. Gleick, Peter. *The World's Water: 1998–1999*. Washington, DC: Island Press, 2000, 264.

7. Ibid., 39 and Gleick, Peter H. "Coping with the Global Fresh Water Dilemma: The State, Market Forces, and Global Governance." *The Global Environment in the Twenty-First Century,* edited by Pamela S. Chase. Tokyo, Japan: United Nations University Press, 2000, 207.

8. *AGENDA*, Magazine of the RWE, Hamburg, Germany, January 2000, 4–8, see http://www.thames-water.com, June 2001.

9. Ibid., 4–8.

10. "Dams and Development: A New Framework for Decision-Making," The Report of the World Commission of Dams, World Commission of Dams, Gland, Switzerland, November 15, 2000, http://www.damsreport.org/wcd_overview.htm, May 2001.

11. Ibid.

12. Silver, Cheryl Simon, and Ruth S. DeFries. *One Earth, One Future: Our Changing Global Environment*. Washington, DC: National Academy Press, 1990, 91.

13. Personal communication with Dennis Nelson, Executive Director, The Watercourse and International Project WET, Montana State University, Bozeman, Montana, August 6, 2001.

14. The Water Education Foundation, http://www.watereducation.org, May 2001.

15. The Groundwater Foundation, http://www.groundwater.org, May 2001.

16. Stockholm International Water Institute, http://www.siwi.org/sws/eng/swsframe.htmll, June 2001.

17. David Stauth, "Water Disputes: Cooperation, Not Conflict," News and Communication Service, Oregon State University, Corvallis, Oregon, http://www.orst.edu/dept/science/srsp99wolf.html, May 2001. Reprinted with permission.

18. "Transboundary Freshwater Dispute Database," Northwest Alliance for Computational Science NACSE, kwysiwyg://63http://terra.geo.orst.edu/users/tfddl, May 2001.

19. Stauth, Oregon State University. Reprinted with permission.

20. Roberts, Wess. *Leadership Secrets of Attila the Hun*. New York: Warner Books, 1985, 57–106.

21. Engel, J. Ronald, and Joan Gibb Engel (Eds.). *Ethics of Environment and Development*. Tucson: University of Arizona Press, 1990, xii.

22. Botzler, Richard G., and Susan J. Armstrong. *Environmental Ethics: Divergence and Convergence*. 2nd ed. Boston: McGraw-Hill, 1998, 45.

23. "Mississippi River Beautification and Restoration Project," kwww.cleanrivers.coml, May 2001.

24. Kilen, Mike, "Mississippi Garbageman Makes a Splash," *Des Moines Register,* Des Moines, Iowa, August 6, 2000, 1E–2E.

25. Living Lands and Water, 2002 Annual Report, http://www.cleanrivers.com, 2002.

26. Kirn, Walter, *Time* magazine, July 10, 2000, http://www.time.com/time/magazine/articles/0,3266,49029,00.htm, May 2001.

27. Personal communication with Chad Pregracke, Mississippi River Beautification and Restoration Project, August 15, 2001.

28. World Meteorological Organization, "The Dublin Statement on Water and Sustainable Development," http://www.wmo.ch/web/homs/icwedece.html, May 2001.

APPENDIX

READING TOPOGRAPHIC MAPS

From the air, the view of the central and western United States is characterized by a grid system of fields, roads, and property lines that often span from horizon to horizon. This checkerboard of land surveying runs in a north-south, east-west configuration and is occasionally bisected by meandering streams, diagonal valleys, and sweeping ridges. The Northwest Ordinance of 1785 provided for the rectangular survey and establishment of the 6-mile-square (9.6 km sq) township grid system.

As early as 1790, the federal government sold public land in Ohio based on a system of sections, townships, and ranges. Prior to that time, colonists had set property boundaries by natural or developed landmarks, such as a grove of trees, roads, a stone fence, or the edge of a stream. Since many of these natural features were more broadly spaced as development moved west, the section, township, and range system was created to divide property into square sections of land that were one mile (1609 m) on each side and contained 640 acres (259 ha) in each section. Townships are grouped into sets of 36 sections (6 miles, or 9.6 km, on each side), with the section numbering system starting in the northeast corner of the township with Section 1 and continuing to the southwest corner of the township and ultimately to Section 36. Townships are bounded by meridians (north-south lines) and by east-west baselines selected by the original land surveyors.

A section can be divided into four equal parts, called quarter sections, with each portion containing 160 acres (640 acres divided by 4 = 160 acres). Each quarter section is identified by its location within the section, such as the northwest quarter, the northeast quarter, the southeast quarter, and the southwest quarter of a particular section. These quarter sections are also often represented as the NW 1/4, NE 1/4, SE 1/4, and SW 1/4 of a section.

Each quarter section can be further divided into four parts, so that within each quarter section there is a NW 1/4, NE 1/4, SE 1/4, and a SW 1/4 of the quarter section. Each of these quarter-quarter sections contains 40 acres (160 acres divided by 4 = 40 acres). The location of a parcel of ground can be identified by its location within the section and quarter section. See Figure A.1.

Topographic maps (from the Greek words *topo* for "place" and *graphos* for "drawn" or "written") are drawn in various scales based on the area to be represented and the dimensions of paper used for the map. The proportion used is called the *scale* of a map and is usually listed as a fraction or a proportion. If the map scale is 1:63,360, a distance of 1 inch on the map will equal one mile on the ground (63,360 inches divided by 12 = 5280 feet = 1 mile). The first number of the map scale is always 1, while the second number varies. A larger second number means the scale of the map is larger so that a more extensive land area can be shown on the map. The problem with larger scale maps is that land features become smaller and many

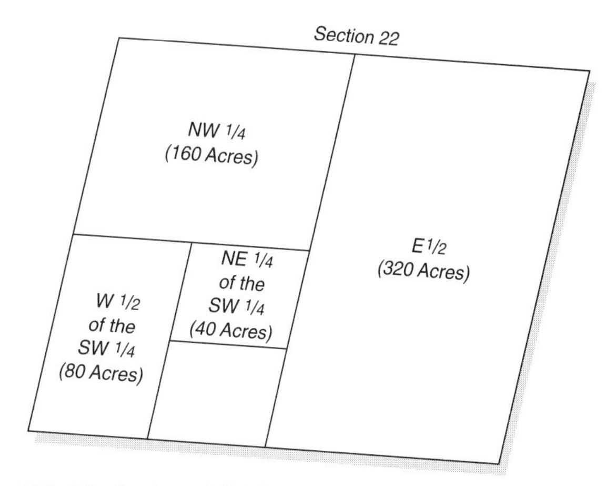

FIG. A.1. Section subdivisions.

developed features are left off. The most common scale used on USGS (U.S. Geological Survey) topographic maps is 1:24,000, which means that one unit on the map (either inches or centimeters) represents 24,000 units on the ground. This scale covers an area of 7.5 minutes of longitude and 7.5 minutes of latitude, and explains why cartographers often call them 7.5-minute quadrangle maps.

Contour lines are drawn on topographic maps to show changes in elevation in regular increments above mean sea level (the average between high and low tide). Each contour represents a single elevation, so contour lines never cross. The shoreline at mean sea level is represented by contour 0, while an elevation 10 feet higher would be represented as contour 10. The one-mile-high marker attached to a stairway of the Colorado State Capitol Building (or the ring of purple seats at Coors Field) in Denver marks contour 5280 feet.

Contours make it possible to measure the height of mountains, elevation changes in watersheds, and slopes of hillsides. Selected contour lines are shaded darker to accentuate changes in elevation at larger intervals, such as at 20-, 40-, or 50-foot gradations. Contour lines that are grouped closely together represent sudden changes in elevation, while contours located far apart (or completely absent) show relatively flat areas. All contour lines eventually form a loop, but the scale of a map may not be small enough to show these connections on a single map. Early contour maps (called *bathymetric maps*) did not show elevation above sea level but instead presented elevations below the water surface of rivers and harbors to assist navigators.

Symbols on a topographic map include blue for water, green for forests or other vegetation, and gray or red for developed areas. Roads, trails, and other boundaries are shown in black, while point symbols are shown to locate wells, springs, water tanks, and other features. All areas of the United States have been mapped. See http://mapping.usgs.gov for national mapping information from the U.S. Geological Survey. This site also includes Landsat satellite photos, links to mapping products, and other cartographic data. Unfortunately, some topographic maps are seriously out of date, particularly in rapidly suburbanizing and urbanizing areas of the country.

SELECTED ENVIRONMENTAL AND CONSERVATION ORGANIZATIONS

THE WILDERNESS SOCIETY

http://www.wilderness.org

The Wilderness Society was formed in 1935 to protect wilderness areas and to help develop a national system of protected lands. The organization is headquartered in Washington, D.C., and has regional offices in Anchorage, Seattle, San Francisco, Boise, Bozeman, Denver, Atlanta, and Boston.

SIDEBAR

Aldo Leopold was born in Burlington, Iowa, in 1887 and loved exploring nearby woods, swamps, and fields. He attended high school in New Jersey, was one of the first graduates from Yale University's new graduate school of forestry, and joined the U.S. Forest Service as a ranger in 1909. He spent time in New Mexico and Wisconsin, and in 1926 he convinced the Forest Service to designate 500,000 acres (202,343 ha) of New Mexico's Gila National Forest as a wilderness area. This was the first such designation in the country and was made an original part of the National Wilderness Protection System in 1964. (1)

Leopold later purchased a rundown farm in south-central Wisconsin along the Wisconsin River where an old chicken coop was converted into a small cabin. Time was spent planting trees, caring for the land, and developing his famous book, *A Sand County Almanac*. Leopold died in 1948 fighting a wildfire near the cabin, but his legacy lives on in his teachings, writings, and the continued health of the "rundown" farm just north of Baraboo, Wisconsin.

Aldo Leopold, one of the original founders of the Wilderness Society, hoped the Society would form a cornerstone for the national movement needed to protect vanishing wilderness areas. Congress eventually passed the Wilderness Act in 1964, with some 9 million acres (3.6 million ha) designated as wilderness. Today federally protected wilderness areas total more than 100 million acres (40.5 million ha) and include high mountain meadows, Alaskan tundra, and fragile desert lands.

SIERRA CLUB

http://www.sierraclub.org

The Sierra Club was founded in 1892 by John Muir and currently has over 600,000 members. Its mission includes exploration, enjoyment, and protection of wild places on Earth, promotion of responsible use of ecosystems and resources, education, and the use of all lawful means to carry out these objectives. The organization has field offices across the country and extensive education programs, and it is actively involved in legislation at the federal level.

John Muir was a staunch opponent of a dam proposed for construction across the Hetch Hetchy Valley near San Francisco in the early 1900s. This pristine valley located in Yosemite National Park, created by the Tuolumne River, was to be dammed to provide additional water supplies for the growing Bay area. Gifford Pinchot (the first director of the U.S. Forest Service) was a close friend of Muir's and was often a hiking partner. However, Pinchot had embraced the "wise use" concept of natural resources—that is, using natural resources for the greatest good of the greatest number of people. In contrast, Muir passionately believed that nature, and the Park in particular, must be protected for its own good. The friendship suffered and was eventually destroyed when the dam was built. Today the valley is inundated by the Hetch Hetchy Reservoir.

Many years, later, in the 1970s, the Sierra Club fell out of favor with the then sitting president, Richard Nixon. Members of the Sierra Club lobbied so aggressively in Washington, D.C., that Nixon ordered the Internal Revenue Service to strip the group of its tax-exempt status. But the organization was not fazed, and today it flourishes around the world. The Sierra Club has its international headquarters in San Francisco.

NATIONAL AUDUBON SOCIETY

http://www.audubon.org

The National Audubon Society was formed in 1905 in response to the slaughter of millions of birds for their plumage. Overhunting during the nineteenth century led to the extinction of numerous species, including the passenger pigeon, the Carolina parakeet, and the great auk. George Bird Grinnell was one of the first to speak out against this senseless destruction of birds. Grinnell was schooled in his early years in New York by the widow of John James Audubon, an avid lover of birds. Later, he published a magazine called *Forest and Stream* (later called *Field and Stream*). In 1886, six years after he received the Ph.D. in Paleontology from Yale, he asked concerned citizens to join him in a new national organization to promote the protection of birds. In only three months, 38,000 people joined the society. Overwhelmed, Grinnell had to disband this group in 1888.

In 1896, Harriet Lawrence Hemenway and Minna B. Hall formed the Massachusetts Audubon Society. To promote their agenda, they refused to wear clothing that included bird feathers or stuffed birds (generally placed on hats), and they wrote newspaper editors and politicians of their concerns. They also gained support from area farmers. Crop yields were falling because of the declining numbers of birds available to eat harmful bugs. A few months later, Audubon societies were formed in Pennsylvania, Texas, and 14 other states.

In 1900, Frank Chapman, an ornithologist with the American Museum of Natural History in New York City, began a Christmas Bird Count, which continues today. Chapman felt it was better to count birds than to shoot them. At the same time William Dutcher, who would later become the first president of the National Audubon Society, began hiring wardens to patrol important bird-nesting sites. Three of these wardens were killed by poachers.

In 1905, the National Association of Audubon Societies for the Protection of Wild Birds and Animals (today known as the Audubon Society) was created with the immediate goal of combating the plume trade. That same year, the Wild Birds Protection Act was passed to prohibit the importation of flamingos (or their feathers) from the Bahamas where they faced decimation.

In 1910, the New York State Audubon Plumage Law was approved and banned the sale of native bird plumes in the state. Later, federal bird sanctuaries and national wildlife refuges were encouraged and developed. In the 1960s and 1970s, the National Audubon Society became involved in implementation of the Clean Water, Wild and Scenic Rivers, and the Endangered Species Acts.

Today, the Audubon Society's efforts include the protection of wetlands and ancient forests in the Pacific Northwest, and the prevention of oil drilling in the Arctic National Wildlife Refuge. The Audubon's main office is located in New York City with a staff of 300 full-time employees, including scientists, educators, sanctuary managers, regional and state directors, and government affairs specialists. The Society has 508 chapters and 550,000 members. (2)

NATIONAL WILDLIFE FEDERATION

http://www.nwf.org

The National Wildlife Federation (NWF) was founded in 1936 by editorial cartoonist J.N. "Ding" Darling. It is the nation's largest member-supported conservation group, with over 4 million members. It was originally called the General Wildlife Federation when it was created in response to the North American Wildlife Conference convened by President Franklin D. Roosevelt in 1936. Its headquarters are located in Reston, Virginia, with a primary purpose of education.

The NWF publishes books, magazines, videos, and other hands-on materials to link children and families back to nature. In addition, the organization educates lawmakers, business leaders, and grassroots activists regarding environmental problems and solutions that are effective and common-sense. Recent activities have included support for the reintroduction of gray wolves into Yellowstone National Park, protection of the Florida Everglades ecosystem, protection of wetlands in the "prairie pothole" region of South Dakota, floodplain management, protection of endangered species, support for reintroduction of the grizzly bear in Idaho and Montana, implementation of global trade rules to protect the environment, and protection of the Copper River Delta of Alaska. (3)

DUCKS UNLIMITED

http://www.ducks.org

Ducks Unlimited (DU), founded in 1937, is the leading waterfowl and wetlands conservation entity in North America. Its primary mission is to fulfill the life-cycle needs of North American waterfowl. Integral to this mission is the reversal of the trend of wetland loss. Over 50 percent of U.S. wetlands have been destroyed or degraded, with continued loss estimated at over 55,000 acres (22,258 ha) per year. Efforts to conserve waterfowl habitats have led to initiatives in the arctic tundra of Alaska, the prairie region of the United States and Canada, the tropical wetlands of Mexico, and innumerable sites in between. Members of DU number over 750,000, including hunters who realize that healthy habitats and intact landscapes are vital for the protection of adequate numbers of waterfowl for sport. DU works with federal, state, and local governments, as well as private landowners, other conservation groups, land trusts, and others to restore and protect wetland, grassland, and other habitats important to waterfowl. DU has headquarters in Memphis, Tennessee, a full-time staff of approximately 600, and an annual budget of over $75 million. (4)

THE NATURE CONSERVANCY

http://www.nature.org

The Nature Conservancy was formed in 1951 and is the world's largest private international conservation group, with over 1 million members and a paid staff of approximately 2800. The Nature Conservancy identifies key conservation targets, such as places, animals, and plants, and then implements programs (such as purchasing land) to reduce or eliminate threats to these targets. The organization currently protects over 12 million acres (4.9 million ha), which is an area greater than Connecticut, Delaware, New Jersey, and Rhode Island combined. Over 61 million acres (24.7 million ha) are protected outside the United States in Mexico, Canada, Central America, South America, Asia Pacific, and the Caribbean. (5)

ENVIRONMENTAL DEFENSE

http://www.environmentaldefense.org

The Environmental Defense (ED) was established in 1967 and has an annual budget of approximately $40 million. Membership is over 300,000 with a staff of about 250, including 75 full-time scientists, economists, and attorneys. Headquarters are in New York City, with regional offices in Washington, D.C., Oakland, California, Boulder, Colorado, Raleigh, North Carolina, and Austin, Texas. (6)

Five years after Rachel Carson's *Silent Spring* was published, volunteer conservationists formed the organization to ban the use of DDT on the marshes of Long Island for mosquito control. The EDF (called the Environmental Defense Fund at the time) filed suit against the Suffolk County Mosquito Control Commission and used scientific data to show that DDT was causing the eggshells of ospreys and other birds to become thin and crack. This was arguably the first attempt to use scientific data in a federal court to stop an environmental problem. Further lawsuits by the EDF in other locations of the country led to a federal ban on the use of DDT in 1972.

ED is currently involved in a broad range of environmental issues but recently has become a leading advocate of economic incentives to solve environmental problems. The organization works closely with grassroots groups, lobbies to influence national environmental policy, and uses litigation when necessary to stop or alter destructive projects.

NATURAL RESOURCES DEFENSE COUNCIL

http://www.nrdc.org

The Natural Resources Defense Council (NRDC) uses the support of over 500,000 members combined with science and the law to launch worldwide campaigns that mobilize citizens to save the planet's most endangered places, such as the Gray Whale nursery in Baja Mexico and the Great Bear Rainforest in British Columbia. This national organization was formed in 1970 and is headquartered in New York City. NRDC also seeks to reduce the disproportionate environmental burdens levied on people of color and those facing social and economic inequities. NRDC's ultimate goal is to create a human lifestyle that can be sustained indefinitely without polluting or depleting the Earth's resources. (7)

TROUT UNLIMITED

http://www.tu.org

Trout Unlimited (TU) is a conservation group made up of 125,000 volunteers with 500 chapters across the country. Its mission is to conserve, protect, and restore North America's trout and salmon fisheries and their watersheds. The organization testifies before Congress, publishes a quarterly magazine, intervenes in federal legal proceedings, and works with local and regional landowners, other conservation groups, municipalities, and regional organizations to protect, enhance, create, and preserve trout habitat. Trout Unlimited has its main office in Arlington, Virginia. (8)

REFERENCES

1. Wilderness Society Homepage, http://www.wilderness.org, April 2001; and personal communication with Ben Beach, The Wilderness Society, June 11, 2001.

2. National Audubon Society Homepage, http://www.audubon.org, October 2001.

3. National Wildlife Federation Homepage, http://www.nwf.org, April 2001.

4. Ducks Unlimited Homepage, http://www.ducks.org, April 2001; and personal communication with Dr. Keith McKnight, Manager of Conservation Programs, Ducks Unlimited, Memphis, Tennessee, August 3, 2001.

5. The Nature Conservancy Homepage, http://www.nature.org, April 2001.

6. Environmental Defense Homepage, http://www.environmentaldefense.org, April 2001; and personal communication with Tom Graff, Environmental Defense, June 12, 2001.

7. Natural Resources Defense Council Homepage, http://www.nrdc.org, April 2001.

8. Trout Unlimited Homepage, http://www.tu.org, April 2001.

PHOTO AND ILLUSTRATION CREDITS

PHOTO CREDITS

Frontmatter Page v: Courtesy Denise Powell. Page vi: Emmett Jordan.

Chapter 1 Page 4: Photo Deutsches Museum, Munchen. Page 6: Reserved to the Ashmolean Museum, Oxford. Not to be reproduced without permission. Page 7: Lou Hanping/China Stock Photo Library. Page 9: Royalty-Free/Corbis Images. Page 10: Courtesy of J. Willard Marriott Library, University of Utah. Page 13: © AP/Wide World Photos. Page 15: Biblioteca-Pinacoteca Ambrosiana. Page 17: U.S. Army Corps of Engineers, Rock Island District. Page 18: David Wells/Corbis Images.

Chapter 2 Page 23: © Art Resource. Page 28: Courtesy Korea Meteorological Administration. Page 29: Courtesy Snow Survey, Natural Resources Conservation Service, USA. Page 32: Courtesy State of Utah, Dept. of Natural Resources, Division of Water Resources. Page 33: David & Peter Turnley/Corbis Images. Page 49: Courtesy Earth Satellite Corporation. Page 50: National Archives of Australia, A1200/L44186.

Chapter 3 Page 61: Courtesy U.S. Army Corps of Engineers. Page 62: Davis Hiser/Stone/Getty Images. Page 71: Courtesy Grace Cech. Page 72: Courtesy U.S. Geological Survey. Page 73: Photo by Willard Price. Page 76: Courtesy April Freier. Page 77: © AP/Wide World Photos.

Chapter 4 Page 83: Courtesy NASA. Page 89: Courtesy Joel Dexter, Illinois State University. Page 90: Courtesy Tom Cech. Page 93: Courtesy German National Tourist Office. Page 94: © A.N.T./Photo Researchers. Page 101: Courtesy European Geophysical Society. Page 105: © Syracuse Newspapers/Jim Commenlucci/The Image Works.

Chapter 5 Page 132: Wellcome Library, London. Page 133 (left): Dennis Kunkel/Phototake. Page 138: Courtesy Frank Gwin, Jr. Page 139: Courtesy Curt Elmore. Page 145: Courtesy Sandra Novotny.

Chapter 6 Page 153: From "Garden Cities of Tomorrow," 1902, by Ebenezer Howard. Page 155: Courtesy Los Angeles Department of Water and Power. Page 156: Courtesy Los Angeles Department of Power and Water. Page 162 (top): Courtesy New York City Department of Environmental Protection. Page 162 (bottom): Courtesy Nancy Cruickshank. Page 165: Don Foley/National Geographic Society. Page 172: Courtesy Bureau of Reclamation, Pacific Northwest Regional Office. Page 175: Courtesy Lindsay Manufacturing Company of Lindsay, Nebraska. Page 176: © 2003 Valmont Irrigation.

Chapter 7 Pages 188,189, 193, & 194: Courtesy Bureau of Reclamation, Pacific Northwest Regional Office. Page 191: Courtesy the Central Nebraska Public Power and Irrigation District. Pages 198, 199 & 200: Courtesy Colin Flahive. Page 203: Courtesy Dr. Ted Nelson. Page 206: Courtesy Tennessee Valley Authority.

Chapter 8 Page 211: Remington, *Fight for the Water Hole*, 1903, oil on canvas. Museum of Fine Arts, Houston; The Hogg Brothers Collection, gift of Miss Ima Hogg. Page 212: Gianni Dagli Orti/Corbis Images. Page 220 (top): U.S. Census Office, Tenth Census, 1880, vols. 16–17, "Reports on the Water-Power of the United States" (Washington, D.C., 1885, 1887), pt. 1, p. 81. Page 220 (bottom): Courtesy Manchester, N.H., Historic Association. Reproduced with permission. Page 224: Courtesy Bureau of Land Management, Colorado State Office. Page 225: Courtesy Denver Public Library, Western History Department. Page 229: Courtesy Lawrence L. Lowry. Page 241: Courtesy John Regan.

Chapter 9 Page 252: Bettman/Corbis Images. Page 254: © AP/Wide World Photos. Page 255: *Water Lily Pool*, by Claude Monet, © Francis G. Mayer/CORBIS. Page 258: From "Exploration of the Colorado River and Its Canyons," by John Wesley Powell. Page 266: Courtesy U.S. Department of Interior, U.S. Fish and Wildlife Service. Page 278: Courtesy Federal Emergency Management Agency.

Chapter 10 Page 291: Courtesy U.S. Army Corps of Engineers. Page 292 (both left and right): © Kent Wood/Photo Researchers. Page 295: Courtesy Bureau of Reclamation. Page 303: Courtesy Chesapeake Bay Foundation. Page 305: Courtesy Mexican Institute of Water Technology. Page 312: Courtesy Ralph Pentland. Page 315: Courtesy Communications Nova Scotia.

Chapter 11 Page 322: Kean Collection/Hulton Archive/Getty Images. Page 324: Yann Arthus-Bertrand/Corbis Images. Pages 325 & 327: Courtesy Dr. Fares Howari. Page 330 (top): Jason Young. Page 330 (bottom): © Peter Yates/Photo Researchers. Page 333: Courtesy Sewerage and Water Board of New Orleans. Page 335: Courtesy Chuck Hamernik. Page 338: Courtesy Clay Houchin.

Chapter 12 Page 349: Reproduced from the Collections of the Library of Congress. Page 356: © AP/Wide World Photos. Page 365: Courtesy Nebraska Game and Parks Commission. Pages 371 & 373: Courtesy U.S. Army Corps of Engineers. Page Page 374: Courtesy Tom Cech.

Chapter 13 Page 381: © The New Yorker Collection 2004 Jack Ziegler from cartoonbank.com. All Rights Reserved. Page 383: Reuters/Corbis Images. Page 390: Courtesy Dr. Ronald Cummings, Environmental Policy Program, Georgia State University.

Chapter 14 Page 398: Courtesy U.S. Department of the Treasury, Internal Revenue Service. Pages 403 & 404: Courtesy Sandow Birk.

Chapter 15 Page 415: Wesley Bocxe/Photo Researchers. Page 419: Courtesy Orange County Water District. Page 425 (top): Courtesy Chad Pregracke and Greg Boll, *Quad-City Times*, Davenport, Iowa. Page 425 (bottom): Courtesy Chad Pregracke. Page 426: Courtesy The Groundwater Foundation.

ILLUSTRATION CREDITS

Chapter 1 Figure 1.1: From *Geography: Realms, Regions, and Concepts,* 10th Edition, by Harm de Blij and Peter O. Muller. Copyright © 2002 H.J. de Blij and John Wiley and Sons, Inc. This map was originally produced in color. Adapted and reprinted by permission of John Wiley and Sons, Inc. (Figure 6.3). Figure 1.2: From D.K. Todd, *Groundwater Hydrology*, Wiley, 1980 (Figure 1.1).

Chapter 2 Figure 2.2: From Raven/Berg, *Environment,* Third Edition, Harcourt College Publishers, 2001 (Figure 6.6) Figure 2.3: From Raven/Berg, *Environment,* Third Edition, Harcourt College Publishers, 2001 (Figure 13.3). Figure 2.4: From Marsh/Grossa, *Environmental Geography,* Second Edition, Wiley, 2002 (Figure 5.8). Figure 2.7: From Marsh/Grossa *Environmental Geography,* Second Edition, Wiley, 2002 (Figure 12.11). Figure 2.10: From Skinner/Porter, *The Blue Planet,* Second Edition, Wiley, 2000 (Figure 13.15). Figure 2.11: From Skinner/Porter, *The Blue Planet,* Second Edition, Wiley, 2000 (Figure 14.4). Figure 2:12: From Marsh/Grossa, *Environmental Geography,* Second Edition, Wiley, 2000 (Figure 12.4).

Chapter 3 Figure 3.3: From Murck/Skinner, *Geology Today*, Wiley, 1999 (Figure 9.24). Figure 3.4: From Skinner/Porter, *The Blue Planet,* Second Edition, Wiley, 2000 (Figure 9.4). Figure 3.9: From Murck/Skinner, *Geology Today*, Wiley, 1999 (Figure 12.9). Figure 3:13: From Murck/Skinner, *Geology Today*, Wiley, 1999 (Figure 12.17). Figure 3.15: From Skinner/Porter, *The Dynamic Earth,* Fourth Edition, Wiley 2000 (Figure 10.8). Figure 3.17, 13.18, & 13.19: GIS Map courtesy of Jake Freier, *GIS Coordinator, Iowa Emergency Management Division*, Des Moines, Iowa.

Chapter 4 Figure 4.2: From Marsh/Grossa, *Environmental Geography,* Second Edition, Wiley, 2002 (Figure 12.15). Figure 4.3: From *Geography: Realms, Regions, and Concepts,* 10th Edition, by Harm de Blij and Peter O. Muller. Copyright © 2002 H.J. de Blij and John Wiley and Sons, Inc. This map was originally produced in color. Adapted and reprinted by permission of John Wiley and Sons, Inc. 2002 (Figure I-6). Figure 4.6: From Skinner/Porter, *The Dynamic Earth,* Fourth Edition, Wiley 2000 (Figure 11.1). Figure 4.7: From Skinner/Porter, *The Dynamic Earth,* Fourth Edition, Wiley, 2000 (Figure 11.10). Figure 4.8: From Skinner/Porter, *The Dynamic Earth,* Fourth Edition, Wiley, 2000 (Figure 11.9). Figure 4.11: From Skinner/Porter/Botkin, *The Blue Planet,* Second Edition, Wiley 1999 (Figure 8.5). Figure 4.16: From Skinner/Porter, *The Blue Planet,* Second Edition, Wiley, 1999 (Figure 9.25).

Chapter 5 Figure 5.1: Courtesy of the *Office of Water, U.S. Environmental Protection Agency*, Washington, D.C. Figure 5.4: From Skinner/Porter, *The Blue Planet,* Second Edition, Wiley 2000 (Figure 16.3). Figure 5.5: from Botkin/Keller, *Environmental Science,* Third Edition, Wiley, 2000 (Figure 19.1). Figure 5.6: From Raven/Berg, *Environment,* Third Edition, Harcourt College Publishers, 2001 (Figure 6.3). Figure 5.13: GIS map courtesy of *Curt Elmore, Ph.D., P.E.,* University of Missouri at Rolla. Figure 5.14: From Marsh/Grossa, *Environmental Geography,* Second Edition, Wiley, 2002 (Figure 13.10).

Chapter 7 Figure 7.1, 7.2: Adapted by U.S. Army Corps of Engineers. Figure 7.14: Adapted by the Tennessee Valley Authority.

Chapter 8 Figure 8.3: Adapted by the *United States Census Office, Tenth Census*, 1880, vols.16–17, "Reports on the Water-Power of the United States" (Washington, D.C., 1885, 1887), pt. 1 p.81. Figure 8.5: Illustration courtesy of the *Bureau of Land Management*, Colorado State Office, Lakewood, Colorado. Figure 8.10: From Botkin/Keller, *Environmental Science,* Third Edition, Wiley, 2000 (Figure 19.1). Figure 8.11, 8.12, & 8.13: GIS maps provided by John Regan, *GIS Manager*, Pima Country Department of Transportation.

Chapter 9 Figure 9.2: Map obtained from the *U. S. Army Corps of Engineers*. Figure 9.6: Map obtained from the *U. S. Bureau of Reclamation*. Figure 9.8: Obtained from the *Federal Emergency Management Agency.*

Chapter 10 Figure 10.1: Courtesy of the *City of Columbia*, Columbia, South Carolina. Figure 10.9, 10.10, 10.11, & 10.12: Courtesy of Dr. Alvaro Aldama, *Director General of the Mexican Institute of Water Technology*, Jiutepec, Morelos, Mexico. Figure 10.13: From *Geography: Realms, Regions, and Concepts,* 10th Edition, by Harm de Blij and Peter O. Muller. Copyright © 2002 H.J. de Blij and John Wiley and Sons, Inc. This map was originally produced in color. Adapted and reprinted by permission of John Wiley and Sons, Inc. (Figure 3.16).

Chapter 12 Figure 12.3: Map courtesy of *South Florida Water Management District*, "Overview of Everglades Construction Projects." Figure 12.4: Courtesy of the *Tennessee Valley Authority*. Figure 12.7, 12.8: Courtesy of the *U. S. Army Corps of Engineers*.

Chapter 13 Figure 13.4: Sourced from the *Murray-Darling Basin Commission*, photo credit Arthur Mostead

Chapter 14 Figure 14.1: From *Geography: Realms, Regions, and Concepts,* 10th Edition, by Harm de Blij and Peter O. Muller. Copyright © 2002 H.J. de Blij and John Wiley and Sons, Inc. This map was originally produced in color. Adapted and reprinted by permission of John Wiley and Sons, Inc. (Figure. 1-10). Figure 14.3: By Linda Sanders, "Water war, closer to end?" *The Post-Searchlight*, May 2, 2001. Figure 14.3: By the "Staff," "Florida says 'No,' water deal goes down the drain," *The Post*, December 22, 2000. Figure 14.6: Courtesy of Sandow Birk of Long Beach, California.

Chapter 15 Figure 15.1: From Marsh/Grossa, *Environmental Geography,* Second Edition, Wiley, 2002 (Figure 1.2).

GLOSSARY

100th Meridian General hydrologic boundary between humid and dry climates in North America. p. 167

100-year flood (Q100) Expected flood event that has a probability of 1 in 100 of occurring in any given year. p. 75

500-year flood (Q500) Expected flood event that has a probability of 1 in 500 of occurring in any given year. p. 75

abutments Sides of a dam structure that are extended into native geologic material for support. p. 184

acenas Spanish word for "watermills." p. 9

acequias Spanish word for "irrigation canals" or "irrigation ditches," which are waterways carved in the ground to transport irrigation water to fields. p. 9

acre-foot Amount of water required to cover 1 acre of land to a depth of 1 foot. p. 73

ACT and ACF Compacts Interstate compacts between Florida, Alabama, and Georgia on the Alabama, Coosa, and Tallapoosa rivers (ACT), and the Apalachicola, Chattahoochee, and Flint rivers (ACF). p. 400

activated sludge Secondary wastewater treatment process that requires constant agitation and aeration to promote bacterial activity. p. 340

adiabatic lapse rate Rate at which atmospheric temperature decreases as altitude increases. p. 42

adiabatic process Process of an air mass rising and expanding without exchanging heat with its surroundings. p. 42

aerobic Condition where ample supplies of oxygen are available. p. 130

aerosols Small liquid droplets that can remain suspended in the air. p. 42

alluvial fan Landform created by the deposition of sediments from a river, and generally found near the mouth of a canyon. p. 68

alluvium Unconsolidated sediments deposited by a stream. p. 89

ambient conditions Existing levels of contamination. p. 143

ammonia (NH_3) Byproduct of decomposition of plants and animals. p. 128

ammonium (NH_4^+) Byproduct of decomposition of plants and animals. p. 128

anaerobic Condition where little or no oxygen is available. p. 130

Anasazi Indians Native American tribe that developed elaborate irrigation systems approximately 1000 years ago in present-day Colorado. p. 9

appurtenances Auxiliary features of a dam such as outlet gates, tunnels, and roadways. p. 186

aqueduct Conduit for carrying large quantities of flowing water, generally constructed of stone in ancient times. p. 4

aquiclude Impermeable beds of geologic material that hinder or prevent groundwater movement. p. 92

aquifer Underground geologic formation through which water can percolate, sometimes very slowly. p. 91

arid Climatic condition where average annual precipitation is under 10 inches (25 cm). p. 26

aridity Permanent, dry climatic condition in a region where average annual precipitation is less than 10 inches (25 cm). p. 48

arsenic Naturally occurring inorganic chemical present in the Earth's crust. p. 121

aspect Direction of exposure of sloping lands in a watershed. p. 59

atmosphere Gaseous portion of a planet. p. 41

auger drilling Process of using a metal bit at the end of a rotating column of pipe to construct a well. p. 105

bank storage Temporary storage of water in the banks of a river. p. 60

barometer Instrument that measures atmospheric pressure. p. 41

bed load Sediment carried by a stream along the bottom of its channel. p. 69

benefit-cost analysis Process of determining the ratio of the present value of project benefits to project costs. p. 196

biological oxygen demand (BOD) Amount of dissolved oxygen demanded by bacteria during the decomposition of organic matter under aerobic conditions. p. 131

Blaney–Criddle Method Mathematical method used to determine evapotranspiration rates. p. 35

braided river Relatively shallow river consisting of numerous intertwining channels. p. 62

Brigham Young Second president of the Church of Jesus Christ of Latter-Day Saints (Mormons), who encouraged development of large-scale irrigation projects in Utah in the mid-1800s. p. 10

Bureau of Land Management (BLM) Federal agency created in 1946 to manage federal public lands. p. 268

Bureau of Reclamation (USBR, Reclamation, or BuRec) Federal agency created in 1902 to construct irrigation projects in the West. p. 257

buttresses Triangular, concrete supports used to distribute water pressure to the foundation of a gravity concrete dam. p. 185

California Doctrine Doctrine of water allocation, which uses aspects of both the Riparian Doctrine and the Doctrine of Prior Appropriation, developed in California in the mid-1800s. p. 224

candidate species Plant or animal species or subspecies that is in danger of becoming threatened or endangered in the near future. p. 354

capillary action Process used by plants to draw water into roots. p. 91

carcinogen Substance that can cause cancer. p. 124

casing Solid piece of pipe, typically steel or PVC plastic, that prevents geologic material from sloughing (falling) into a completed well. p. 106

center pivot sprinkler Type of sprinkler system that moves across a field in a circular motion. p. 175

charcas Spanish word for "reservoirs." p. 9

check dam Barrier in a farm ditch that creates a small reservoir for syphon tubes to obtain water. p. 172

Chesapeake Bay Commission Regional government agency created to improve water quality in Chesapeake Bay. p. 303

chlorination Use of chlorine gas to disinfect drinking water. p. 334

cirque lake Lake created behind the debris left by a melting glacier. p. 64

Class A Evaporation Pan Pan used to measure evaporation rates at a given location. p. 34

Clean Water Act Amendments (1972) Federal legislation that included provisions to protect wetlands. pp. 255, 270

Clean Water Act Amendments (1977) Federal legislation that increased regulation of toxic pollutants. p. 270

climate Sum of all statistical weather information that helps describe a place or region. p. 37

cloud seeding Introduction of dry ice or silver iodide into clouds to enhance precipitation or to retard hail development. p. 46

coagulation Step in the water treatment process that causes suspended material to join together for easier removal. p. 332

Code Napoléon Compilation of French law in 1804 ordered by Napoléon I. p. 218

Code of Hammurabi Ancient Babylonian law created by King Hammurabi. p. 212

Coffin* v. *Left Hand Ditch Company Landmark Colorado Supreme Court case that eliminated the use of the Riparian Doctrine in Colorado. p. 226

cold front Weather front where a cold air mass mixes beneath a warmer air mass. p. 43

Colorado Doctrine Concept of strict adherence to the Doctrine of Prior Appropriation developed in Colorado in the mid-1800s. p. 226

Colorado River Basin Salinity Control Act (1974) Federal legislation to reduce salinity of water in the Colorado River at the border with Mexico. p. 122

Comprehensive Environmental Response, Compensation, and Liability Act (Superfund) of 1980 Federal legislation to regulate accidental spills and disposal of hazardous wastes. p. 271

concrete arch dam Curved, concrete structure that distributes water pressure to the abutments. p. 185

condensation The process through which water vapor turns to water liquid. p. 36

cone of depression Cone-shaped depression in the groundwater table caused by the drawdown of a pumping well. p. 108

confined aquifer Aquifer that is overlain by a confining bed of geologic material. p. 92

confluence Location where a tributary joins the main river channel. p. 60

conglomerate Sedimentary rock composed of rounded and gravel-sized particles. p. 86

conservancy district Local government agency that provides water management services. p. 296

consolidated rock Tightly bound geologic formation composed of sandstone, limestone, granite, or other rock. p. 92

consumptive use (CU) Amount of water transpired and retained within a plant or animal during a growing season. p. 36

contour lines Lines on a topographic map which show a constant elevation. p. 58

convection heating Heating of the atmosphere from the latent heat of the land surface. p. 43

Coriolis effect Deflective force of the Earth's rotation on all free-moving objects, including the atmosphere and oceans. p. 37

Correlative Rights Principle Legal doctrine that requires riparian landowners to share all water in a stream. p. 229

Council on Environmental Quality (CEQ) Agency created in the executive branch of the federal government to oversee implementation of NEPA. pp. 270, 352

crest Top of a dam. p. 184

crop yield Total amount of a crop harvested from an acre (or hectare) of land. p. 169

Cryptosporidiosis (Crypto) Disease caused by the ingestion of *Cryptosporidium parvum*. p. 135

Cryptosporidium parvum Microscopic parasite found in wastewater. p. 135

cubic feet per second (cfs) Volume of water equal to one cubic foot of water moving past a given point in one second. p. 70

cubic meters per second (cms) Volume of water equal to one cubic meter of water moving past a given point in one second. p. 70

cumuliform clouds Billowy, individual cloud masses that often have flat bases. p. 43

dam axis Total length of a dam. p. 184

dam foundation Excavated land surface at the toe of a dam. p. 184

dam tender Individual in charge of dam operations. p. 185

Darcy's Law Groundwater movement equation developed by Henry Darcy in the mid-1800s. p. 100

DDT Pesticide developed to kill pests. p. 124

dead capacity (dead storage) Storage capacity of a reservoir that cannot be released by gravity. p. 185

delivery canal Artificial channel used to carry irrigation water from a river or reservoir. p. 171

delta Landform created by the deposition of sediments at the mouth of a river. p. 68

dendrochronology Study of tree rings to determine historical climatic data. p. 40

denitrification Natural process where nitrate changes into nitrite and then into atmospheric nitrogen. p. 130

desalination Removal of salts and other minerals from sea water. p. 323

Desert Land Act (1877) Federal legislation that authorized the sale of public lands. p. 259

dew point (frost point) State at which the rate of evaporation and condensation are equal. p. 42

diligence Legal demonstration of continued efforts toward completion of a water project that requires a water right. p. 231

discharge Amount of water in a river, pipe, or other conduit that passes a given point during a given period of time. p. 70

dissolved load Portion of a stream's sediment load carried in solution. p. 68

dissolved oxygen (DO) Existence of soluble oxygen in water; related to pressure and temperature. p. 117

Doctrine of Prior Appropriation Doctrine of water allocation that states the right to use water is separate from other property rights. The first person to withdraw water and put it to beneficial use holds the better right. p. 223

Doppler Radar Technology that measures change in wave length caused by the relative motion of precipitation, tornados, and other severe weather events. p. 28

downstream Direction toward the confluence with a larger stream or river. p. 60

drawdown Lowering of the groundwater table caused by pumping of groundwater from wells. p. 108

drip irrigation Highly efficient (but expensive) method of irrigation that uses buried pipe to deliver water to individual plants p. 177

drought Period of extended dry weather that is abnormal for a region. p. 48

drought indices Numerical method to assess the severity of below-average precipitation in a region. p. 49

dryland farming Farming methods in arid or semiarid regions where no irrigation water is applied to crops. p. 169

earthen embankment (earthfill) dam Structure composed primarily of compacted earth material. p. 185

effluent (gaining) river A river, or reach of river, that gains water through the inflow of groundwater. p. 63

El Niño Spanish term for the effects of the natural warming of water in the Pacific Ocean off the coast of South America. p. 45

endangered species Plant or animal (vertebrate or invertebrate) species in danger of becoming extinct. p. 354

Endangered Species Act (ESA) of 1973 Landmark federal legislation that gave the USFWS responsibility for the protection of endangered species and associated habitat. p. 265, 354

English Common Law Body of law developed from the Justinian Code. p. 215

Environmental Impact Statement (EIS) Process required by NEPA to assess the impacts of all federal projects or other projects requiring a federal permit. p. 270

ephemeral stream Stream that is normally dry but may quickly fill with water during a storm event. p. 63

epilimnion Layer of warmer water in a lake. p. 65

Escherichia coli (E. coli) Microorganism that indicates the presence of pathogenic organisms from animal waste. p. 134

eutrophic Water that contains excess quantities of nutrients. p. 131

eutrophic lake An "old" lake that has excessive organic material which inhibits or prevents the growth of aquatic species. p. 65

eutrophication Aging of a surface water body caused by excess nutrients. p. 131

evaporation Process whereby water changes from a liquid to a gas. p. 34

evapotranspiration (ET) Sum of evaporation and transpiration from a plant. p. 35

exchange Legal mechanism that allows the diversion of water upstream of a given point of diversion as long as a similar quantity of water is replaced downstream with no injury to intervening water users. p. 233

face Exposed surface of a dam. p. 182

farm ditch (field ditch) Smaller artificial channel used to deliver irrigation water from a delivery canal to individual farm fields. p. 171

fate and transport Movement and residence of pollutants. p. 136

fecal coliform Microorganism that indicates the presence of animal waste. p. 134

Federal Emergency Management Agency (FEMA) Federal agency that provides assistance during and after natural and human-caused disasters. p. 276

Federal Energy Regulatory Commission (FERC) Federal agency that regulates nonfederal hydroelectric projects. p. 274

Federal Insecticide, Fungicide, and Rodenticide Act (FIFRA) of 1972 Federal law that regulates the use of pesticides. p. 125

Federal Land Policy and Management Act (1976) Federal legislation that reorganized management of federal lands by the BLM. p. 269

federal reserved water rights Rights obtained by the federal government through establishment of national parks or national monuments. p. 239

Federal Water Pollution Control Act (1956) Federal legislation that provided funds for pollution studies and construction of wastewater treatment plants. p. 269

Federal Water Pollution Control Act (1972) (Clean Water Act Amendments) Landmark federal legislation to protect all waters of the United States. p. 270

filtration Mechanical straining process, typically using sand, gravel, charcoal, or other media in the drinking water treatment process. p. 332

fines Very small particles from weathered or eroded sedimentary rocks similar in size to talcum powder. p. 86

firm yield Amount of water that can be stored in a reservoir with reasonable certainty during various hydrologic events. p. 186

"first in time, first in right" Legal concept used to allocate scarce water resources in dry climates. p. 223

Fish and Wildlife Coordination Act (1958) Federal legislation which required that wildlife conservation must be considered in any water resources development project. p. 265

fissure Rock fracture that has separated and moved apart. p. 89

flash flood Flood that occurs with little or no warning. p. 48

flexible plastic pipe Disposable irrigation pipe used to deliver irrigation water to a field. Similar to gated-pipe. p. 173

flocculation Growth of coagulated particles during the water treatment process. p. 332

Flood Control Act (1928) Major federal legislation to improve levees and channels along the Mississippi River and its tributaries. p. 252

Flood Control Act (1936) Federal legislation that expanded the flood-control efforts of the USACE. p. 253

Flood Control Act (1944) Federal legislation that again expanded the flood-control efforts of the USACE. p. 253

flood-control capacity (flood pool) All space above the dead capacity and inactive capacity if the reservoir is used for flood-control purposes. p. 186

flood-control dam Dam constructed for the purpose of protecting downstream property and human life from flood damage. p. 182

flood-control district Local agency that constructs flood protection projects. p. 289

Flood Insurance Act (1968) Federal legislation that encourages state and local governments to promote sound land-use planning in floodplains. p. 277

floodplain Flat, low-lying lands adjacent to a river and subject to periodic flooding. p. 67

flow meter Electronic device used to measure water velocity. p. 71

fluid drilling Process of using water or air at the end of a column of pipe to construct a well. p. 106

fluoridation Process of adding fluoride to drinking water to prevent tooth decay. p. 334

fluoride Chemical comound frequently added during the drinking water process to improve the dental health of water uses. p. 123

fluvial material (alluvial deposits) Sediments deposited by flowing rivers. p. 68

fracture Hairline break or rupture of rock where no apparent movement has occurred. p. 89

fractured aquifer Type of aquifer found in consolidated rock that contains usable quantities of groundwater in fissures or joints. p. 92

freeboard Difference in elevation between the dam crest and the top of the maximum water surface of a reservoir. p. 186

French Common Law Body of law developed from Roman, Teutonic, and Visigothic law. p. 215

fungicides Type of pesticide used to kill fungus. p. 125

furrow irrigation Gravity irrigation method whereby irrigation water is delivered to individual crop rows (furrows) in a field. p. 171

gage height Elevation of a water surface on a staff gage. p. 71

Gallatin Report (1808) Report to Congress which identified a plan for federal funding of the construction of national roads and canals. pp. 15, 249

gated-pipe Plastic or metal irrigation pipe that uses small slide-gates to deliver irrigation water to individual furrows in a field. p. 173

General Survey Act (1824) Legislation that authorized a survey of all roads and canals of national importance. p. 249

Giardia (Giardiasis) Diarrheal illness caused by *Giardia lamblia*. p. 135

Giardia lamblia Flagellated protozoan responsible for giardiasis. p. 135

glacial outwash Sediments (generally well-sorted sand and gravel) deposited by the meltwater of a glacier. p. 89

glacial till Glacial material (generally unsorted sand, silt, clay, gravel, and boulders) deposited directly by receding ice sheets. p. 89

global trade winds Two belts of wind that blow almost constantly from the east in the tropics. p. 37

gradient Slope of a river, typically measured in feet per mile or meters per kilometer. p. 63

gravity concrete dam Solid, concrete structure that uses its mass for stability. p. 185

gravity irrigation Common method of delivering water to crops through the use of gravity. p. 171

greenhouse gases Life-maintaining gases in the Earth's atmosphere. p. 41

groundwater Water contained in interconnected pores of geologic material below the land surface. p. 84

groundwater hydrology Study of the characteristics, movement, and occurrence of water found beneath the land surface. p. 85

groundwater management district (GMD) Local government agency in Kansas that provides water management services. p. 298

groundwater mining Process of pumping more water from an aquifer than is replaced through natural or artificial recharge. p. 96

groundwater recharge Downward movement of water from the land surface into and through upper soil layers. p. 90

groundwater table Top of the saturated zone in a groundwater system. p. 90

gulf stream Ocean currents that affect weather and climate around the world. p. 38

gyres Giant oceanic current circles p. 38

hardness Total concentration of calcium, iron, and magnesium in water. p. 119

headgate Device used to regulate the flow of water from a delivery canal. p. 171

headwaters Beginning of a river. p. 60

herbicides Type of pesticide used to kill unwanted plants and weeds. p. 125

high barometric pressure Area where air near the ground flows away from a region of high pressure in a clockwise motion. p. 46

Hohokam Indians Native American tribe that developed large-scale irrigation projects approximately 1200 years ago in present-day Arizona. p. 9

Homestead Act (1862) Landmark federal legislation that allowed U.S. citizens to acquire public lands. pp. 10, 268

Horace Greeley Newspaper editor and founder of the *New York Tribune* in 1841. Promoted the saying "Go West, Young Man." p. 11

humidity Amount of water moisture in the air. p. 42

hydraulic conductivity Coefficient that describes the rate at which water can move through a permeable medium. p. 98

hydraulic gradient Slope of the top of a groundwater table. p. 98

hydraulic head The height of a column of water that can be supported by water pressure at the point of measurement. p. 98

hydroelectric power generation Process of using moving water to generate electricity. p. 182

hydrograph Graph that shows changes in surface water or groundwater elevations over time. p. 72

hydrologic cycle Circulation of water from rivers, lakes, and oceans through the atmosphere, and back to the land surface and oceans. p. 24

hypolimnion Layer of cooler water in a lake. p. 65

hyporheic zone Area directly beneath a river. p. 60

inactive capacity (inactive storage) Quantity of water in a reservoir, located just above the dead storage zone, typically used for recreation. p. 185

indicator organism Organism used to identify the presence of infectious microorganisms. p. 134

influent (losing) river A river, or reach of river, that loses water by percolation into the ground. p. 63

inorganic chemical Chemicals that do not contain carbon. p. 119

insecticides Type of pesticide used to kill insects and other arthropods. p. 125

intakes Pipes used to withdraw adequate quantities of water from a river or reservoir. p. 330

integrated pest management (IPM) Use of economically reasonable practices to control pest damage with the least possible harm to the environment. p. 126

interflow Lateral movement of water, just below the land surface, during and immediately following a precipitation event. p. 59

intermittent stream Stream that generally flows only after a storm event or during wet seasons. p. 63

Internal Revenue Service (IRS) water depletion allowance Federal ruling that allows irrigators to receive a credit on federal income taxes if groundwater is declining beneath their property. p. 399

interstate compact Water allocation or water quality protection agreement between states. p. 235

intertropical convergence zone (ITCZ) Low-pressure zone at the equator where northeast and southeast trade winds flow together; characterized by strong upward air motion and heavy rainfall. p. 37

irrigation Artificial application of water to crops. p. 167

irrigation district Local government agency that provides irrigation water to irrigators. p. 294

irrigation efficiency Ratio between the total amount of irrigation water applied to a crop and the total amount of water consumed by the crop. p. 173

Irwin* v. *Phillips Landmark California Supreme Court case that defined the concept of a priority system of water use. p. 223

isobars Curved lines on a weather map that represent differences in air pressure. p. 45

isotherms Linear method used to graphically represent weather data. p. 41

jeopardy opinion Official federal opinion that a proposed action will have a significant adverse effect on a threatened or endangered species. p. 355

jet stream Very strong winds at high altitudes created by differences in air pressure gradients. p. 46

jetties Structures extending into a river to reduce water velocity near a shore. p. 202

joint use capacity (active conservation pool or active storage capacity) Quantity of water in a reservoir, located just above the inactive storage zone, used for a variety of purposes such as irrigation, power generation, and municipal use. p. 186

junior appropriator The owner of a water right whose right was acquired after other water right holders on the same stream. p. 223

Justinian Code Set of Roman law developed by Roman Emperor Flavius Petrus Sabbatius Justinianus, known as the Roman Emperor Justinian I. p. 213

karst Geologic terrain where significant solution of carbonate rocks has occurred due to flowing groundwater. p. 87

kettle lake Depressions created by blocks of buried glacial ice that later melted. p. 65

La Niña Spanish term for the effects of the natural cooling of waters in the Pacific Ocean off the coast of South America. p. 45

lake Any body of surface water of reasonable size, other than an ocean. pp. 31, 64

lake turnover Natural process of the hypolimnion replacing the epilimnion. p. 66

lateral sprinkler system Type of sprinkler system that moves across a field in a straight-line motion. p. 175

leach The process of minerals dissolving from soil. p. 121

lead Naturally occurring inorganic chemical. p. 120

lentic Habitat that generally contains nonmoving water, such as lakes. p. 64

levee district Local agency that constructs levees along flood-prone waterways. p. 289

lift Total distance between the land surface and groundwater that exists when pumping groundwater through a well. p. 107

limnetic zone Area of a lake toward open water where sunlight cannot penetrate to the lake bottom. p. 65

limnology Science of the characteristics and behavior of lakes. p. 64

littoral zone Area of a lake near its shore which provides adequate sunlight to promote shallow-rooted plant growth. p. 65

live capacity Amount of all storage capacity of a reservoir that can be released by gravity. p. 186

locks Step-like structures used to raise or lower vessels along rivers with steep gradients. p. 14

Louisiana Swamp Land Act (1849) Federal legislation that encouraged the draining and cultivation of swamps in Louisiana. p. 254

low barometric pressure Area where air near the ground flows into a region of low pressure in a counterclockwise motion. p. 46

lysimeter Field device containing soil and vegetation used to measure evapotranspiration rates. p. 35

Maryland Mill Act (1669) Law enacted by the Maryland General Assembly to encourage construction of watermills. p. 216

Massachusetts Mill Act (1714) Law enacted by the Massachusetts Legislature which allowed construction of mills with little or no compensation to upstream flood-damaged landowners. p. 216

maximum contaminant level (MCL) Highest concentration of a solute permissible in a public water supply as determined by the federal government. p. 124

maximum water surface (maximum water pool) Highest acceptable elevation for stored water in a reservoir. p. 186

meanders Loop-like bends in a river. p. 61

mesotrophic lake A "middle-aged" lake with adequate organic material to support a wide variety of aquatic species. p. 65

metals Elements found naturally in the Earth's crust. p. 119

microorganism Very small organisms found in all water. p. 133

mill raceway Channel of flowing water from a river to the waterwheel of a mill. p. 17

mineral Naturally occurring inorganic crystalline material. p. 121

mineralization/ammonification Process of organic matter decomposing in the presence of oxygen, which includes ammonification. p. 129

miner's inch Unit of water measurement developed by prospectors in the 1800s. p. 228

Mining Law (1872) Landmark federal legislation that set standards for prospecting and mining operations. p. 268

Missouri River Basin Association (MRBA) Regional government agency created to coordinate Missouri River planning activities. p. 304

monitored natural attenuation (MNA) Process of monitoring a contaminated aquifer to determine if conditions improve naturally over time. p. 137

municipal water department Local agency that provides drinking water to residents. p. 285

mutual ditch and irrigation company Private company that constructs and operates irrigation systems. p. 293

National Environmental Policy Act (NEPA) of 1970 Landmark federal legislation that directs all federal agencies to foster improved protection of the environment. pp. 270, 352

National Flood Insurance Program Established to provide federally backed flood insurance to communities. p. 277

National Irrigation Congress National effort to promote the federal funding of irrigation projects. p. 259

National Marine Fisheries Service (NMFS or NOAA Fisheries) Federal agency responsible for management of most marine mammals. p. 276

National Park Service (Park Service or NPS) Federal agency created in 1916 to manage the national park system. p. 266

National Pollutant Discharge Elimination System Permits (NPDES) Permit Federal permit required to discharge wastewater into a navigable stream. pp. 271, 341

National Rivers and Harbors Congress (1901) National effort to promote the federal funding of navigation along rivers. p. 250

natural levee Elevated landforms that parallel some rivers, deposited by sediments of previous floods. p. 67

natural organic chemicals (NOC) Chemicals that contains carbon and occurs naturally from decomposition of plants or animals. p. 124

Natural Resources Conservation Service (NRCS) Federal agency that protects soil and water resources in agricultural areas. p. 272

natural resources district (NRD) Local government agency in Nebraska that provides water and soil management services. p. 297

negative benefit-cost ratio A benefit-cost ratio less than 1:1. p. 196

nematacides Type of pesticide used to kill nematodes. p. 125

Nilometer Device used in ancient Egypt to measure the stage of the Nile River. p. 73

nitrate (NO_3^-) Salt formed as an end-product of the aerobic stabilization of organic nitrogen. p. 127

nitrification Natural process where organic nitrogen changes into ammonia, then into nitrite, and finally into nitrate. p. 130

nitrite (NO_2^-) Salt formed by the action of bacteria on ammonia and organic nitrogen. p. 127

nitrogen (N) Common nutrient that supports plant growth. p. 127

nitrogen cycle Natural process of converting nitrogen gas of the atmosphere into usable forms of nutrients for plants and animals. p. 128

nitrogen fixation Natural process of changing nitrogen gas into organic form. p. 128

nitrogen gas (N_2) Form of nitrogen abundant in the atmosphere. p. 127

nonpoint source pollution Contamination discharged from broad and difficult to identify sources. p. 113

nonvolatile organic chemicals (NVOC) Heavier compounds, including PCBs and DDT, which can settle to the bottom of rivers and lakes. p. 124

noria Ancient Egyptian waterwheel used for irrigation. p. 16

nutrients Substances that provide energy and growth to plants and animals. p. 127

oligotrophic lake A "young" lake with little or no organic material on its bottom. p. 65

open meeting law State law that requires government agencies to conduct business in full view of the general public. p. 299

open records law State law that requires most records of a government agency to be available to the general public for inspection. p. 299

organic chemicals Chemicals that contains carbon. p. 124

orientation General direction of the course of a river. p. 59

orographic lifting Weather phenomenon created when moisture-laden air meets a mountain or mountain range and the air is forced to ascend. This rising air cools adiabatically, often resulting in clouds and precipitation. p. 43

outlet Opening used to release water from a reservoir to a river or canal. p. 184

outlet gate Metal door (headgate) used to control the release of water from a reservoir. p. 184

overland flow Runoff water that flows on the land surface rather than percolating into the ground. p. 59

oxbow lakes Curved lakes created when a river cuts off a meander. p. 62

ozone gas Very potent, and effective germicide used in the final stage of the treatment of drinking water. p. 334

Palmer Index First drought index developed in the United States. p. 49

pan coefficient Correction factor used to eliminate excess heat retained in the metal of an evaporation pan. p. 34

parapet wall Typically a low, concrete wall along the crest of a dam. p. 184

Penman Equation Mathematical method used to determine evaporation rates from plants and soil. p. 35

perched aquifer Localized zone of saturation above the main water table created by an underlying layer of impermeable material. p. 92

percolate Surface water that infiltrates underground (also called groundwater recharge). p. 59

percolation Naturally occurring downward movement of groundwater from original precipitation or surface water. p. 90

percussion drilling Process of using hydraulic pressure to hit a sharp rod at the end of a column of pipe to construct a well. p. 106

permeability Measure of the ability of geologic material to transmit water. p. 98

pesticides Synthetic organic chemicals used to kill pests. p. 125

pH Common measure of the acidity or alkalinity of a solution. p. 117

phosphorus (P) Common nutrient found in soil and water. p. 128

phreatophyte Plant with a high rate of water use typically found along a river corridor. p. 35

piping Internal erosion of an earthen dam caused by the undesired movement of reservoir water through a dam. p. 192

plume An underground zone of contamination. p. 115

pluvial lake Lake formed during a period of increased precipitation. p. 65

point of diversion Location of the removal of water from a stream or an aquifer by a ditch, canal, or well. p. 233

point source pollution Contamination discharged through a pipe or other identifiable location. pp. 113, 114

pollution Change in water quality that renders it unfit for a given use. p. 112

pore space (void) Openings between geologic material found underground. p. 96

porosity Ratio of void spaces in a geologic formation to the total volume of the formation. p. 96

positive benefit-cost ratio A benefit-cost ratio greater than 1:1. p. 196

potable water Water delivered to a consumer that can be safely used for drinking, cooking, and washing. p. 323

potential evapotranspiration Maximum amount of evapotranspiration that would occur if soil moisture was unlimited. p. 35

potentiometric map Contour map of the potentiometric surface of an aquifer. p. 103

potentiometric surface Elevation to which groundwater will rise in a well. p. 103

precipitation Process of water in the atmosphere returning to the Earth's surface in liquid or solid form. p. 25

present value (time value of money or discount rate) Current value of money with the benefits of accruing interest added. p. 196

primary treatment Initial treatment of wastewater that includes screening, chemical treatment, grit removal, and settling. p. 338

private good The right of water to be developed, used, traded, and sold for economic productivity and financial gain. p. 384

privatization Practice by local governments to use private firms to deliver public services. p. 384

Probable Maximum Flood (PMF) Maximum surface water flow in a drainage area expected from a Probable Maximum Precipitation event. p. 76

Probable Maximum Precipitation (PMP) Greatest amount of precipitation that is reasonably expected from a single storm event at a given location. p. 76

profundal zone Area at a lake bottom where rooted plants cannot grow due to a lack of sunlight. p. 65

public good In reference to water resources, the right to water for all people. p. 383

Pueblo water rights Water allocation system often used in Spanish communities. p. 215

qanat Underground water delivery system, consisting of a long tunnel with vertical shafts, generally found in Afghanistan and other Middle Eastern countries. p. 2

radionuclides Naturally occurring isotopes from uranium and radon. p. 328

rain shadow Region of relatively low precipitation that occurs downwind of a mountain or mountain range. p. 44

rating curve Graph that depicts the discharge of water over a variety of river elevations or stages. p. 71

Rational Formula Mathematical formula used to determine overland flow. p. 69

reasonable and prudent alternatives Alternatives that can be implemented that will not adversely jeopardize a threatened or endangered species. p. 355

Reasonable Use Principle Legal doctrine that allows a riparian landowner to divert and use any quantity of water as long as it does not interfere with the reasonable use of other riparian landowners. p. 229

Reclamation Act (1902) Federal legislation that created the Reclamation Service, later renamed the Bureau of Reclamation. p. 260

Recopilación de leyes de los Reynos de las Indias Set of Spanish laws developed in 1680. p. 214

recorder Device used to measure and record the stage of flowing water. p. 72

recreational river Federal designation for a river that is accessible by road and may have some development. p. 351

Refuse Act (1899) Section of the Rivers and Harbors Act (1899) that gave the USACE authority to enforce pollution protection programs along navigable rivers. p. 249

reservoir Human-made body of water used to store surface water runoff for flood-control or other purposes. pp. 31, 65

residence time Period of time that groundwater remains in an aquifer. p. 102

Resource Conservation and Recovery Act (1976) Federal legislation to regulate land disposal of hazardous wastes. p. 270

reverse osmosis (RO) Water treatment process that uses membranes to remove inorganic ions, turbidity, organic material, bacteria, and viruses from drinking water. p. 325

Riparian Doctrine (Common Law of Water) Framework for water allocation law developed within the Justinian Code. p. 213

rip-rap Randomly placed layers of concrete, rock, or other impervious material used to control erosion caused by moving water. p. 185

Rivers and Harbors Act (1899) Legislation that gave the USACE authority to regulate construction activities along navigable rivers. p. 249

rodenticide Type of pesticide used to kill rodents. p. 125

root zone Vertical area of soil that contains a plant's root system. p. 91

Rule of Absolute Ownership (English Rule) Groundwater doctrine which holds that landowners have the right to use all groundwater beneath their property. p. 234

Rule of Capture Doctrine of groundwater use which provides a landowner unrestricted use of water obtained from beneath their property. p. 234

runoff Amount of water that flows along the land surface after a storm event or from melting snow in the spring. p. 30

Safe Drinking Water Act (SDWA) of 1974 Landmark federal legislation to protect public drinking water supplies. pp. 270, 328

Safe Drinking Water Act Amendments (1996) Federal legislation that promotes the protection of source water drinking supplies. pp. 271, 328

safe harbor Federal incentives to nonfederal property owners to protect habitat for threatened or endangered species. p. 355

salinity Proportion of dissolved salts to pure water. pp. 121, 174

salt Compound of sodium and chloride found naturally in soil and water. p. 121

saltation Transportation of larger sediments through a series of leaps or bounces along the bottom of a river channel. p. 69

salvaged water Legal process in some states that allows a water right holder to sell a portion of that right if certain conservation practices are taken. p. 233

sandstone Sedimentary rock composed of sand-sized particles. p. 86

saqia Elaborate ancient animal-powered waterwheel with multiple buckets. p. 6

saturated thickness Total water-bearing thickness of an aquifer. p. 92

saturated zone Area below the land surface where all open spaces between sediments and rock are filled with water. p. 90

scale Relationship between the size of map features and the actual dimensions on the ground. p. 58

scenic river Federal designation for a river with shorelines generally not accessible by roads. p. 351

scoping process Federal procedure that solicits public input into proposed activities or projects that could affect threatened or endangered species. p. 355

secondary treatment Second step of wastewater treatment that involves aerobic oxidation of nutrients. p. 339

Section 404 Federal legislation that gave the USEPA and the USACE authority to protect wetlands. p. 255

sediment Unconsolidated particles created by weathering and erosion of rock, chemical precipitation from solution in water, or secretions of organisms, and transported by wind, water, or glaciers. p. 67

sediment yield Amount of sediments carried from a watershed by a river. p. 69

sedimentary rock Weathered and eroded igneous and metamorphic rock. p. 86

sedimentation Settling of sediments from water as velocity decreases. p. 67

seiches Difference in water elevations in a lake. p. 67

semiarid Climatic condition in which average annual precipitation is between 10 and 20 inches (25 and 51 cm). p. 37

senior appropriator The owner of a water right whose right was acquired prior to other water right holders on the same stream. p. 223

set Maximum number of syphon tubes that can be placed in a "checked" ditch. p. 173

sewage lagoons Sewage collection system that utilizes natural bacterial processes to treat wastewater. p. 338

shadouf Water-lifting device consisting of a long pole with a bucket at one end and counterbalanced with rocks at the other end. p. 6

shale Sedimentary rock commonly known as clay. p. 86

siltstone Sedimentary rock composed of particles smaller than grains of sand. p. 86

sinkhole Geologic formations where water has dissolved limestone, carbonate rock, or salt beds to create large holes at the land surface. p. 87

snow core Sample of snow obtained in a snow tube. p. 28

snow course Area used for the collection of snow cores, and generally along a predetermined line. p. 29

snow pillow Device used to measure snow depth and water equivalents. p. 29

snow tube Metal pipe used to obtain snow samples for measurement purposes. p. 29

soil Combination of mineral and organic matter, water, and air, which supports plant growth. p. 90

soil moisture A thin film of water that surrounds each particle of material (conglomerate, sand, clay, siltstone) in the unsaturated zone. p. 90

sorting Process whereby geologic material of various sizes is moved by wind or water and deposited into areas based on size and weight. p. 68

source water Untreated water used for public drinking water supplies. p. 141

southern oscillation Changing pattern of atmospheric pressure along the equator in the Pacific Ocean. p. 45

specific capacity Ratio obtained by dividing the pumping capacity (yield) of a well by the drawdown created by a pumping well. p. 108

specific yield Ratio of the volume of groundwater that will be yielded from an aquifer by gravity drainage to the total volume of rock or soil in an aquifer. p. 101

spillway Wide chute or conduit used to bypass excess water from a reservoir past a dam. p. 184

spring Flow of groundwater that emerges naturally at the land surface. p. 93

sprinkler irrigation Mechanized method of applying irrigation water to crops through sprinkler heads. p. 174

staff gage Metal ruler used to determine the stage of a river or lake. p. 71

stage Maximum water level in a river or lake above a set reference point. p. 71

stage-capacity curve (rating curve) Graphic representation of reservoir storage volumes at various water depths. p. 74

state water agency Agencies of state government that provide water management services. p. 301

stationary fronts Parallel weather fronts with little or no movement. p. 43

stratiform clouds Low sheets or layers of clouds that cover much or all of the sky. p. 43

stream depletion Reduction in the flow of a stream or river caused by groundwater pumping in the vicinity. p. 108

stream depletion factor Mathematical calculation to determine the effect of well pumping on the flow of water in a river. p. 108

sublimation Process whereby water changes from a solid to a gas without passing through the liquid state. p. 35

surcharge capacity (temporary storage capacity or temporary flood pool) Zone of temporary storage in a reservoir, and located above the joint use capacity and flood-control capacity, if a dam is used for flood control. p. 186

surface water hydrology Study of the occurrence, distribution, and chemistry of water, which is open to the atmosphere and subject to surface runoff. p. 56

Surface Water Supply Index (SWSI) Weighted index to assess quantities of surface water in a region. p. 50

suspended load Fine sediments carried within a body of moving water. p. 68

Swamp Lands Act (1850) Federal legislation that encouraged the draining and cultivation of swamps in other public-land states. p. 254

Swamp Lands Act (1860) Federal legislation that encouraged the draining and cultivation of swamps in Oregon and Minnesota. p. 254

synthetic organic chemicals (SOC) Chemicals that contains carbon and is generally developed in a laboratory. p. 124

syphon tubes Metal or plastic tubes used to divert water from a farm ditch to a crop. p. 172

tambour Ancient auger-type water delivery device. p. 6

Taylor Grazing Act (1934) Federal legislation that regulated cattle grazing on federal lands. p. 268

tertiary treatment Third step of wastewater treatment that involves addition of chemicals, chlorination, and use of ultraviolet lights. p. 340

Thales Ancient Greek philosopher and scientist. p. 23

thalweg Imaginary line that follows the deepest points in a river. p. 60

thermal spring Heated groundwater that naturally flows to the land surface. p. 95

thermocline Layer of water with a rapid change of temperature in the vertical direction. p. 65

threatened species Plant or animal species or subspecies that is likely to become endangered in the near future. p. 354

time value of money (discount rate) Interest rate used to determine the present value of a sum of money. p. 196

toe Bottom of a dam where the structure meets that natural land surface. p. 184

topographic map USGS map that shows land and water features, elevation, and structures in various scales of detail. p. 58

total capacity Amount of all storage capacity of a reservoir, including the dead and inactive storage pools. p. 186

total coliform Coliform bacteria used as an indicator of the sanitary quality of a water body. p. 134

Total Maximum Daily Load (TMDL) Program Federal standards developed to protect water quality in a stream. p. 143

Tragedy of the Commons Human perspective that often leads to resource mismanagement and conflict. p. 410

transmissivity Rate at which groundwater moves laterally through the saturated thickness of an aquifer with a hydraulic gradient of 0. p. 99

transpiration Process of a plant releasing water vapor through its stomata. p. 35

tributaries Smaller streams that combine to eventually form a river. p. 60

trickling filters Rectangular or circular beds filled with coarse material, such as rock or gravel, that promotes biological treatment of wastewater p. 339

turbidity Measure of fine suspended matter in water. p. 118

Tyler* v. *Wilkinson Landmark U.S. Supreme Court case that defined the Riparian Doctrine in the United States. p. 219

U.S. Army Corps of Engineers (Corps or USACE) Oldest water resources management agency in the United States. p. 248

U.S. Department of Agriculture (USDA) Federal agency created in 1862 to assist the nation's farmers with agricultural issues. p. 272

U.S. Department of Interior (DOI or Interior) Federal agency created in 1849 to administer public lands. p. 257

U.S. Environmental Protection Agency (USEPA) Federal regulatory agency created in 1970 to protect the earth, air, land, and water of the United States. p. 269

U.S. Fish & Wildlife Service (USFWS) Only federal agency with the primary responsibility to conserve, protect, and enhance fish, wildlife, and plants. p. 264

U.S. Forest Service (USFS) Federal agency that manages the country's national forests. p. 273

U.S. Geological Survey (USGS) Federal scientific agency created in 1879. p. 262

ultraviolet (UV) systems Electromagnetic radiation that effectively kills bacteria and other microorganisms in water. p. 334

unconfined aquifer Aquifer with no confining bed of material between the saturated zone and the land surface. p. 92

unconsolidated rock Loosely bound geologic formation composed of sands and gravels. p. 92

upstream Direction toward the headwaters of a river. p. 60

usufructory property right The right to use something without actual ownership of the property. p. 219

vadose zone (unsaturated zone) Area between the land surface and the top of an aquifer where openings in soil, sediment, and rock are not saturated with water, but are filled primarily with air. p. 90

velocity Speed of moving water. p. 68

virga Atmospheric condition where precipitation evaporates before reaching the ground. p. 25

volatile organic chemicals (VOC) Lightweight compounds often found in solvents and plastics, that vaporize and evaporate easily. p. 124

warm front Weather front where a warm air mass overrides a retreating cool air mass. p. 43

wastewater Irrigation water that is applied to a crop but is not consumed. p. 173

water and sewer district Local agency that treats and disposes wastewater. p. 288

water banking Use of a local clearinghouse for the sale or lease of water within a defined set of rules. p. 390

"water-diamond" paradox Water is very necessary to life yet relatively inexpensive, while (paradoxically) diamonds are relatively inessential to life yet extremely expensive. p. 382

water equivalent Amount of liquid contained in a snow core. p. 29

water management district (WMD) Local government agency in Florida that provides water management services. p. 299

water marketing The sale or lease of water in a market-based system. p. 386

Water Pollution Control Act (1948) Federal legislation that provided funds to states for water quality protection. p. 269

Water Quality Act (1965) Federal legislation that created the first federal water quality standards. p. 269

water right The right to use water based on seniority and flow rate. p. 223

watershed Fundamental hydrologic unit of land that contributes surface water to a stream. p. 56

Watershed Protection Approach (WPA) Federal program to develop partnerships between organizations to protect water quality in watersheds. p. 143

weather The state of the atmosphere at any given time. p. 41

weather modification Any change in local weather events caused by humans. p. 46

well screen Pipe with slots or holes attached to the end of a well casing that allows groundwater to move into a well casing. p. 106

Wellhead Protection Program (WHPP) Federal program to protect groundwater quality used as a source of drinking water. p. 141

wetland Natural or artificial areas of shallow water or saturated soils that contain or could support hydric plants. p. 33

Wild and Scenic Rivers Act (1968) Federal legislation to protect rivers from development. p. 350

wild flood irrigation Gravity irrigation method whereby irrigation water is allowed to spread broadly across fields such as alfalfa or grains that do not have furrows (crop rows). p. 171

wild river Federal designation for a river untouched by development. p. 351

wind Horizontal movement of air. p. 45

Winters Doctrine (Indian Reserved Rights) A U.S. doctrine which holds that the federal government legally reserved water rights for Native Americans when reservations were established. p. 239

yazoo stream Tributary stream that flows parallel to a main river. A natural levee separates the two river systems. p. 67

INDEX